Advanced High Voltage Power Device Concepts

B. Jayant Baliga

Advanced High Voltage Power Device Concepts

Springer

B. Jayant Baliga
Department of Electrical and Computer Engineering
North Carolina State University
Raleigh, NC
USA
bjbaliga@ncsu.edu

ISBN 978-1-4614-0268-8 e-ISBN 978-1-4614-0269-5
DOI 10.1007/978-1-4614-0269-5
Springer New York Dordrecht Heidelberg London

Library of Congress Control Number: 2011932228

Printed on acid-free paper

Springer is part of Springer Science+Business Media (www.springer.com)

The author would like to dedicate this book to his wife, Pratima, for her unwavering support throughout his career devoted to the enhancement of the performance and understanding of power semiconductor devices.

Preface

The adverse impact on the environment resulting from carbon emissions into the atmosphere is widely accepted worldwide. The carbon emission can be reduced by taking two approaches. The first approach is based upon energy conservation. Energy conservation can be achieved without compromising the standard of life in our society by improving the efficiency for the management and distribution of power. Power semiconductor devices are recognized as a key component for achieving this goal [1]. It is estimated that at least 50% of the electricity used in the world is controlled by power devices. With the wide spread use of electronics in the consumer, industrial, lighting, and transportation sectors, power devices have a major impact on the economy because they determine the cost and efficiency of systems. The second approach to mitigating carbon emissions is by the development of renewable energy sources such as wind power and solar power. These installations require power electronic inverters to convert the generated power to a well-regulated 60 Hz AC power that can be distributed to consumers and the industry. Due to the relatively high power levels involved, the power semiconductor devices used in these applications must have high voltage and current handling capability.

In the 1950s, the power rectifiers and thyristors were introduced to replace the existing vacuum tubes. The solid state devices offered much smaller size, improved ruggedness, and greater efficiency. Over the last six decades, the power ratings of thyristors have steadily grown. The current handling capability has been increased from 100 A to over 4,000 A by using larger diameter silicon wafers while the blocking voltage capability has simultaneously been increased from 200 V to more than 8,000 V by using higher resistivity silicon produced with the neutron transmutation doping process. These devices have been primarily used in HVDC power transmission and distribution systems.

The complexity and power losses in the commutation circuits required with thyristors motivated the development of Gate Turn-Off (GTO) thyristors in the 1960s. These devices found favor in large motor drive applications such as in steel mills and electric trains (traction). The power ratings of GTOs grew steadily in the last five decades to reach a current handling capability of 4,000 A with a blocking voltage capability of 6,000 V [1].

The large drive current needed for silicon GTOs encouraged the development of the insulated gate bipolar transistor (IGBT) [2] in the 1980s. During the last three decades, the IGBT has become the dominant device used in all medium and high power electronic systems in the consumer, industrial, transportation, and military systems, and even found applications in the medical sector. The U.S. Department of Energy has estimated that the implementation of IGBT-based variable speed drives for controlling motors is producing an energy savings of over 2 quadrillion btus per year, which is equivalent to 70 GW of power. This energy savings eliminates the need for generating electricity from 70 coal-fired power plants resulting in reducing carbon dioxide emissions by over one trillion pounds each year. The power ratings of IGBTs have been increased to a current handling capability of over 1,000 A with blocking voltages of 6,000 V. They are now being applied to not only high power motor drives for traction [3] (Shinkansen bullet train) but also for HVDC power distribution [4].

With on-going investments in renewable energy sources such as wind and solar power that utilize power semiconductor device in inverters, it is anticipated that there will be an increasing need for technologists trained in the discipline of designing and manufacturing power semiconductor devices. My recently published textbook [5] provides a comprehensive analysis of the basic power rectifier, transistor, and thyristor structures. In 2009, the textbook was complemented with a monograph on *Advanced Power Rectifier Concepts* to familiarize students and engineering professionals with diodes that exhibit improved performance attributes. In 2010, the textbook was complemented with a monograph on 'Advanced Power MOSFET Concepts' to familiarize students and engineering professionals with switches that exhibit improved performance attributes.

This monograph introduces the reader to advanced MOS-gated power thyristor concepts that enable improvement of performance of these high voltage structures. The voltage ratings for the devices discussed here range from 5,000 V to 20,000 V. For the convenience of readers, analysis of the basic thyristor structures, with the same voltage ratings as the novel device structures, has been included in the monograph to enable comparison of the performance. As in the case of the textbook, analytical expressions that describe the behavior of the advanced power thyristor structures have been rigorously derived using the fundamental semiconductor Poisson's, continuity, and conduction equations in this monograph. The characteristics of IGBTs have also been included in this book because they have displaced thyristors in many high power motor drive and power transmission systems. The electrical characteristics of all the power devices discussed in this book can be computed using these analytical solutions as shown by typical examples provided in each section. In order to corroborate the validity of these analytical formulations, I have included the results of two-dimensional numerical simulations in each section of the book. The simulation results are also used to further elucidate the physics and point out two-dimensional effects whenever relevant. Due to increasing interest in the utilization of wide band-gap semiconductors for power devices, the book includes the analysis of silicon carbide structures.

In the first chapter, a broad introduction to potential applications for high voltage power devices is provided. The electrical characteristics for ideal power switches are then defined and compared with those for typical devices. The second and third chapters provide analyses of the silicon and silicon carbide power thyristors. The analysis includes the blocking characteristics, the on-state voltage drop, and switching behavior. The silicon Gate Turn-Off thyristor structure is then discussed in Chap. 4. The fifth chapter is devoted to silicon IGBT structures to provide a benchmark. Any alternate silicon or silicon carbide device technology must outperform the commonly used silicon IGBT systems today.

The analysis of silicon carbide MOSFETs and IGBTs is provided in Chaps. 6 and 7. The much larger breakdown field strength for 4H-SiC allows increasing the doping concentration in the drift region by a factor of 200 times while shrinking the thickness of the drift region by one-order of magnitude. This makes 5 kV and 10 kV silicon carbide MOSFETs with low on-resistance feasible. For even higher blocking voltages, the silicon carbide IGBT structure needs to be developed. However, the silicon carbide MOSFET and IGBT structures must be designed to shield the gate oxide from the much larger electric fields prevalent in silicon carbide to avoid rupture. In addition, the base region must be shielded to avoid reach-through breakdown. The on-state voltage drop of these devices becomes limited by the channel resistance and buffer layer design.

The eighth and ninth chapters discuss the MOS-Controlled Thyristor (MCT) structure and the Base-Resistance Controlled Thyristor (BRT) structure, which utilize MOS-gate control of the turn-on and turn-off of the thyristor. The tenth chapter describes the Emitter Switched Thyristor (EST) which also utilizes an MOS-gate structure to control the turn-on and turn-off of the thyristor while allowing construction with the IGBT process. This device has the added feature of a good safe operating area.

The final chapter provides a comparison of all the high voltage power device structures discussed in this book. The performance of all the devices is compared over a wide range of blocking voltages to provide a broader view.

I am hopeful that this monograph will be useful for researchers in academia and to product designers in the industry. It can also be used for the teaching of courses on solid state devices as a supplement to my textbook [5].

Raleigh, NC
March 2011

B. Jayant Baliga

References

1. H. Akagi, "Utility Applications of Power Electronics in Japan", IEEE Industrial Electronics, Control, and Instrumentation Conference, Vol. 2, pp. 409–416, 1997.

2. B.J. Baliga, "How the Super-Transistor Works", Scientific American Magazine, Special Issue on 'The Solid-State-Century', pp. 34-41, January 22, 1988.
3. H. Akagi, "Large Static Converters for Industry and Utility Applications", Proc. IEEE, Vol. 89, pp. 976–983, 2001.
4. M.P. Bahram, "HVDC Transmission Overview", IEEE Transmission and Distribution Conference and Exposition, pp. 1–7, 2008.
5. B.J. Baliga, "Fundamentals of Power Semiconductor Devices", Springer Scientific, New York, 2008.

Contents

Chapter 1
Introduction

Power devices are required for applications that operate over a broad spectrum of power levels as shown in Fig. 1.1 [1]. Based upon this figure, the applications can be broken down into several categories. The first category is applications that require low operating current (typically less than 1 A) levels. These applications, such as display drives, usually require a large number of transistors that must be capable of blocking up to 300 V. The small size of the low-current transistors allows their integration on a single chip with control circuits to provide a cost-effective solution.

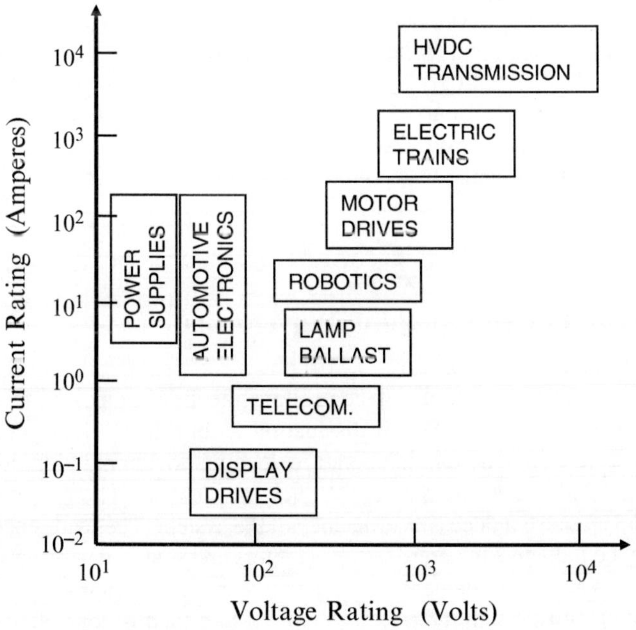

Fig. 1.1 Applications for power devices

B.J. Baliga, *Advanced High Voltage Power Device Concepts*,
DOI 10.1007/978-1-4614-0269-5_1, © Springer Science+Business Media, LLC 2011

The second category is applications where the operating voltage of the power circuit is relatively small (<100 V). Typical examples are automotive electronics and power supplies used in desktop computers and laptops. Silicon power metal–oxide–semiconductor field-effect transistor (MOSFET) structures offer the best performance for these applications because of their low on-resistance and fast switching speed. A variety of power MOSFET structures, which enable enhanced performance for these applications, are discussed in a companion monograph [2].

The third category is applications with high operating voltages (above 200 V). Typical examples are lamp ballasts, consumer appliances that utilize motors, and electric vehicle drives. The on-resistance of conventional silicon power MOSFET structures is too large to serve these applications. Consequently, these applications utilize silicon insulated gate bipolar transistors (IGBTs). The silicon IGBT combines the physics of the MOSFET structure with the physics of the bipolar transistor structure.

The fourth category is applications with very high operating voltages (>5,000 V). Typical examples are high power motor control in steel mills and for traction (electric trains), and in power transmission and distribution. This monograph discusses high voltage power devices suitable for these applications. At the 5,000 V level, the silicon power thyristor and gate-turn-off (GTO) thyristor were extensively utilized until the turn of the century. Since then, the voltage and current ratings of IGBTs have been scaled such that they have now replaced the silicon GTO for traction applications and have even found favor in power distribution systems [3]. In addition, power MOSFET structures built using silicon carbide as the base material have been shown to exhibit very promising characteristics for applications that require blocking voltages of up to 5,000 V [4]. Furthermore, silicon carbide IGBTs with voltage ratings of 15 kV are under investigation for use in smart grid applications [5]. Consequently, this monograph includes the discussion of the power MOSFET and IGBT structures that are specially configured to obtain a high performance from silicon carbide.

1.1 Typical Power Switching Waveforms

Power devices must be capable of controlling the flow of power to loads with minimum power dissipation. Power dissipation within the devices generates heat that causes a rise in junction temperature leading to degradation in electrical characteristics and sometimes destructive thermal runaway. The power dissipation within the devices also reduces the efficiency of the system. The loads encountered in systems may be inductive in nature (such as motors and solenoids), resistive in nature (such as heaters and lamp filaments), or capacitive in nature (such as transducers and LCD displays). Most often, the power delivered to a load is controlled by turning on a power switch on a periodic basis to generate pulses of current that can be regulated by a control circuit.

Typical waveforms for the power delivered through a power switch are shown in Fig. 1.2. During each switching cycle, the switch remains on for a time t_6 to t_1 and maintains an off-state for the remainder of the period T. This produces pulses of current that flow through the circuit as controlled by the turning on of power switches. The voltage drop during the on-state (V_F in the figure) produces power dissipation in a typical power switch. Similarly, during the off-state, the leakage current (I_L in the figure) in the typical power switch is finite resulting in power dissipation, although this leakage current is usually small for the devices discussed in this monograph. In addition, it is assumed that the typical power switch makes the transition between the on-state and off-state with finite time intervals (t_1 to t_3 and t_4 to t_6 in the figure) resulting in switching power losses as well.

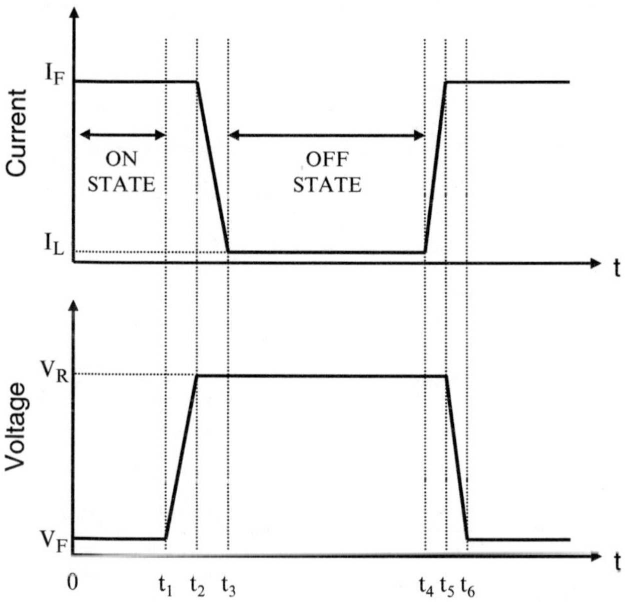

Fig. 1.2 Typical switching waveforms for power delivery

The most appropriate power device for any application produces the lowest combination of power losses so as to maximize the efficiency for the application. In addition, the choice of the power devices is dictated by the overall cost including the gate control circuit and any snubber circuits required for fail-safe operation of the device. The complex and expensive gate drive and snubber circuits for silicon GTOs have motivated the development of silicon IGBTs. The development of silicon carbide power MOSFETs is motivated by reduced power losses in the on-state and during switching. The development of devices with larger blocking voltage capability is motivated by reducing the number of devices connected in series to serve the very high operating voltages in power distribution systems.

The cost of the system is greatly reduced when fewer devices are connected in series because of the expensive level-shifting of gate drive signals and the need for voltage sharing networks.

1.2 Typical High Voltage Power Device Structures

The high voltage (>5,000 V) power device structures that are discussed in this monograph can be classified into four categories: (a) power thyristors, (b) power MOSFETs, (c) IGBTs, and (d) MOS-gated thyristors. Within these categories, the devices can be fabricated from either silicon or silicon carbide. Since silicon high voltage (>5,000 V) power MOSFETs have extremely high on-resistances, only silicon carbide power MOSFET structures are included in this monograph.

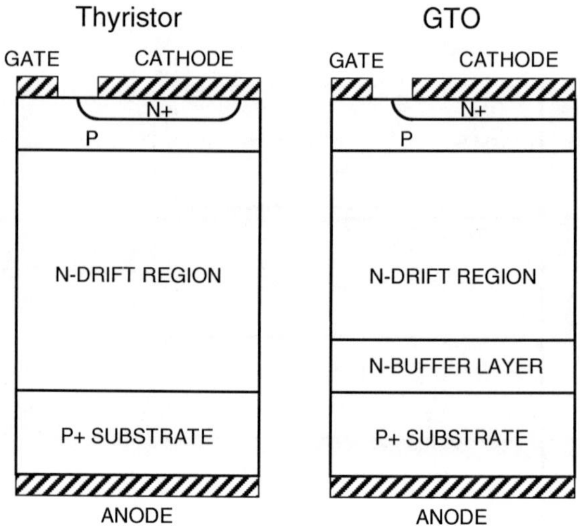

Fig. 1.3 Typical silicon power thyristor structures

Typical power thyristor structures are illustrated in Fig. 1.3. The thyristor structure has high voltage blocking capability in both the first and third quadrants of operation, i.e., when both positive and negative bias is applied to the anode terminal. The device can be triggered to the on-state by the application of a small current at the gate terminal when operating in the first quadrant. These features are attractive for power applications with large AC power supply voltages. The gate-turn-off (GTO) thyristor structure is tailored for power circuits operating from a DC power source, such as electric trains. The device can support a high voltage only in the first quadrant. A large gate current is required to turn off the device, making the gate control circuit bulky and expensive.

MOS-gated power thyristor structures were proposed and demonstrated in the 1980s and 1990s to simplify the gate drive requirements for power thyristors while maintaining their desirable low on-state voltage drop when compared with IGBTs. The three basic MOS-gated thyristor structures that have been investigated are illustrated in Fig. 1.4. The first MOS-gated operation of a four-layer thyristor structure was proposed and demonstrated in 1979 for turning on the device [6]. The MOS-controlled thyristor (MCT) was the earliest device structure proposed and demonstrated with the capability for turning off the thyristor in the first quadrant [7, 8]. This device is difficult to fabricate because of the triple diffusions (P^+, N-base, and P-base regions under the gate) required to form the structure. The base resistance controlled thyristor (BRT) structure, which is compatible with the IGBT process, was proposed in the 1980s [9] and demonstrated [10] in the 1990s with thyristor turn-off capability. These device structures require snubber circuits because they do not exhibit significant safe-operating-area. To address this shortcoming, an MOS-gated thyristor structure, called the emitter switched thyristor (EST), with turn-off capability and good safe-operating-area was proposed in the 1980s [11] and demonstrated [12] in the 1990s.

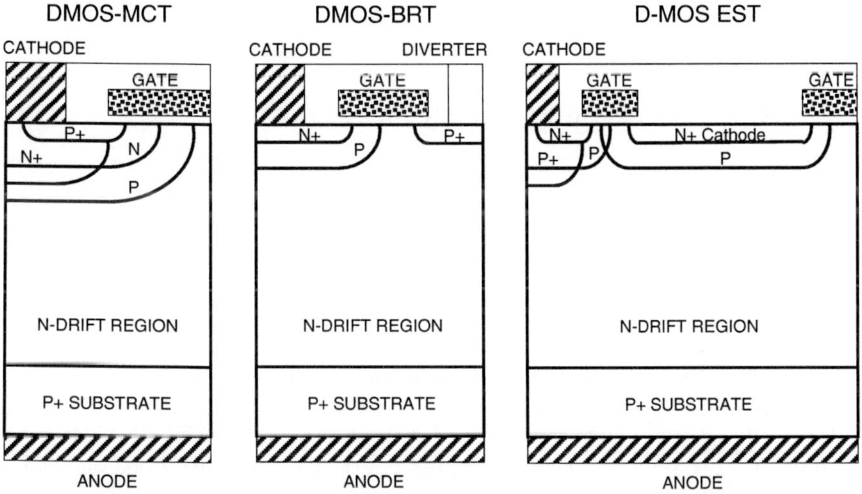

Fig. 1.4 MOS-gated power thyristor structures

The silicon power MOSFET structure has a very high specific on-resistance when designed for operation at high voltages (>5,000 V) [1, 2]. However, the high electric field within silicon carbide drift regions enables achieving a low specific on-resistance for the drift region while taking advantage of the superior switching behavior of unipolar devices [4]. One critical problem with the development of silicon carbide power MOSFETs is the rupture of the gate oxide due to the high electric field in the drift region. The shielded-gate planar power MOSFET structures were proposed [13] and demonstrated [14] in the 1990s to solve this problem. In order to reduce the resistance contributed from the channel, the accumulation-mode structure was also proposed and demonstrated to take advantage of the larger mobility for electrons in accumulation layers. These silicon carbide power MOSFET structures are illustrated in Fig. 1.5.

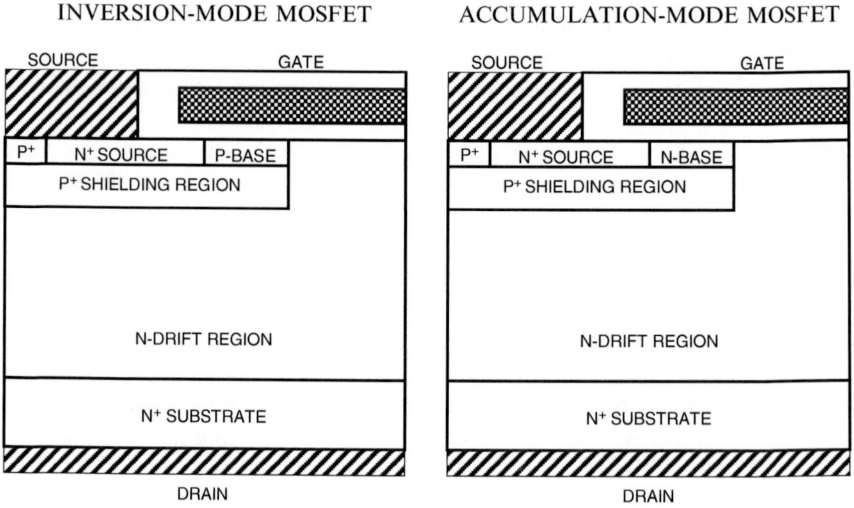

Fig. 1.5 Silicon carbide planar power MOSFET structures

1.3 Revised Breakdown Models for Silicon

In the textbook [1], the breakdown voltage for silicon devices was analyzed by using the Fulop's power law relating the impact ionization coefficient to the electric field. The Fulop's power law [15] for impact ionization in silicon is given by:

$$\alpha_{\mathrm{F}}(\mathrm{Si}) = 1.80 \times 10^{-35} E^7 \tag{1.1}$$

The values for the impact ionization coefficient obtained by using this equation are compared with the impact ionization coefficients measured for electrons and holes in silicon [16] as represented by Chynoweth's equation in Fig. 1.6. It can be observed that Fulop's power law falls between that for electrons and holes and consequently underestimates the values for the impact ionization coefficients for electrons. This results in the prediction of larger breakdown voltages than in actual devices when performing the analytical calculations as pointed out in the textbook.

A better prediction of breakdown in silicon devices using analytical models can be achieved by improving the match between the power law and the measured data for impact ionization coefficients for electrons and holes in silicon. The proposed Baliga's power law for impact ionization in silicon [2] is given by:

$$\alpha_{\mathrm{B}}(\mathrm{Si}) = 3.507 \times 10^{-35} E^7 \tag{1.2}$$

From Fig. 1.6, it can be observed that this equation provides a larger value for the impact ionization coefficients, which will result in reducing the breakdown voltage.

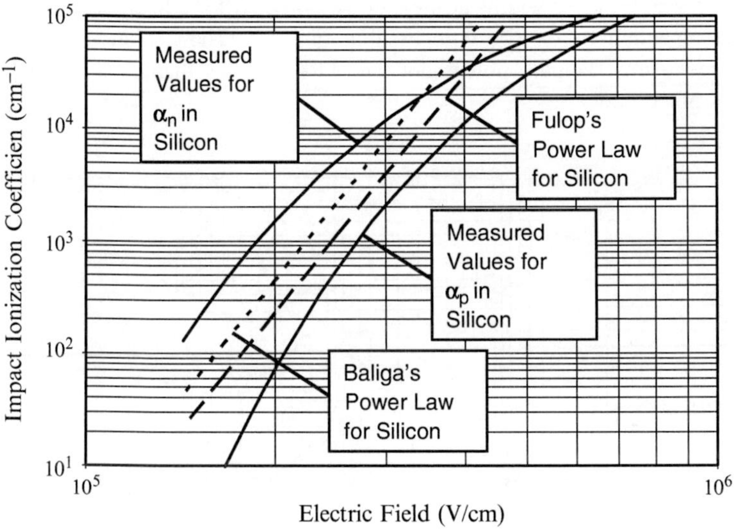

Fig. 1.6 Impact ionization coefficients for silicon

In the case of one-dimensional parallel-plane junctions, discussed in Chap. 3 in the textbook [1], the electric field takes a triangular distribution in the lightly doped side of the P-N junction given by

$$E(x) = -\frac{qN_{\mathrm{D}}}{\varepsilon_{\mathrm{S}}}(W_{\mathrm{D}} - x) \tag{1.3}$$

where W_{D} is the depletion layer width, and N_{D} is the doping concentration on the lightly doped side of the junction. The breakdown voltage in this case is determined by the ionization integral becoming equal to unity:

$$\int_0^{W_{\mathrm{D}}} \alpha \, \mathrm{d}x = 1 \tag{1.4}$$

Substituting Eq. 1.2 into the above equation with the distribution given by Eq. 1.3, an expression for the depletion layer width at breakdown can be obtained:

$$W_{\mathrm{PP,B}}(\mathrm{Si}) = 2.404 \times 10^{10} N_{\mathrm{D}}^{-7/8} \tag{1.5}$$

In contrast, the expression for the depletion layer width at breakdown obtained by using Fulop's power law is given by:

$$W_{\mathrm{PP,F}}(\mathrm{Si}) = 2.67 \times 10^{10} N_{\mathrm{D}}^{-7/8} \tag{1.6}$$

The depletion layer widths at breakdown obtained for silicon devices by using the above equations can be compared in Fig. 1.7. The depletion layer widths computed using Baliga's power law are 11% smaller than those predicted by Fulop's power law.

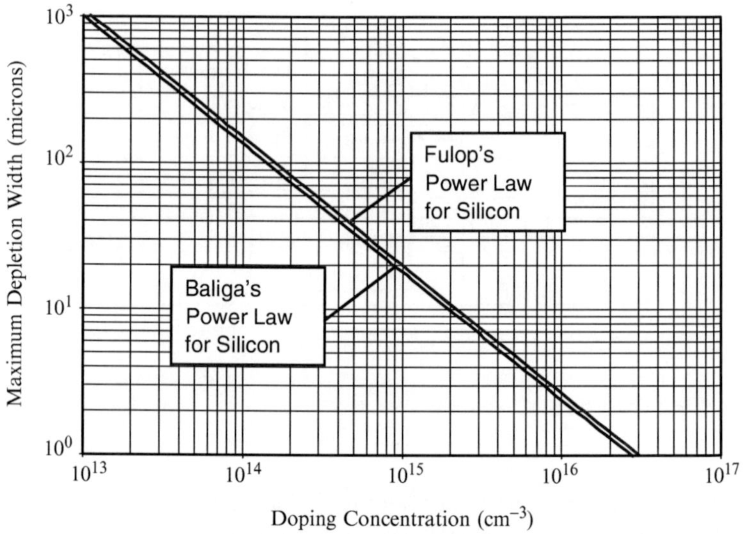

Fig. 1.7 Depletion layer width at breakdown in silicon for the one-dimensional parallel-plane junction

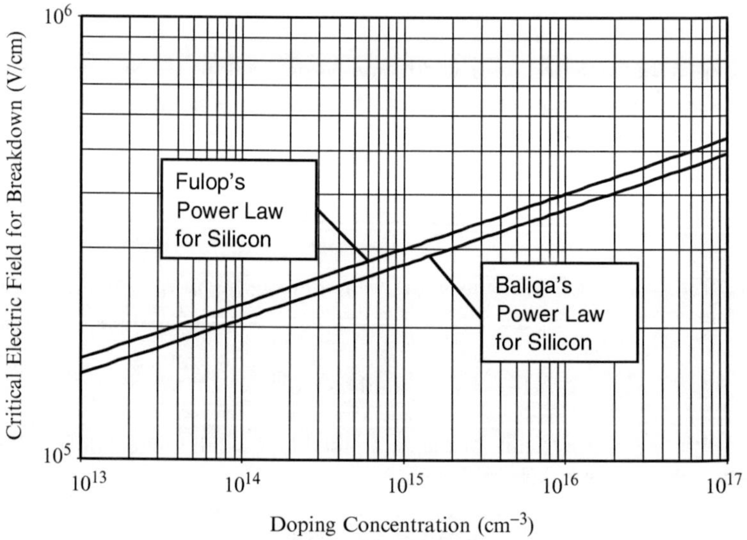

Fig. 1.8 Critical electric fields for breakdown in silicon for the one-dimensional parallel-plane junction

The maximum electric field located at the P-N junction for the one-dimensional parallel-plane case is given by:

$$E_M = \frac{qN_D}{\varepsilon_S} W_D \tag{1.7}$$

The critical electric field for breakdown of the one-dimensional parallel-plane junction can be obtained by substituting the depletion layer width at breakdown into the above equation. In the case of Baliga's power law, the critical electric field for breakdown of the one-dimensional parallel-plane junction is given by:

$$E_{C,1D,B}(\text{Si}) = 3,700 N_D^{1/8} \tag{1.8}$$

In contrast, in the case of Fulop's power law, the critical electric field for breakdown of the one-dimensional parallel-plane junction is given by:

$$E_{C,1D,F}(\text{Si}) = 4,010 N_D^{1/8} \tag{1.9}$$

The critical electric fields for breakdown of the one-dimensional parallel-plane junction obtained for silicon devices by using the above equations can be compared in Fig. 1.8. The critical electric fields for breakdown of the one-dimensional parallel-plane junction computed using Baliga's power law are 8.4% smaller than those predicted by Fulop's power law.

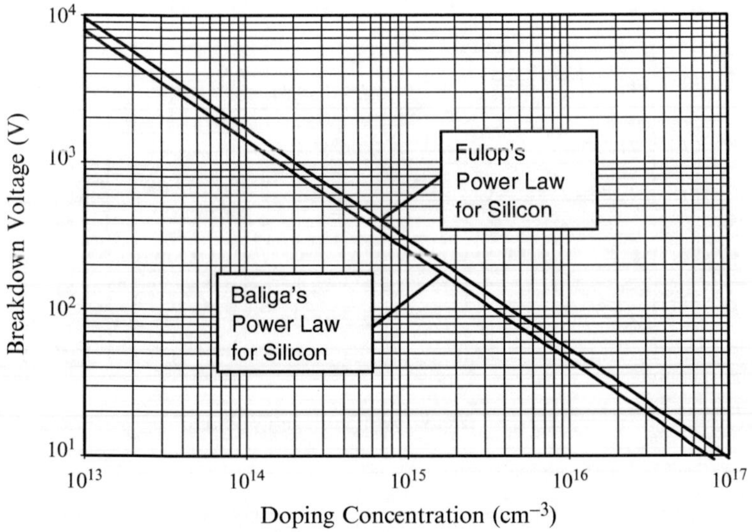

Fig. 1.9 One-dimensional parallel-plane breakdown voltages in silicon

The one-dimensional parallel-plane breakdown voltage for abrupt P-N junctions in silicon can be computed using the critical electric field and the depletion layer width at breakdown:

$$BV_{PP} = \frac{1}{2} E_{C,1D} W_{PP} \tag{1.10}$$

Using the equations derived above for the critical electric field and the depletion layer width at breakdown with Baliga's power law for the impact ionization coefficients, the breakdown voltage of the one-dimensional parallel-plane junction is given by:

$$BV_{PP,B}(Si) = 4.45 \times 10^{13} N_D^{-3/4} \tag{1.11}$$

In contrast, in the case of Fulop's power law, the breakdown voltage of the one-dimensional parallel-plane junction is given by:

$$BV_{PP,F}(Si) = 5.34 \times 10^{13} N_D^{-3/4} \tag{1.12}$$

The breakdown voltage of the one-dimensional parallel-plane junction obtained for silicon devices by using the above equations can be compared in Fig. 1.9. The breakdown voltages for the one-dimensional parallel-plane junction computed using Baliga's power law are 20% smaller than those predicted by Fulop's power law.

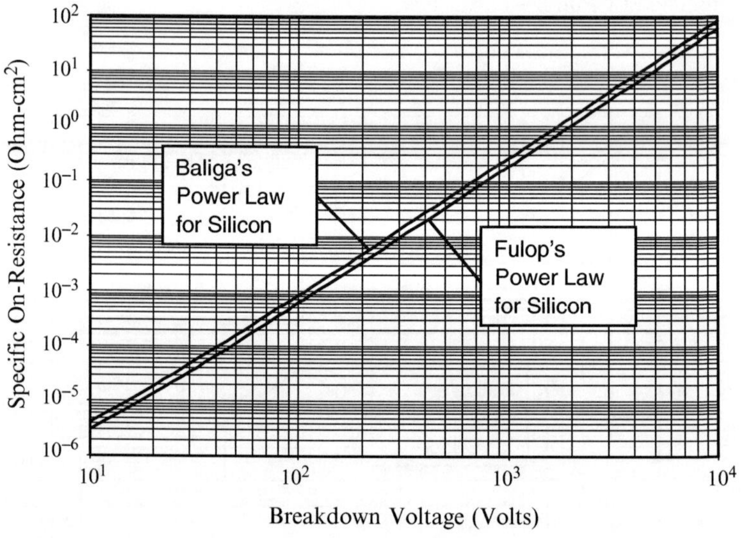

Fig. 1.10 Ideal specific on-resistances for silicon

The ideal specific on-resistance is defined as the resistance per unit area for the drift region with the doping concentration and thickness corresponding to each breakdown voltage. This resistance can be computed using the following equation:

$$R_{\text{on,sp}}(\text{Ideal}) = \frac{W_{\text{PP}}}{q\mu_n N_{\text{D}}} \tag{1.13}$$

It is important to include the dependence of the mobility on the doping concentration when computing the ideal specific on-resistance. The ideal specific on-resistance for silicon devices is slightly larger when the Baliga's power law for the impact ionization is utilized as compared with Fulop's power law, as shown in Fig. 1.10. At breakdown voltages above 40 V, the mobility in silicon can be assumed to be independent of the doping concentration. The ideal specific on-resistance obtained by using Baliga's power law for the impact ionization coefficients is then given by:

$$R_{\text{on,sp,B}}(\text{Si}) = 8.37 \times 10^{-9}\,\text{BV}^{2.5} \tag{1.14}$$

In contrast, the ideal specific on-resistance obtained by using Fulop's power law for the impact ionization coefficients is given by:

$$R_{\text{on,sp,F}}(\text{Si}) = 5.93 \times 10^{-9}\,\text{BV}^{2.5} \tag{1.15}$$

The ideal specific on-resistances for silicon devices computed using Baliga's power law are 40% larger than those predicted by Fulop's power law.

The revised information provided above based upon using Baliga's power law for the impact ionization coefficients is intended to bring the analytical calculations of breakdown voltages in silicon devices more in-line with the results of numerical simulations. This information can be used to compute the doping concentration and thickness of the drift region to achieve a desired breakdown voltage for power devices.

1.4 Typical High Voltage Applications

Power devices are commonly used to control power flow to loads. Two typical examples for the application of high voltage (>5,000 V) power devices are provided in this section to emphasize the characteristics of importance from an application standpoint. The first example is in variable speed motor drives. This application is popular for traction (electric train) applications. The second example is in the power transmission and distribution used to deliver power from the generation site to homes and factories.

1.4.1 Variable-Frequency Motor Drive

Electric trains require operation of motors over a wide range of frequencies in order to alter the speed of the locomotive. The most commonly used topology converts the constant frequency input AC power to a DC bus voltage and then uses an inverter stage to produce the variable frequency output power [17] that controls the speed of the motor. The circuit diagram for a three-phase motor drive system is shown in Fig. 1.11. Six IGBTs are used with six fly-back rectifiers in the inverter stage to deliver the variable frequency power to the motor windings. A pulse-width-modulation (PWM) scheme is used to generate the variable frequency AC voltage waveform that is fed to the motor windings [18].

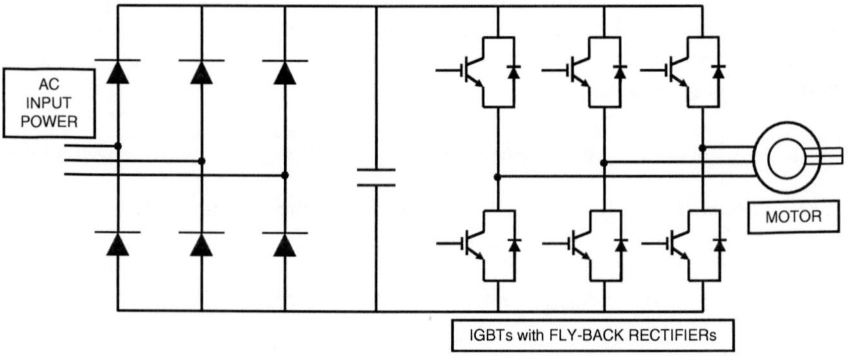

Fig. 1.11 Variable frequency motor drive circuit

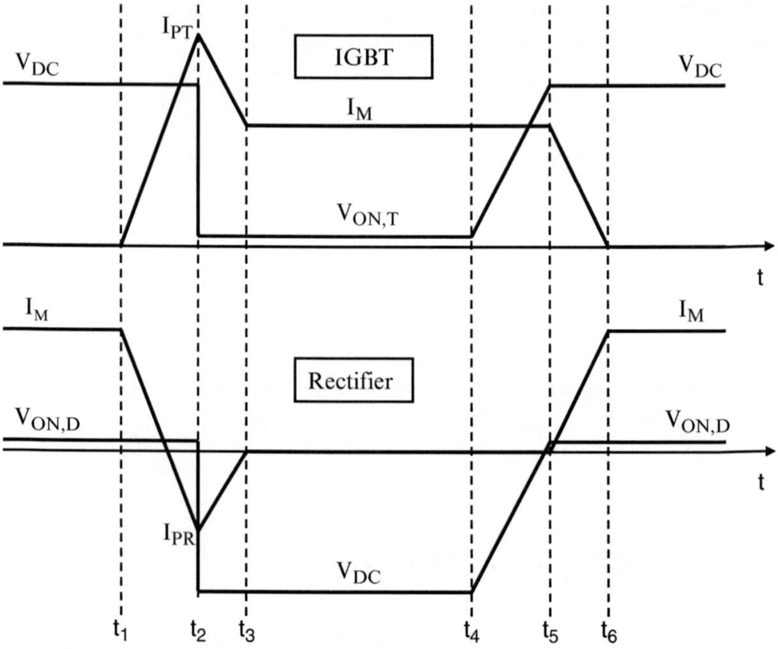

Fig. 1.12 Linearized waveforms for the PWM motor drive circuit

During each cycle of the PWM period, the current in the motor winding can be considered to remain approximately constant. This allows linearization of the waveforms for the current and voltage experienced by the IGBTs and the rectifiers. Typical waveforms for the transistor and the fly-back diode are illustrated in Fig. 1.12. The large reverse recovery current typically observed in silicon P-i-N rectifiers during the time interval from t_1 to t_3 produces high power dissipation not only in the diodes but also in the transistors [1]. This power loss can be eliminated by replacing the silicon P-i-N rectifiers with silicon carbide Schottky rectifiers.

With the availability of high voltage silicon carbide Schottky rectifiers, the power loss in the motor control application becomes dominated by the on-state and turn-off losses in the IGBTs [19]. The development of alternate high performance, high voltage devices can reduce these power losses, making the motor drive more efficient. The characteristics of various device structures discussed in this monograph are therefore compared with those of the silicon GTO and IGBT.

1.4.2 High Voltage Direct Current (HVDC) Power Transmission and Distribution

Power transmission is generally performed over long distances because of the geographical separation between the energy source and the end user. The power sources are typically hydroelectric dams, nuclear power stations, or coal-fired power plants. There is also an increasing interest in the generation of electricity from renewable energy sources such as wind power and solar power. Typical end users are residences, offices, and factories. The power from the source can be transmitted to the end user by using either an AC or DC power transmission system. DC power transmission is favored over AC power transmission because of cable charging losses in an AC system which limit the power transmission distance. This is particularly true for submarine cables but is also applicable to landlines [3]. The cost of AC power transmission is also increased because more AC lines are needed to deliver the same power over the same distance due to system stability limitations, the need for intermediate switching stations, and the need for reactive power compensation. Further, back-to-back converters with asynchronous HVDC links act as an effective firewall against propagation of cascading outages [20].

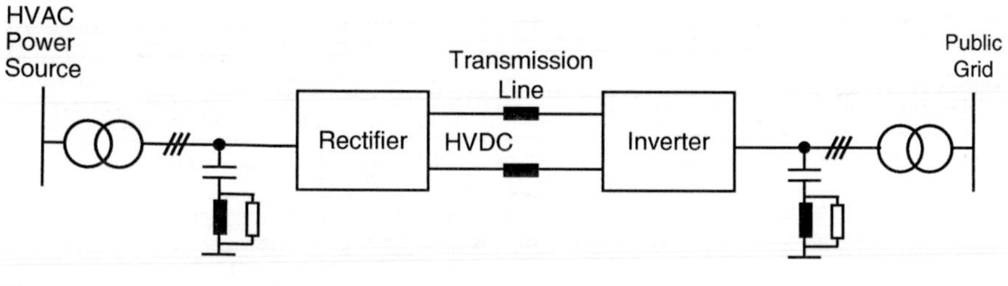

Fig. 1.13 HVDC power transmission system

A typical HVDC power transmission system is illustrated in Fig. 1.13. The power is transmitted at very high voltages (above 100 kV) in order to reduce the current on the cables. Large currents in cables require more copper which adds to the cost and weight. Since power semiconductor devices are unable to withstand such high voltages, it is necessary to connect many devices in series to satisfy the system requirements. In addition, for higher power levels, many devices may have to be connected in parallel as well. The series and parallel combination of power devices comprises an HVDC valve. The most common configuration for modern overhead HVDC transmission lines is bipolar because it provides two independent DC circuits each capable of operating at half capacity [20]. Two basic converter topologies are used in modern HVDC transmission systems: conventional line-commutated, current-source converters (CSC) based upon thyristor valves and self-commutated, voltage source converters (VSC) based upon IGBT valves. Each valve consists of a large number of series-connected thyristors or IGBTs to sustain the desired DC voltage rating. In the case of current source converters with thyristor valves, a Graetz bridge configuration is used, allowing six commutations or switching operations per period. Self-commutated, voltage source converters using IGBTs are preferred because they allow independent rapid control of both active and reactive power. Reactive power can also be controlled at each end of the transmission line providing total flexibility in network design.

In order to ensure equal voltage distribution between the power devices that are connected in series, it is necessary to include passive components with the power devices. The passive networks add to the cost and increase power losses in the power transmission system. The triggering of thyristors connected in series is complicated by the level-shifting of the gate drive signals. It is, therefore, advantageous to reduce the number of power devices that are connected in series by increasing the blocking voltage capability for each device. In the case of silicon thyristors, the blocking voltage has been increased up to 10 kV, with further increase inhibited by the basic properties of the material as shown in this monograph. Higher voltage thyristors are possible by utilizing silicon carbide as the semiconductor as discussed in this monograph.

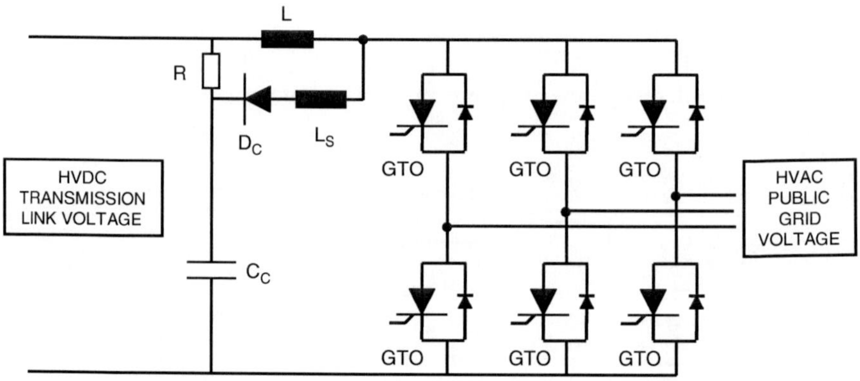

Fig. 1.14 GTO-valve with snubber circuit

The self-commutated, voltage source converters can be constructed using GTOs. In this case it is necessary to include a snubber circuit, as illustrated in Fig. 1.14, which controls the rate of rise in the current through the device when it is turned on. The inductance (L) is used to limit the rate of rise of current when one of the GTOs is turned on. This is necessary to control the reverse recovery of the antiparallel diodes and avoid very high reverse recovery currents that would damage the devices. The energy stored in the inductance (L) is subsequently dissipated in resistance (R). The clamping capacitor (C_C) is used to minimize the voltage overshoot during the transients. These additional components increase the cost and degrade the efficiency of the GTO-based VSC inverters [21].

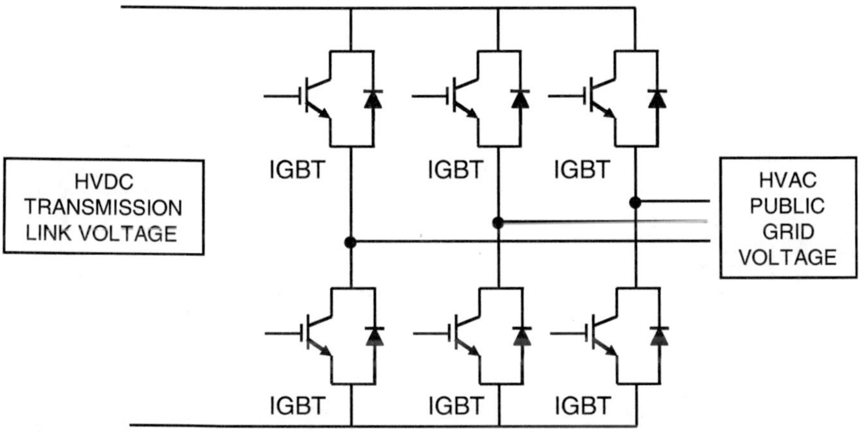

Fig. 1.15 IGBT-valve

The self-commutated, voltage source converters can be constructed using IGBTs without the snubbers required for GTOs, as illustrated in Fig. 1.15. The rate of rise of the current in the IGBT can be controlled by tailoring the gate drive voltage waveform without any ancillary components. This allows controlling the reverse recovery of the antiparallel rectifiers without the snubbers. The reduced passive components in the IGBT-based VSC inverters reduce system cost. The IGBT-based H-bridge configuration, shown in Fig. 1.15, is remarkably similar to that used for motor control application as discussed in the previous section. Consequently, voltage source converters with pulse width modulation have evolved. The high commutation frequency used in these VSCs allows elimination of the converter transformer leading to the concept being called "HVDC Light" [22]. This approach is attractive for connecting groups of windmills to an overall network [23] and for power transmission from offshore wind farms [24].

With increasing awareness of the need to mitigate carbon emissions by the generation of electricity from renewable energy sources, the smart grid concept has been gaining popularity. A smart grid integrates energy sources with end users in residential communities and industrial parks. In order to balance the variable power generation from renewable sources with variable demand from end users, the smart grid also incorporates energy storage capability. In addition, it is necessary to make the smart grid fault tolerant to provide high power quality. The development of the technology infrastructure for the smart grid is being pursued in an NSF-sponsored engineering research center titled “FREEDM: Future Renewable Electric Energy Delivery and Management System” [25]. The new system, shown in Fig. 1.16, envisions a plug-and-play capability for all the components consisting of the distributed energy storage devices (DESD), distributed renewable energy sources (DRER), and power loads. This capability is made possible by utilization of the intelligent energy management (IEM) units as the interface between the components in the power network and power grid. The system also contains intelligent fault management (IFM) units to prevent the network from going down during localized faults within the network.

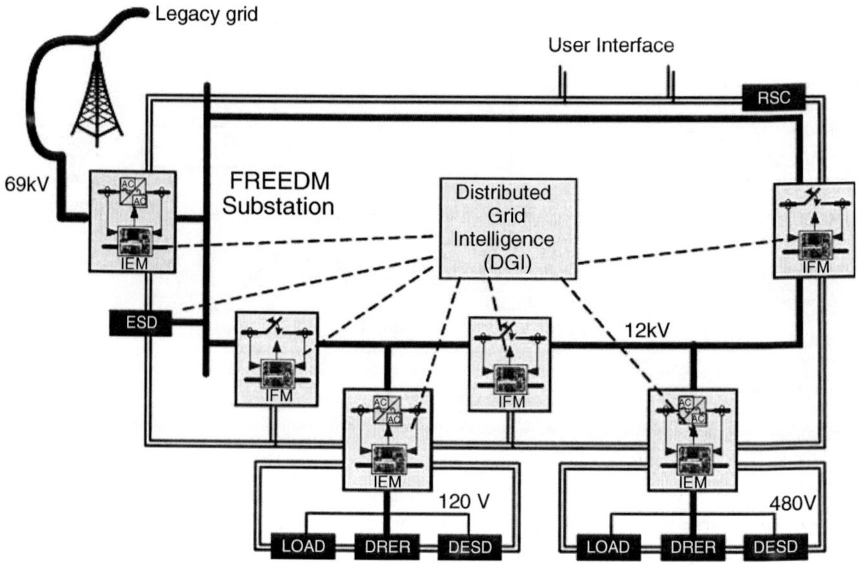

Fig. 1.16 Smart grid concept

The centerpiece of the intelligent energy management (IEM) unit is the solid-state transformer (SST). Unlike the conventional AC-to-AC 60-Hz transformer that has been the workhorse of the power community for more than a century, the SST is designed to operate at a high (typically 10 kHz) frequency to reduce the size and weight of the magnetic material. However, this requires conversion of the very high (12 kV AC) power distribution voltage to a very high (18 kV) DC bus voltage using an AC-to-DC rectifier stage, as shown in Fig. 1.17, followed by a DC-to-AC inverter stage to convert the DC power to 10 kHz AC power that can be fed to the primary of the high-frequency transformer. The transformer reduces the high voltage AC power to a lower voltage suitable for delivery to consumers and the industry. In the example shown in Fig. 1.17, the secondary side AC output voltage of the SST is assumed to be 120 V, suitable for delivery of power to residences. In order to achieve this outcome, the high-frequency voltage on the secondary side of the high-frequency transformer is first converted to a 200 V DC bus using the AC-to-DC rectifier stage. The DC voltage is then converted to 120 V AC voltage using the DC-to-AC inverter stage. The power control algorithms used in the SST can be used for power factor correction and to provide fail-safe operation, which are significant advantages when compared with the conventional transformer.

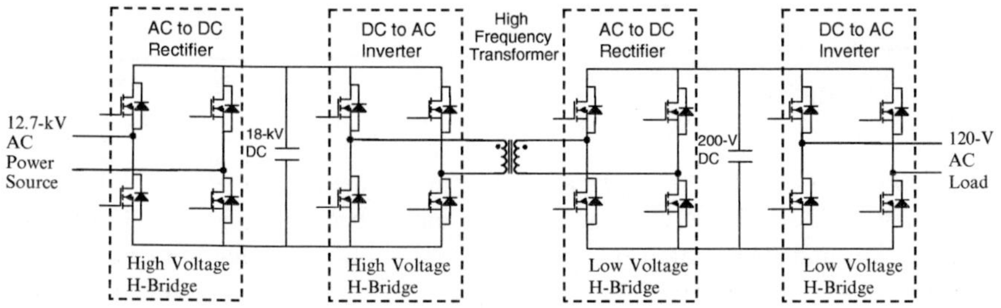

Fig. 1.17 The solid-state transformer (SST)

In order for the IEM concept to be viable, the efficiency of the SST must approach the very high efficiency of the conventional 60-Hz transformers. The efficiency of the SST is determined by power losses in the power switches and rectifiers. Due to the relatively high operating frequency, switching power losses are particularly important although the on-state losses cannot be ignored. In Fig. 1.17, the power switches are all shown as power MOSFETs as representative devices. However, the optimum commercially available silicon power devices are 6 kV IGBTs for the primary side of the transformer. These devices would have to be used in an interleaved topology on the primary side because their voltage rating is insufficient to withstand a peak 18 kV AC voltage. The interleaved topology greatly increases the number of transistors and rectifiers required for the SST which adds to the power losses and cost. An alternate option is to utilize 18 kV 4H-SiC power transistors. The best transistor for this application has been determined to be the IGBT structure [26], although the 18 kV SiC power MOSFET is still a contender, in spite of its large on-resistance, due to its superior switching losses.

On the secondary side of the transformer, the optimum silicon power devices are either 600 V IGBTs, 600 V COOLMOS transistors, or 600 V GD-MOSFETs [2]. In the future, power MOSFET structures with reduced specific on-resistance may be possible by using GaN HEMT structures [27]. The HEMT structure creates a very high sheet charge electron density between the gate and the drain, resulting in reduced specific on-resistance. The cost for these devices is reduced, making their manufacturability viable, by the growth of the GaN layer on silicon substrates.

The reverse recovery behavior of the fly-back rectifiers used across each of the power switches also plays a major role in determination of the power losses.

On the primary side, 5 kV silicon P-i-N rectifiers can be used with the 6 kV IGBTs if the interleaved topology is used. Alternately, 18 kV rectifiers must be developed using silicon carbide as the base material. Although high voltage 4H-SiC P-i-N rectifiers have been reported, their reverse recovery switching losses are a problem at the high (10 kHz) operating frequency of the SST. The switching losses are much smaller for the 4H-SiC JBS rectifier but it has a very high on-state voltage drop. The most promising option is the high voltage 4H-SiC MPS rectifier which has an on-state voltage drop similar to the P-i-N rectifier but a much lower reverse recovery power loss [19].

1.5 Conclusions

The applications for high voltage (>5,000 V) power devices have been reviewed in this chapter. These devices are required in motor control for traction, and for power transmission and distribution applications. The application requirements for the devices have been reviewed in this chapter. Various power device structures suitable for these applications are discussed in detail in subsequent chapters of this monograph.

References

1. B. J. Baliga, "Fundamentals of Power Semiconductor Devices", Springer Science, New York, 2008.
2. B. J. Baliga, "Advanced Power MOSFET Concepts", Springer Science, New York, 2009.
3. H. Akagi, "Large Static Converters for Industry and Utility Applications", Proceedings of the IEEE, Vol. 89, pp. 976–983, 2001.
4. B. J. Baliga, "Silicon Carbide Power Devices", World Scientific Press, Singapore, 2005.
5. J. Wang, et al, "Smart Grid Technologies: Development of 15-kV SiC IGBTs and Their Impact on Utility Applications", IEEE Industrial Electronics Magazine, pp. 16–23, June 2009.
6. B. J. Baliga, "Enhancement and Depletion Mode Vertical Channel MOS-Gated Thyristors", Electronics Letters, Vol. 15, pp. 645–647, 1979.
7. V. A. K. Temple, "MOS Controlled Thyristors", IEEE International Electron Devices Meeting, Abstract 10.7, pp. 282–285, 1984.
8. M. Stoisiek and H. Strack, "MOS-GTO - A Turn-off Thyristor with MOS Controlled Emitter Shorts", IEEE International Electron Devices Meeting, Abstract 6.5, pp. 158–161, 1985.
9. B. J. Baliga, "Gated Base Controlled Thyristor", U.S. Patent 5,099,300, Filed June 14, 1990, Issued March 24, 1992.
10. M. Nandakumar, et al, "A New MOS-Gated Power Thyristor Structure with Turn-off achieved by Controlling the Base Resistance", IEEE Electron Device Letters, Vol. EDL-12, pp. 227–229, 1991.
11. D. N. Pattanayak and B. J. Baliga, "Monolithically Integrated Insulated Gate Semiconductor Device", U.S. Patent 4,847,671, Issued July 11, 1989.
12. M. S. Shekar, et al, "Characteristics of the Emitter Switched Thyristor", IEEE Transactions on Electron Devices, Vol. ED-38, pp. 1619–1623, 1991.

13. B. J. Baliga, "Silicon Carbide Semiconductor Devices having Buried Silicon Carbide Conduction Barrier Layers Therein", U.S. Patent 5,543,637, Issued August 6, 1996.
14. P. M. Shenoy and B. J. Baliga, "The Planar 6 H-SiC ACCUFET", IEEE Electron Device Letters, Vol. EDL-18, pp. 589–591, 1997.
15. W. Fulop, "Calculation of Avalanche Breakdown in Silicon P-N Junctions", Solid State Electronics, Vol. 10, pp. 39–43, 1967.
16. R. Van Overstraeten and H. DeMan, "Measurements of the Ionization Rates in Diffused Silicon P-N Junctions", Solid State Electronics, Vol. 13, pp. 583–608, 1970.
17. J. D. van Wyk, "Power Electronic Converters for Drives", pp. 81–137, in 'Power Electronics and Variable Frequency Drives', Edited by B.K. Bose, IEEE Press, 1997.
18. J. Holtz, "Pulse Width Modulation for Electronic Power Conversion", pp. 138–208, in 'Power Electronics and Variable Frequency Drives', Edited by B.K. Bose, IEEE Press, 1997.
19. B. J. Baliga, "Advanced Power Rectifier Concepts", Springer-Science, New York, 2009.
20. M.P. Bahrman, "HVDC Transmission Overview", IEEE Transmission and Distribution Conference and Exposition, pp. 1–7, 2008.
21. E.I. Carroll, "Power Electronics for Very High Power Applications", Power Engineering Journal, Vol. 13, pp. 81–87, April, 1999.
22. V.F. Lescale, "Modern HVDC: State of the Art and Development Trends", Proceedings of the International Conference on Power Systems and Technology, Vol. 1, pp. 446–450, 1998.
23. S.K. Olsen and O. Tonnesen, "Power Applications for Superconducting Cables in Denmark", IEEE Transactions on Applied Superconductivity, Vol. 9, pp. 1285–1288, 1999.
24. T. Volker, et al, "New HVDC Concept for Power Transmission from Off-shore Wind Farms", EPE Seminar on 'Wind Power to the Grid', EPE-WECS, pp. 1–6, 2008.
25. A. Huang and B. J. Baliga, "FREEDM System", IEEE International Symposium on Power Semiconductor Devices and ICs, Abstract A-3, pp. 9–12, 2009.
26. W. Sung, et al, "Design and Investigation of the Frequency Capability of 15 kV 4 H-SiC IGBTs", IEEE International Symposium on Power Semiconductor Devices and ICs, pp. 271–274, 2009.
27. F. Medjdoub, et al, "Low On-Resistance High Breakdown Normally Off AlN/GaN/AlGaN DHFET on Si Substrate", IEEE Electron Device Letters, Vol. 31, pp. 111–113, 2010.

Chapter 2
Silicon Thyristors

As discussed in the textbook [1], the power thyristor was developed as a replacement for the thyratron, a vacuum tube used for power applications prior to the advent of solid-state devices. The simple construction of these structures using P-N junctions enabled commercialization of devices in the 1950s. These devices were found to be attractive from an applications viewpoint because they eliminated the need for the cumbersome filaments required in vacuum tubes and were much more rugged and smaller in size. The power thyristor provides both forward and reverse voltage blocking capability, making it well suited for AC power circuit applications. The device can be triggered from the forward-blocking off-state to the on-state by using a relatively small gate control current. Once triggered into the on-state, the thyristor remains stable in the on-state even without the gate drive current. In addition, the device automatically switches to the reverse-blocking off-state upon reversal of the voltage in an AC circuit. These features greatly simplify the gate control circuit, relative to that required for the power transistor, reducing its cost and size.

Due to significant interest in the development of solid-state devices for the control of motors operating from DC power sources, a structure called the *gate turn-off thyristor* (abbreviated as GTO), was also developed in the 1960s. In this device, the structure is modified to enable the switching of the device from the on-state to the off-state while operating in the first quadrant. This is performed by the application of a large reverse gate drive current, akin to that used for turning off the bipolar power transistor. In spite of the bulky and expensive gate control circuits required for the GTO, it was widely adopted for the control of motors in traction (electric street-cars and electric locomotives) applications until recently. The scaling of the power handling capability of the insulated gate bipolar transistor (IGBT) to handle very high (megawatt) power levels in the twenty-first century has resulted in the displacement of these devices by the IGBT in traction applications.

B.J. Baliga, *Advanced High Voltage Power Device Concepts*,
DOI 10.1007/978-1-4614-0269-5_2, © Springer Science+Business Media, LLC 2011

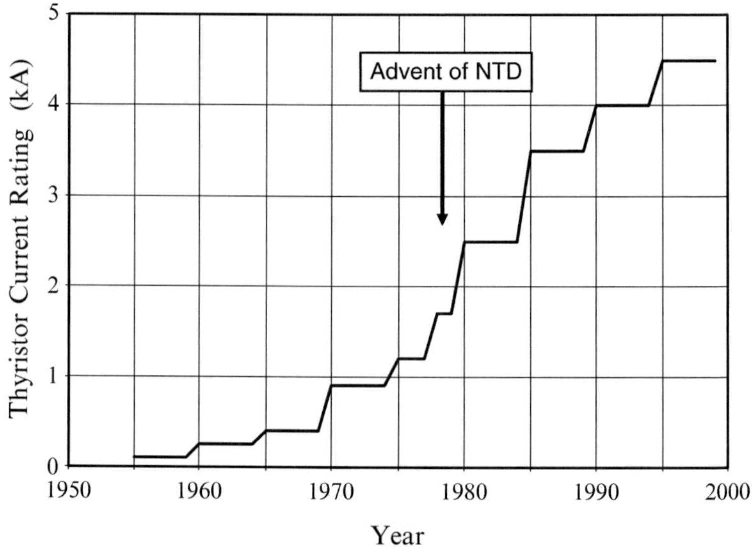

Fig. 2.1 Growth in current ratings for the thyristor

The ability to control operation between the blocking and on states for a thyristor by using a third terminal was first reported in the 1950s [2, 3]. The extensive application potential for these devices to home appliances and power distribution systems generated strong interest in making improvements in the power ratings for the thyristors. The growth of the current handling capability for the power thyristors is charted in Fig. 2.1. Starting with a modest current of 100 A in the 1950s, the current rating has been scaled to approach 5,000 A for a single device. These high current levels are required for power distribution systems, such as high voltage DC transmission (HVDC) networks. From the figure, it can be observed that the most rapid increase in the current handling capability took place at the end of the 1970s. This outcome can be traced to the development of the neutron-transmutation-doping (NTD) process in the mid-1970 time frame. Using the NTD process, it became possible to obtain larger diameter silicon wafers with uniform properties enabling the observed scaling of the current handling capability of thyristors.

There was a concomitant increase in the voltage blocking capability for the thyristor as illustrated in the chart in Fig. 2.2. Beginning with devices capable of operating upto a few hundred volts in the 1950s, the voltage rating for thyristors has been escalated to 8,000 V. The increase in the voltage rating had to be accomplished by the availability of higher resistivity silicon wafers. This was initially achieved by the development of the float-zone process. However, the resistivity variation produced by this process was inadequate for utilization in the large diameter wafers desired to increase the current ratings. The NTD process was instrumental in providing the breakthrough required to create large diameter silicon wafers with low n-type doping concentration and high uniformity in the resistivity across the wafers. Consequently, a substantial gain in the voltage rating occurred in the late 1970s after the commercial availability of NTD silicon as indicated in the chart.

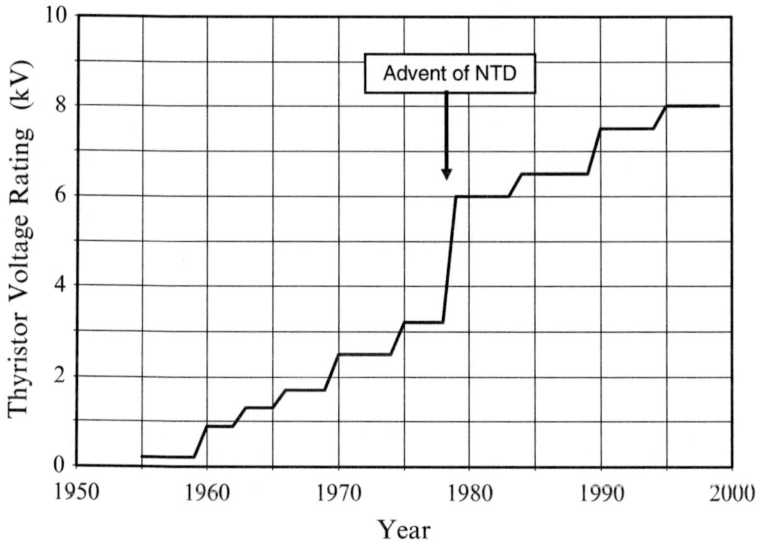

Fig. 2.2 Growth in voltage ratings for the thyristor

Today, single thyristors are available with the capability to block over 8,000 V and conduct 5,000 A in the on-state. Consequently, a single thyristor device can control 40 MW of power. Such devices are attractive for power distribution networks to reduce the total number of devices required in a HVDC station. The reduction in the number of devices connected in series and parallel provides the added benefits of a smaller number of other components that are needed in the system to ensure proper voltage and current distribution between the multiple thyristors. Light-triggered thyristors have also been developed to enable stacking them in a series string to hold off the very high voltages (in excess of 100 kV) that are commonplace for power distribution.

The basic structure and operation of the thyristor were discussed in the textbook. The thyristor contains two coupled bipolar transistors that provide an internal positive feedback mechanism that allows the device to sustain itself in the on-state. This internal feedback mechanism makes it difficult to turn off the structure by external means. In order to enable operation at elevated temperatures, it is necessary to short-circuit the emitter and base regions of the thyristor. The impact of this on the gate control current and switching behavior was also analyzed in the textbook. Analytical models were provided in the textbook for all the operating modes of the thyristor, including the switching transient for the GTO. These models are applicable to the devices with high (>5,000 V) blocking voltage capability discussed in this monograph. The characteristics of high voltage silicon carbide thyristors are also considered in the next chapter for blocking voltage rating of 20 kV because there are no high performance silicon devices available with blocking voltages above 10 kV.

2.1 Power Thyristor Structure and Operation

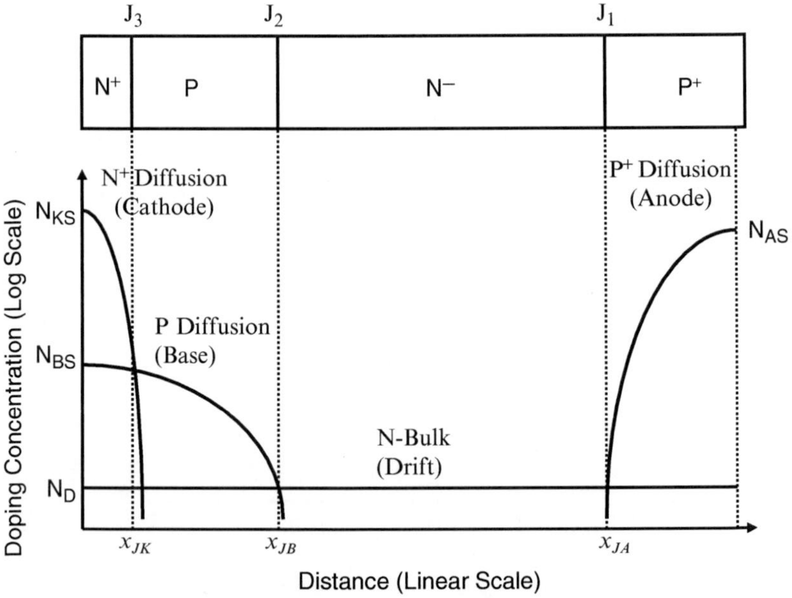

Fig. 2.3 The power thyristor structure and its doping profile

The basic structure for an N^+-P-N-P^+ power thyristor is illustrated in Fig. 2.3. The structure is usually constructed by starting with a lightly doped N-type silicon wafer whose resistivity is chosen based upon the blocking voltage rating for the device. The anode P^+ region is formed by the diffusion of dopants from the backside of the wafer to a junction depth x_{JA}. The P-base and N^+ cathode regions are formed by the diffusion of dopants from the front of the wafer to a depth of x_{JB} and x_{JK}, respectively. Electrodes are formed on the front side of the wafer to contact the cathode and P-base regions, and on the backside of the wafer to contact the anode region. No contact electrode is usually attached to the N-drift (N-base) region.

In order to achieve such high voltage blocking capability in both forward and reverse operating quadrants, the P^+ anode/N-drift junction and the P-base/N-drift junction must have a highly graded doping profile. This increases the blocking voltage capability due to voltage supported on the more highly doped side of the junction and allows the utilization of the positive and negative bevels at the edges to suppress premature breakdown at the junction termination. The highly graded junctions can be produced by using gallium and aluminum as the p-type dopants instead of boron, the more commonly used dopant for integrated circuits as well as other power devices discussed in the book. These dopants have much larger diffusion coefficients than boron, allowing the production of large junction depths (20–100 μm) in a reasonable processing time. Gallium is used as the dopant for the P^+ anode region because of its high solid solubility. Aluminum is used as the dopant for the P-base region due to its lower solid solubility. The doping concentration for the P-base region of the thyristor must be in the range of $1 \times 10^{16} - 1 \times 10^{17}$ cm^{-3} to obtain a reasonable gain for the internal N-P-N bipolar transistor and for enabling edge termination with a negative bevel for the P-base/N-drift junction.

The output characteristics for the thyristor structure are illustrated in Fig. 2.4. The thyristor structure contains three P-N junctions that are in series as indicated in Fig. 2.3. When a negative bias is applied to the anode terminal of the device, the P^+ anode/N-drift junction (J_1) and the N^+ cathode/P-base junction (J_3) become reverse biased, while the P-base/N-drift junction (J_2) becomes forward biased. Due to high doping concentrations on both sides of the N^+ cathode/P-base junction (J_3), it is capable of supporting less than 50 V. Consequently, most of negative bias applied to the anode terminal is supported by the P^+ anode/N-drift junction (J_1). The reverse blocking voltage capability for the device is determined by the doping concentration and thickness of the N-drift region. Note that an open-base bipolar transistor is formed within the thyristor structure between junctions J_1 and J_2. Consequently, the breakdown voltage is not determined by the avalanche breakdown voltage but by the open-base transistor breakdown voltage. The width of the N-drift region between these two junctions must be carefully optimized to maximize the blocking voltage capability and minimize the on-state voltage drop.

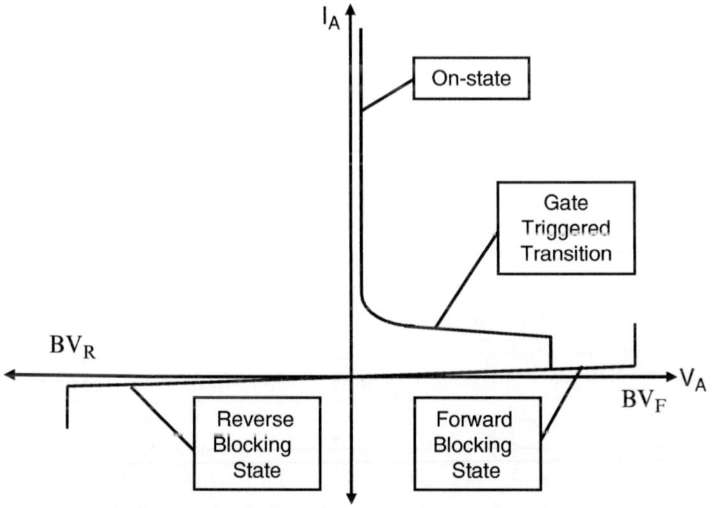

Fig. 2.4 Output characteristics of the power thyristor structure

When a positive bias is applied to the anode terminal of the thyristor, the P^+ anode/N-drift junction (J_1) and the N^+ cathode/P-base junction (J_3) become forward biased while the junction (J_2) between the P-base region and the N-drift region becomes reverse biased. The applied positive bias is mostly supported across the N-drift region. As in the case of reverse blocking operation, the blocking voltage capability is determined by open-base transistor breakdown rather than avalanche breakdown. The reverse and forward blocking capability for the thyristor structure must be approximately equal, making it suitable for use in AC power circuits. This is achieved by using cathode shorts to reduce the gain of the N-P-N transistor at low leakage current levels. A cross section of the thyristor with the cathode short is illustrated in Fig. 2.5.

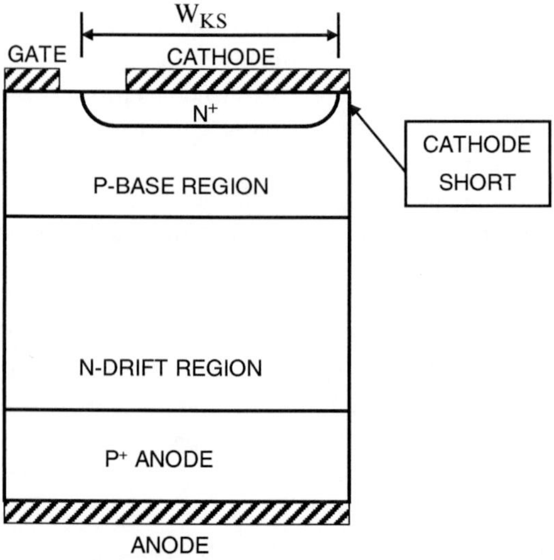

Fig. 2.5 Thyristor structure with cathode short

Current flow through the thyristor can be induced in the first quadrant of operation by using a current supplied through the gate terminal to trigger the device into its on-state. The gate current forward biases the N^+ cathode/P-base junction (J_3) to initiate the injection of electrons. The injected electrons trigger a positive feedback mechanism produced by the two coupled bipolar transistors within the thyristor structure. The first bipolar transistor is an N-P-N transistor formed between the N^+ cathode/P-base/N-drift regions while the second bipolar transistor is a P-N-P transistor formed between the P^+ anode/N-drift/P-base regions. Once current flow is initiated through the transistors, they are able to provide the base drive current for each other by a process referred to as *regenerative action*. In this process, the collector current of the N-P-N transistor provides the base drive current for the P-N-P transistor and the collector current of the P-N-P transistor provides the base drive current for the N-P-N transistor. The regenerative action inherent within the thyristor structure allows stable operation of the device in its on-state without any external gate drive current. This is one of its advantages when compared with the bipolar transistors.

Once the thyristor is operating in its on-state, the *i–v* characteristics can be shown to become similar to that for a P-i-N rectifier, resulting in the anode current increasing exponentially with the on-state voltage drop (or anode–cathode voltage). Consequently, thyristors can be designed with very high voltage blocking capability with low on-state voltage drops making them excellent power devices for circuits used in power distribution systems.

The power thyristor can be switched from its on-state to the blocking state by reversing the bias applied to the anode electrode. The reverse bias applied to the anode electrode forces the thyristor to undergo a reverse-recovery process similar to that observed in a P-i-N rectifier. Once the thyristor has entered the reverse blocking mode, a positive voltage can once again be applied to the anode without turning on the device until a gate control signal is applied.

2.2 5,000-V Silicon Thyristor

The design and characteristics of the 5,000-V symmetric blocking silicon thyristor structure are discussed in this section. The design parameters for the N-base region required to achieve this blocking voltage are first analyzed. Using the optimum N-base width, the blocking characteristics for the device are then obtained as a function of the lifetime in the drift region. The on-state characteristics for the device are obtained for various lifetime values as well. The switching behavior of the thyristor structure is obtained by observation of the turn-on process and the reverse recovery process.

2.2.1 Blocking Characteristics

The physics for blocking voltages in the first and third quadrants by thyristors is discussed in detail in the textbook. The device can have equal or symmetric forward and reverse blocking capability if cathode shorts are utilized to suppress the gain of the N-P-N transistor when small leakage currents are flowing in the structure. The reverse blocking voltage is supported across the P^+ anode/N-drift junction with a depletion layer extending mostly within the N-drift region. The maximum electric field occurs at the P^+ anode/N drift junction (J_1). The forward blocking voltage is supported across the P-base/N-drift junction, with a depletion layer extending mostly within the N-drift region. The maximum electric field occurs at the P-base/N-drift junction (J_2).

The breakdown voltage for the thyristor in both blocking modes is governed by the open-base transistor breakdown phenomenon [4]. According to open-base-transistor breakdown [1], the anode current will increase very rapidly when the common base current gain of the P-N-P bipolar transistor within the thyristor structure approaches unity. As the anode bias is increased, the width of the un-depleted portion of the N-drift region becomes smaller, producing an increase in the base transport factor (α_T). Concurrently, the maximum electric field at the blocking junction becomes larger leading to an increase in the multiplication coefficient. Both phenomena produce an increase in the common base current gain with increasing anode bias until it becomes equal to unity resulting in open-base transistor breakdown.

The open-base transistor breakdown condition is given by:

$$\alpha_{PNP} = (\gamma_E \cdot \alpha_T)_{PNP} M = 1 \tag{2.1}$$

The injection efficiency of the P-base/N-drift and P^+ anode/N-drift junctions is close to unity because of the relatively high doping concentration in the P^+ anode and P-base regions and the low doping concentration of the N-drift region. The magnitude of the other two terms in the above equation is a function of the anode bias. The base transport factor is determined by the width (l) of the un-depleted portion of the N-drift region:

$$\alpha_T = \frac{1}{\cosh(l/L_P)} \tag{2.2}$$

with

$$l = W_N - \sqrt{\frac{2\varepsilon_S V_A}{qN_D}} \tag{2.3}$$

where V_A is the applied bias to the anode electrode. As the anode bias increases, the width of the un-depleted portion of the N-drift region shrinks resulting in an increase in the base transport factor.

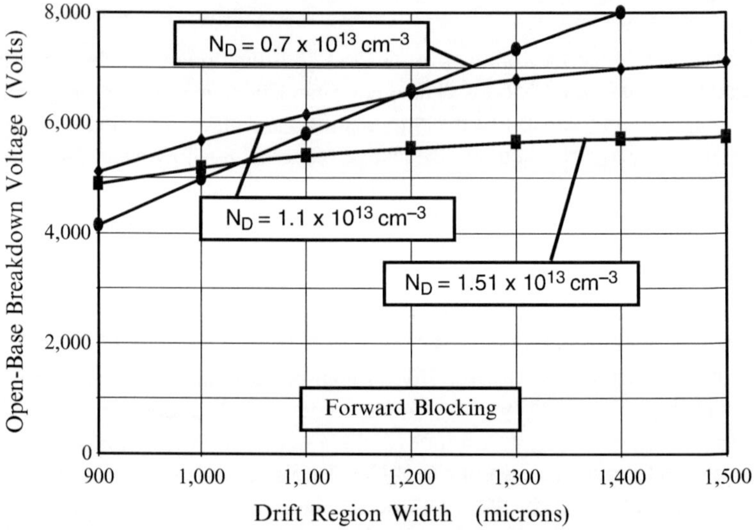

Fig. 2.6 Open-base breakdown voltage for the 5,000-V power thyristor structure

During forward blocking, the multiplication factor is determined by the anode bias relative to the avalanche breakdown voltage of the P-base/N-drift junction ($BV_{PP,F}$):

$$M = \frac{1}{1 - \left(V_{\rm A}/{\rm BV}_{\rm PP,F}\right)^n} \tag{2.4}$$

where $n = 6$ for the case of a P^+/N diode. The multiplication coefficient also increases with increasing anode bias. The open-base transistor breakdown voltage is determined by the anode voltage at which the multiplication factor becomes equal to the reciprocal of the base transport factor.

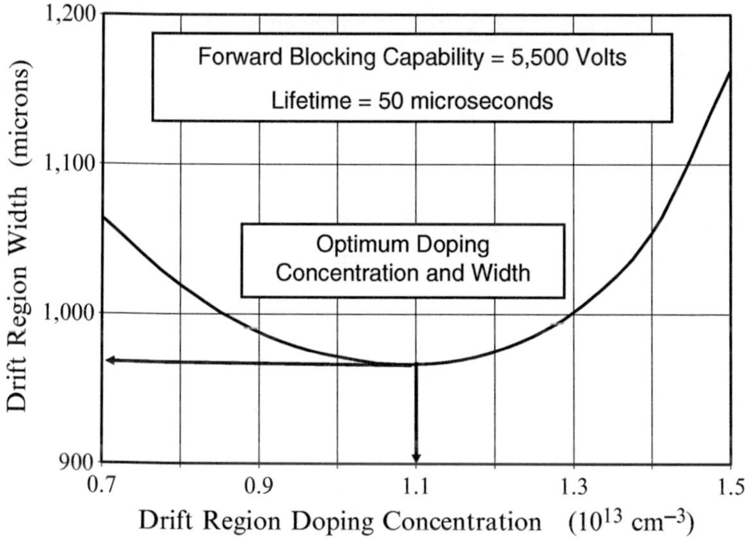

Fig. 2.7 Optimum width and doping concentration of the drift region for the 5,000-V power thyristor structure

Consider the case of a power thyristor that must have a reverse breakdown voltage of 5,500 V to achieve a blocking voltage rating of 5,000 V. In the case of avalanche breakdown, there is a unique value of 1.62×10^{13} cm^{-3} for the drift region doping concentration to obtain this blocking voltage. However, in the case of open-base transistor breakdown, many combinations of the drift region doping concentration and width can be used to obtain this blocking voltage capability. This is illustrated in Fig. 2.6 where the open-base breakdown voltage is plotted as a function of the drift region width for three cases of the drift region doping concentration. A lifetime of 50 μs was used in the N-drift region for this analysis. It can be observed from Fig. 2.6 that the open-base breakdown voltage becomes equal to 5,500 V at a drift region width of 1,070 μm for a drift region doping concentration of 0.7×10^{13} cm^{-3}. In this case, the base transport factor becomes close to unity under breakdown conditions. When the doping concentration of drift region is increased to 1.1×10^{13} cm^{-3}, the drift region thickness is reduced to 970 μm to achieve the same open-base breakdown voltage of 5,500 V. The drift region thickness increases to 1,200 μm, if the doping concentration of drift region is increased to 1.5×10^{13} cm^{-3}, to achieve the same open-base breakdown voltage of 5,500 V. In this case, the multiplication coefficient becomes large under open-base breakdown conditions. These examples demonstrate that there is an optimum drift region doping concentration to obtain a minimum drift region width to achieve an open-base breakdown voltage of 5,500 V. The location of the optimum design with a width of 970 μm and doping concentration of 1.1×10^{13} cm^{-3} is illustrated for this case in Fig. 2.7.

The leakage current for the thyristor is determined by the space charge generation current produced by the thermally generated carriers inside the depletion region. The space generation current is amplified by the gain of the open-base PNP transistor. Consequently:

$$J_L = \frac{J_{SCG}}{1 - \alpha_{PNP}} \tag{2.5}$$

The space charge generation current increases with increasing anode bias voltage due to the enlargement of the depletion region.

$$J_{SCG} = \frac{qW_D n_i}{\tau_{SC}} = \frac{n_i}{\tau_{SC}}\sqrt{\frac{2q\varepsilon_S V_A}{N_D}} \tag{2.6}$$

As the anode bias approaches the breakdown voltage, the multiplication factor becomes larger than unity, producing a more rapid increase in the leakage current.

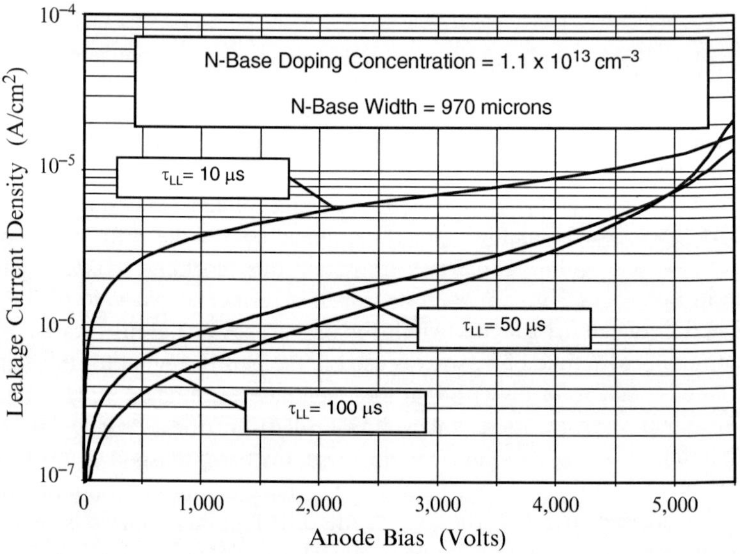

Fig. 2.8 Leakage current for the 5,000-V power thyristor structure

The leakage current density computed for the 5-kV thyristor structure by using the above analytical model is shown in Fig. 2.8. The structure had an optimized N-base doping concentration of 1.1×10^{13} cm^{-3} and optimized width of 970 μm at a lifetime of 50 μs. The impact of changing the lifetime on the leakage current is shown in the figure. The leakage current density increases when the lifetime is reduced.

Simulation Results

In order to gain insight into the physics of operation for the power thyristor under voltage blocking conditions, the results of two-dimensional numerical simulations for the optimized structure are described here. The total width of the structure, as shown by the cross section in Fig. 2.5, was 1,000 μm (area $= 1 \times 10^{-5}$ cm^{-2}) with a cathode finger width of 980 μm. The breakdown voltage obtained by using the optimized N-base region doping concentration and width from the analytical model yielded a breakdown voltage above 7,000 V. For the case of an N-drift region doping concentration of 1.5×10^{13} cm^{-3} and width of 1,070 μm, the breakdown voltage was found to be about 6,000 V. The P-base region had a Gaussian doping profile with a surface concentration of 5×10^{17} cm^{-3} and a depth of 30 μm. The N^+ cathode region had a Gaussian doping profile with a surface concentration of 1×10^{20} cm^{-3} and a depth of 10 μm. The P^+ anode region had a Gaussian doping profile with a surface concentration of 1×10^{20} cm^{-3} and a depth of 50 μm. The resulting doping profile is shown in Fig. 2.9.

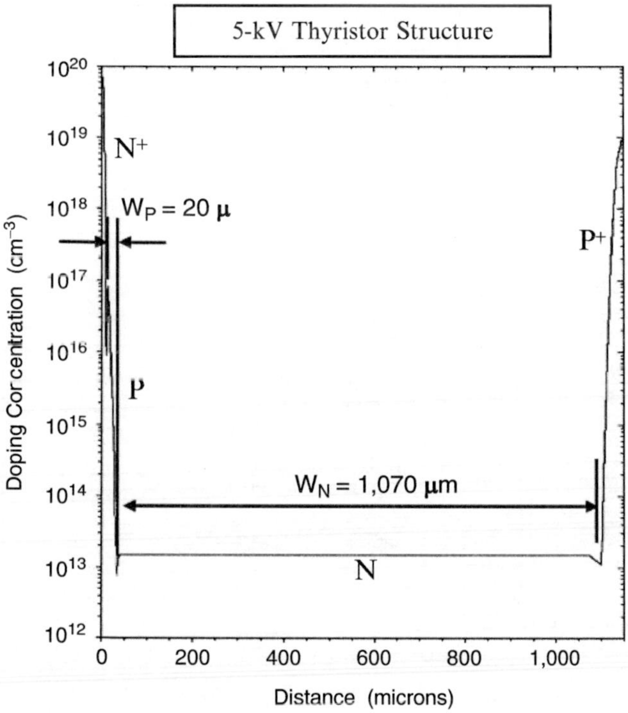

Fig. 2.9 Doping profile for the simulated 5-kV power thyristor structure

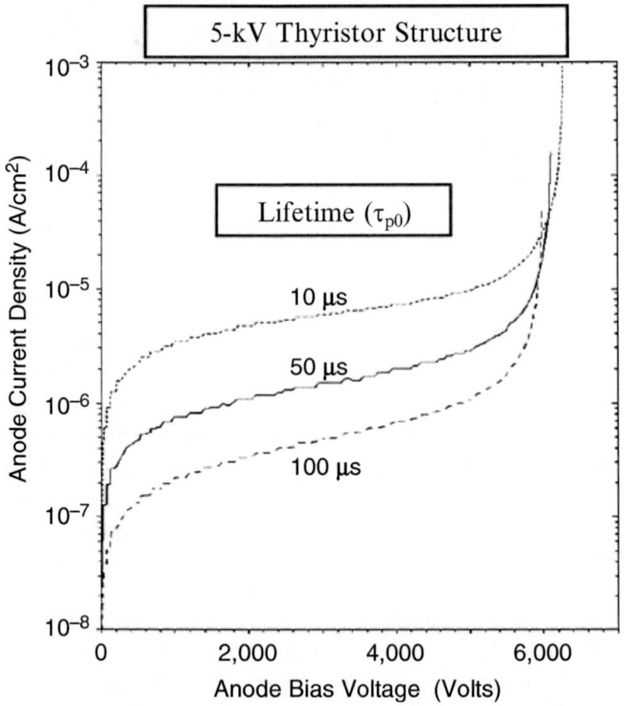

Fig. 2.10 Forward blocking characteristics for the 5-kV power thyristor structure

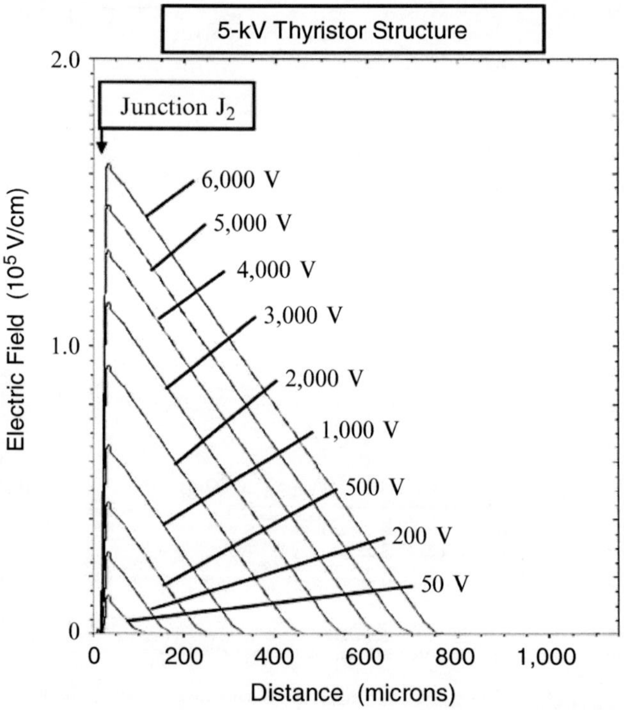

Fig. 2.11 Electric field profiles in the forward blocking mode for the 5-kV power thyristor structure

The forward blocking characteristics for the thyristor structure are shown in Fig. 2.10 for the case of various lifetime (τ_{p0}, τ_{n0}) values in the N-base region. The breakdown voltage is indicated by an abrupt increase in the anode current. At room temperature (300 K), the forward blocking voltage obtained with the simulations is about 6,000 V. The value predicted by the analytical model is smaller than this value because the analytical model does not account for the voltage being supported in the P-base region due to its graded doping profile. The leakage current density increases with decreasing lifetime as predicted by the analytical model (see Fig. 2.8) due to enhanced space charge generation.

The voltage is primarily supported in the thyristor within the N-drift region. This is illustrated in Fig. 2.11 where the electric field profiles are shown during operation in the forward blocking mode at various positive anode bias voltages. It can be observed that the P-base/N-drift junction (J_2) becomes reverse biased during the forward blocking mode with the depletion region extending toward the right-hand side with increasing anode bias. Some extension of the depletion region is observed within the P-base region due to its graded, diffused doping profile. This allows the thyristor structure to support a larger voltage than predicted by the analytical model which is based upon an abrupt junction assumption.

2.2.2 *On-State Characteristics*

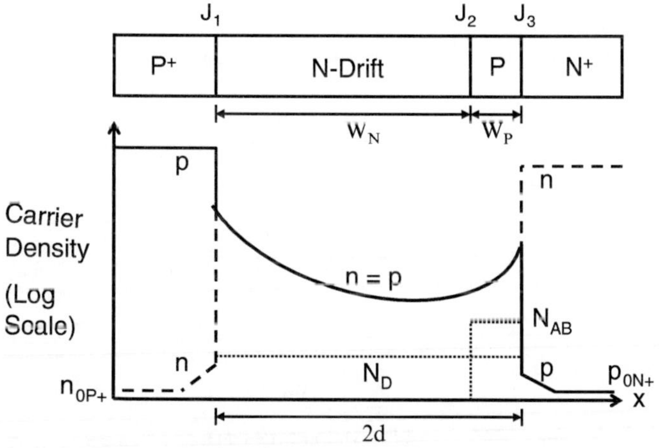

Fig. 2.12 Carrier distribution within the power thyristor structure in the on-state

One of the attributes of the thyristor structure is its excellent forward conduction characteristic even when designed to support large voltage levels. The thyristor structure can be triggered from the forward blocking mode at the anode supply voltage (V_{AS}) into the forward conduction mode by the application of a small gate current to initiate the turn-on process [1]. This gate current serves to increase the current gain of the N-P-N transistor, despite the presence of cathode shorts, until the combined gain of the integral N-P-N and P-N-P transistors within the thyristor structure is sufficient to sustain its regenerative action. The thyristor then operates in its on-state with a forward conduction characteristic similar to that observed for a P-i-N rectifier. In the on-state, strong conductivity modulation of the N-drift region occurs due to high level injection of holes, allowing the thyristor to carry high current levels with a low on-state voltage drop.

The power thyristor can be treated as a P-i-N rectifier for analysis of its forward conduction characteristics. In this case, it is assumed that the junction J_2 is strongly forward biased resulting in high level injection in not only the N-base region but also the P-base region as illustrated in Fig. 2.12. The thyristor can then be regarded as a P-i-N rectifier between the P^+ anode and N^+ cathode regions. In this context, the N^+ cathode region is referred to as a *remote emitter* because it provides electrons to the N-base region through the intervening P-base region.

The electron and hole concentrations within the N-base and P-base regions take a catenary distribution in accordance with the analysis for the P-i-N rectifier in the textbook:

$$n(x) = p(x) = \frac{\tau_{HL} J_A}{2qL_a}\left[\frac{\cosh(x/L_a)}{\sinh(d/L_a)} - \frac{\sinh(x/L_a)}{2\cosh(d/L_a)}\right] \tag{2.7}$$

The distance "d" for the thyristor structure is given by:

$$d = \frac{W_N + W_P}{2} \tag{2.8}$$

as indicated in the figure. A minimum on-state voltage drop occurs for the thyristor structure when the ambipolar diffusion length (L_a) is equal to the distance "d" (see textbook analysis for the P-i-N rectifier in Chap. 5).

The on-state *i–v* characteristic for the thyristor in the high level injection regime of operation is given by [1]:

$$V_{ON} = \frac{2kT}{q}\ln\left[\frac{J_T d}{2qD_a n_i F(d/L_a)}\right] \tag{2.9}$$

where

$$F\left(\frac{d}{L_a}\right) = \frac{(d/L_a)\tanh(d/L_a)}{\sqrt{1 - 0.25\tanh^4(d/L_a)}}e^{-\frac{qV_M}{2kT}} \tag{2.10}$$

The i-region (or middle-region) voltage drop can be computed using the following approximations. For d/L_a ratios of up to 2, an asymptote A given by:

$$V_M = \frac{2kT}{q}\left(\frac{d}{L_a}\right)^2 \tag{2.11}$$

provides a good fit. For d/L_a ratios of greater than 2, an asymptote B given by:

$$V_M = \frac{3\pi kT}{8q} e^{(d/L_a)} \tag{2.12}$$

provides a good fit.

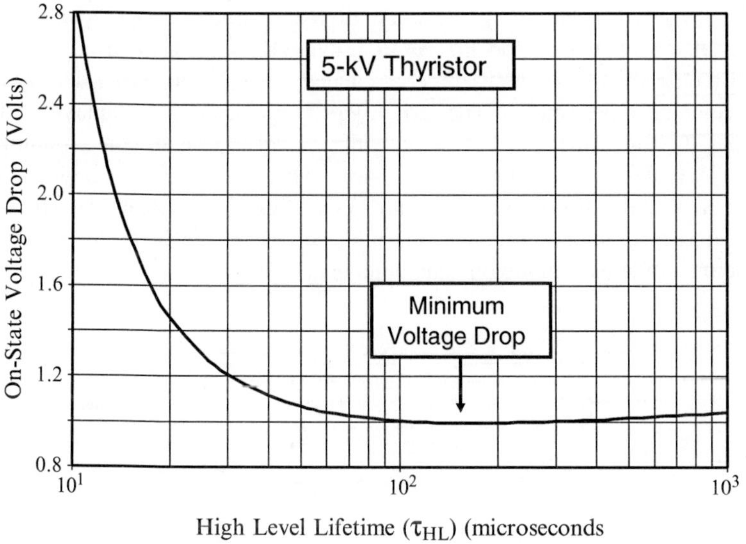

Fig. 2.13 On-state voltage drop for the 5-kV power thyristor structure

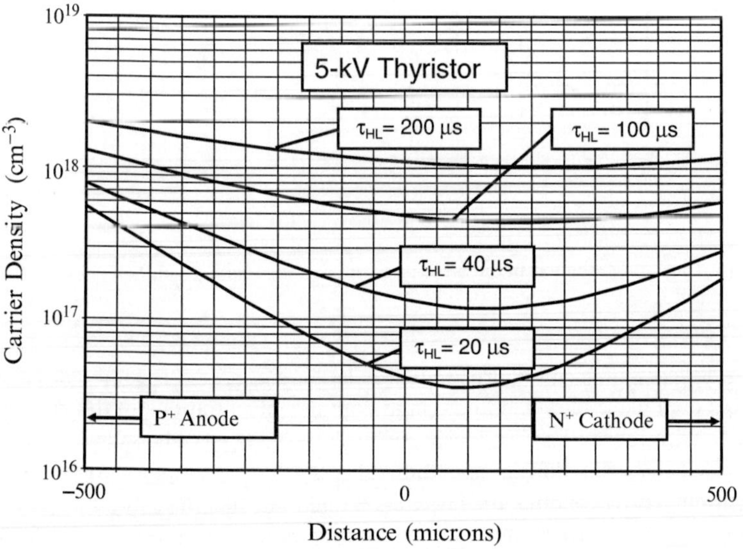

Fig. 2.14 Carrier distribution in the 5-kV power thyristor structure with middle region recombination dominant

The on-state voltage drop (at an on-state current density of 100 A/cm^2) computed for the 5-kV power thyristor structure by using Eq. 2.9 is provided in Fig. 2.13 for various values for the high-level lifetime in the drift region. This thyristor structure had the optimized N-base region width of 970 μm and a P-base width of 25 μm. As expected, the on-state voltage drop has a minimum value when the $[d/L_a]$ value is equal to unity. The minimum on-state voltage drop is found to be 0.993 V at a high-level lifetime of 160 μs.

The carrier distribution in the 5-kV thyristor structure as predicted by the analytical model based upon Eq. 2.7 is provided in Fig. 2.14 for various high-level lifetime values. As the lifetime is reduced, the injected carrier density in the drift region becomes smaller. The carrier density predicted by the analytical model is larger than in actual devices because the model is based upon assuming that recombination occurs only in the drift region. In actual devices, recombination also occurs in the end region of the device as discussed in Sect. 5.1.4 of the textbook [1]. End-region recombination reduces the injected carrier density in the drift region producing an increase in the on-state voltage drop as well.

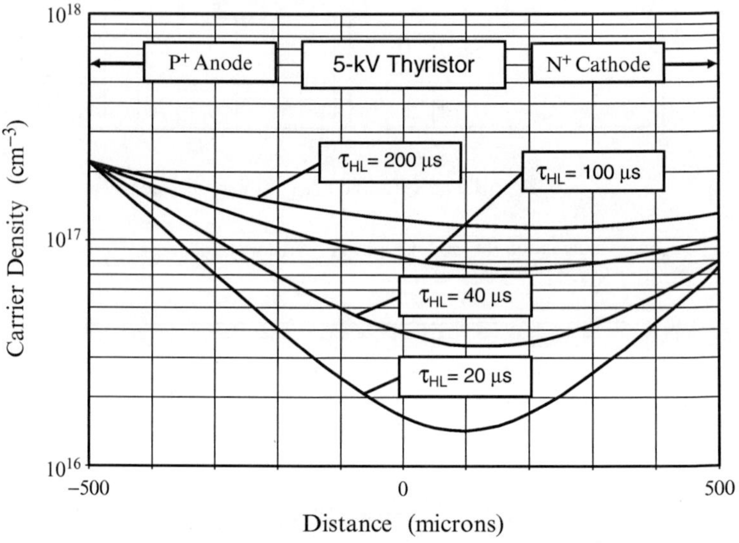

Fig. 2.15 Carrier distribution in the 5-kV power thyristor structure with end-region recombination dominant

When the lifetime in the drift region is large, end-region recombination begins to take a dominant role. Due to the high doping concentrations in the end regions, the injected minority carrier density in these regions is well below the majority carrier density even during operation at very high on-state current densities. The current corresponding to the end regions can therefore be analyzed using low-level injection theory under the assumption of a uniform doping concentration in these regions [1]. The current flow through the device in the on-state under these conditions is described by:

$$J_{P+} = J_{SP+}\left[\frac{n(-d)}{n_{ieP+}}\right]^2 \tag{2.13}$$

for the P$^+$ anode region and:

$$J_{N+} = J_{SN+}\left[\frac{n(+d)}{n_{ieN+}}\right]^2 \tag{2.14}$$

for the N$^+$ cathode region. In these equations, n_{ieP+} and n_{ieN+} are the effective intrinsic carrier concentrations in the P$^+$ and N$^+$ end regions including the influence of band-gap narrowing. From these equations, it can be concluded that the carrier concentration in the drift region will increase as the square root of the current density if the end-region recombination becomes dominant. Under these circumstances, the middle region voltage drop is no longer independent of the current density resulting in an increase in the total on-state voltage drop.

When end-region recombination is dominant, the electron and hole concentrations within the N-base and P-base regions take a catenary distribution with a smaller concentration at the boundaries as given by Eqs. 2.13 and 2.14:

$$n(x) = p(x) = K_E\left[\frac{\cosh(x/L_a)}{\sinh(d/L_a)} - \frac{\sinh(x/L_a)}{2\cosh(d/L_a)}\right] \tag{2.15}$$

The constant K_E can be obtained by using Eq. 2.13 with $x = -d$ in the above equation. The results obtained by using this approach are shown in Fig. 2.15 for the case of various values of lifetime. A saturation current density of 4×10^{13} A/cm^2 was utilized for the plots. It can be observed that the carrier concentration is reduced by an order of magnitude when compared with the plots in Fig. 2.14.

Simulation Results

The results of two-dimensional numerical simulations for the 5-kV silicon thyristor structure are described here. The total width of the structure, as shown by the cross section in Fig. 2.5, was 1,000 μm (area $= 1 \times 10^{-5}$ cm^{-2}) with a cathode finger width of 980 μm. The N-base region had an optimized doping concentration of 1.5×10^{13} cm^{-3} and width of 1,070 μm. The on-state characteristics were obtained by using a gate drive current of 2×10^{-8} A/μm using various values for the lifetime in the drift region. The characteristics obtained from the numerical simulations are shown in Fig. 2.16. The small anode current observed at anode biases below 0.5 V is associated with the gate drive current. It can be observed that the on-state voltage drop increases as expected with reduction of the lifetime (τ_{p0}, τ_{n0}) indicated in the figure. The on-state voltage drop at a lifetime value of 50 μs is found to be 1.758 V at an on-state current density of 100 A/cm^2. This value is much larger than predicted by the analytical model because it does not account for recombination in the end regions.

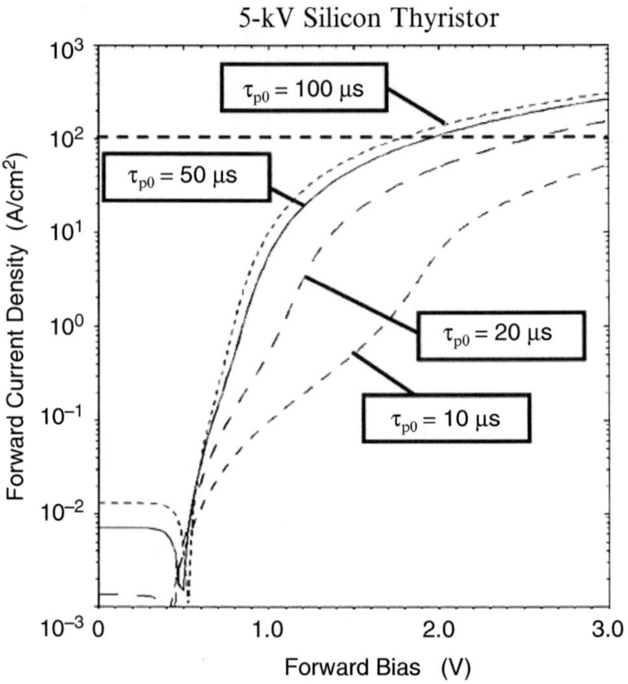

Fig. 2.16 On-state characteristics of the 5-kV power thyristor structure

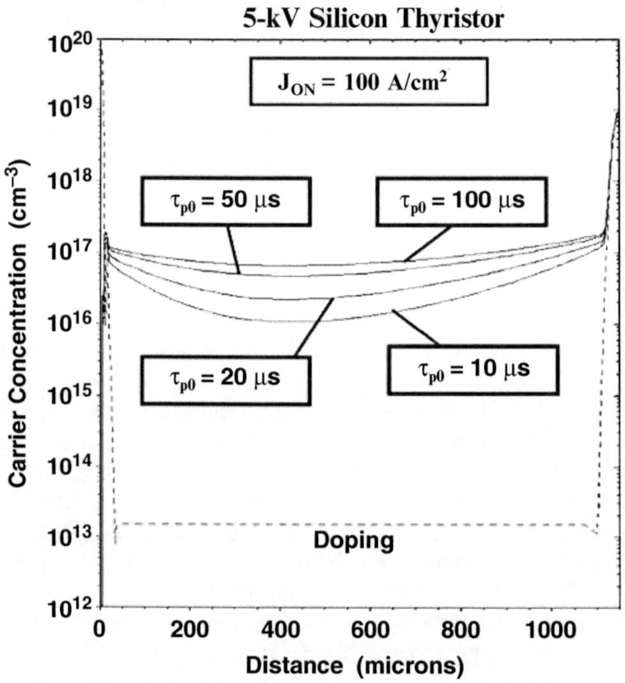

Fig. 2.17 On-state carrier distribution in the 5-kV power thyristor structure

The low on-state voltage drop for the 5-kV thyristor structure is due to the large number of carriers injected into the drift region producing a drastic reduction of its resistance. This is illustrated in Fig. 2.17 where the injected carrier density is shown for four cases of the lifetime (τ_{p0}, τ_{n0}) in the drift region of the thyristor structure. It can be observed that the injected carrier density is four orders of magnitude larger than the doping concentration for the case of a lifetime of 100 μs. The injected carrier density is reduced by an order of magnitude in the middle of the drift region when the lifetime is reduced to 10 μs. The predictions of the analytical model (see Fig. 2.14) have the same general characteristics but the injected carrier density is much smaller in the numerical simulations when compared with the analytical model that neglects the end-region recombination. When the end-region recombination is taken into account, the carrier densities predicted by the analytical model (see Fig. 2.15) are similar to those observed in the numerical simulations.

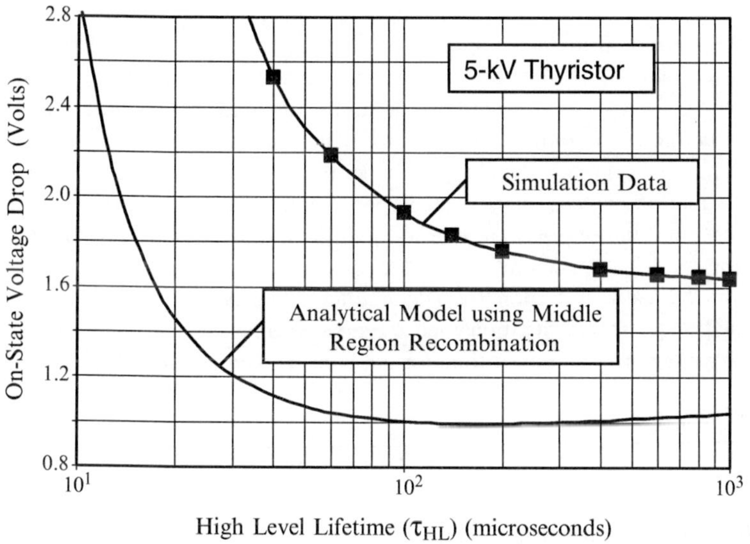

Fig. 2.18 On-state voltage drop for the 5-kV power thyristor structure obtained using numerical simulations

The reduction of the injected carrier density in the middle region with smaller lifetime leads to an increase in the on-state voltage drop. The on-state voltage drop for the 5-kV thyristor obtained using numerical simulations can be compared with that obtained using the analytical model without end-region recombination in Fig. 2.18. The on-state voltage drop obtained using numerical simulations is much larger than that predicted by the analytical model based upon middle-region recombination. Further, the minimum on-state voltage drop occurs at a significantly larger value for the lifetime in the drift region. It can be concluded that the analytical model is very optimistic and not reliable for predicting the on-state voltage drop, indicating that the impact of end-region recombination is very important for the silicon 5-kV thyristor structure.

2.2.3 Turn-On

The thyristor can be triggered from the forward blocking mode to the on-state by the application of a gate drive current. The gate drive current flows from the gate terminal to the first row of cathode shorts. The voltage drop in the P-base region due to the gate current flow forward biases the N^+ emitter/P-base junction (J_3) at the edge of the cathode closest to the gate terminal producing the injection of electrons from the cathode region. This does not immediately produce anode current flow. The injected electrons diffuse through the P-base region in a finite time interval referred to as the *base transit time*. Once the electrons cross the P-base/N-base junction (J_2), they immediately promote the injection of holes from the P^+ anode/N-base junction (J_1) in order to preserve charge neutrality in the N-base region. The injection of carriers at the P^+ anode/N-base junction initiates current flow through the device. Consequently, the anode current begins to flow after a *delay-time* interval, which is equal to the transit time for the N-P-N transistor. The delay time is typically 50 ns in duration.

After injection from the P^+ anode/N-base junction, the holes diffuse through the N-base region until they are collected at the P-base/N-base junction (J_2). The base transit time for the N-base region is a strong function of the initial anode bias voltage before turning on the thyristor [1]. The transit time can be limited by the diffusion of holes in the neutral region under the assumption that the applied anode bias is supported across only the depletion region. During the turn-on process, a high concentration of holes are injected into the N-drift region from the P^+ anode region and a high concentration of electrons are injected into the N-drift region from the N^+ cathode region (remote emitter). The depletion region cannot be sustained under these conditions and the anode voltage is distributed throughout the N-base region. The transit time for the holes is then given by [1]:

$$t_{\mathrm{t,PNP}} = \frac{W_{\mathrm{N}}}{v_{\mathrm{P}}} = \frac{W_{\mathrm{N}}^2}{\mu_{\mathrm{P}} V_{\mathrm{A}}} \tag{2.16}$$

which is much smaller than the transit time determined by the diffusion process.

An analytical model for the increase in the anode current during the turn-on transient for a one-dimensional thyristor structure was derived based upon charge control principles discussed in the textbook [1]. This analysis takes into consideration the internal feedback mechanism between the N-P-N and P-N-P transistors within the thyristor structure to determine the growth of the stored charge within the N-base region and the P-base region. Due to relatively short time for the turn-on transient when compared with the lifetime, the recombination within the N-base region and P-base region can be assumed to be negligible during this analysis. The increase in the anode current during the turn-on process is given by:

$$J_{\mathrm{A}}(t) = \frac{J_{\mathrm{G}}}{\alpha_{\mathrm{PNP}}} \left(e^{t/\sqrt{t_{\mathrm{t,NPN}} t_{\mathrm{t,PNP}}}} - 1 \right) \tag{2.17}$$

Based upon this equation, it can be concluded that the anode current will grow exponentially with time after the delay phase. The time constant for the rise in anode current is observed to be the geometric mean of the transit times for the N-P-N and P-N-P transistors.

The *rise-time* is defined as the time taken for the anode current density to increase to the on-state value. Using Eq. 2.17, the rise-time can be obtained:

$$t_R = \sqrt{t_{t,NPN} \cdot t_{t,PNP}} \ln\left(\frac{\alpha_{PNP} J_{A,SS}}{J_G} + 1\right) \tag{2.18}$$

where $J_{A,SS}$ is the steady-state (on-state) anode current density. The rise-time is determined by the transit times for the internal N-P-N and P-N-P transistors within the thyristor structure.

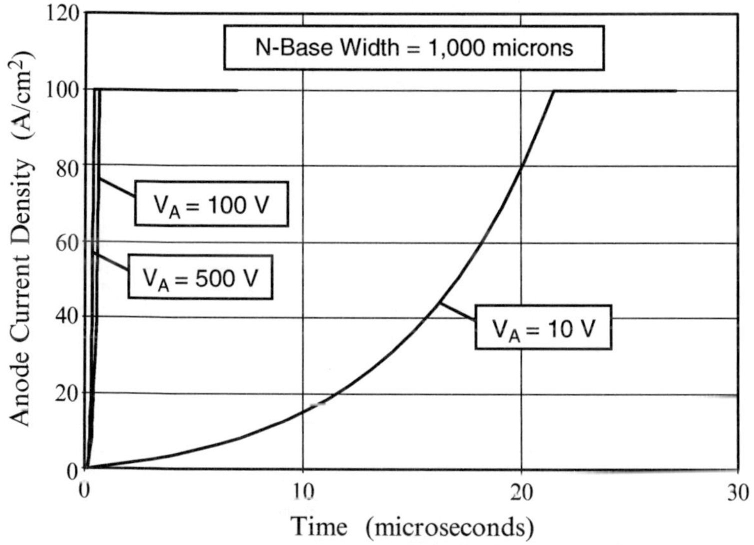

Fig. 2.19 Turn-on transient for a one-dimensional 5-kV thyristor

The rise in the anode current with time is shown in Fig. 2.19 for the case of a one-dimensional thyristor structure with a P-base region width of 25 μm and an N-base region width of 1,000 μm. The transit time for the N-P-N transistor is calculated to be 120 ns. For the case of an anode bias of 10 V, the current reaches its steady-state value with a rise-time of 21 μs. However, when the anode voltage is increased to 100 V, the transit time for the P-N-P transistor is greatly reduced to 0.2 μsdue to the enhanced drift current. Consequently, the rise-time for the anode current also decreases to only 0.5 μs. A further increase in the anode voltage to 500 V produces a reduction of the rise-time to only 0.26 μs.

Simulation Results

In order to gain further insight into the physics of turn-on for the 5-kV power thyristor structure, the results of two-dimensional numerical simulations for the structure (described in the previous section) are discussed here. The total width of the structure is 1,000 μm (area = 1×10^{-5} cm^{-2}) with a cathode finger width of 980 μm. In order to turn on the thyristor, the gate drive current was abruptly increased from 0 to 2×10^{-6} A/μm (0.2 A/cm^2) while the anode voltage was maintained fixed with a load resistance in series with the thyristor structure.

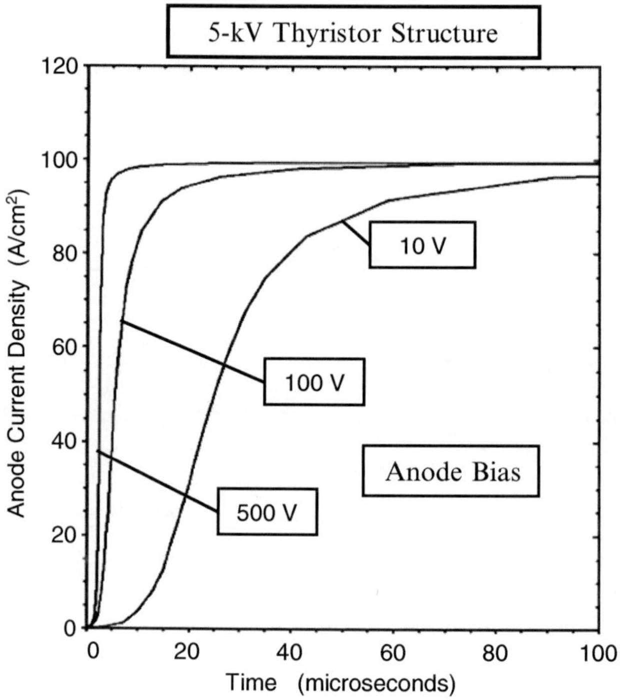

Fig. 2.20 Turn-on characteristics for the 5-kV power thyristor structure

The turn-on characteristics for the 5-kV power thyristor structure are shown in Fig. 2.20 for three values of the anode bias. In each case the load resistance was altered to obtain the same anode current density (100 A/cm^2) after the thyristor is in the on-state. In all cases, there is a short delay time of about 100 ns, which is consistent with the delay time computed by using the base transit time of the N-P-N transistor. The anode current then increases exponentially with time as described by the analytical Eq. 2.17. For the case of an anode bias of 10 V, the anode current increases exponentially for about 35 μs. This behavior is consistent with the analytical model for turn-on based upon the diffusion of carriers through the N-base region. When the anode bias is increased to 100 and 500 V, the anode current density increases much more rapidly. This behavior is also consistent with the analytical model for turn-on based upon the drift of carriers through the N-base region. In all cases, a gradual increase in the anode current density is also observed at the end of the turn-on transient due to the current spreading across the emitter width in the two-dimensional structure used for the numerical simulations.

2.2.4 *Reverse Recovery*

The switching of the thyristor from the on-state to the reverse blocking state produces substantial power dissipation. The high concentration of minority carriers stored within the N-drift region due to high-level injection conditions in the on-state, must be removed before the device can support a high reverse bias voltage. When the anode voltage crosses zero, the anode current continues to flow in the reverse direction with an approximately constant rate of change or constant [d*i*/d*t*] until the P^+ anode/N-base junction is able to support voltage. Once the anode voltage reaches the supply voltage, the anode current decreases to zero in an exponential manner after this junction becomes reverse biased. This behavior is similar to the reverse recovery process for the P-i-N power rectifier.

During the reverse recovery process, most of the power dissipation occurs when the anode current decays to zero after reaching the peak reverse recovery current because the anode voltage supported during this time is large (the supply voltage). The anode current waveform during the decay is described by:

$$J_A(t) = J_{PR} e^{-t/\tau_{RR}} \tag{2.19}$$

where J_{PR} is the peak reverse recovery current density and τ_{RR} is a characteristic decay time constant during the reverse recovery. The energy loss per reverse recovery transient can be obtained by integrating the instantaneous power dissipation:

$$E_{RR} = \int_0^{\infty} V_S J_{PR} e^{-t/\tau_{RR}} \mathrm{d}t = V_S J_{PR} \tau_{RR} \tag{2.20}$$

If switching power loss is assumed to be dominant, the maximum operating frequency for the thyristor, as limited by a power dissipation of P_D, is given by:

$$f_{Max} = \frac{P_D}{E_{RR}} \tag{2.21}$$

Simulation Results

In order to gain further insight into the physics of turn-off for the power thyristor under a constant [d*i*/d*t*], the results of two-dimensional numerical simulations for the 5-kV thyristor structure (described in the previous sections) are discussed here. The total width of the structure used for the numerical simulations was 1,000 μm (area = 1×10^{-5} cm^{-2}) with a cathode finger width of 980 μm between the gate contact and cathode short. The thyristor was switched from on-state operation at a current density of 100 A/cm^2 with a constant [d*i*/d*t*] (rate of 150 A/cm^2 in 10 μs). This reverse ramp rate was applied until the anode voltage reached −1,000 V. The anode voltage was then held constant, allowing the anode current to decay to zero.

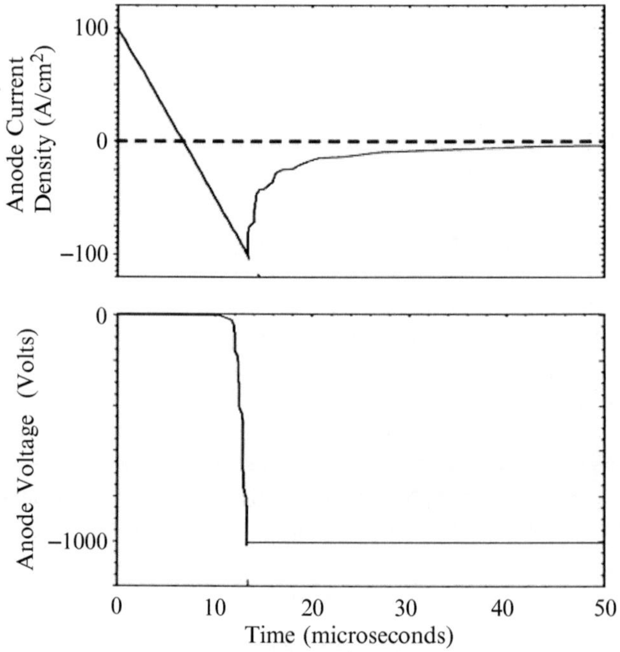

Fig. 2.21 Reverse recovery characteristics for the 5-kV thyristor structure with constant [d*i*/d*t*] applied to the anode electrode

The anode current and voltage waveforms for the thyristor are shown in Fig. 2.21. As expected, these waveforms are very similar to those exhibited by the P-i-N rectifier [1]. The anode current flows in the reverse direction until the anode voltage becomes equal to the reverse bias voltage and then decreases exponentially to zero. These waveforms confirm that the thyristor structure can be analyzed by using the reverse recovery analysis discussed in the textbook for the P-i-N rectifier. In the case of the baseline simulation structure with a lifetime (τ_{p0}, τ_{n0}) of 50 μs, the reverse recovery process requires a total time interval of about 50 μs.

2.2.5 *Summary*

The results of the numerical simulations discussed above indicate that the silicon 5-kV thyristor structure requires a minority carrier lifetime (τ_{p0}, τ_{n0}) of at least 50 μs to achieve a low on-state voltage drop. For such a large lifetime value, the reverse recovery process takes over 50 μs. From the waveforms in Fig. 2.21, it can be observed that most of the power dissipation occurs during the decay of the anode current after it reaches the peak reverse recovery current density. For the conditions used during the numerical simulations, the peak reverse recovery current (J_{PR}) is found to be 100 A/cm^2 and the time constant for the reverse recovery (t_{RR}) is found to be 3.25 μs. Using a power supply voltage of 1,000 V, the energy loss per switching event is found to be 0.325 J/cm^2 by using Eq. 2.20. If the maximum power dissipation due to switching is limited to 100 W/cm^2, the maximum operating frequency for the 5-kV thyristor is then found to be 300 Hz. This demonstrates that the 5-kV silicon thyristor cannot be operated at high frequencies.

2.3 10,000-V Silicon Thyristor

The design and characteristics of the silicon 10-kV symmetric blocking thyristor structure are discussed in this section. The design parameters for the N-base region required to achieve this blocking voltage are first analyzed. Using the optimum N-base width, the blocking characteristics for the device are then obtained as a function of the lifetime in the drift region. The on-state characteristics for the device are obtained for various lifetime values as well. The switching behavior of the thyristor structure is obtained by observation of the turn-on process and the reverse recovery process.

2.3.1 Blocking Characteristics

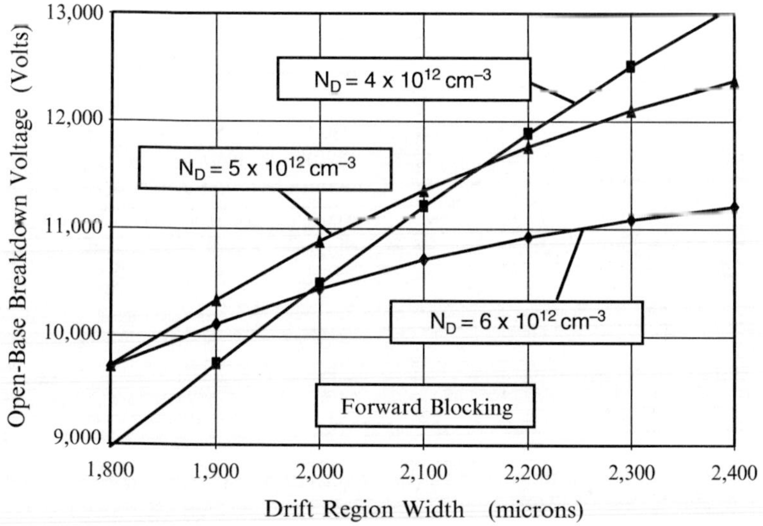

Fig. 2.22 Open-base breakdown voltage for the 10-kV power thyristor structure

The physics for blocking voltages for the 10-kV silicon symmetric blocking thyristor is identical to that for the 5-kV device described in the previous section. The open-base transistor breakdown condition is given by Eq. 2.1. Consider the case of a power thyristor that must have a reverse breakdown voltage of 11,000 V to achieve a blocking voltage rating of 10,000 V. In the case of avalanche breakdown, there is a unique value of 6.45×10^{12} cm^{-3} for the drift region doping concentration to obtain this blocking voltage. However, in the case of open-base transistor breakdown, many combinations of the drift region doping concentration and width can be used to obtain this blocking voltage capability. This is illustrated in Fig. 2.22 where the open-base breakdown voltage is plotted as a function of the drift region width for three cases of the drift region doping concentration. A lifetime of 100 μs was used in the N-drift region for this analysis. It can be observed from Fig. 2.22 that the open-base breakdown voltage becomes equal to 11,000 V at a drift region width of 2,070 μm for a drift region doping concentration of 4×10^{12} cm^{-3}. In this case, the base transport factor becomes close to unity under breakdown conditions. When the doping concentration of drift region is increased to 5×10^{12} cm^{-3}, the drift region thickness is reduced to 2,020 μm to achieve the same open-base breakdown voltage of 11,000 V. The drift region thickness increases to 2,240 μm, if the doping concentration of drift region is increased to 6×10^{12} cm^{-3}, to achieve the same open-base breakdown voltage of 11,000 V. In this case, the multiplication coefficient becomes large under open-base breakdown conditions. These examples demonstrate that there is an optimum drift region doping concentration to obtain a minimum drift region width to achieve an open-base breakdown voltage of 11,000 V. The location of the optimum design with a width of 2,020 μm and doping concentration of 4.7×10^{12} cm^{-3} is illustrated for this case in Fig. 2.23.

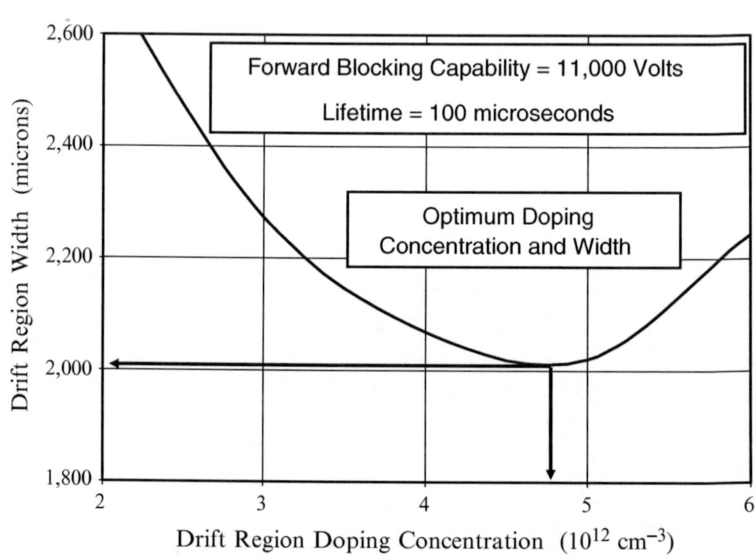

Fig. 2.23 Optimum width and doping concentration of the drift region for the 10-kV power thyristor structure

Simulation Results

In order to gain insight into the physics of operation for the 10-kV silicon power thyristor under voltage blocking conditions, the results of two-dimensional numerical simulations for the optimized structure are described here. The total width of the structure, as shown by the cross section in Fig. 2.5, was 1,000 μm (area = 1×10^{-5} cm^{-2}) with a cathode finger width of 980 μm. The P-base region had a Gaussian doping profile with a surface concentration of 5×10^{17} cm^{-3} and a depth of 30 μm. The N$^+$ cathode region had a Gaussian doping profile with a surface concentration of 1×10^{20} cm^{-3} and a depth of 10 μm. The P$^+$ anode region had a Gaussian doping profile with a surface concentration of 1×10^{20} cm^{-3} and a depth of 50 μm. An optimum N-drift region doping concentration of 5×10^{12} cm^{-3} and width of 2,000 μm was used for the baseline device structure. The doping profile is similar to that shown in Fig. 2.9.

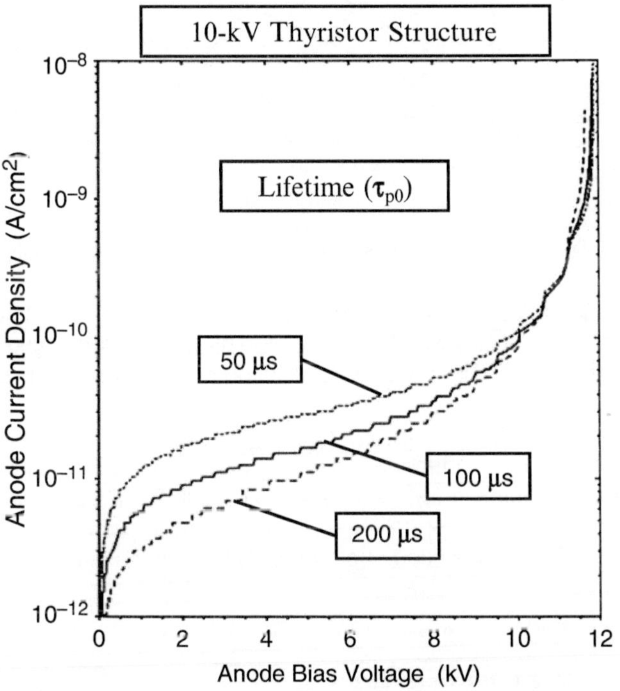

Fig. 2.24 Forward blocking characteristics for the 10-kV power thyristor structure

The forward blocking characteristics for the thyristor structure are shown in Fig. 2.24 for the case of various lifetime (τ_{p0}, τ_{n0}) values in the N-base region. The breakdown voltage is indicated by an abrupt increase in the anode current. At room temperature (300 K), the forward blocking voltage obtained with the simulations is about 11,000 V in agreement with the value predicted by the analytical model. The leakage current density increases with decreasing lifetime as predicted by the analytical model due to enhanced space charge generation.

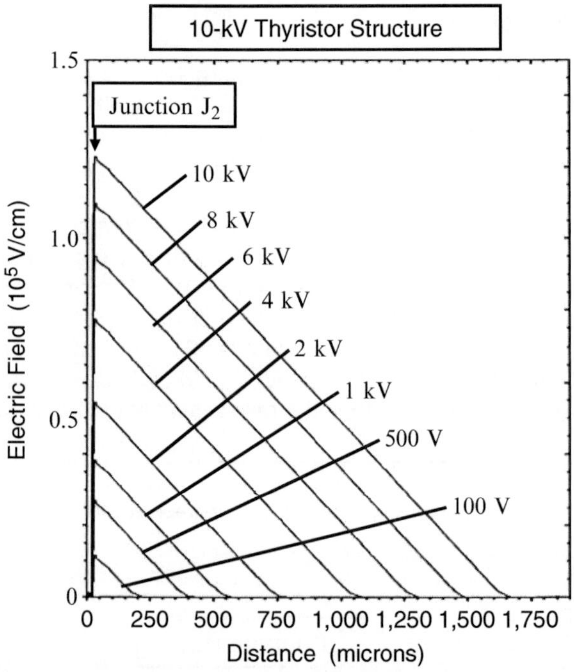

Fig. 2.25 Electric field profiles in the forward blocking mode for the 5-kV power thyristor structure

The voltage is primarily supported in the 10-kV thyristor structure within the N-drift region. This is illustrated in Fig. 2.25 where the electric field profiles are shown during operation in the forward blocking mode at various positive anode bias voltages. It can be observed that the P-base/N-drift junction (J_2) becomes reverse biased during the forward blocking mode, with the depletion region extending toward the right-hand side with increasing anode bias. Very little extension of the depletion region is observed within the P-base region. Consequently, the 10-kV thyristor structure supports voltage that is well predicted by the analytical model which is based upon an abrupt junction assumption.

2.3.2 *On-State Characteristics*

The on-state *i–v* characteristic for the 10-kV thyristor structure is described by Eq. 2.9, with a carrier distribution described by Eq. 2.7 if middle region recombination is assumed to be dominant. The on-state voltage drop (at an on-state current density of 50 A/cm^2) computed for the 10-kV power thyristor structure by using Eq. 2.9 is provided in Fig. 2.26 for various values for the high-level lifetime in the drift region. This thyristor structure had the optimized N-base region width of 2,000 μm and a P-base width of 25 μm. As expected, the on-state voltage drop has a minimum value when the $[d/L_a]$ value is equal to unity. The minimum on-state voltage drop is found to be 1.03 V at a high-level lifetime of 500–1,000 μs.

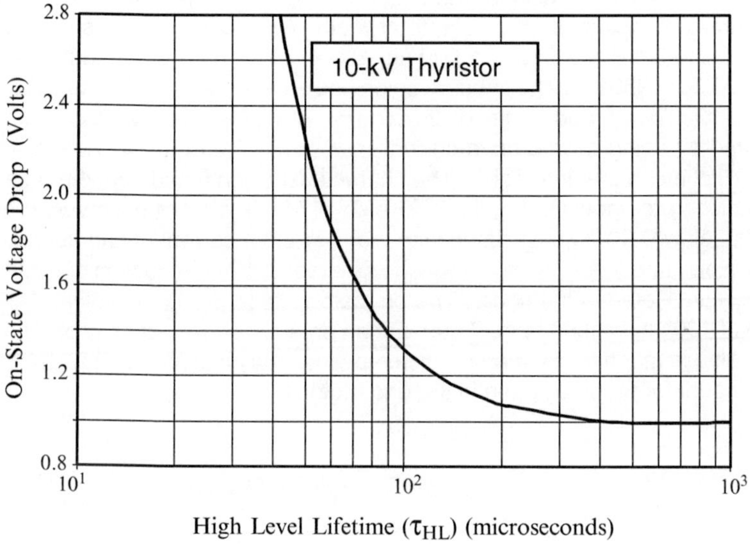

Fig. 2.26 On-state voltage drop for the 10-kV power thyristor structure

Simulation Results

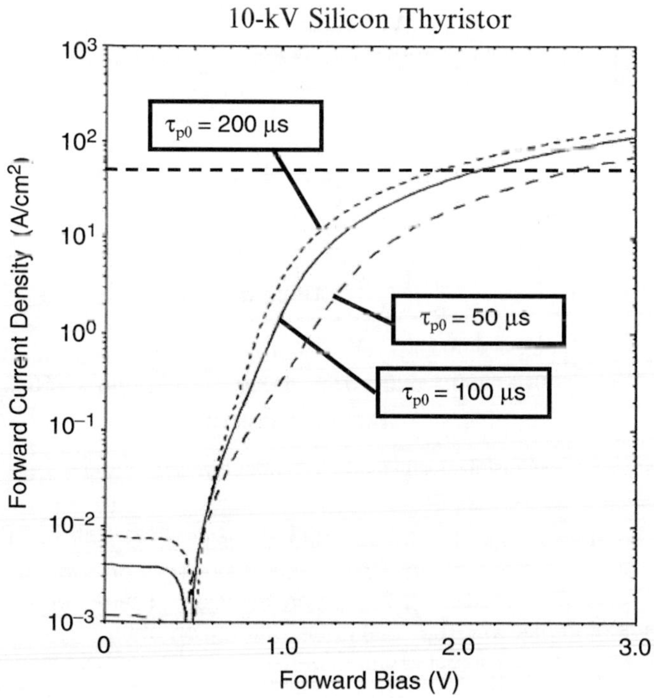

Fig. 2.27 On-state characteristics of the 10-kV power thyristor structure

The results of two-dimensional numerical simulations for the 10-kV silicon thyristor structure are described here. The total width of the structure, as shown by the cross section in Fig. 2.5, was 1,000 μm (area = 1×10^{-5} cm^{-2}) with a cathode finger width of 980 μm. The N-base region had an optimized doping concentration of 5×10^{12} cm^{-3} and width of 2,000 μm. The on-state characteristics were obtained by using a gate drive current of 2×10^{-8} A/μm using various values for the lifetime in the drift region. The characteristics obtained from the numerical simulations are shown in Fig. 2.27. The small anode current observed at anode biases below 0.5 V is associated with the gate drive current. It can be observed that the on-state voltage drop increases as expected with reduction of the lifetime (τ_{p0}, τ_{n0}) indicated in the figure. The on-state voltage drop at a lifetime (τ_{p0}, τ_{n0}) value of 100 μs is found to be 2.099 V at an on-state current density of 50 A/cm^2. This value is much larger than predicted by the analytical model because it does not account for recombination in the end regions.

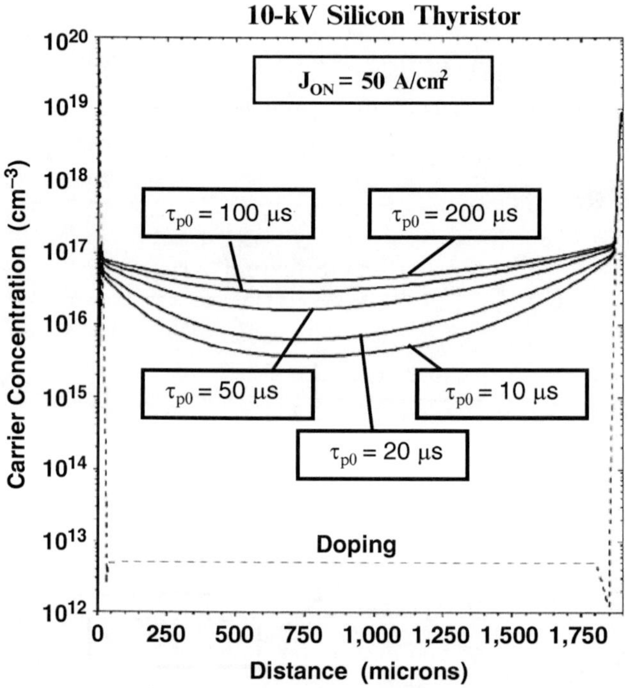

Fig. 2.28 On-state carrier distribution in the 10-kV power thyristor structure

The low on-state voltage drop for the 10-kV thyristor structure is due to the large number of carriers injected into the drift region producing a drastic reduction of its resistance. This is illustrated in Fig. 2.28 where the injected carrier density is shown for five cases of the lifetime (τ_{p0}, τ_{n0}) in the drift region of the thyristor structure. It can be observed that the injected carrier density is four orders of magnitude larger than the doping concentration for the case of a lifetime of 200 μs. The injected carrier density is reduced by more than an order of magnitude in the middle of the drift region when the lifetime is reduced to 10 μs. As in the case of the 5-kV thyristor structure, the injected carrier density is much smaller in the numerical simulations when compared with the analytical model that neglects the end-region recombination.

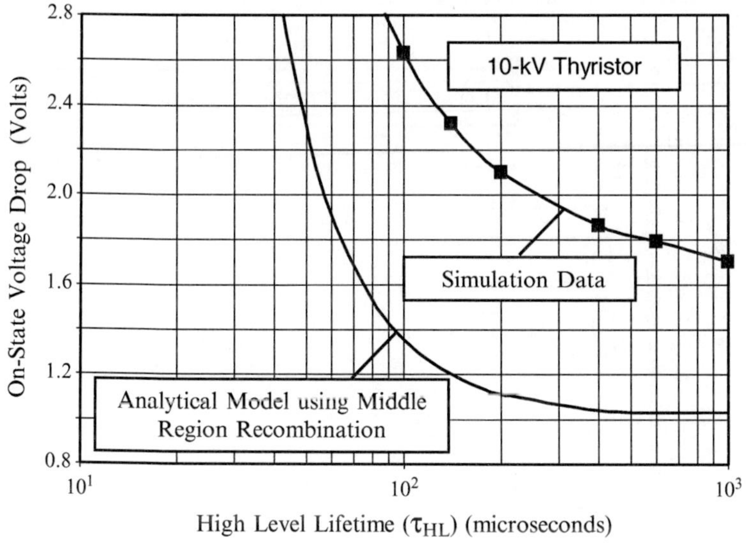

Fig. 2.29 On-state voltage drop for the 10-kV power thyristor structure obtained using numerical simulations

The reduction of the injected carrier density in the middle region with smaller lifetime leads to an increase in the on-state voltage drop. The on-state voltage drop for the 10-kV thyristor obtained using numerical simulations can be compared with that obtained using the analytical model without end-region recombination in Fig. 2.29. The on-state voltage drop obtained using the numerical simulations are much larger than those predicted by the analytical model based upon middle-region recombination. Further, the minimum on-state voltage drop occurs at a significantly larger value for the lifetime in the drift region. It can be concluded that the analytical model is very optimistic and not reliable for predicting the on-state voltage drop, indicating that the impact of end-region recombination is very important for the silicon 10-kV thyristor structure.

2.3.3 *Turn-On*

The 10-kV thyristor can be triggered from the forward blocking mode to the on-state by the application of a gate drive current. The analytical model for the increase in the anode current during the turn-on transient for a one-dimensional thyristor structure was discussed in the previous section. The increase in the anode current during the turn-on process is described by Eq. 2.17. The time constant for the rise in anode current is observed to be the geometric mean of the transit times for the N-P-N and P-N-P transistors. The transit time for the holes through the N-base region of the 10-kV thyristor is much greater than in the 5-kV device because of the larger base width. This produces an increase in the rise-time for the anode current as illustrated in Fig. 2.30 for three values of the anode voltage.

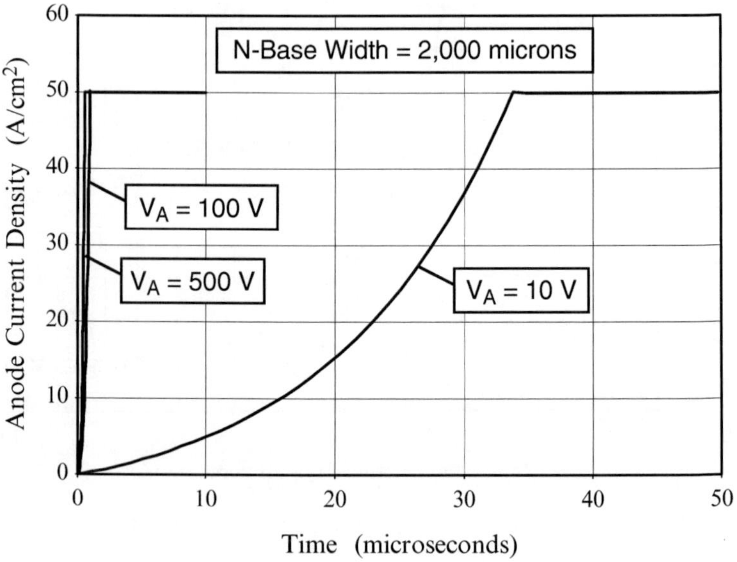

Fig. 2.30 Turn-on transient for a one-dimensional 10-kV thyristor

The rise in the anode current with time is shown in Fig. 2.30 for the case of a one-dimensional thyristor structure with a P-base region width of 25 μm and an N-base region width of 2,000 μm. The transit time for the N-P-N transistor is calculated to be 120 ns. For the case of an anode bias of 10 V, the current reaches its steady-state value with a rise-time of 34 μs. However, when the anode voltage is increased to 100 V, the transit time for the P-N-P transistor is greatly reduced due to the enhanced drift current. Consequently, the rise-time for the anode current also decreases to only 0.95 μs. A further increase in the anode voltage to 500 V produces a reduction of the rise-time to only 0.56 μs.

Simulation Results

In order to gain further insight into the physics of turn-on for the 10-kV power thyristor structure, the results of two-dimensional numerical simulations for the structure (described in the previous section) are discussed here. The total width of the structure is 1,000 μm (area $= 1 \times 10^{-5}$ cm^{-2}) with a cathode finger width of 980 μm. In order to turn on the thyristor, the gate drive current was abruptly increased from 0 to 2×10^{-6} A/μm (0.2 A/cm^2) while the anode voltage was maintained fixed with a load resistance in series with the thyristor structure.

The turn-on characteristics for the 10-kV power thyristor structure are shown in Fig. 2.31 for three values of the anode bias. In each case the load resistance was altered to obtain the same anode current density (50 A/cm^2) after the thyristor is in the on-state. In all cases, there is a short delay time of about 100 ns, which is consistent with the delay time computed by using the base transit time of the N-P-N transistor. The anode current then increases exponentially with time as described by the analytical Eq. 2.17. For the case of an anode bias of 10 V, the anode current increases exponentially for about 80 μs. When the anode bias is increased to 100 and 500 V, the anode current density increases much more rapidly. This behavior is consistent with the analytical model for turn-on based upon the drift of carriers through the N-base region. In all cases, a gradual increase in the anode current density is also observed at the end of the turn-on transient due to the current spreading across the emitter width in the two-dimensional structure used for the numerical simulations. The analytical model describes the correct trends with increasing anode voltage but is not in quantitative agreement with the results of the numerical simulations. This is due to current spreading across the cell structure which produces a different current density than assumed for the one-dimensional analytical model.

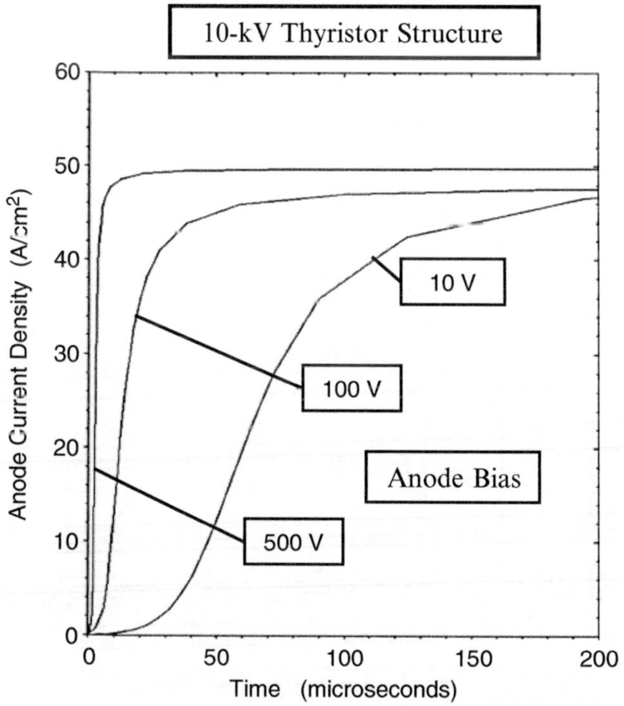

Fig. 2.31 Turn-on characteristics for the 10-kV power thyristor structure

2.3.4 Reverse Recovery

The reverse recovery behavior for the 10-kV thyristor structure can be expected to be similar to that previously described in Sect. 2.2.4 for the 5-kV device structure. The larger stored charge in the 10-kV device structure increases the reverse recovery current and consequently the power dissipation during this transient. This reduces the maximum operating frequency for the 10-kV device when compared with the 5-kV device.

Simulation Results

In order to gain further insight into the physics of turn-off for the power thyristor under a constant [d*i*/d*t*], the results of two-dimensional numerical simulations for the 10-kV thyristor structure (described in the previous sections) are discussed here. The total width of the structure used for the numerical simulations was 1,000 μm (area $= 1 \times 10^{-5}$ cm^{-2}) with a cathode finger width of 980 μm between the gate contact and cathode short. The thyristor was switched from on-state operation at a current density of 50 A/cm^2 with a constant [d*i*/d*t*] (rate of 75 A/cm^2 in 10 μs). This reverse ramp rate was applied until the anode voltage reached −1,000 V. The anode voltage was then held constant, allowing the anode current to decay to zero.

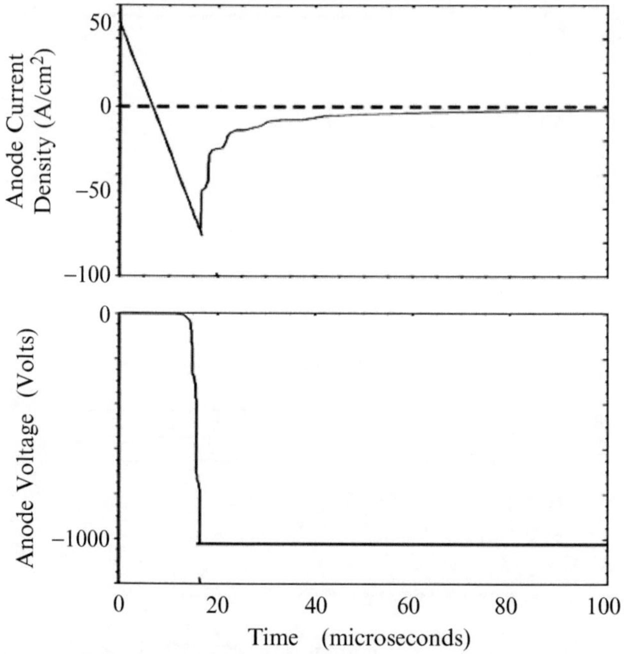

Fig. 2.32 Reverse recovery characteristics for the 10-kV thyristor structure with constant [d*i*/d*t*] applied to the anode electrode

The anode current and voltage waveforms for the thyristor are shown in Fig. 2.32. As expected, these waveforms are very similar to those exhibited by the P-i-N rectifier. The anode current flows in the reverse direction until the anode voltage becomes equal to the reverse bias voltage and then decreases exponentially to zero. These waveforms confirm that the thyristor structure can be analyzed by using the reverse recovery analysis discussed in the textbook for the P-i-N rectifier. In the case of the baseline simulation structure with a lifetime (τ_{p0}, τ_{n0}) of 100 μs, the reverse recovery process requires a total time interval of about 100 μs.

2.3.5 *Summary*

The results of the numerical simulations discussed above indicate that the silicon 10-kV thyristor structure requires a minority carrier lifetime (τ_{p0}, τ_{n0}) of at least 100 μs to achieve a low on-state voltage drop. For such a large lifetime value, the reverse recovery process takes over 100 μs. From the waveforms in Fig. 2.32, it can be observed that most of the power dissipation occurs during the decay of the anode current after it reaches the peak reverse recovery current density. For the conditions used during the numerical simulations, the peak reverse recovery current (J_{PR}) is found to be 74 A/cm^2 and the time constant for the reverse recovery (t_{RR}) is found to be 6.75 μs. Using a power supply voltage of 1,000 V, the energy loss per switching event is found to be 0.50 J/cm^2 by using Eq. 2.20. If the maximum power dissipation due to switching is limited to 100 W/cm^2, the maximum operating frequency for the 10-kV thyristor is then found to be 200 Hz. This demonstrates that the 10-kV silicon thyristor cannot be operated at frequencies as high as the 5-kV device.

2.4 Conclusions

The design and characteristics of the 5-kV and 10-kV silicon thyristor structures have been analyzed in this chapter. Such high voltage ratings require extremely high resistivity silicon wafers with relatively large thickness. The large thickness of the drift region limits the ability to reduce the lifetime in the drift region to improve the switching speed as determined by the reverse recovery process. Typical silicon devices are found to be limited to an operating frequency of less than 300 Hz for the 5-kV device and 200 Hz for the 10-kV device.

References

1. B.J. Baliga, "Fundamentals of Power Semiconductor Devices", Springer-Science, 2008.
2. I.M. Mackintosh, "The Electrical Characteristics of Silicon p-n-p-n Triodes", Proc. IRE, Vol. 46, pp. 1229, 1958.
3. R.W. Aldrich and N. Holonyak, "Multiterminal p-n-p-n Switches", Proc. IRE, Vol. 46, pp. 1236, 1958.
4. A. Herlet, "The Maximum Blocking Capability of Silicon Thyristors", Solid State Electronics, Vol. 8, pp. 655–671, 1965.

Chapter 3
Silicon Carbide Thyristors

The basic structure and operation of the thyristor are discussed briefly in Chap. 2 and in more detail in the textbook [1]. The thyristor contains two coupled bipolar transistors that provide an internal positive feedback mechanism that allows the device to sustain itself in the on-state. Analytical models were provided in Chap. 2 for all the operating modes of the thyristor. These models are applicable to the silicon carbide devices discussed in this chapter. The motivation for the development of thyristors from silicon carbide originates from the high on-state voltage drop and slow switching speed of the high-voltage silicon devices. In Chap. 2, it was demonstrated that even a 10-kV silicon thyristor structure has a relatively high on-state voltage drop (close to 3 V) even when the high-level lifetime in its drift region is 100 μs. In the case of silicon carbide devices, the width of the drift region can be greatly reduced (about ten times) when compared with a silicon device with the same voltage rating. This allows obtaining devices with much faster switching speed.

The development of silicon carbide-based thyristor structures has been pursued due to the anticipated faster switching speed and higher operating temperature capability. The first symmetrical blocking silicon carbide thyristors were demonstrated in 1996 with a blocking voltage capability of 700 V [2]. These devices exhibited an on-state voltage drop of 3.9 V. Subsequent efforts to develop SiC thyristors have utilized the asymmetric device structure which is more suitable for the gate-turn-off (GTO) thyristor devices. Silicon carbide GTO structures are not discussed in this book because of their very high turn-off gate drive currents [3]. Silicon carbide GTO devices with a chip size of 1 cm by 1 cm were reported in 2005 with an asymmetric blocking capability of 1,770 V [4]. Devices with the capability to support 9 kV during the forward blocking mode have been more recently reported [5]. The on-state characteristics of asymmetric blocking thyristors with forward blocking capability ranging from 8 to 20 kV have also been reported [6]. The 8-kV devices had a good on-state voltage drop of 3.5 V but the on-state voltage drop for the 20-kV devices was very high at over 10 V.

The characteristics of high-voltage silicon carbide thyristors are considered in this chapter for a blocking voltage rating of 20 kV because there are no high

B.J. Baliga, *Advanced High Voltage Power Device Concepts*,
DOI 10.1007/978-1-4614-0269-5_3, © Springer Science+Business Media, LLC 2011

performance silicon devices available with blocking voltages above 10 kV. These devices have potential applications in power distribution systems. For this application, the most important device characteristic, beside the high operating voltage capability, is its on-state voltage drop because the switching frequency is very low. Consequently, the analysis in this chapter focuses only on the on-state voltage drop of the 20-kV devices.

3.1 SiC Thyristor Structure

The basic structure for the silicon carbide power thyristor is illustrated in Fig. 3.1. Although the silicon carbide thyristor comprises the same four-layer structure as the silicon device, the doping profiles of all the regions are uniform because it is usually constructed with epitaxially grown layers due to the very low diffusion rates for dopants in silicon carbide. Further, the silicon carbide device is fabricated using a P-type drift region because of the larger diffusion length for electrons and due to the availability of highly doped, low resistivity N-type substrates. Low resistivity P-type silicon carbide substrates are not available for the fabrication of the complementary structure.

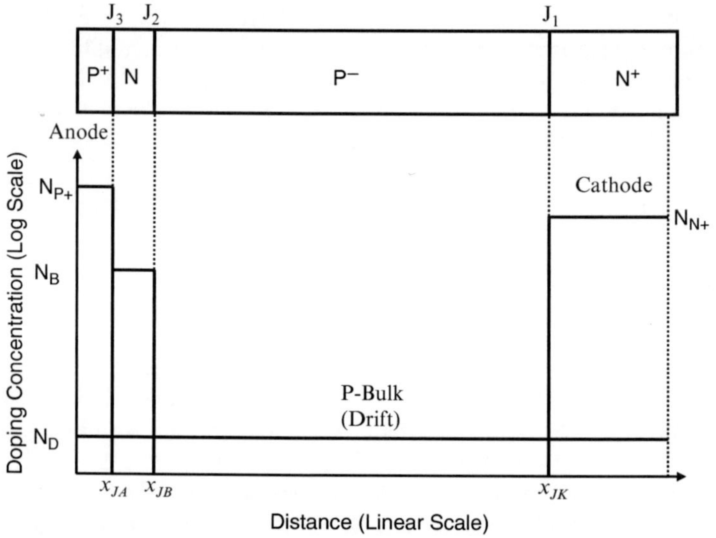

Fig. 3.1 SiC power thyristor structure and its doping profile

The silicon carbide thyristor structure is usually constructed by starting with a highly doped N-type substrate (as cathode region) on which the lightly doped P-type drift region is first formed by epitaxial growth. The P-type drift region has a doping concentration and thickness optimized for the desired blocking voltage capability of the thyristor. An N-type base region is then epitaxially grown on the drift region with a typical doping concentration of 1×10^{17} cm^{-3} and thickness of 2 μm [5]. The P$^+$ anode region is then formed by epitaxial growth with a typical doping concentration of 2×10^{19} cm^{-3} and thickness of 1 μm [5]. The doping profile for the silicon carbide thyristor structure is shown in Fig. 3.1 with uniformly doped regions and abrupt junctions. The size of the anode regions are defined by etching through the upper P$^+$ layer.

The basic *i–v* characteristics of the silicon carbide thyristor structure are similar to those shown in Chap. 2 (see Fig. 2.4). The physics of operation of the silicon carbide thyristor is also basically the same as that for the silicon counterpart. However, the lifetime for free carriers in silicon carbide has been found to be relatively small. Recent improvements in process technology for silicon carbide have produced material with lifetime values above 1 μs [7].

3.2 20-kV Silicon Baseline Thyristor

The design and characteristics of the 20-kV symmetric blocking silicon thyristor structure are discussed in this section to serve as a baseline for comparison with the 20-kV silicon carbide thyristor. The design parameters for the N-base region required to achieve this blocking voltage are first analyzed. Using the optimum N-base width, the blocking characteristics for the device are first obtained. The on-state characteristics for the device are then described for various lifetime values. It is demonstrated in this section that the silicon 20-kV thyristor structure has a very high on-state voltage drop. This has motivated the development of the silicon carbide thyristor structure.

3.2.1 Blocking Characteristics

The physics for blocking voltages in the first and third quadrants by thyristors is discussed in detail in the textbook [1]. The breakdown voltage for the thyristor in both blocking modes is governed by the open-base transistor breakdown phenomenon [8]. According to open-base-transistor breakdown, the anode current will increase very rapidly when the common base current gain of the P-N-P bipolar transistor within the thyristor structure approaches unity. As the negative anode bias is increased, the width of the un-depleted portion of the N-drift region becomes smaller producing an increase in the base transport factor (α_T). Concurrently, the maximum electric field at the blocking junction becomes larger leading to an increase in the multiplication coefficient. Both phenomena produce an increase in the common base current gain with increasing anode bias until it becomes equal to unity resulting in open-base transistor breakdown. The equations governing the blocking voltage capability were already provided in Chap. 2.

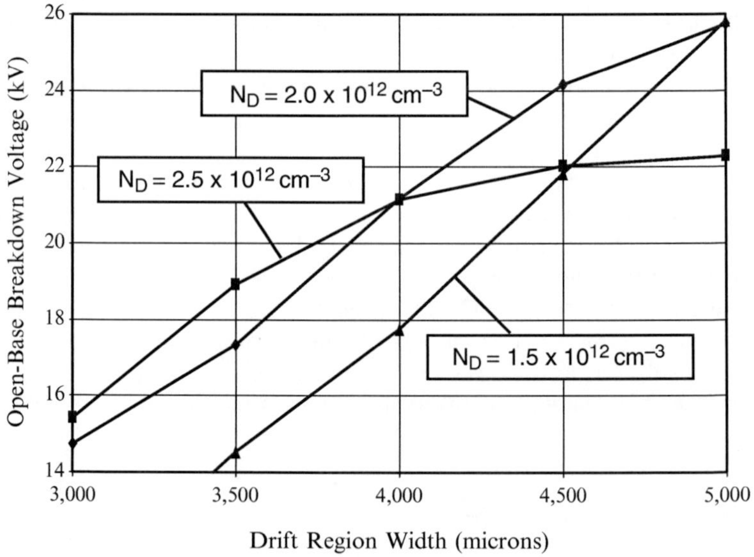

Fig. 3.2 Open-base breakdown voltage for the 20-kV silicon thyristor structure

Consider the case of a silicon symmetric power thyristor that must have a breakdown voltage of 22 kV to achieve a blocking voltage rating of 20 kV. In the case of avalanche breakdown, there is a unique value of 2.56×10^{12} cm^{-3} for the drift region doping concentration to obtain this blocking voltage. However, in the case of open-base transistor breakdown, many combinations of the drift region doping concentration and width can be used to obtain this blocking voltage capability. This is illustrated in Fig. 3.2 where the open-base breakdown voltage is plotted as a function of the drift region width for three cases of the drift region doping concentration. A lifetime of 100 μs was used in the N-drift region for this analysis. It can be observed from Fig. 3.2 that the open-base breakdown voltage becomes equal to 22 kV at a drift region width of 4,550 μm for a drift region doping concentration of 1.5×10^{12} cm^{-3}. In this case, the base transport factor becomes close to unity under breakdown conditions. When the doping concentration of drift region is increased to 2.0×10^{12} cm^{-3}, the drift region thickness is reduced to 4,150 μm to achieve the same open-base breakdown voltage of 22 kV. The drift region thickness increases to 4,500 μm, if the doping concentration of drift region is increased to 2.5×10^{12} cm^{-3}, to achieve the same open-base breakdown voltage of 22 kV. In this case, the multiplication coefficient becomes large under open-base breakdown conditions. These examples demonstrate that there is an optimum drift region doping concentration to obtain a minimum drift region width to achieve an open-base breakdown voltage of 22 kV. The location of the optimum design with a width of 4,080 μm and doping concentration of 2.2×10^{12} cm^{-3} is illustrated for this case in Fig. 3.3.

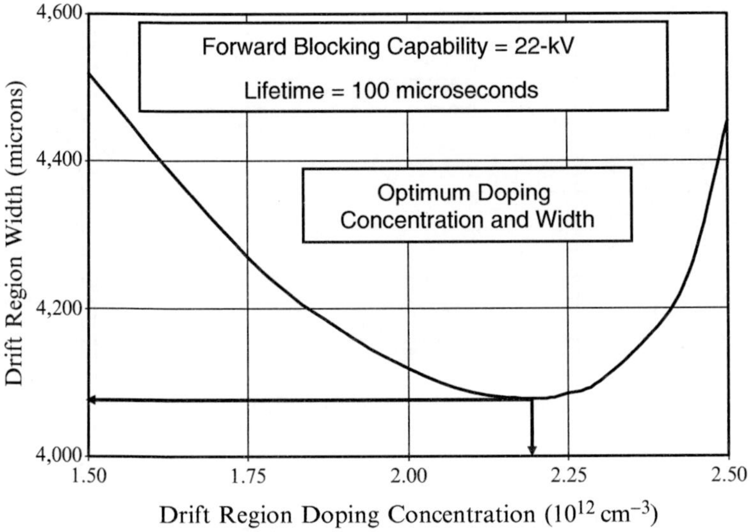

Fig. 3.3 Optimum width and doping concentration of the drift region for the 20-kV silicon symmetric power thyristor structure

Simulation Results

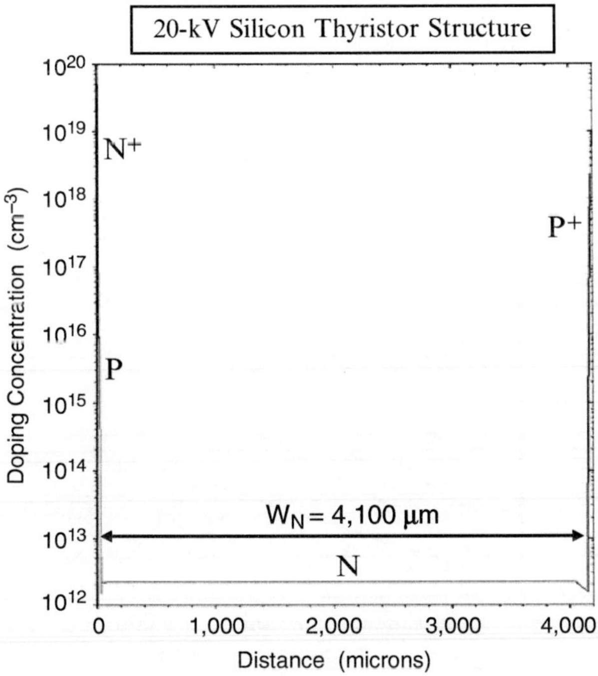

Fig. 3.4 Doping profile for the simulated 20-kV silicon thyristor structure

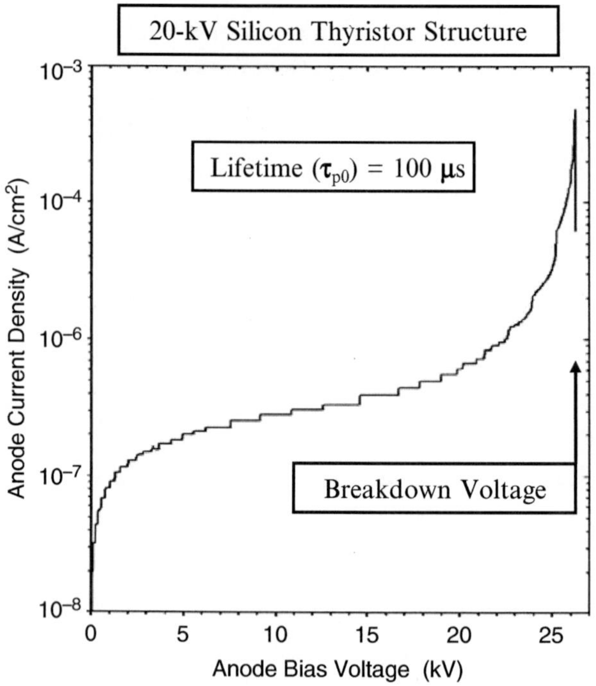

Fig. 3.5 Forward blocking characteristics for the 20-kV silicon thyristor structure

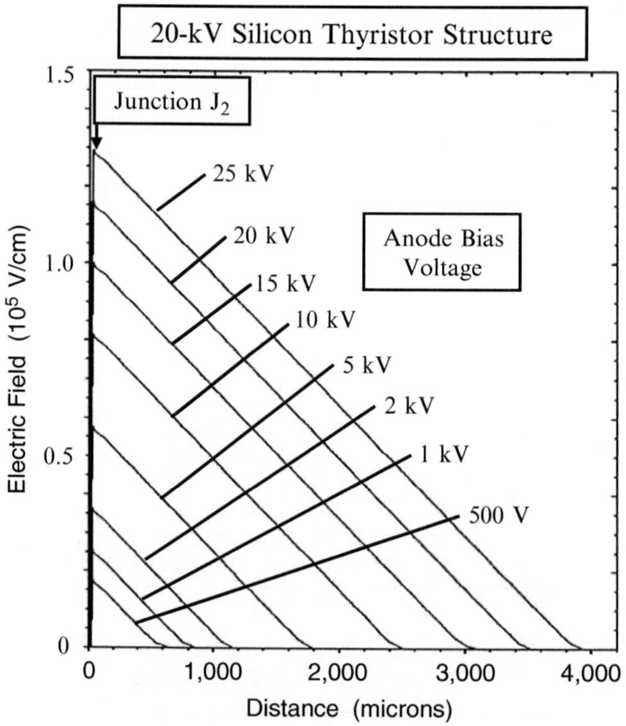

Fig. 3.6 Electric field profiles in the forward blocking mode for the 20-kV silicon power thyristor structure

The doping profile for the 20-kV silicon thyristor structure used for the numerical simulations is provided in Fig. 3.4. The N-type drift region has a doping concentration of 2.2×10^{12} cm^{-3} and thickness of 4,100 μm based upon the optimum values obtained from the analytical model (see Fig. 3.3). The depths and doping profiles for the P^+ anode region, the P-base region, and the N^+ cathode region were chosen to be the same as those for the 5-kV silicon thyristor structure (see Chap. 2).

The forward blocking characteristics for the thyristor structure are shown in Fig. 3.5 for the case of a lifetime (τ_{p0}, τ_{n0}) value of 100 μs in the N-base region. The leakage current increases rapidly for anode bias voltages above 22 kV due to the increasing gain of the open-base PNP transistor. The breakdown voltage is indicated by an abrupt increase in the anode current at an anode bias of 25 kV.

The voltage is primarily supported in the silicon symmetric thyristor structure within the N-drift region. This is illustrated in Fig. 3.6 where the electric field profiles are shown during operation in the forward blocking mode at various positive anode bias voltages. It can be observed that the P-base/N-drift junction (J_2) becomes reverse biased during the forward blocking mode, with the depletion region extending toward the right-hand side with increasing anode bias. The depletion region extends through most of the N-base region at an anode bias of 20 kV indicating an optimum design.

3.2.2 *On-State Characteristics*

One of the attributes of the thyristor structure is its excellent forward conduction characteristic even when designed to support large voltage levels. In the on-state, the thyristor structure operates like a P-i-N rectifier with strong conductivity modulation of the N-drift region due to high level injection of holes. However, for the case of the silicon thyristor structure designed to support 20 kV in the blocking modes, the drift region thickness becomes extremely large (4,100 μm) as shown in the previous section. In order to obtain good conductivity modulation of the drift region with a low on-state voltage drop, it is necessary to achieve a (d/L_a) ratio of close to unity. Based upon a diffusion coefficient of 15 cm^2/s under high-level injection conditions, the high-level lifetime required to achieve a low on-state voltage drop is found to be 2.8 ms. It is not possible to achieve such an extremely large value for the high-level lifetime in silicon. Typical high-level lifetime values even for very high resistivity silicon wafers are in the 100-μs range.

The on-state *i–v* characteristic for the thyristor in the high level injection regime of operation is given by [1]:

$$V_{\mathrm{ON}} = \frac{2kT}{q} \ln\left[\frac{J_{\mathrm{T}} d}{2qD_{\mathrm{a}} n_i F(d/L_{\mathrm{a}})}\right] \tag{3.1}$$

where

$$F\left(\frac{d}{L_a}\right) = \frac{(d/L_a)\tanh(d/L_a)}{\sqrt{1 - 0.25\tanh^4(d/L_a)}} e^{-\frac{qV_M}{2kT}} \tag{3.2}$$

The i-region (or middle-region) voltage drop can be computed using the following approximations. For d/L_a ratios of up to 2, an asymptote A given by:

$$V_M = \frac{2kT}{q}\left(\frac{d}{L_a}\right)^2 \tag{3.3}$$

provides a good fit. For d/L_a ratios of greater than 2, an asymptote B given by

$$V_M = \frac{3\pi kT}{8q} e^{(d/L_a)} \tag{3.4}$$

provides a good fit.

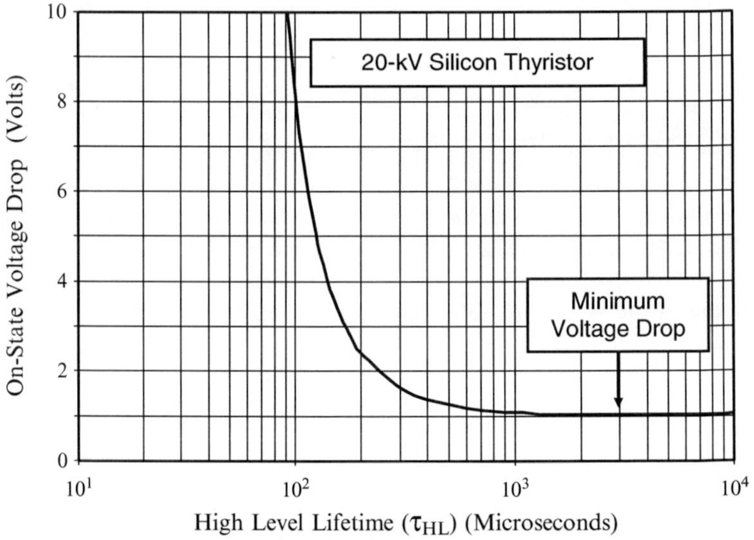

Fig. 3.7 On-state voltage drop for the 20-kV power thyristor structure

The on-state voltage drop (at an on-state current density of 100 A/cm^2) computed for the 20-kV silicon thyristor structure by using Eq. 3.1 is provided in Fig. 3.7 for various values for the high-level lifetime in the drift region. This thyristor structure had the optimized N-base region width of 4,100 μm and a P-base width of 25 μm. As expected, the on-state voltage drop has a minimum value when the high-level lifetime is 3,000 μs at which point the [d/L_a] value becomes equal to unity. The minimum on-state voltage drop is found to be 1.03 V. The on-state voltage drop increases very rapidly when the high-level lifetime is reduced below 500 μs. For practical high-level lifetime values of 100 μs, the on-state voltage drop for the

20-kV silicon thyristor structure is found to be 8 V. This relatively large on-state voltage drop provides motivation for the development of silicon carbide 20-kV thyristors even though the diode knee voltage for silicon carbide devices is 3 V.

Simulation Results

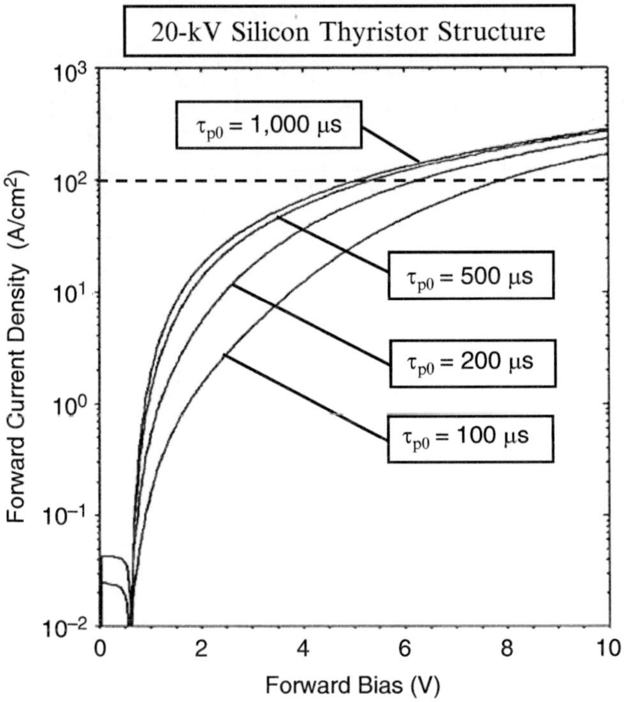

Fig. 3.8 On-state characteristics of the 20-kV silicon thyristor structure: lifetime dependence

The results of two-dimensional numerical simulations for the 20-kV silicon thyristor structure are described here. The total width of the structure, as shown by the cross-section in Fig. 2.5, was 1,000 μm (Area = 1×10^{-5} cm^{-2}) with a cathode finger width of 980 μm. The N-base region had an optimized doping concentration of 2.2×10^{12} cm^{-3} and width of 4,100 μm as shown in Fig. 3.4. The on-state characteristics were obtained by using a gate drive current of 2×10^{-8} A/μm using various values for the lifetime in the drift region. The characteristics obtained from the numerical simulations are shown in Fig. 3.8. The small anode current observed at anode biases below 0.5 V is associated with the gate drive current. It can be observed that the on-state voltage drop increases as expected with reduction of the lifetime (τ_{p0}, τ_{n0}) indicated in the figure. The on-state voltage drop at a lifetime value of 1,000 μs is found to be 5.12 V at an on-state current density of 100 A/cm^2. This value is much larger than predicted by the analytical model because it does not account for recombination in the end regions. The on-state voltage drop increases to 7.98 V when the lifetime is reduced to 200 μs. The data obtained from the numerical simulations can be compared with the values obtained by using the analytical model with middle region recombination in Fig. 3.9.

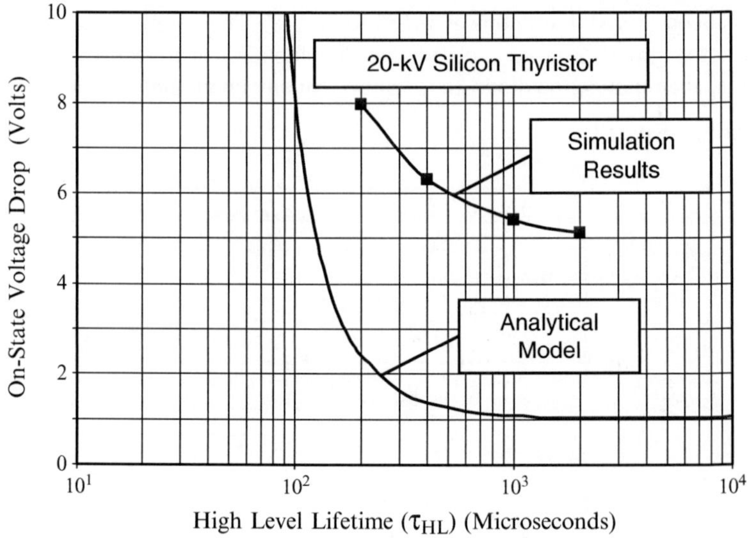

Fig. 3.9 On-state voltage drop for the 20-kV silicon thyristor structure

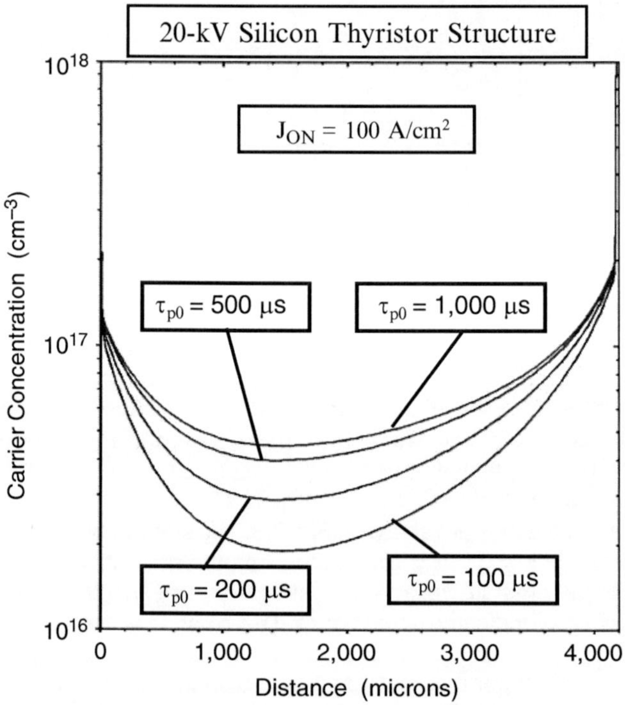

Fig. 3.10 On-state carrier distribution in the 20-kV silicon thyristor structure

The relatively high on-state voltage drop for the 20-kV silicon thyristor structure is due to reduced carrier density in the middle of the very thick drift region. This is illustrated in Fig. 3.10 where the injected carrier density is shown for four cases of the lifetime (τ_{p0}, τ_{n0}) in the drift region of the thyristor structure. It can be observed that the injected carrier density is four orders of magnitude larger than the doping concentration. In spite of this, the on-state voltage drop becomes large due to large thickness of the drift region required to support the very high blocking voltage for this structure.

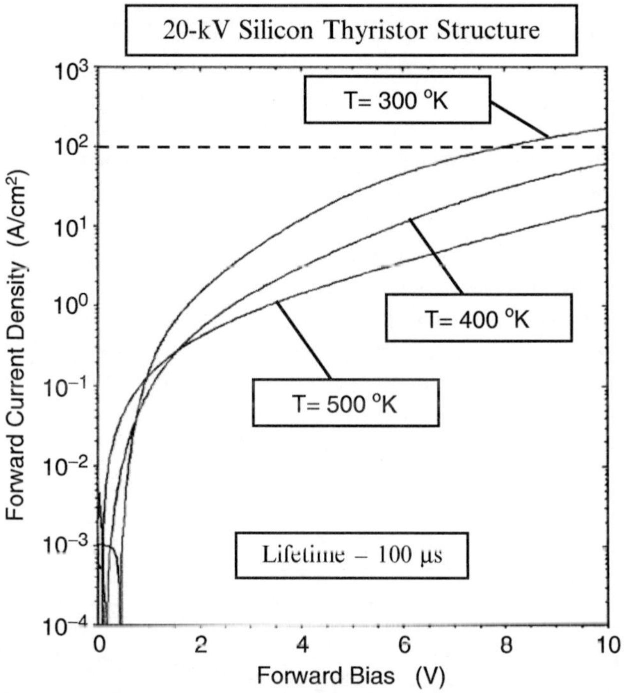

Fig. 3.11 On-state characteristics of the 20-kV silicon thyristor structure: temperature dependence

The on-state *i–v* characteristics of the 20-kV silicon thyristor structure are shown in Fig. 3.11 for various temperatures. The lifetime (τ_{p0}, τ_{n0}) was assumed to be 100 μs during these simulations. It can be observed that the on-state voltage drop at an on-state current density of 100 A/cm^2 increases rapidly with increasing temperature. Based upon the results of the numerical simulations, it can be concluded that the silicon thyristor designed with a blocking voltage capability of 20 kV has poor on-state characteristics.

3.3 20-kV Silicon Carbide Thyristor

The design and characteristics of the 20-kV symmetric blocking silicon carbide thyristor structure are discussed in this section. The design parameters for the N-base region required to achieve this blocking voltage are first analyzed. The on-state characteristics for the device are then obtained for various lifetime values.

3.3.1 Blocking Characteristics

The physics for blocking voltages in the first and third quadrants by the silicon carbide thyristor (Fig. 3.1) is similar to that for the silicon devices. The breakdown voltage for the thyristor in both blocking modes is governed by the open-base transistor breakdown phenomenon. According to open-base-transistor breakdown [1], the anode current will increase very rapidly when the common base current gain of the N-P-N bipolar transistor within the thyristor structure approaches unity. As the negative anode bias is increased, the width of the un-depleted portion of the P-drift region becomes smaller producing an increase in the base transport factor (α_T). Concurrently, the maximum electric field at the blocking junction becomes larger leading to an increase in the multiplication coefficient. Both phenomena produce an increase in the common base current gain with increasing anode bias until it becomes equal to unity resulting in open-base transistor breakdown. The equations governing the blocking voltage capability were already provided in Chap. 2.

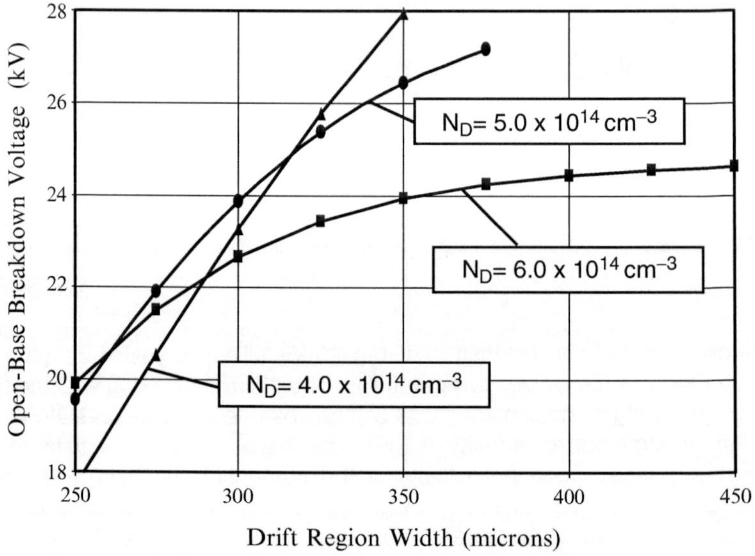

Fig. 3.12 Open-base breakdown voltage for the 20-kV silicon carbide thyristor structure

Consider the case of a silicon carbide symmetric thyristor structure that must have a breakdown voltage of 22 kV to achieve a blocking voltage rating of 20 kV. In the case of avalanche breakdown, there is a unique value of 7.02×10^{14} cm^{-3} for the drift region doping concentration to obtain this blocking voltage. However, in the case of open-base transistor breakdown, many combinations of the drift region doping concentration and width can be used to obtain this blocking voltage capability. This is illustrated in Fig. 3.12 where the open-base breakdown voltage is plotted as a function of the drift region width for three cases of the drift region doping concentration. A lifetime of 1.0 μs was used in the P-drift region for this analysis. It can be observed from Fig. 3.12 that the open-base breakdown voltage becomes equal to 22 kV at a drift region width of 290 μm for a drift region doping concentration of 4.0×10^{14} cm^{-3}. In this case, the base transport factor becomes close to unity under breakdown conditions. When the doping concentration of drift region is increased to 5.0×10^{12} cm^{-3}, the drift region thickness is reduced to 275 μm to achieve the same open-base breakdown voltage of 22 kV. The drift region thickness increases to 285 μm, if the doping concentration of drift region is increased to 6.0×10^{14} cm^{-3}, to achieve the same open-base breakdown voltage of 22 kV. In this case, the multiplication coefficient becomes large under open-base breakdown conditions. These examples demonstrate that there is an optimum drift region doping concentration to obtain a minimum drift region width to achieve an open-base breakdown voltage of 22 kV.

The location of the optimum design with a width of 275 μm and doping concentration of 5.0×10^{14} cm^{-3} is illustrated for this case in Fig. 3.13. The optimum thickness of the drift region for the 20-kV silicon carbide thyristor is much (a factor of 15 times) smaller than that required for the silicon device. This allows the device to operate with much lower lifetime values which are feasible in silicon carbide material.

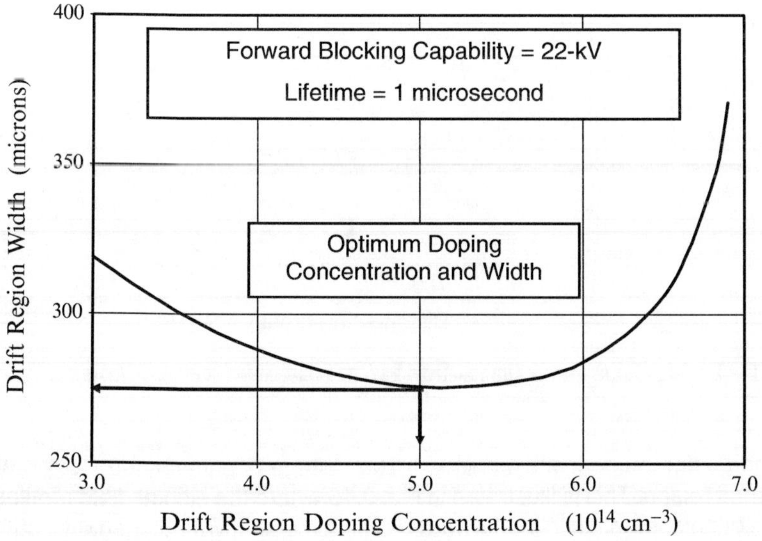

Fig. 3.13 Optimum width and doping concentration of the drift region for the 20-kV silicon carbide symmetric power thyristor structure

Simulation Results

The doping profile for the 20-kV silicon carbide thyristor structure used for the numerical simulations is provided in Fig. 3.14. This device structure has a P-type drift region because this allows utilization of available N^+ substrates as the initial starting material during device fabrication. The N^+ substrate acts as the cathode region of the silicon carbide thyristor structure. The P-type drift region has a doping concentration of 5.0×10^{14} cm^{-3} and thickness of 275 μm based upon the optimum values obtained from the analytical model (see Fig. 3.13). The N^+ cathode region has a thickness of 10 μm and its doping concentration is 1×10^{19} cm^{-3}. The N-base region has a thickness of 2 μm and its doping concentration is 2×10^{17} cm^{-3}. The P^+ anode region has a thickness of 1 μm and its doping concentration is 2×10^{19} cm^{-3}. All of the regions have a uniform doping concentration because the silicon carbide device structures are usually fabricated using sequential epitaxial growth. The doping concentrations and thicknesses used for the simulations are consistent with the values for the device parameters published in the literature [5].

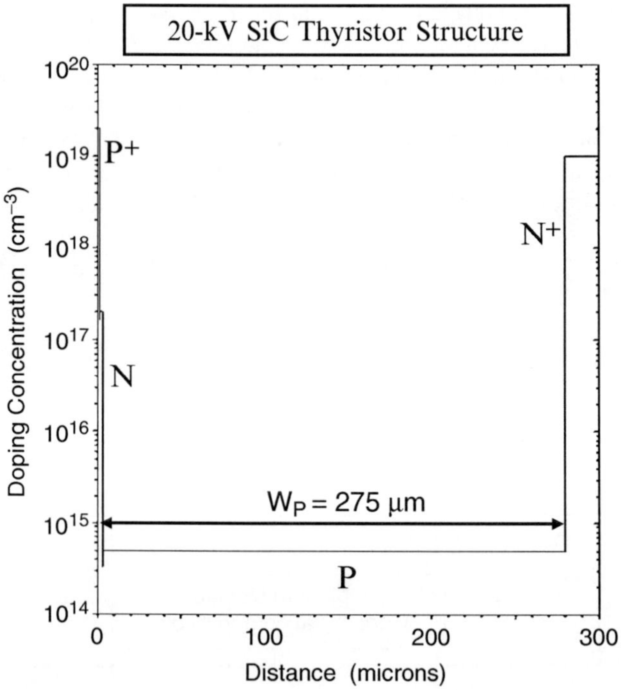

Fig. 3.14 Doping profile for the simulated 20-kV silicon carbide thyristor structure

The forward blocking characteristics for the 20-kV SiC thyristor structure were obtained using numerical simulations. The voltage is primarily supported within the P-drift region. This is illustrated in Fig. 3.15 where the electric field profiles are shown during operation in the forward blocking mode at various negative cathode bias voltages. It can be observed that the N-base/P-drift junction (J_2) becomes reverse biased during the forward blocking mode with the depletion region extending toward the right-hand side with increasing anode bias. The depletion region extends through most of the P-base region at an anode bias of 20 kV indicating an optimum design. It is worth pointing out that the maximum electric field is ten times larger than that for silicon devices.

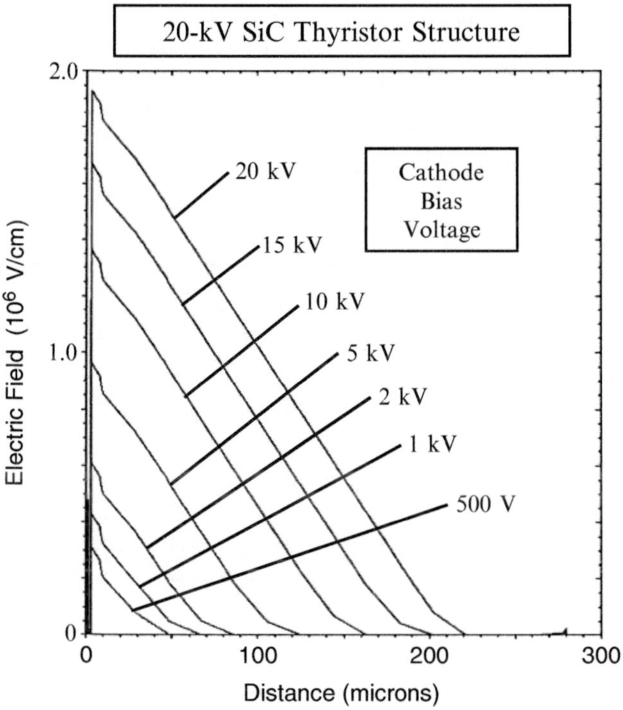

Fig. 3.15 Electric field profiles in the forward blocking mode for the 20-kV silicon carbide power thyristor structure

3.3.2 On-State Characteristics

One of the attributes of the silicon carbide thyristor structure is its excellent forward conduction characteristic even when designed to support large voltage levels. This is mainly due to the much smaller drift region width to support the voltage in a silicon carbide structure. In the on-state, the silicon carbide thyristor structure operates like a P-i-N rectifier with strong conductivity modulation of the N-drift region due to high level injection of electrons into the P-type drift region. As in the case of silicon devices, in order to obtain good conductivity modulation of the drift region with a low on-state voltage drop, it is necessary to achieve a (d/L_a) ratio of close to unity. Based upon a diffusion coefficient of 15 cm^2/s under high-level injection conditions, the high-level lifetime required to achieve a low on-state voltage drop is found to be 12.8 μs for a drift region thickness of 275 μm. Typical lifetime values measured in silicon carbide epitaxial layers are in the 1–4 μs range. These values are sufficiently close to the optimum lifetime value to allow the 20-kV silicon carbide thyristor to operate with a low on-state voltage drop.

The on-state *i–v* characteristic for the thyristor in the high level injection regime of operation is given by [1]:

$$V_{ON} = \frac{2kT}{q} \ln\left[\frac{J_T d}{2qD_a n_i F(d/L_a)}\right] \quad (3.5)$$

In the case of silicon carbide, the intrinsic concentration (n_i) is much smaller than for silicon leading to a high "knee-voltage" before high current flow commences. The function $F(d/L_a)$ can be computed using the same equations as for the silicon structure.

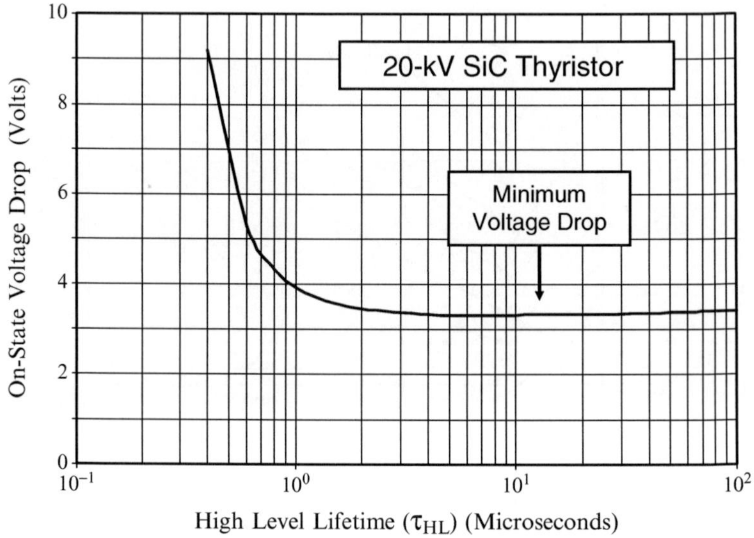

Fig. 3.16 On-state voltage drop for the 20-kV SiC thyristor structure

The on-state voltage drop (at an on-state current density of 100 A/cm^2) computed for the 20-kV silicon carbide thyristor structure by using Eq. 3.5 is provided in Fig. 3.16 for various values for the high-level lifetime in the drift region. This thyristor structure had the optimized P-base region width of 275 μm and an N-base width of 2 μm. As expected, the on-state voltage drop has a minimum value when the high-level lifetime is 13 μs at which point the $[d/L_a]$ value becomes equal to unity. The minimum on-state voltage drop is found to be 3.32 V. The on-state voltage drop increases very rapidly when the high-level lifetime is reduced below 1 μs. For practical high-level lifetime values of 1–4 μs, the on-state voltage drop for the 20-kV silicon thyristor structure is found to be less than 3.5 V.

Simulation Results

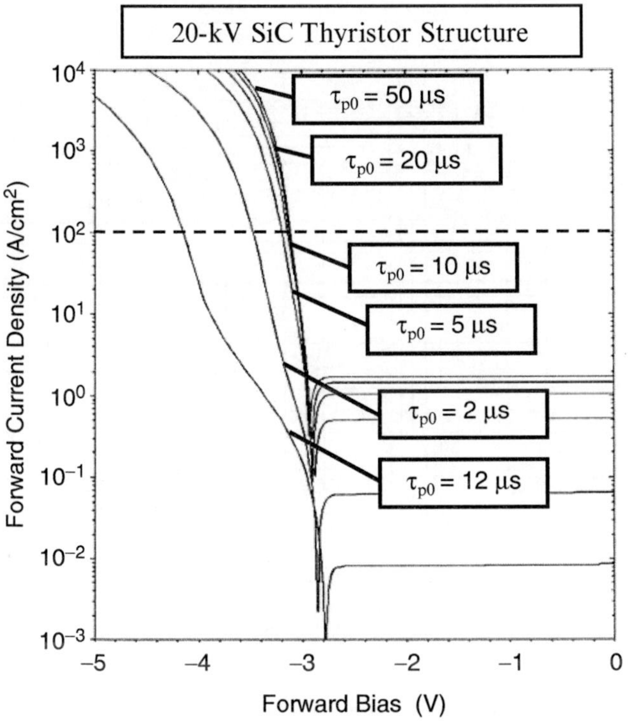

Fig. 3.17 On-state characteristics of the 20-kV SiC thyristor structure: lifetime dependence

The results of two-dimensional numerical simulations for the 20-kV silicon carbide thyristor structure are described here. The total width of the structure, as shown by the cross section in Fig. 2.5, was 1,000 μm (area = 1×10^{-5} cm^{-2}) with a cathode finger width of 980 μm. The P-base region had an optimized doping concentration of 5.0×10^{14} cm^{-3} and width of 275 μm as shown in Fig. 3.13. The on-state characteristics were obtained by using a gate drive current of 2×10^{-5} A/μm (corresponding to a gate current density of 2 A/cm^2) using various values for the lifetime in the drift region. The characteristics obtained from the numerical simulations are shown in Fig. 3.17. The small anode current observed at anode biases below 3 V is associated with the gate drive current.

It can be observed that the on-state voltage drop increases as expected with reduction of the lifetime (τ_{p0}, τ_{n0}) indicated in the figure. The on-state voltage drop at a relatively large lifetime value of 10 μs is found to be 3.13 V at an on-state current density of 100 A/cm^2. This value is in very good agreement with the on-state voltage drop predicted by the analytical model as shown in Fig. 3.18. It can be concluded from Fig. 3.18 that the on-state voltage drop predicted by the simple analytical model based upon recombination in the middle region works quite well for silicon carbide thyristors for a broad range of lifetime values. The on-state voltage drop for 20-kV silicon carbide thyristor obtained using the numerical simulations increases to 6.27 V when the high-level lifetime is reduced to 0.6 μs.

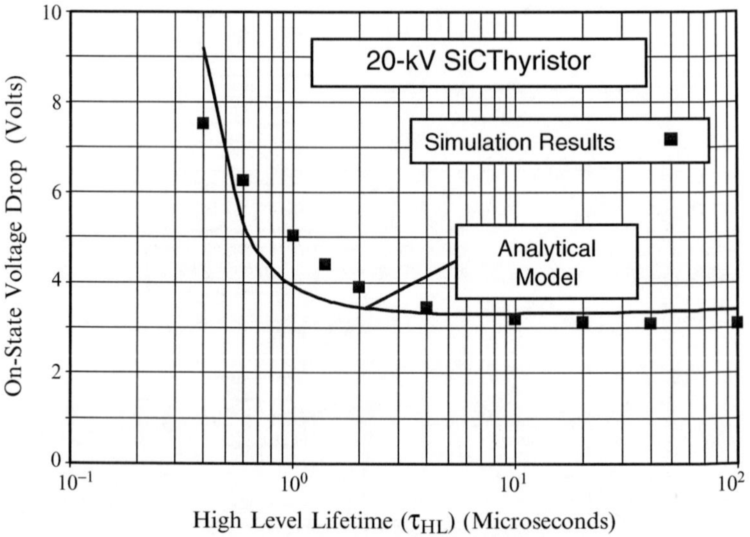

Fig. 3.18 On-state voltage drop for the 20-kV SiC thyristor structure

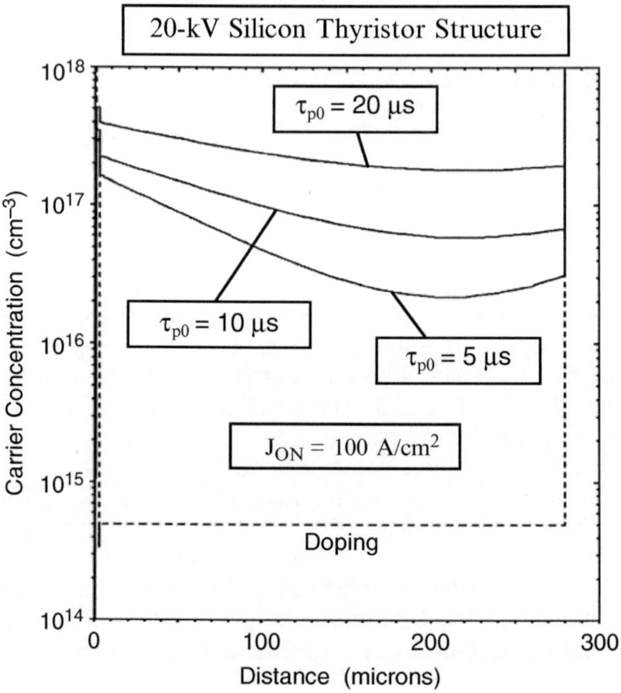

Fig. 3.19 On-state carrier distribution in the 20-kV SiC thyristor structure

The increase in the on-state voltage drop for the 20-kV silicon carbide thyristor structure can be correlated with the injected carrier density. The injected electron concentration in the P-base region is shown in Fig. 3.19 for three cases of the lifetime (τ_{p0}, τ_{n0}) in the drift region of the thyristor structure. It can be observed that the injected carrier density is three orders of magnitude larger than the doping concentration when the lifetime is large (20 μs). It is reduced by an order of magnitude when the lifetime is reduced to 5 μs.

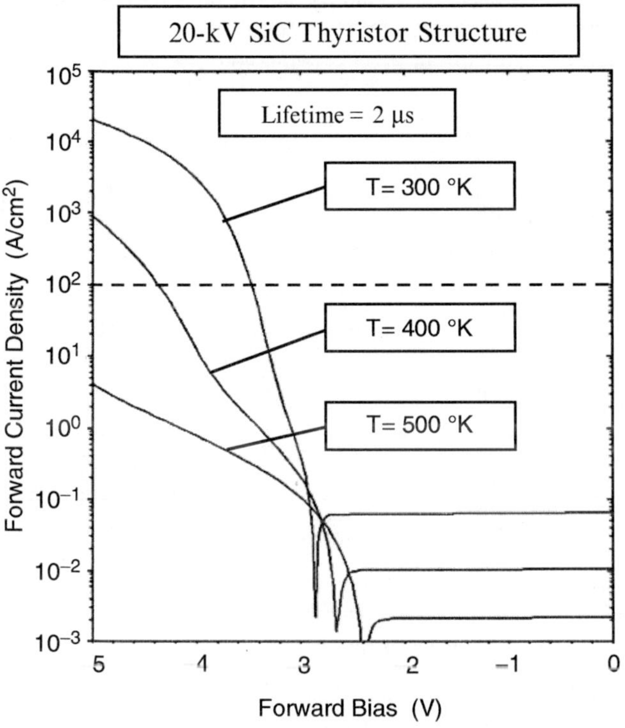

Fig. 3.20 On-state characteristics of the 20-kV silicon carbide thyristor structure: temperature dependence

The on-state *i–v* characteristics of the 20-kV silicon carbide thyristor structure are shown in Fig. 3.20 for various temperatures. The lifetime (τ_{p0}, τ_{n0}) was assumed to be 2 μs during these simulations. It can be observed that the on-state voltage drop at an on-state current density of 100 A/cm^2 increases rapidly with increasing temperature. Based upon the results of the numerical simulations, it can be concluded that the silicon carbide thyristor designed with a blocking voltage capability of 20 kV can be operated at up to only 400 K (127°C). At above 400 K, the silicon carbide thyristor structure has difficulty in latching-up to sustain the regenerative action that can produce a low on-state voltage drop.

3.4 Conclusions

The results of the numerical simulations provided in this chapter demonstrate that the silicon 20-kV thyristor structure has a very high on-state voltage drop due to its extremely wide drift region required to support the voltage. In contrast, the 20-kV silicon carbide thyristor has an attractive on-state voltage drop due to its much smaller drift region width because of the very high electric fields that can be supported by this semiconductor material. The low on-state voltage drop for the silicon carbide thyristor can be obtained only if the lifetime in the drift region is more than 1 μs. Such lifetime values are possible in epitaxial layers grown using low defect density with current process technology.

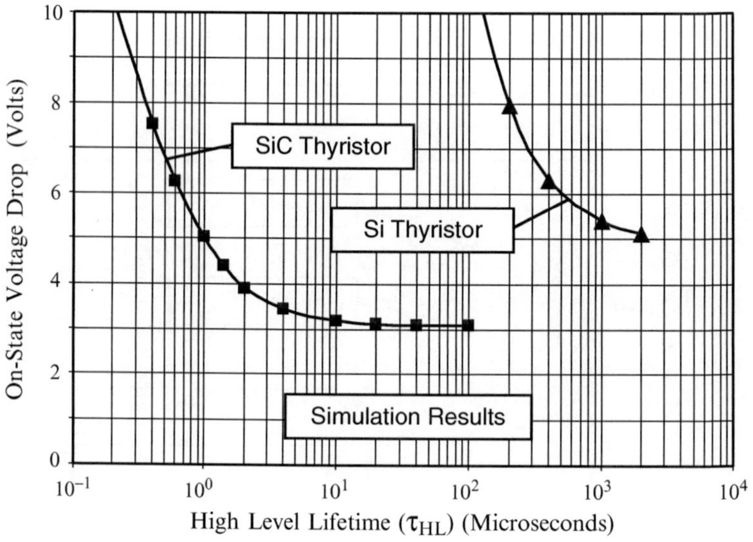

Fig. 3.21 Comparison of 20-kV silicon and SiC thyristors

The on-state voltage drops for the 20-kV silicon carbide thyristors are compared with those for the silicon thyristor structure in Fig. 3.21. It can be seen that the 20-kV silicon thyristor is not feasible due to its very high on-state voltage drop when the high-level lifetime is 100 μs, which is the best achievable value for this material. In contrast, the 20-kV silicon carbide thyristor has a sufficiently low on-state voltage drop of about 3–4 V even when the high-level lifetime takes a much smaller range of values, which are practical for epitaxial layers grown using current process technology. In comparison with an on-state voltage drop of about 8 V for the silicon device when the lifetime has a practical value of 100 μs, the silicon carbide thyristor has an on-state voltage drop close to 3 V. Consequently, on-state power losses can be reduced by a factor of 2.7 times which makes the device attractive for power distribution applications.

References

1. B. J. Baliga, "Fundamentals of Power Semiconductor Devices", Springer-Science, 2008.
2. J. W. Plamour, et al, "4 H-SiC High Temperature Power Devices", Third International Conference on High Temperature Electronics, Vol. 2, p. XVI-9, 1996.
3. A. Elasser, et al, "Silicon Carbide GTOs: Static and Dynamic Characterization", IEEE Industry Applications Conference, Vol. 1, pp. 359–364, 2001.
4. A. Agarwal, et al, "The First Demonstration of the 1 cm x 1 cm SiC Thyristor Chip", IEEE International Symposium on Power Semiconductor Devices and ICs, Abstract HV-P1, pp. 195–198, 2005.
5. A. Agarwal, et al, "9-kV, 1 cm × 1 cm SiC Super GTO Technology Development for Pulse Power", IEEE Pulsed power Conference, pp. 264–269, 2009.
6. G. G. Walden and J. A. Cooper, "On-State Characteristics of SiC Thyristors for the 8–20 kV Regime", IEEE Device Research Conference, pp. 91–92, 2009.
7. J.A. Cooper and A. Agarwal, "SiC Power Switching Devices – The Second Electronics Revolution?", Proceedings of the IEEE, Vol. 90, pp. 956–968, 2002.
8. A. Herlet, "The Maximum Blocking Capability of Silicon Thyristors", Solid State Electronics, Vol. 8, pp. 655–671, 1965.

Chapter 4
Silicon GTO

As discussed in the previous chapter, the thyristor structure contains a set of coupled transistors that provide a regenerative action during the conduction of current in the on-state. These devices are designed for operation in AC circuits where the anode voltage cycles between positive and negative values. The regenerative action is disrupted whenever the anode voltage reverses from positive to negative. The turn-off of the device then occurs with a reverse recovery process to establish blocking voltage capability. Such device structures are not suitable for applications in DC circuits unless expensive commutation circuits [1] are added to reverse the anode voltage polarity. The development of a thyristor structure that can be designed to turn on and turn off current flow under control by a gate signal in a DC circuit was motivated by this need. Such thyristors have been named gate turn-off (GTO) thyristors. The GTO is turned on in the same manner as the thyristor structures described in the previous chapter, while the turn-off for the GTO is accomplished by the application of a large reverse gate current. The gate current must be sufficient to remove stored charge from the P-base region and disrupt the regenerative action of the internal coupled transistors.

The basic operating principles and characteristics for the GTO have been described in the textbook [2] and reviewed in the literature [3]. In this chapter, the performance of the high voltage (5 and 10 kV) silicon GTO structure is discussed for purposes of comparison with the other devices covered in the book.

4.1 Basic Structure and Operation

Although similar to the conventional thyristor structure, the GTO structure does not contain cathode-shorts because the width (W_{KS}) of the cathode region for the GTO structure is made much shorter than for the conventional thyristor to facilitate turning-off of the anode current. In addition, since the GTO structure is

B.J. Baliga, *Advanced High Voltage Power Device Concepts*,
DOI 10.1007/978-1-4614-0269-5_4, © Springer Science+Business Media, LLC 2011

intended for use in DC circuits, its reverse blocking capability does not have to match the forward blocking capability allowing the use of the asymmetric GTO structure illustrated in Fig. 4.1. In the asymmetric structure, an N-buffer layer is added in the N-base region adjacent to the P^+ anode region. The N-buffer layer has a much larger doping concentration than the lightly doped portion of the N-base region. These changes result in a trapezoidal shape for the electric field profile as illustrated on the right-hand side of the figure. The same forward blocking capability can be achieved for the asymmetrical GTO structure with a smaller net thickness for the N-base region than necessary for the symmetrical structure. This enables reduction of the on-state voltage drop. The presence of the N-buffer layer also reduces the current gain of the P-N-P transistor which improves the turn-off gain of the GTO.

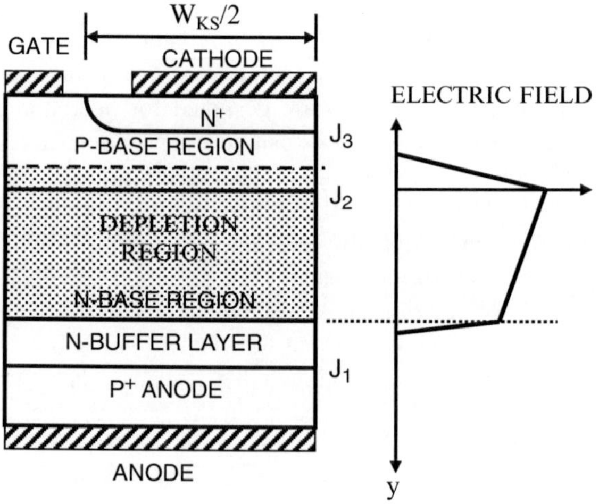

Fig. 4.1 Asymmetric gate turn-off thyristor structure

The output characteristics for the asymmetric GTO structure are illustrated in Fig. 4.2. The asymmetrical structure can support a large voltage (BV_F) in the forward blocking mode but only a relatively small voltage ($BV_{R,AS}$) in the reverse blocking mode. The GTO can be triggered into the on-state while operating in the forward blocking mode by the application of a small gate current. Once the device enters its regenerative mode of operation, it can sustain the on-state current flow without any gate drive signal. The device can be turned off without reversing the anode voltage by the application of a large reverse gate current. The *i–v* trajectory during the turn-off transient is shown by the dashed lines in the case of an inductive load. During this transient, the current crowds toward the center of the cathode fingers because portions close to the gate contact are turned off first. The increase in the local current density can lead to destructive failure.

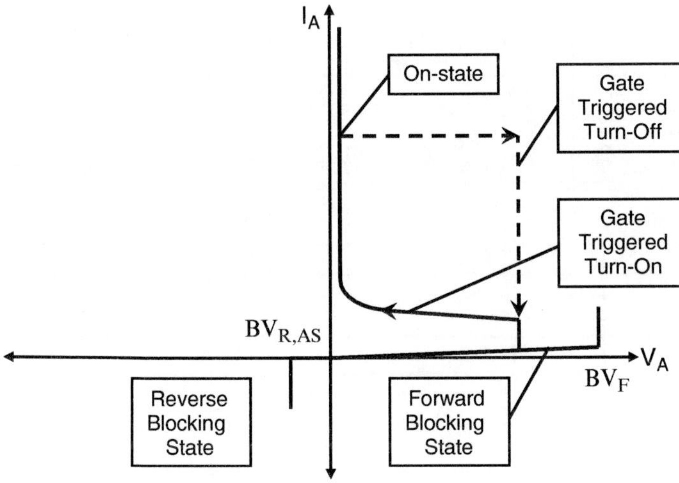

Fig. 4.2 Output characteristics of the GTO structure

4.2 5,000-V Silicon GTO

The design and characteristics of the 5,000-V asymmetric silicon gate turn-off thyristor structure are discussed in this section. The design parameters for the N-base region required to achieve this blocking voltage are first analyzed. Using the optimum N-base width, the blocking characteristics for the device are then obtained as a function of the lifetime in the drift region. The on-state characteristics for the device are obtained for various lifetime values as well. The gate controlled turn-off behavior of the silicon GTO structure is analyzed including the effect of the lifetime in the drift region, the buffer layer concentration, and the transparent emitter design.

4.2.1 Blocking Characteristics

The physics for blocking voltages in the first and third quadrants by the GTO structure is discussed in detail in the textbook [2]. When a positive bias is applied to the anode terminal of the asymmetric GTO structure, the P-base/N-base junction (J_2) becomes reverse biased while the junction (J_1) between the P^+ anode region and the N-base region becomes forward biased. The forward blocking voltage is supported across the P-base/N-base junction (J_2) with a depletion layer extending mostly within the N-base region as illustrated in Fig. 4.3. Here, a one-dimensional view of only the portion under the N^+ cathode region is illustrated. However, the current flowing into the P-base region is shunted to the gate contact because the gate terminal is shorted to the cathode terminal during the forward blocking mode of operation. Consequently, the N-P-N transistor is assumed to be inactive.

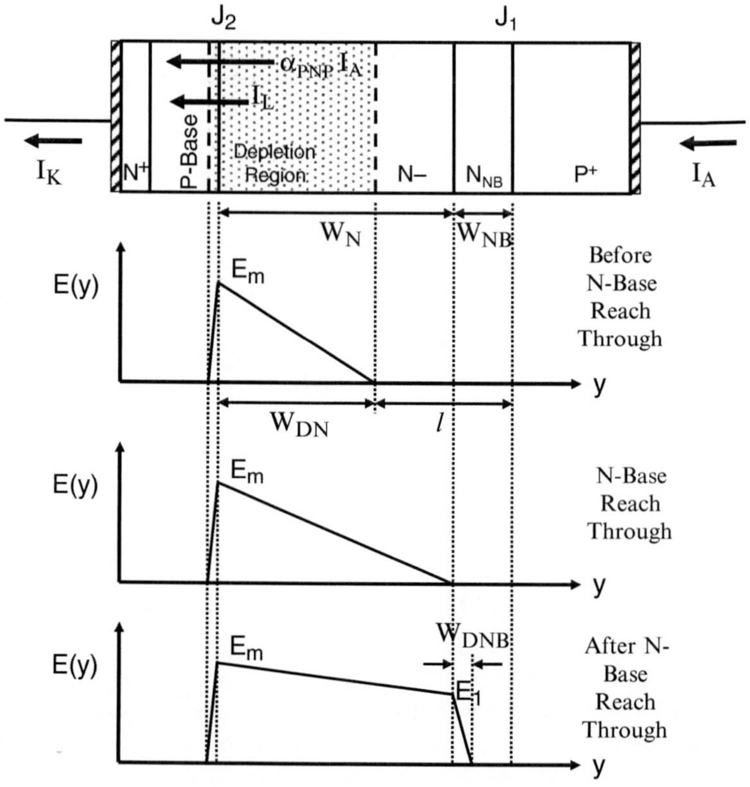

Fig. 4.3 Electric field distribution during the forward blocking mode in the GTO structure

In the forward blocking mode, the maximum electric field occurs at the P-base/N-base junction (J_2). On the one hand, if the width of the N-base region is very large when compared with the depletion layer width, the blocking voltage capability is limited by avalanche breakdown when the maximum electric field (E_m) becomes equal to the critical electric field. This corresponds to the multiplication coefficient (M) becoming equal to infinity. The avalanche breakdown voltage is given by:

$$\mathrm{BV_{PP}} = 4.45 \times 10^{13} N_D^{-3/4} \tag{4.1}$$

where N_D is the doping concentration in the lightly doped portion of the N-base region. However, a very large width for the N-base region is unacceptable because it increases the on-state voltage drop of the thyristor.

On the other hand, if the width of the N-base region is made small when compared with the depletion layer width, the depletion region can extend through the entire lightly doped portion of the N-base region as illustrated in the middle of Fig. 4.3, while the maximum electric field at the junction is well below the critical electric field for breakdown. The depletion region reaches through the lightly doped portion of the N-base region at a collector bias given by:

$$V_{RT} = \frac{qN_D}{2\varepsilon_S} W_N^2 \tag{4.2}$$

where W_N is the width of the lightly doped portion of the N-base region.

Due to the presence of the N-buffer layer in the asymmetric GTO structure, the device can continue to support voltages after the depletion region reaches through the lightly doped portion of the N-base region. The electric field distribution for this case is illustrated at the bottom of Fig. 4.3. (The vertical axes for the three electric field plots in the figure have different scales to allow displaying the field distributions. The slope of the electric field with distance within the lightly doped portion of the N-base region is the same for all three cases as determined by the doping concentration in the region.) The electric field has a trapezoidal shape typical of punch-through structures. As discussed in the textbook [2] the avalanche breakdown voltage for the punch-through structure is given by:

$$BV_{PT} = E_C W_N - \frac{qN_D}{2\varepsilon_S} W_N^2 \tag{4.3}$$

where E_C is the critical electric field for breakdown corresponding to the doping concentration N_D in the lightly doped portion of the N-base region.

The actual breakdown voltage for the asymmetric GTO structure in the forward blocking mode falls below the punch-through breakdown voltage because it is governed by the open-base transistor breakdown phenomenon. In order to analyze this phenomenon for the asymmetric GTO structure, consider the currents flowing at the boundary of the depletion region, as illustrated in Fig. 4.3. The current consists of the leakage current due to the generation process within the depletion region and the collector current amplified by the current gain of the P-N-P transistor. Based upon the application of Kirchhoff's Current Law to the thyristor structure in the absence of a gate current:

$$I_A = \alpha_{PNP} I_A + I_L = I_K \tag{4.4}$$

leading to the relationship:

$$I_A = \frac{I_L}{(1 - \alpha_{PNP})} \tag{4.5}$$

From this expression, it can be concluded that the anode current will increase very rapidly when the common base current gain of the P-N-P bipolar transistor within the asymmetric GTO structure approaches unity. When the positive anode bias is increased, the width of the un-depleted portion of the N-base region becomes smaller producing an increase in the base transport factor (α_T). Concurrently, the maximum electric field at junction J_2 becomes larger leading to an increase in the multiplication coefficient. Both phenomena produce an increase in the common base current gain of the P-N-P transistor with increasing collector bias.

For the asymmetric GTO structure, the emitter injection efficiency becomes smaller than unity due to the high doping concentration of the N-buffer layer. The emitter injection efficiency for the P^+ anode/N-buffer junction (J_1) can be obtained by using an analysis similar to that described in the textbook for the bipolar power transistor [2]:

$$\gamma_E = \frac{D_{pNB} L_{nA} N_{AA}}{D_{pNB} L_{nA} N_{AA} + D_{nA} W_{NB} N_{DNB}} \tag{4.6}$$

where D_{pNB} and D_{nA} are the diffusion coefficients for minority carriers in the N-buffer and P^+ anode regions, N_{AA} and L_{nA} are the doping concentration and diffusion length for minority carriers in the P^+ anode region, and N_{DNB} and W_{NB} are the doping concentration and width of the N-buffer layer. In determining the diffusion coefficients and the diffusion length, it is necessary to account for the high doping concentrations in the P^+ anode region and N-buffer layer. In addition, the lifetime within the highly doped P^+ anode region is reduced due to heavy doping effects, which shortens the diffusion length.

Based upon the above analysis, the open-base transistor breakdown condition for the asymmetric GTO structure is given by:

$$\alpha_{PNP} = (\gamma_E \cdot \alpha_T)_{PNP} M = 1 \tag{4.7}$$

Before the advent of reach-through of the lightly doped portion of the N-base region, the base transport factor (α_T) is determined by the width (l) of the un-depleted portion of the N-drift region (see Fig. 4.3):

$$\alpha_T = \frac{1}{\cosh(l/L_p)} \tag{4.8}$$

with

$$l = W_N - \sqrt{\frac{2\varepsilon_S V_A}{q N_D}} \tag{4.9}$$

if the width of the buffer layer is assumed to be small. As the forward bias increases, the width of the un-depleted portion of the N-base region shrinks resulting in an increase in the base transport factor. However, when the collector bias exceeds the reach-through voltage (V_{RT}), the electric field is truncated by the high doping concentration of the N-buffer layer making the un-depleted width equal to the width of the N-buffer layer. The base transport factor is then given by

$$\alpha_T = \frac{1}{\cosh(W_{NB}/L_{p,NB})} \tag{4.10}$$

which is independent of the collector bias. Here, $L_{p,NB}$ is the diffusion length for holes in the N-buffer layer. This analysis neglects the depletion region extension (shown as W_{DNB} in Fig. 4.3) within the N-buffer layer.

The diffusion length for holes ($L_{p,NB}$) in the N-buffer layer depends upon the diffusion coefficient and the minority carrier lifetime in the N-buffer layer. The diffusion coefficient varies with the doping concentration in the N-buffer layer based upon the concentration dependence of the mobility. In addition, the minority carrier lifetime has been found to be dependent upon the doping concentration [4]. It has been empirically observed that the low-level lifetime decreases with increasing doping concentration [4, 5]. This can be modeled by using the relationship:

$$\frac{\tau_{LL}}{\tau_{p0}} = \frac{1}{1 + (N_D/N_{REF})} \tag{4.11}$$

where N_{REF} is a reference doping concentration [6] whose value will be assumed to be 5×10^{16} cm^{-3}. The reduction of the low-level lifetime indicated by this model must be taken into account for computing the diffusion length of minority carriers when the doping concentration in the N-buffer layer is increased.

The multiplication factor for a P-N junction is given by:

$$M = \frac{1}{1 - (V_A/BV_{PP})^n} \tag{4.12}$$

with a value of $n = 6$ for the case of a P^+/N junction and the avalanche breakdown voltage of the P-base/N-base junction (BV_{PP}) *without the punch-through phenomenon*. In order to apply this formulation to the punch-through case relevant to the asymmetric GTO structure, it is necessary to relate the maximum electric field at the junction for the two cases. The electric field at the interface between the lightly doped portion of the N-base region and the N-buffer layer is given by:

$$E_1 = E_m - \frac{qN_D W_N}{\varepsilon_S} \tag{4.13}$$

The voltage supported by the device is given by:

$$V_A = \left(\frac{E_m + E_1}{2}\right) W_N = E_m W_N - \frac{qN_D}{2\varepsilon_S} W_N^2 \tag{4.14}$$

From this expression, the maximum electric field is given by:

$$E_m = \frac{V_A}{W_N} + \frac{qN_D W_N}{2\varepsilon_S} \tag{4.15}$$

The corresponding equation for the non-punch-through case is:

$$E_{\mathrm{m}} = \sqrt{\frac{2qN_{\mathrm{D}}V_{\mathrm{NPT}}}{\varepsilon_{\mathrm{S}}}} \tag{4.16}$$

Consequently, the non-punch-through voltage that determines the multiplication coefficient "M" corresponding to the applied collector bias "V_{A}" for the punch-through case is given by:

$$V_{\mathrm{NPT}} = \frac{\varepsilon_{\mathrm{S}}E_{\mathrm{m}}^2}{2qN_{\mathrm{D}}} = \frac{\varepsilon_{\mathrm{S}}}{2qN_{\mathrm{D}}}\left(\frac{V_{\mathrm{A}}}{W_{\mathrm{N}}} + \frac{qN_{\mathrm{D}}W_{\mathrm{N}}}{2\varepsilon_{\mathrm{S}}}\right)^2 \tag{4.17}$$

The multiplication coefficient for the asymmetric GTO structure can be computed by using this non-punch-through voltage:

$$M = \frac{1}{1 - (V_{\mathrm{NPT}}/\mathrm{BV}_{\mathrm{PP}})^n} \tag{4.18}$$

The multiplication coefficient increases with increasing collector bias. The open-base transistor breakdown voltage (and the forward blocking capability of the asymmetric IGBT structure) is determined by the collector voltage at which the multiplication factor becomes equal to the reciprocal of the product of the base transport factor and the emitter injection efficiency.

The silicon GTO structure must have a forward blocking voltage of 5,500 V for a 5,000-V rated device. In the case of avalanche breakdown, there is a unique value of $1.62 \times 10^{13}\ \mathrm{cm}^{-3}$ for the drift region to obtain this blocking voltage. However, in the case of the asymmetric IGBT it is advantageous to use a much lower doping concentration for the lightly doped portion of the N-base region in order to reduce its width. The strong conductivity modulation of the N-base region during on-state operation favors a smaller thickness for the N-base region independent of its original doping concentration. A doping concentration of $5 \times 10^{12}\ \mathrm{cm}^{-3}$ will be assumed for the N-base region.

The doping concentration of the N-buffer layer must be sufficiently large to prevent reach-through of the electric field to the P^+ collector region. Doping concentrations above $1 \times 10^{16}\ \mathrm{cm}^{-3}$ are sufficient to accomplish this goal. For the baseline device structure, an N-buffer layer doping concentration of $1.2 \times 10^{17}\ \mathrm{cm}^{-3}$ will be assumed. In this case, the emitter injection efficiency computed using Eq. 4.6 is 0.741 for a doping concentration of $1 \times 10^{19}\ \mathrm{cm}^{-3}$ in the P^+ anode region. When the device is close to breakdown, the entire N-base region is depleted, and the base transport factor computed by using Eq. 4.10 in this case is 0.628. In computing these values, a lifetime of 5 μs was assumed for the N-base region resulting in a lifetime of 2.5 μs in the N-buffer layer due to the scaling according to Eq. 4.11. Based upon Eq. 4.7, open-base transistor breakdown will then occur when the multiplication coefficient becomes equal to 2.15 for the above values for the injection efficiency and base transport factor.

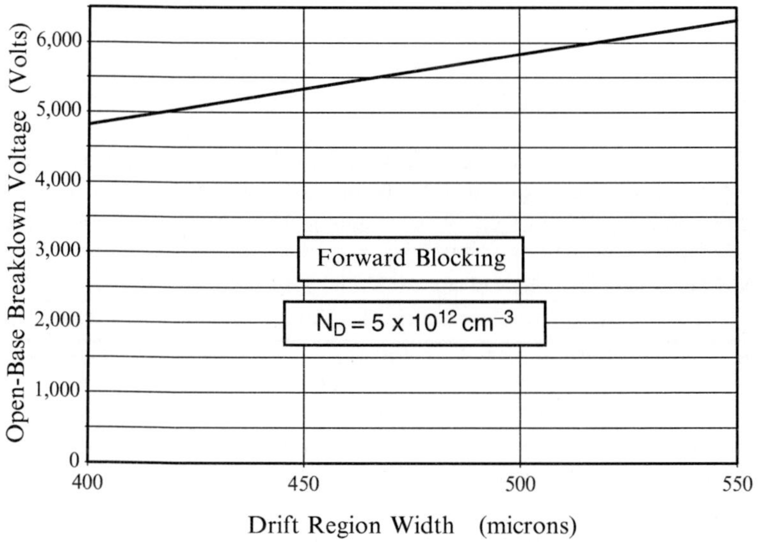

Fig. 4.4 Open-base breakdown voltage for the asymmetric GTO structure in the forward blocking mode

The forward blocking capability for the silicon asymmetric GTO structure can be computed by using Eq. 4.7 for various widths for the N-base region. The analysis requires determination of the voltage V_{NPT} by using Eq. 4.17 for each width of the N-base region. The resulting values for the forward blocking voltage are plotted in Fig. 4.4. From this graph, the N-base region width required to obtain a forward blocking voltage of 5,500-V is 470 μm. This width can be slightly reduced when taking into account the voltage supported within the P-base region due to its graded doping profile.

4.2.2 Leakage Current

The leakage current in forward blocking mode is produced by space-charge-generation within the depletion region. In the case of the asymmetric GTO structure in the forward blocking mode, the space-charge-generation current at the reverse biased deep P^+/N-base junction J_2 is amplified by the gain of the internal P-N-P transistor:

$$J_L = \frac{J_{SCG}}{(1 - \alpha_{PNP})} \tag{4.19}$$

The space-charge-generation current density is given by:

$$J_{SCG} = \frac{qW_D n_i}{\tau_{SC}} = \frac{n_i}{\tau_{SC}} \sqrt{\frac{2q\varepsilon_S V_A}{N_D}} \tag{4.20}$$

at low collector bias voltages *before the depletion region in the lightly doped portion of the N-base regions reaches-through to the interface between the lightly doped portion of the N-base region and the N-buffer layer*. The space-generation current increases with increasing collector bias in this regime of operation for the asymmetric GTO structure. Concurrently, the current gain (α_{PNP}) of the P-N-P transistor is also a function of the collector bias voltage because the base transport factor increases when the collector bias increases. Prior to the complete depletion of the lightly doped portion of the N-base region, the multiplication factor remains close to unity. It is therefore sufficient to account for the increase in the base transport factor with collector bias as given by Eqs. 4.8 and 4.9.

For the case of the silicon asymmetric GTO structure with a width of 450 μm for the lightly doped portion of the N-base region with a doping concentration of $5 \times 10^{12}\ \text{cm}^{-3}$, the entire lightly doped portion of the N-base region is completely depleted at a reach-through voltage of 780 V according to Eq. 4.2. Once the lightly doped portion of the N-base region becomes completely depleted, the electric field becomes truncated at the interface between the lightly doped portion of the N-base region and the N-buffer layer as illustrated at the bottom of Fig. 4.3. The space-charge generation width then becomes independent of the anode bias because the depletion width in the N-buffer layer is small. Under these bias conditions, the base transport factor also becomes independent of the anode bias as given by Eq. 4.10. Consequently, the leakage current becomes independent of the anode bias until the onset of avalanche multiplication.

The transport of minority carriers through the N-base region of the P-N-P transistor occurs through the N-buffer layer and the lightly doped portion of the N-base region that has not been depleted by the applied collector bias. The base transport factor is therefore given by:

$$\alpha_{\mathrm{T}} = \alpha_{\mathrm{T,N-buffer}}\alpha_{\mathrm{T,N-base}} \tag{4.21}$$

The base-transport factor associated with the N-buffer layer can be obtained from the decay of the hole current within the N-buffer layer as given by low-level injection theory [2]:

$$\alpha_{\mathrm{T,N-buffer}} = \frac{J_{\mathrm{p}}(W_{\mathrm{NB^-}})}{J_{\mathrm{p}}(x_{\mathrm{N}})} = \frac{\gamma_{\mathrm{E}} J_{\mathrm{C}} e^{-W_{\mathrm{NB}}/L_{\mathrm{p,NB}}}}{\gamma_{\mathrm{E}} J_{\mathrm{C}}} = e^{-W_{\mathrm{NB}}/L_{\mathrm{p,NB}}} \tag{4.22}$$

where $J_{\mathrm{p}}(x_N)$ is the hole current density at the P^+ collector/N-buffer layer junction (J_1), $J_{\mathrm{p}}(W_{\mathrm{NB-}})$ is the hole current density at the interface between the lightly doped portion of the N-base region and the N-buffer layer; W_{NB} is the width of the N-buffer layer, and $L_{\mathrm{p,NB}}$ is the minority carrier diffusion length in the N-buffer layer. The base transport factor for the lightly doped portion of the N-base region under low-level injection conditions appropriate for computation of the leakage current is:

$$\alpha_{\mathrm{T,N-base}} = \frac{1}{\cosh\left[(W_{\mathrm{N}} - W_{\mathrm{D}})/L_{\mathrm{p,N}}\right]} \tag{4.23}$$

where W_N is the width of the lightly doped portion of the N-base region and $L_{p,N}$ is the minority carrier diffusion length in the lightly doped portion of the N-base region. In this expression, the depletion width W_D prior to punch-through is given by:

$$W_D = \sqrt{\frac{2\varepsilon_S V_A}{qN_D}} \tag{4.24}$$

As the anode bias voltage increases, the base transport factor for the P-N-P transistor increases until it becomes equal to that for the N-buffer layer.

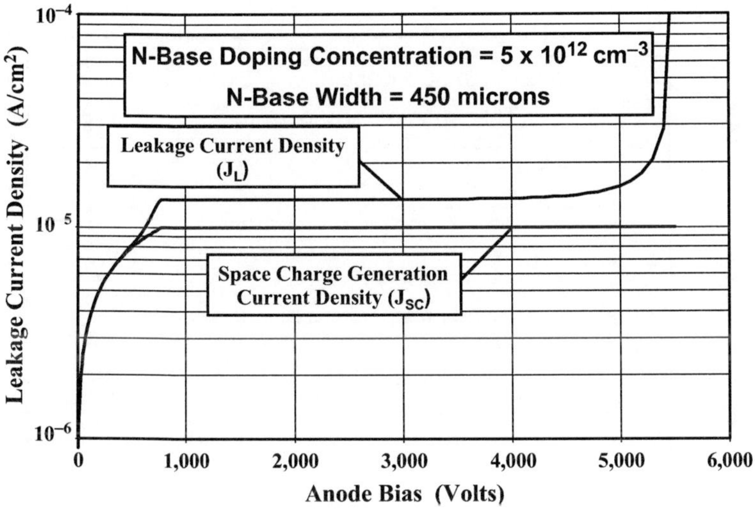

Fig. 4.5 Leakage current for the 5-kV asymmetric GTO structure

Consider the case of an asymmetric GTO structure that is designed with a forward blocking capability of 5,000 V. This would be satisfied by using an N-base region with a lightly doped portion having a doping concentration of 5×10^{12} cm^{-3} and width of 450 μm. The N-buffer layer will be assumed to have a doping concentration of 1.2×10^{17} cm^{-3} and thickness of 30 μm (corresponding to the baseline device structure used for the numerical simulations). The leakage current computed by using the above analysis is shown in Fig. 4.5 for the case of a lifetime (τ_{p0}, τ_{n0}) of 5 μs in the lightly doped portion of the N-drift region. This corresponds to a space-charge-generation lifetime (τ_{SC}) of 10 μs if the recombination center is located at mid-gap. In performing the analysis, the reduction of the lifetime with increasing doping concentration in the N-buffer layer was taken into account by using Eq. 4.11. The space-charge-generation current is also included in the figure for comparison purposes. It can be seen that the space-charge-generation leakage current increases with anode bias voltage due to the expansion of the width of the depletion region until 780 V. This corresponds to a reach-through voltage of 780 V obtained by using Eq. 4.2. The space-charge-generation leakage current then becomes independent of the anode bias voltage. The leakage current for the asymmetric GTO structure is larger than the space-charge-generation current due to the current gain of the P-N-P transistor. For the case of an N-buffer layer doping of 1.2×10^{1} cm^{-3}, the current gain of the P-N-P transistor is 0.261 after the lightly doped portion of the N-base region becomes completely depleted. The leakage current density for this case is 1.35×10^{-5} A/cm^2. When the anode voltage increases above 5,000 V, the multiplication factor starts to increase rapidly producing a breakdown voltage of about 5,500 V.

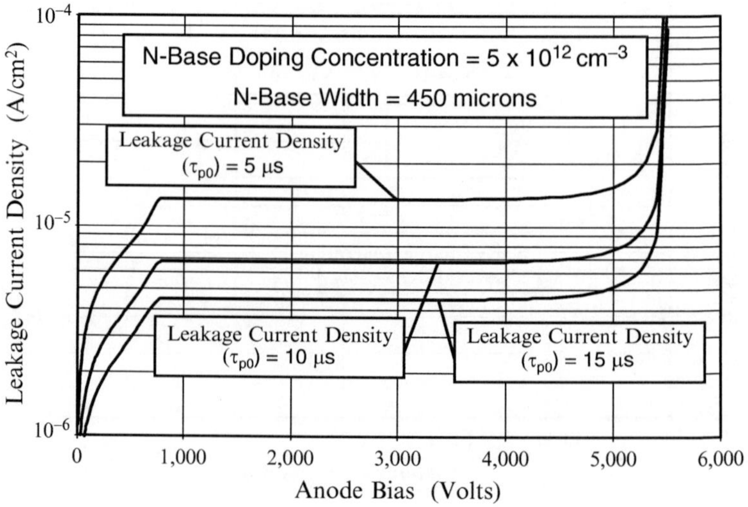

Fig. 4.6 Leakage current for the 5-kV asymmetric GTO structure: lifetime dependence

The leakage current computed by using the above analysis for the silicon 5-kV GTO structure is shown in Fig. 4.6 for the case of three lifetime (τ_{p0}, τ_{n0}) values in the lightly doped portion of the N-drift region. This corresponds to space-charge-generation lifetimes (τ_{SC}) of 10, 20, and 30 μs if the recombination center is located at mid-gap. In performing the analysis, the reduction of the lifetime with increasing doping concentration in the N-buffer layer was taken into account by using Eq. 4.11. In all three cases, the leakage current increases with increasing anode bias voltage until the depletion region reaches through the lightly doped portion of the N-drift region at 780 V, and then becomes independent of the anode bias. The leakage current density is observed to increase when the lifetime is reduced. The breakdown voltage is essentially independent of the lifetime in the N-drift region.

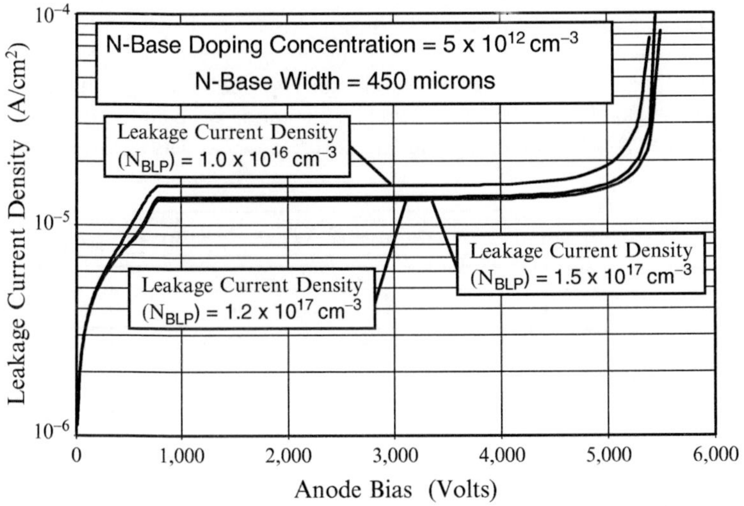

Fig. 4.7 Leakage current for the 5-kV asymmetric GTO structure: buffer layer doping dependence

The leakage current computed by using the above analysis for the silicon 5-kV GTO structure is shown in Fig. 4.7 for the case of three buffer layer doping concentration (N_{BLP}) values. The lifetime (τ_{p0}, τ_{n0}) was assumed to be 5 μs for all three cases. In performing the analysis, the reduction of the lifetime with increasing doping concentration in the N-buffer layer was taken into account by using Eq. 4.11. The leakage current decreases with increasing buffer layer doping concentration due to a reduction of the injection efficiency, the base transport factor, and consequently, the gain of the P-N-P transistor. The breakdown voltage increases slightly with increase in the buffer layer doping concentration.

Simulation Example

In order to gain insight into the physics of operation for the 5-kV asymmetric GTO structure under voltage-blocking conditions, the results of two-dimensional numerical simulations for several structures are described here. The simulations were performed using a cell with the structure shown in Fig. 4.1. This cell had a width of 100 μm (area $= 1.0 \times 10^{-6}$ cm^{-2}) with a cathode width (W_{KS}) of 180 μm. The asymmetric GTO structure used for the simulations was formed by diffusions performed into a uniformly doped N-type drift region with a doping concentration of 5×10^{12} cm^{-3}. A lifetime (τ_{p0}, τ_{n0}) of 5 μs was used for the baseline device. The N-buffer layer was formed by diffusion from the anode side with a depth of 55 μm. For the baseline device structure, the surface concentration of the N-type diffusion was adjusted to achieve a peak doping concentration of 1.2×10^{17} cm^{-3} in the buffer layer. The P-base region was formed with a Gaussian doping profile with a surface concentration of 5×10^{17} cm^{-3} and a depth of 30 μm. The N$^+$ cathode region was formed with a Gaussian doping profile with a surface concentration of 1×10^{20} cm^{-3} and a depth of 10 μm. The doping profile in the vertical direction through the N^+ cathode region is shown in Fig. 4.8 indicating the net width of the lightly doped portion of the N-base region is 410 μm after accounting for the diffusions. The peak doping concentration of the P-base region is 8×10^{16} cm^{-3} and its thickness is 23 μm. The peak doping concentration of the N-buffer layer is 1.2×10^{17} cm^{-3} and its thickness is 30 μm.

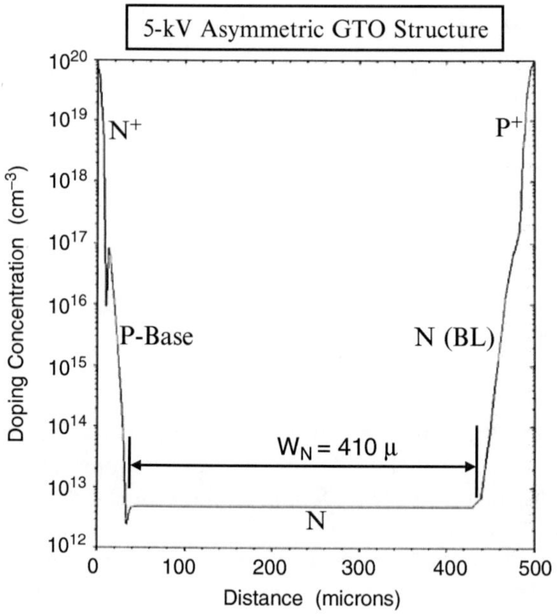

Fig. 4.8 Doping profile for the simulated baseline asymmetric 5-kV GTO structure: lifetime in N-base region

The forward blocking capability of the silicon GTO structures was obtained by increasing the anode bias while maintaining the gate electrode at zero volts. The characteristics obtained for three lifetime (τ_{p0}) values are provided in Fig. 4.9. In all cases, the leakage current increases rapidly with increasing anode bias voltage until about 780 V as predicted by the analytical model (see Fig. 4.6). This occurs due to the increase in the space-charge-generation volume and the increase in the current gain (α_{PNP}) of the open base P-N-P transistor until the anode bias becomes equal to the reach-through voltage obtained using the analytical solution given by Eq. 4.2. The leakage current then becomes independent of the collector voltage until close to the breakdown voltage. This behavior is well described by the analytical model (see Fig. 4.6). The leakage current density obtained using the analytical model is within a factor of 2 of the values derived from the numerical simulations for all cases. Thus, the simple analytical theory provides a very good qualitative and quantitative description of the leakage current behavior as a function of both the anode bias voltage and the lifetime in the N-drift region.

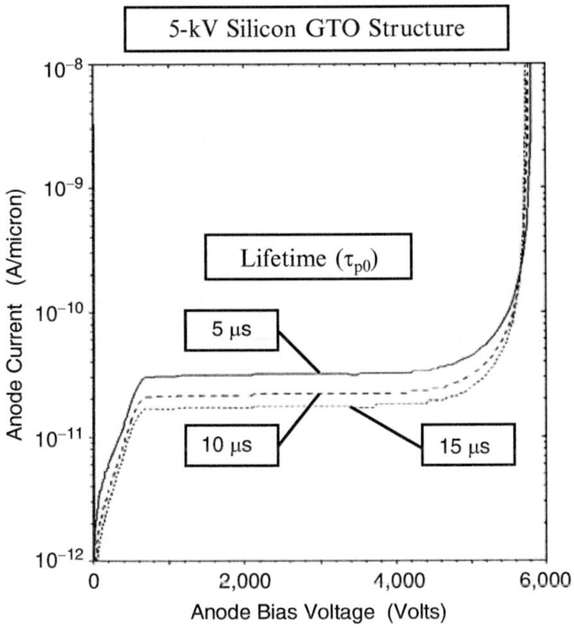

Fig. 4.9 Forward blocking characteristics for the asymmetric GTO structure

The leakage currents obtained for the asymmetric silicon GTO structure using the numerical simulations are provided in Fig. 4.10 for three doping concentrations in the N-buffer layer. A lifetime of 5 μs (τ_{p0}) was used in the lightly doped portion of the N-base region for all these cases. It can be observed that the leakage current decreases when the N-buffer layer doping concentration is increased due to the reduction of the emitter injection efficiency and base transport factor of the P-N-P transistor. The reduced current gain of the P-N-P transistor also results in a small increase in the open-base breakdown voltage. The behavior obtained by using the simple analytical model for the leakage current (see Fig. 4.7) is in good qualitative and quantitative agreement with these simulation results.

The voltage is primarily supported within the lightly doped portion of N-base region in the asymmetric GTO structure during operation in the forward blocking mode. This is illustrated in Fig. 4.11 where the electric field profiles are shown during operation in the forward blocking mode at several anode voltages. It can be observed that the P-Base/N-base junction (J_2) becomes reverse biased during the forward blocking mode with the depletion region extending toward the right-hand side with increasing (positive) collector bias. The electric field has a triangular shape until the entire lightly doped portion of the N-base region becomes completely depleted. This occurs at a collector bias of about 800 V in agreement with the value obtained using the analytical solution (see Eq. 4.2). The electric field profile then takes a trapezoidal shape due to the high doping concentration in the N-buffer layer.

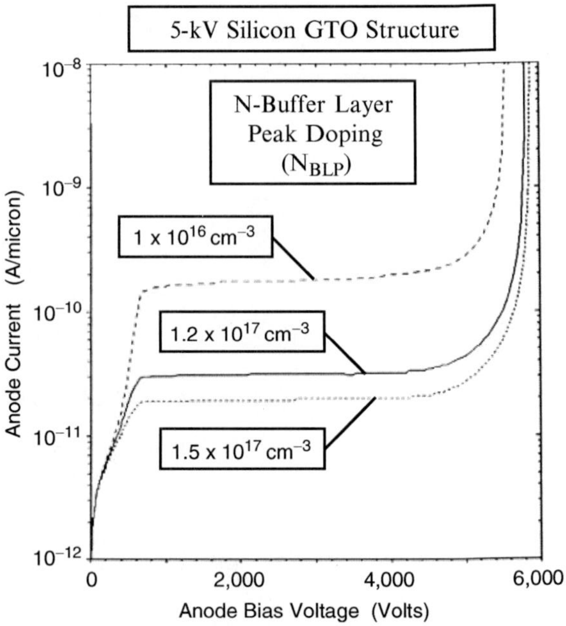

Fig. 4.10 Forward blocking characteristics for the asymmetric GTO structure

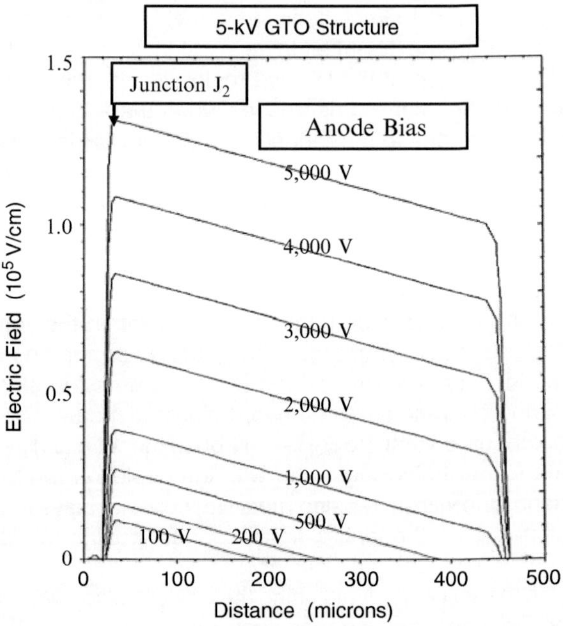

Fig. 4.11 Electric field profiles in the 5-kV asymmetric GTO structure

4.2.3 On-State Voltage Drop

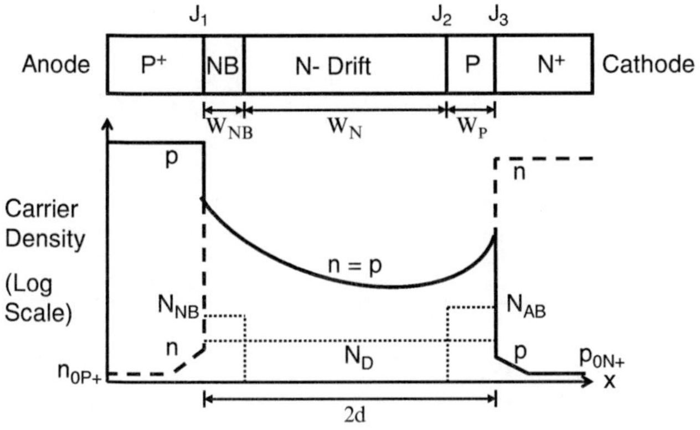

Fig. 4.12 Carrier distribution within the asymmetric GTO structure in the on-state

The GTO structure has excellent forward conduction characteristic even when designed to support large voltage levels because it operates like a thyristor. The device can be triggered from the forward-blocking mode at the anode supply voltage (V_{AS}) into the forward-conduction mode by the application of a small gate current to initiate the turn-on process [2]. The GTO structure then operates in its on-state with a forward conduction characteristic similar to that observed for a P-i-N rectifier. In the on-state, strong conductivity modulation of the N-drift region occurs due to high level injection of holes allowing the thyristor to carry high current levels with a low on-state voltage drop.

The GTO structure can be treated as a P-i-N rectifier for analysis of its forward conduction characteristics. In this case, it is assumed that the junction J_2 is strongly forward-biased resulting in high level injection in not only the N-base region but also the P-base region and the N-buffer layer as illustrated in Fig. 4.12. The GTO structure can then be regarded as a P-i-N rectifier between the P^+ anode and N^+ cathode regions. The electron and hole concentrations within the N-base and P-base regions take a catenary distribution in accordance with the analysis for the P-i-N rectifier in the textbook [2]:

$$n(x) = p(x) = \frac{\tau_{HL} J_A}{2qL_a}\left[\frac{\cosh(x/L_a)}{\sinh(d/L_a)} - \frac{\sinh(x/L_a)}{2\cosh(d/L_a)}\right] \tag{4.25}$$

The distance "*d*" for the asymmetric GTO structure is given by:

$$d = \frac{W_{NB} + W_N + W_P}{2} \tag{4.26}$$

as indicated in the figure. A minimum on-state voltage drop occurs for the thyristor structure when the ambipolar diffusion length (L_a) is equal to the distance "*d*" (see textbook analysis for the P-i-N rectifier in Chap. 5).

The on-state *i–v* characteristic for the GTO structure in the high-level injection regime of operation is given by [2]:

$$V_{ON} = \frac{2kT}{q}\ln\left[\frac{J_T d}{2qD_a n_i F(d/L_a)}\right] \quad (4.27)$$

where

$$F\left(\frac{d}{L_a}\right) = \frac{(d/L_a)\tanh(d/L_a)}{\sqrt{1 - 0.25\tanh^4(d/L_a)}}e^{-\frac{qV_M}{2kT}} \quad (4.28)$$

The i-region (or middle-region) voltage drop can be computed using the following approximations. For d/L_a ratios of up to 2, an asymptote A given by:

$$V_M = \frac{2kT}{q}\left(\frac{d}{L_a}\right)^2 \quad (4.29)$$

provides a good fit. For d/L_a ratios of greater than 2, an asymptote B given by:

$$V_M = \frac{3\pi kT}{8q}e^{(d/L_a)} \quad (4.30)$$

provides a good fit.

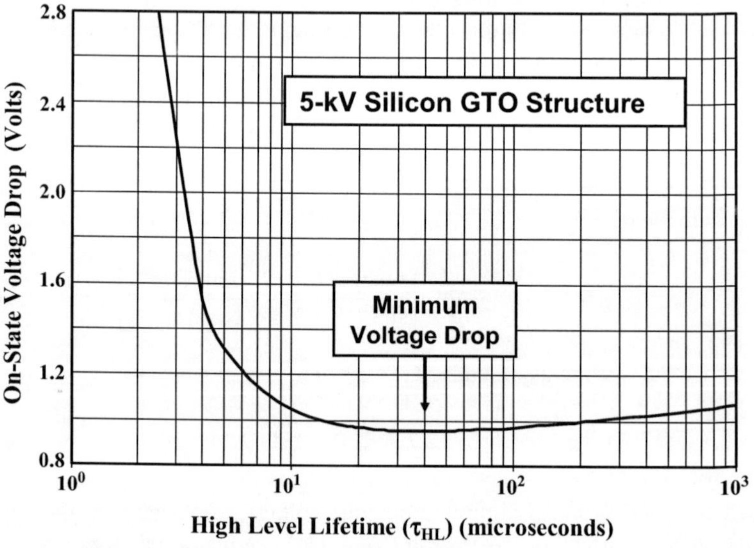

Fig. 4.13 On-state voltage drop for the 5-kV GTO structure

The on-state voltage drop (at an on-state current density of 100 A/cm^2) computed for the 5-kV silicon GTO structure by using Eq. 4.27 is provided in Fig. 4.13 for various values for the high-level lifetime in the drift region. This thyristor structure had the optimized N-base region width of 420 μm, N-buffer layer width of 30 μm, and a P-base width of 20 μm. As expected, the on-state voltage drop has a minimum value when the [d/L_a] value is equal to unity. The minimum on-state voltage drop is found to be 0.94 V at a high-level lifetime of 40 μs. This lifetime value is four times smaller than the optimum high-level lifetime for the symmetric blocking 5-kV thyristor structure (see Chap. 2) allowing the asymmetric GTO to switch at higher frequencies. The on-state voltage drop of the asymmetric GTO structure begins to rise rapidly only when the high-level lifetime is reduced below 4 μs.

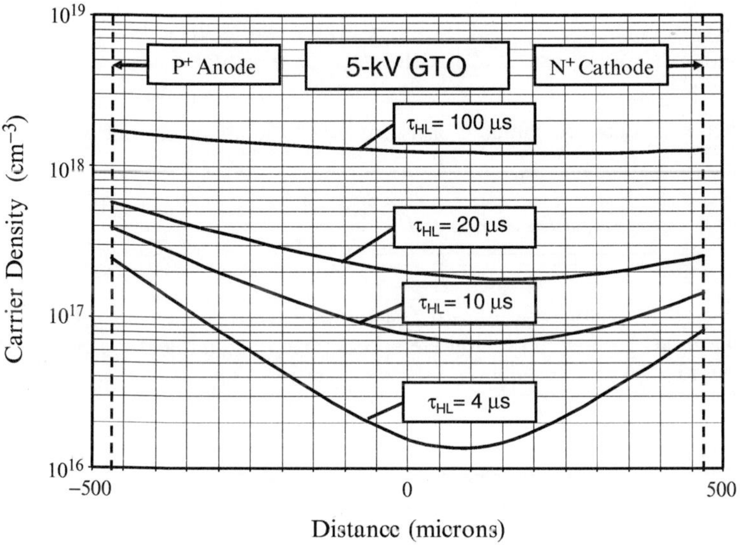

Fig. 4.14 Carrier distribution in the 5-kV GTO structure when middle region recombination is dominant

The carrier distribution in the 5-kV thyristor structure, as predicted by the analytical model based upon Eq. 4.25, is provided in Fig. 4.14 for various high-level lifetime values. As the lifetime is reduced, the injected carrier density in the drift region becomes smaller. The carrier density for each lifetime value is smaller for the 5-kV symmetric thyristor structure because the asymmetric GTO structure has a smaller thickness for the drift region. The carrier density predicted by the analytical model is larger than in actual devices because the model is based upon assuming that recombination occurs only in the drift region. In actual devices, recombination also occurs in the end region of the device as discussed in Sect. 5.1.4 of the textbook [2]. End-region recombination reduces the injected carrier density in the drift region producing an increase in the on-state voltage drop as well.

When the lifetime in the drift region is large, end-region recombination begins to take a dominant role. The theory for the carrier distribution in the drift region when end-region recombination is dominant was provided in Chap. 2. When end-region recombination is dominant, the electron and hole concentrations within the N-base and P-base regions take a catenary distribution with a smaller concentration at the boundaries:

$$n(x) = p(x) = K_{\mathrm{E}}\left[\frac{\cosh(x/L_{\mathrm{a}})}{\sinh(d/L_{\mathrm{a}})} - \frac{\sinh(x/L_{\mathrm{a}})}{2\cosh(d/L_{\mathrm{a}})}\right] \tag{4.31}$$

The constant K_{E} can be obtained by using Eq. 4.31 with $x = -d$ in the above equation:

$$n(-d) = K_{\mathrm{E}}\left[\frac{\cosh(-d/L_{\mathrm{a}})}{\sinh(d/L_{\mathrm{a}})} - \frac{\sinh(-d/L_{\mathrm{a}})}{2\cosh(d/L_{\mathrm{a}})}\right] \tag{4.32}$$

When end-region recombination is dominant, the anode current is given by:

$$J_{\mathrm{A,ON}} = J_{\mathrm{P+}} + J_{\mathrm{N+}} = J_{\mathrm{SP+}}\left[\frac{n(-d)}{n_{\mathrm{ieP+}}}\right]^2 + J_{\mathrm{SN+}}\left[\frac{n(+d)}{n_{\mathrm{ieN+}}}\right]^2 \tag{4.33}$$

Under the assumption that the saturation current densities and intrinsic concentrations for the end region are approximately equal, and the concentrations $n(+d)$ and $n(-d)$ are also approximately equal, it can be shown that:

$$K_{\mathrm{E}} = n_{\mathrm{ieP+}}\sqrt{\frac{J_{\mathrm{A,ON}}}{2J_{\mathrm{SP+}}}\left[\frac{2\sinh(2d/L_{\mathrm{a}})}{1 + 3\cosh(2d/L_{\mathrm{a}})}\right]} \tag{4.34}$$

The average carrier concentration in the drift region can be obtained by using the value at the ends, i.e., $n(-d) = n(+d)$ and the value $n(0)$ at the midpoint:

$$n_{\mathrm{a}} = p_{\mathrm{a}} = \frac{n(-d) + n(0)}{2} \tag{4.35}$$

The carrier concentration at the midpoint can be derived using Eqs. 4.31 and 4.34:

$$n(0) = \frac{K_{\mathrm{E}}}{\sinh(d/L_{\mathrm{a}})} \tag{4.36}$$

Substituting Eqs. 4.32 and 4.36 into Eq. 4.35 yields:

$$n_a = \frac{n_{ieP+}}{2}\sqrt{\frac{J_{A,ON}}{2J_{SP+}}}\left\{1 + \frac{2\sinh(2d/L_a)}{\sinh(d/L_a)[1 + 3\cosh(2d/L_a)]}\right\} \tag{4.37}$$

This equation can be written as:

$$n_a = K_{av}\sqrt{J_{A,ON}} \tag{4.38}$$

where

$$K_{av} = \frac{n_{ieP+}}{2\sqrt{2J_{SP+}}}\left\{1 + \frac{2\sinh(2d/L_a)}{\sinh(d/L_a)[1 + 3\cosh(2d/L_a)]}\right\} \tag{4.39}$$

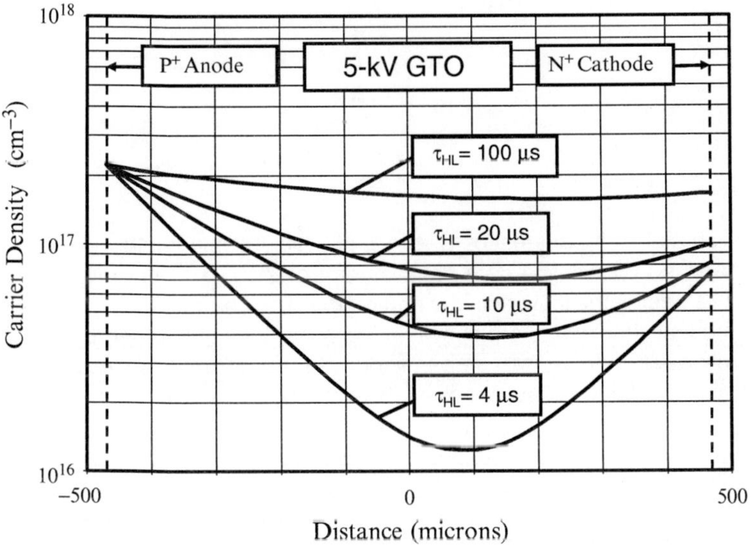

Fig. 4.15 Carrier distribution in the 5-kV power thyristor structure with end-Region recombination dominant

The results obtained by using this approach are shown in Fig. 4.15 for the case of various values of lifetime. A saturation current density of 4×10^{13} A/cm^2 was utilized for the plots. It can be observed that the carrier concentration is reduced by an order of magnitude when compared with the plots in Fig. 4.14.

Simulation Results

The results of two-dimensional numerical simulations for the 5-kV asymmetrical silicon GTO structure are described here. The total width of the structure, as shown by the cross section in Fig. 4.1, was 100 μm (area = 1×10^{-6} cm^{-2}) with a cathode finger width (W_{KS}) of 180 μm. The doping profile for the baseline device structure was already shown in Fig. 4.8.

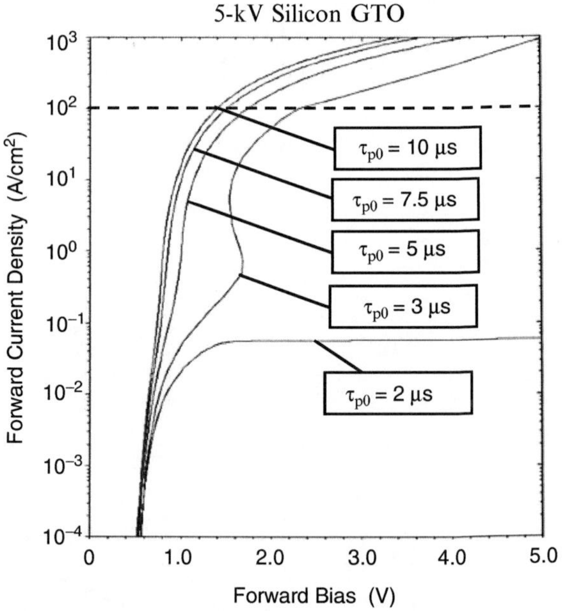

Fig. 4.16 On-state characteristics of the 5-kV asymmetric GTO structure

The on-state characteristics of the 5-kV silicon asymmetric GTO structure were obtained by using a gate drive current of 2×10^{-8} A/μm for the case of various values for the lifetime in the drift region. The characteristics obtained from the numerical simulations are shown in Fig. 4.16. It can be observed that the on-state voltage drop increases as expected with reduction of the lifetime (τ_{p0}, τ_{n0}) indicated in the figure. A snapback is observed in the *i–v* characteristics when the lifetime is reduced to 3 μs and the device does not latch up when the lifetime is reduced to 2 μs for the chosen gate drive current. The on-state voltage drop at a lifetime value of 10 μs is found to be 1.403 V at an on-state current density of 100 A/cm^2. This value is larger than predicted by the analytical model based upon middle region recombination.

The low on-state voltage drop for the 5-kV asymmetric GTO structure is due to the large number of carriers injected into the drift region producing a drastic reduction of its resistance. This is illustrated in Fig. 4.17 where the injected carrier density is shown for four cases of the lifetime (τ_{p0}, τ_{n0}) in the drift region of the GTO structure. It can be observed that the injected carrier density is four orders of magnitude larger than the doping concentration even for the case of a lifetime of 10 μs. The injected carrier density is reduced by an order of magnitude in the middle of the drift region when the lifetime is reduced to 3 μs. The predictions of the analytical model (see Fig. 4.14) have the same general characteristics but the injected carrier density is much smaller in the numerical simulations when compared with the analytical model that neglects the end-region recombination. When the end-region recombination is taken into account, the carrier densities predicted by the analytical model (see Fig. 4.15) are similar to those observed in the numerical simulations.

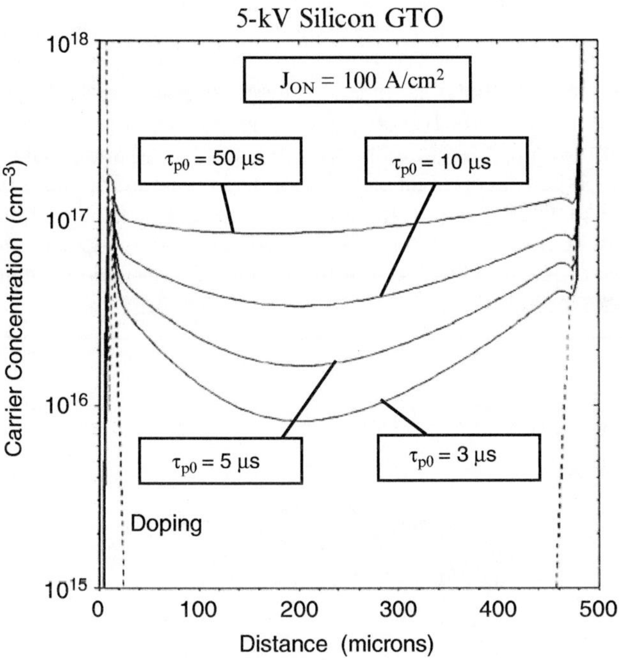

Fig. 4.17 On-state carrier distribution in the 5-kV asymmetric GTO structure

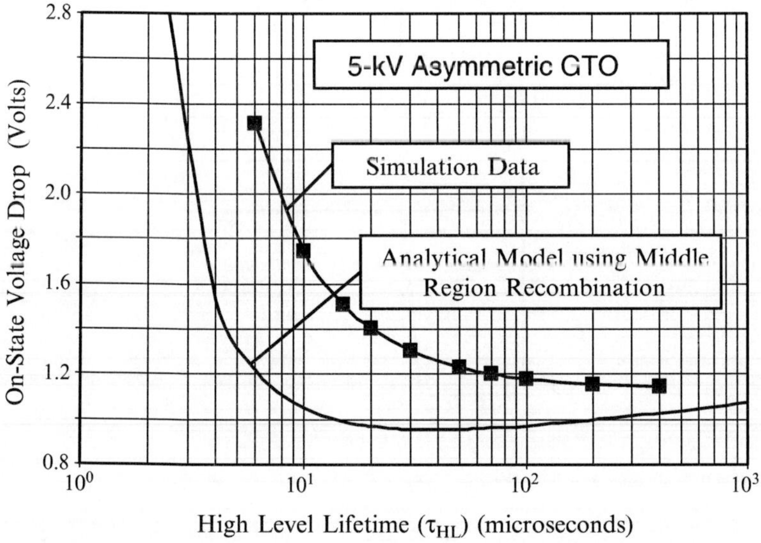

Fig. 4.18 On-state voltage drop for the 5-kV asymmetric GTO structure obtained using numerical simulations

The reduction of the injected carrier density in the middle region with smaller lifetime leads to an increase in the on-state voltage drop. The on-state voltage drop for the 5-kV asymmetric GTO structure obtained using numerical simulations can be compared with that obtained using the analytical model without end-region recombination in Fig. 4.18. The on-state voltage drop obtained using the numerical simulations is much larger than that predicted by the analytical model based upon middle-region recombination. Further, the minimum on-state voltage drop occurs at a significantly larger value for the lifetime in the drift region. It can be concluded that the analytical model is optimistic and not reliable for predicting the on-state voltage drop indicating that the impact of end-region recombination is very important for the silicon 5-kV asymmetric GTO structure.

4.2.4 Turn-Off Characteristics

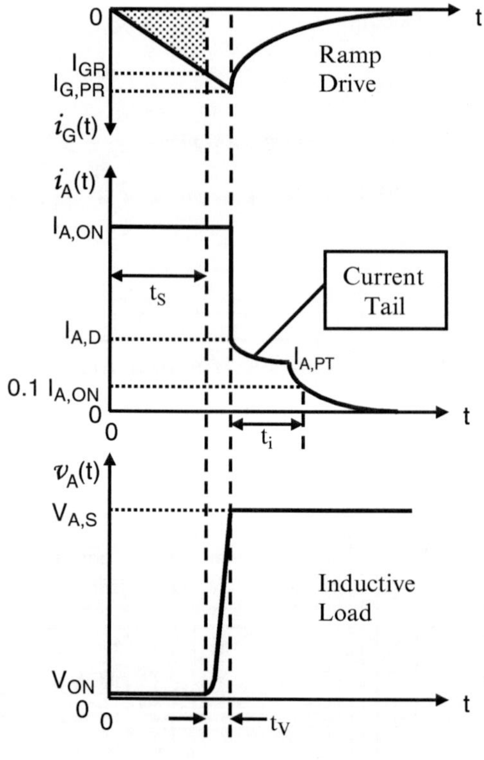

Fig. 4.19 Turn-off characteristics of the GTO structure

As discussed in the textbook [2], the GTO structure is usually turned off by the application of a gate current that increases with time in the reverse direction (ramp drive) as shown in the upper part of Fig. 4.19. The regenerative action within the thyristor does not cease until a time interval called the *storage time* (t_S). After the storage time interval, the anode voltage increases until it reaches the DC power supply voltage. Subsequently, the anode current exhibits an abrupt decrease followed by a long slow decay of the current. This slow decay of the anode current is referred to as the *current tail*. For the symmetric blocking GTO structure discussed in the textbook, the current tail occurs with a single time constant. In the case of the asymmetric blocking GTO structure discussed in this book, the current tail occurs with two time constants as discussed in detail below. Due to high anode voltage and current values during the voltage rise-time and the current tail-time, a large switching power loss occurs that limits the frequency of operation of the GTO structure.

4.2.4.1 Storage Time

The storage time for the GTO structure can limit its operating frequency. A simple two-dimensional analysis for the storage time during the turn-off of the GTO structure can be performed by assuming that the reverse gate current is used to remove the charge stored within the P-base region. If a linear GTO structure topology is assumed with a depth "Z" orthogonal to the cross section, the total stored charge in the P-base region is given by:

$$Q_{\mathrm{S,PB}} = qW_{\mathrm{P}}\frac{W_{\mathrm{K}}}{2}Zn_{\mathrm{a}} \tag{4.40}$$

where the average carrier concentration within the P-base region due to the on-state current flow is given by:

$$n_{\mathrm{a}} = K_{\mathrm{av}}\sqrt{J_{\mathrm{A,ON}}} \tag{4.41}$$

if end-region recombination is dominant. Combining the above equations:

$$Q_{\mathrm{S,PB}} = qW_{\mathrm{P}}\left(\frac{W_{\mathrm{K}}Z}{2}\right)K_{\mathrm{av}}\sqrt{J_{\mathrm{A,ON}}} \tag{4.42}$$

In the case of a ramp drive with an anode current density ramp rate "a" (A/cm^2-s) during the turn-off transient, the charge removed by the gate current (as illustrated by the shaded area in Fig. 4.19) is given by:

$$Q_{\mathrm{R}} = \frac{1}{2}I_{\mathrm{GR}}t_{\mathrm{S,R}} = \frac{1}{2}a\left(\frac{W_{\mathrm{K}}Z}{2}\right)t_{\mathrm{S,R}}^2 \tag{4.43}$$

Equating this to the stored charge in the P-base region given by Eq. 4.42 provides the storage time:

$$t_{\mathrm{S,R}} = \sqrt{\frac{2qW_{\mathrm{P}}K_{\mathrm{av}}\sqrt{J_{\mathrm{A,ON}}}}{a}} \tag{4.44}$$

For a GTO structure with a P-base width of 20 μm and an N-base width of 450 μm (including the N-buffer layer) with a high-level lifetime of 5 μs, the storage time obtained by using the above equation is 2.9 μs if the ramp rate for the gate is 10 A/cm^2-μs and the on-state current density is 100 A/cm^2.

4.2.4.2 Voltage Rise-Time

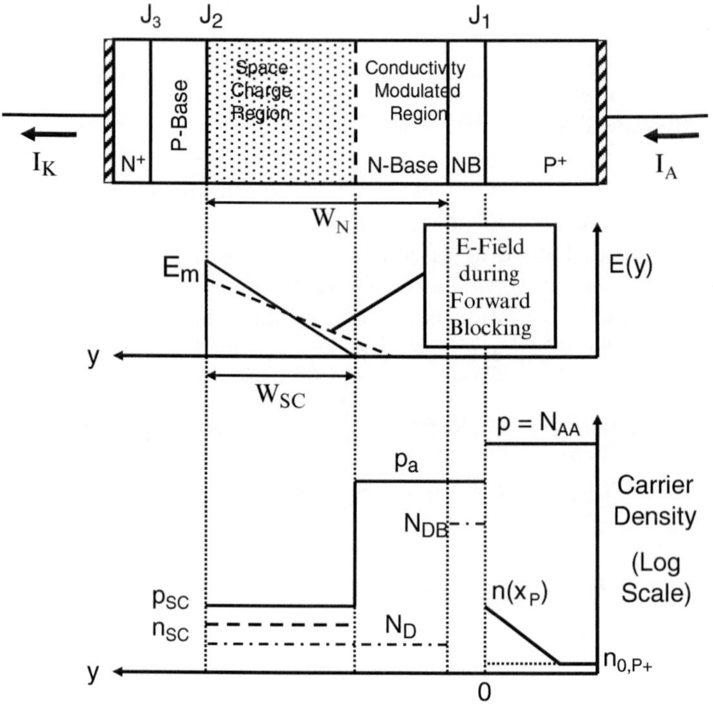

Fig. 4.20 Electric field and free carrier distribution during turn-off of the GTO structure

Once the stored charge in the P-base region has been removed by the reverse gate drive current, the P-base/N-base junction (J_2) begins to support voltage across a space-charge region as shown in Fig. 4.20. Unlike the depletion region formed across this junction under the steady-state forward blocking mode, the space-charge region formed during the turn-off of the GTO contains a large concentration of holes due to the transport of these carriers from the stored charge remaining within the N-base region. In addition, the space-charge region contains electrons due to injection of carriers from the N$^+$ cathode region from the portion of the thyristor that has not yet turned off. Since the electric field in the space-charge region is large, it can be assumed that the holes and electrons are transported at their saturated drift velocity ($v_{sat,p}$, $v_{sat,n}$) in this region.

The hole concentration (p_{SC}) within the space-charge region is related to the on-state anode current density that continues to flow during the voltage rise-time interval:

$$p_{SC} = \frac{J_{A,ON}}{qv_{sat,p}} \tag{4.45}$$

As an example, the hole concentration within the space-charge region is 8.3×10^{13} cm^{-3} at an anode current density of 100 A/cm^2 if an average saturated velocity of 7.5×10^6 cm/s is assumed. The electron concentration in the space-charge region is related to the cathode current multiplied by the common base current gain (α_{NPN}) of the NPN transistor within the thyristor structure. Consequently:

$$n_{SC} = \frac{\alpha_{NPN} J_{K,ON}}{qv_{sat,n}} = \frac{\alpha_{NPN} J_{A,ON}}{qv_{sat,n}} \tag{4.46}$$

As an example, the electron concentration within the space-charge region is 4.2×10^{13} cm^{-3} at an anode current density of 100 A/cm^2 (assuming an average saturated velocity of 7.5×10^6 cm/s) if the common base current gain (α_{NPN}) of the NPN transistor is 0.5 for the typical doping levels for the N^+ cathode and P-base regions and the cathode current density is equal to the anode current density in the on-portion of the thyristor. Since these concentrations are much larger than the doping concentration in the lightly doped portion of the N-base region (5×10^{12} cm^{-3}), the presence of the holes and electrons must be accounted for when deriving the width of the space-charge region.

The width of the space-charge region (W_{SC}) at any point in time is determined by the solution of Poisson's equation with a net charge given by the sum of the positive charge from the ionized donors, the positive charge from the holes, and the negative charge from the electrons being transported through the region. The anode voltage at any time during the voltage rise-time interval is then related to the space-charge region width by:

$$W_{SC}(t) = \sqrt{\frac{2\varepsilon_S V_A(t)}{q(N_D + p_{SC} - n_{SC})}} \tag{4.47}$$

Taking the time derivative of this expression yields:

$$\frac{dW_{SC}(t)}{dt} = \sqrt{\frac{\varepsilon_S}{2q(N_D + p_{SC} - n_{SC})V_A(t)}} \frac{dV_A}{dt} \tag{4.48}$$

As the space-charge region expands, it extracts some of the remaining stored charge at its boundary. The anode current density is also related to this extraction of the stored charge:

$$J_{A,ON} = qp_a \frac{dW_{SC}(t)}{dt} \tag{4.49}$$

where p_a is the concentration of the holes in the stored charge region. Using Eq. 4.48:

$$J_{A,ON} = qp_a\sqrt{\frac{\varepsilon_S}{2q(N_D + p_{SC} - n_{SC})V_A(t)}}\frac{dV_A}{dt} \tag{4.50}$$

This equation can be rewritten in the form:

$$\sqrt{\frac{2(N_D + p_{SC} - n_{SC})J_{A,ON}^2}{q\varepsilon_S p_a^2}}dt = \frac{dV_A}{\sqrt{V_A(t)}} \tag{4.51}$$

Integration of this expression yields:

$$\sqrt{\frac{2(N_D + p_{SC} - n_{SC})J_{A,ON}^2}{q\varepsilon_S p_a^2}}t = 2\sqrt{V_A(t)} \tag{4.52}$$

The voltage transient during the rise-time interval is then obtained:

$$V_A(t) = \frac{(N_D + p_{SC} - n_{SC})J_{A,ON}^2}{2q\varepsilon_S p_a^2}t^2 \tag{4.53}$$

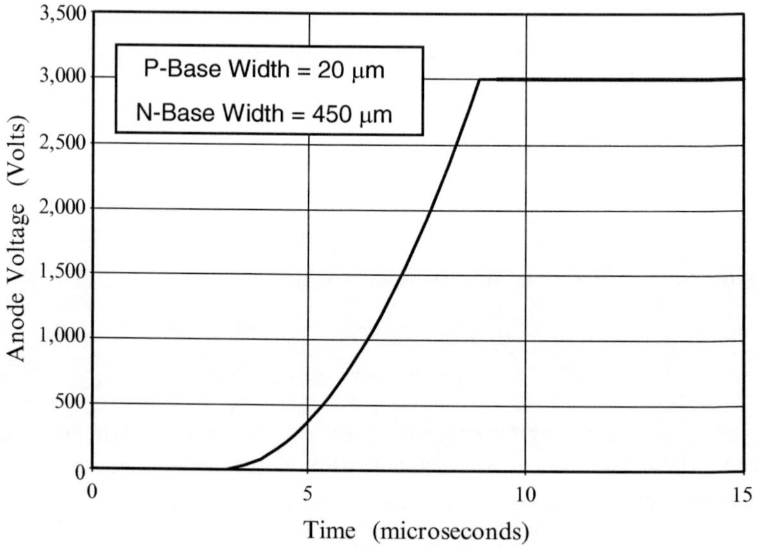

Fig. 4.21 Anode voltage transient during turn-off for the 5-kV asymmetric GTO structure

The average hole concentration (p_a) in the stored charge region is determined by the free carrier distribution within the GTO during the on-state. When end-region recombination is dominant, the average hole concentration is given by Eq. 4.41. Substituting this into Eq. 4.53 yields:

$$V_A(t) = \frac{(N_D + p_{SC} - n_{SC})J_{A,ON}}{2q\varepsilon_S K_{av}^2} t^2 \tag{4.54}$$

According to this solution for the voltage transient, the anode voltage will rise as the square of time. This solution is valid for the asymmetric GTO structure while the space-charge region is expanding through the lightly doped portion of the drift region. If the space-charge region extends through the entire lightly doped portion of the N-drift region before the anode voltage reaches the supply voltage, the anode voltage will increase abruptly to the supply voltage at that time.

The voltage transient obtained by using the above one-dimensional analysis is illustrated in Fig. 4.21 for the case of the 5-kV asymmetrical GTO structure with a P-base width of 20 μm, and an N-base width of 450 μm including the N-buffer layer. The doping concentration of the N-base region was assumed to be 5×10^{12} cm^{-3} with a high-level lifetime of 10 μs. The device was assumed to be turned off from an initial on-state current density of 100 A/cm^2 leading to a hole concentration within the space-charge region of 8.3×10^{13} cm^{-3}. The electron concentration in the space-charge region is assumed to be 4.2×10^{13} cm^{-3} corresponding to a common base current gain of 0.5 for the NPN transistor. It can be observed that the anode voltage reaches 3,000 V in 6.0 μs (t_V) after the end of the storage time interval. For this asymmetric GTO device structure, the space-charge region width is only 320 μm (compared to 420 μm for the lightly doped portion of the N-base region) when the anode voltage reaches a supply voltage of 3,000 V due to the presence of the high concentration of holes and electrons. Consequently, the voltage does not change abruptly during the transient.

The time taken for the anode voltage to reach a supply voltage ($V_{A,S}$), defined as the *voltage rise-time* t_V, can be derived from Eq. 4.55:

$$t_V = K_{av}\sqrt{\frac{2q\varepsilon_S V_{A,S}}{(N_D + p_{SC} - n_{SC})J_{A,ON}}} \tag{4.55}$$

The value for the voltage rise-time given in the previous paragraph can be computed by using this expression. It is worth pointing out that the voltage rise-time is dependent upon the lifetime in the N-base region through the term K_{av}. In the case of switching an inductive load with a clamping diode, the anode voltage becomes constant at the supply voltage after the rise-time interval.

4.2.4.3 Current Fall-Time

Once the anode voltage reaches the supply voltage, the anode current decreases with a sharp drop in value followed by a current tail as illustrated in Fig. 4.19.

The sudden drop in the anode current is associated with the cessation of cathode current flow at the end of the voltage rise-time [2]. Just prior to the turn-off of the thyristor, the base drive current for the P-N-P transistor ($I_{B,PNP}$) due to the electrons supplied by the N-P-N transistor is given by ($\alpha_{NPN}.I_K$). This base drive current for the P-N-P transistor produces an anode current given by:

$$I_{AT} = \beta_{PNP} I_{B,PNP} = \left(\frac{\alpha_{PNP}}{1 - \alpha_{PNP}}\right) \alpha_{NPN} I_K \tag{4.56}$$

When the cathode current is interrupted by the cessation of regenerative action during the turn-off of the GTO structure, the anode current is abruptly reduced by the above amount leading to the sudden fall in the anode current. The anode current at the start of the current tail is therefore given by:

$$I_{A,D} = I_{A,ON} - I_{AT} = I_{A,ON} - \left(\frac{\alpha_{PNP}}{1 - \alpha_{PNP}}\right) \alpha_{NPN} I_K \tag{4.57}$$

with the cathode current related to the anode current by:

$$I_K = I_A - I_{GR} = I_A - a \cdot \left(t_{S,R} + t_V\right) \tag{4.58}$$

where I_{GR} is the reverse gate drive current at the end of the voltage-rise transient. For a typical common-base current gain of 0.5 for the N-P-N transistor and 0.60 for the P-N-P transistor, a 70% fall in the anode current is predicted by this equation, resulting in an initial tail current of 30% of the initial on-state anode current.

The decay of the anode current after the initial abrupt reduction is governed by the recombination of the excess holes and electrons that are trapped within the N-base region and the N-buffer layer. The minority carrier density in the N-buffer layer is initially larger than its doping concentration. In the case of typical diffused buffer layers, the doping concentration in the buffer layer varies from a low value of the lightly doped portion of the N-base region (5×10^{12} cm^{-3}) to a high concentration of about 1×10^{17} cm^{-3}. Consequently, the N-buffer layer operates partly with high-level injection and partly with low-level injection conditions.

Model Using High-Level Injection in Buffer Layer

In this model, it will be assumed that the N-buffer layer operates under high-level injection conditions during the first phase of the current tail. In the absence of diffusion, the continuity equation for holes in the N-base region and N-buffer layer is given by:

$$\frac{d\delta p_N}{dt} = -\frac{\delta p_N}{\tau_{HL}} \tag{4.59}$$

where δp_{N} is the excess hole concentration in the N-base region and N-buffer layer. The solution for this equation is:

$$\delta p_{\mathrm{N}}(t) \approx p_{\mathrm{N}}(t) = p_{\mathrm{a}} e^{-t/\tau_{\mathrm{HL}}} \tag{4.60}$$

because high-level injection conditions prevail in the N-base region and N-buffer layer, with the initial concentration of holes in the stored charge region being equal to the average hole concentration due to the injection of carriers in the on-state. Since this concentration (p_{a}) is given by Eq. 4.41 when recombination in the end regions is dominant:

$$p_{\mathrm{N}}(t) = K_{\mathrm{av}} \sqrt{J_{\mathrm{A,ON}}} e^{-t/\tau_{\mathrm{HL}}} \tag{4.61}$$

The anode current flow that supports the recombination of carriers within the stored charge region can be analyzed by examination of the carrier distribution on both sides of the P$^+$ anode/N-base junction (J$_1$). The high concentration of electrons in the N-base region and N-buffer layer produces the injection of electrons into the P$^+$ anode region [2]. The anode current is produced by the diffusion of the injected electrons in the P$^+$ anode side of the junction:

$$J_{\mathrm{A}}(t) = \frac{qD_{\mathrm{nP+}} p_{\mathrm{N}}^2(t)}{L_{\mathrm{nP+}} N_{\mathrm{AA}}} = \frac{qD_{\mathrm{nP+}} K_{\mathrm{av}}^2 J_{\mathrm{A,ON}}}{L_{\mathrm{nP+}} N_{\mathrm{AA}}} e^{-2t/\tau_{\mathrm{HL}}} \tag{4.62}$$

This equation indicates that the anode current varies as the square of the carrier density in the stored charge region during the anode current tail. Consequently, the anode current tail decreases exponentially with time with a time constant of one half of the high-level lifetime, even though the free carrier density in the stored charge region is decreasing exponentially with time with a time constant equal to the high-level lifetime. The equation can be written in the form:

$$J_{\mathrm{A}}(t) - J_{\mathrm{A,D}} e^{-2t/\tau_{\mathrm{HL}}} \tag{4.63}$$

where

$$J_{\mathrm{A,D}} = \frac{qD_{\mathrm{nP+}} K_{\mathrm{av}}^2 J_{\mathrm{A,ON}}}{L_{\mathrm{nP+}} N_{\mathrm{AA}}} \tag{4.64}$$

In the case of the asymmetrical GTO structure, the abrupt reduction of the anode current density during the initial current fall time produces a decrease in the hole concentration in the space-charge region. At the same time, the cathode current is interrupted leading to a rapid reduction of the electron concentration to zero within the space-charge region. The reduced hole concentration produces a decrease in the net positive charge within the space-charge region. Consequently, the space-charge region width increases even though the anode voltage is held constant until it expands through the entire N-base region. Following the abrupt reduction of the anode current, the holes in the N-base region are removed by a combination the recombination process and the expansion of the space-charge region. Once the space-charge region punches-through to the N-buffer layer, the excess holes are removed by recombination in the buffer layer. This process occurs at a faster rate due to the lower lifetime in the buffer layer because its doping concentration is larger than in the N-base region.

The width of the space-charge region during the initial portion of the anode current tail is given by:

$$W_{SC}(t) = \sqrt{\frac{2\varepsilon_S V_{A,S}}{q[N_D + p_{SC}(t)]}} \tag{4.65}$$

because the electron current has ceased. The hole concentration in the space-charge region is given by:

$$p_{SC} = \frac{J_A(t)}{q v_{sat,p}} \tag{4.66}$$

Combining these equations yields:

$$J_A(t) = \frac{2 v_{sat,p} \varepsilon_S V_{A,S}}{W_{SC}^2} - q v_{sat,p} N_D \tag{4.67}$$

The space-charge region punches-through to the N-buffer layer at a time (t_{PT}) when its width becomes equal to the width (W_N) of the lightly doped portion of the N-base region. The corresponding anode current density is:

$$J_{A,PT} = \frac{2 v_{sat,p} \varepsilon_S V_{A,S}}{W_N^2} - q v_{sat,p} N_D \tag{4.68}$$

During the initial portion of the anode current tail, the current decays in accordance with Eq. 4.63. The time (t_{PT}) at which the space-charge region punches-through to the N-buffer layer is the time taken for the anode current to decay from $J_{A,D}$ to $J_{A,PT}$:

$$t_{PT} = \frac{\tau_{HL}}{2} \ln\left[\frac{J_{A,D}}{J_{A,PT}}\right] \tag{4.69}$$

Once the space-charge region extends through the entire width of the lightly doped portion of the N-base region, its width cannot increase any further due to the high doping in the N-buffer layer. The remaining stored charge in the N-buffer layer is then removed by recombination at a rate corresponding to the lifetime in the N-buffer layer. The current transient is then described by:

$$J_A(t) = J_{A,PT} e^{-2t/\tau_{BL}} \quad (4.70)$$

where τ_{BL} is the lifetime in the N-buffer layer. The lifetime in the buffer layer is smaller than the corresponding value in the lightly doped portion of the N-base region:

$$\tau_{BL} = \frac{\tau_{LL}}{1 + (N_{D,BL}/N_{REF})} \quad (4.71)$$

where τ_{LL} is the low-level lifetime in the lightly doped portion of the N-base region, $N_{D,BL}$ is the doping concentration of the buffer layer, and N_{REF} is a reference doping concentration (typically $5 \times 10^{16}\,cm^{-3}$). The current fall time (t_i), defined as the time taken for the anode current to reach one tenth of the on-state value, is then given by:

$$t_i = t_{PT} + \frac{\tau_{BL}}{2} \ln\left[\frac{10 J_{A,PT}}{J_{A,ON}}\right] \quad (4.72)$$

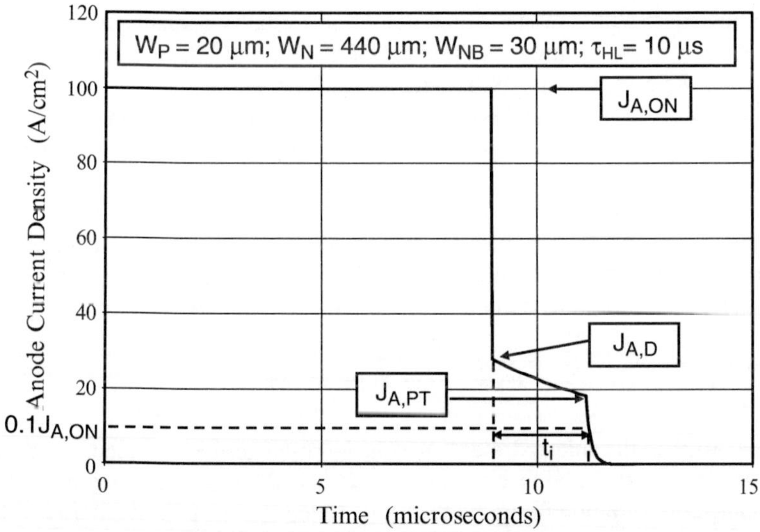

Fig. 4.22 Anode current transient during turn-off for the GTO structure under high-level injection conditions in the buffer-layer

Consider the case of a GTO structure with P-base, N-base, and N-buffer layer widths of 20, 440, and 30 µm, respectively, with a high-level lifetime of 10 µs in the N-base region, N-buffer layer doping concentration of 1×10^{17} cm^{-3}, and an effective anode doping concentration of 5×10^{18} cm^{-3}. The diffusion coefficient for electrons (D_{nP+}) in the anode region for this doping concentration is 3.5 cm^2/V-s. The diffusion length for electrons (L_{nP+}) in the anode region is then 0.7 μm if the low-level recombination lifetime in this region is 1.35 ns. Using these values in Eq. 4.64, with an on-state anode current density of 100 A/cm^2, yields an anode current density ($J_{A,D}$) of 28 A/cm^2 at the beginning of the anode current tail. The anode current density when the space-charge region punches-through to the N-buffer layer is found to be 18 A/cm^2 based upon Eq. 4.68.

The anode current waveform predicted by the above equations is shown in Fig. 4.22. It can be seen that the anode current undergoes an abrupt reduction followed by a slow decay in magnitude until the space-charge region extends through the entire lightly doped portion of the N-base region. After this, the anode current decays much more rapidly to zero. Using the values given in the previous paragraph, the current turn-off time is found to be 2.25 μs. This is a relatively long time interval during which there is a substantial anode current density flowing through the GTO structure while its anode voltage is high. Consequently, the high power dissipation associated with the current tail limits the frequency of operation for the GTO structure.

Model Using Low-Level Injection in Buffer Layer

In this model, it will be assumed that the N-buffer layer operates under low-level injection conditions during the first phase of the current tail. However, the recombination of the carriers in the N-base region during the first phase of the current tail will be assumed to occur under high-level injection conditions. Consequently, the decay of the hole concentration is described by Eq. 4.62. The anode current flow that supports the recombination of carriers within the stored charge region can be analyzed by examination of the carrier distribution on both sides of the P$^+$ anode/N-base junction (J_1). The high concentration of electrons in the N-base region and N-buffer layer produces the injection of electrons into the P$^+$ anode region [2]. The carrier concentrations on both sides of the anode junction are related by:

$$n_A(0,t)N_{AA} = p_N(t)N_{D,BL} \tag{4.73}$$

where $n_A(0,t)$ is the injected electron concentration on the P$^+$ anode side of the junction, N_{AA} is the doping concentration of the P$^+$ anode region, and $N_{D,BL}$ is the doping concentration of the buffer layer. Using Eq. 4.61 for the decay of holes in the N-base region yields:

$$n_A(0,t) = \frac{N_{D,BL}}{N_{AA}} K_{av}\sqrt{J_{A,ON}}e^{-t/\tau_{HL}} \tag{4.74}$$

The anode current is produced by the diffusion of the injected electrons in the P$^+$ anode side of the junction:

$$J_A(t) = \frac{qD_n n_A(0,t)}{L_{nP+}} = \frac{qD_n N_{D,BL} K_{av} \sqrt{J_{A,ON}}}{L_{nP+} N_{AA}} e^{-t/\tau_{HL}} \tag{4.75}$$

This equation indicates that the anode current tail decreases exponentially with time with a time constant equal to the high-level lifetime in the N-base region.

In the case of the asymmetrical GTO structure, the abrupt reduction of the anode current density during the initial current fall time produces a decrease in the hole concentration in the space-charge region. At the same time, the cathode current is interrupted leading to a rapid reduction of the electron concentration to zero within the space-charge region. The reduced hole concentration produces a decrease in the net positive charge within the space-charge region. Consequently, the space-charge region width increases even though the anode voltage is held constant until it expands through the entire N-base region. Following the abrupt reduction of the anode current, the holes in the N-base region are removed by a combination of the recombination process and the expansion of the space-charge region. Once the space-charge region punches-through to the N-buffer layer, the excess holes are removed by recombination in the buffer layer. This process occurs at a faster rate due to the lower lifetime in the buffer layer because its doping concentration is larger than in the N-base region.

The width of the space-charge region during the initial portion of the anode current tail is given by:

$$W_{SC}(t) = \sqrt{\frac{2\varepsilon_S V_{A,S}}{q[N_D + p_{SC}(t)]}} \tag{4.76}$$

because the electron current has ceased. The hole concentration in the space-charge region is given by:

$$p_{SC} = \frac{J_A(t)}{q v_{sat,p}} \tag{4.77}$$

Combining these equations yields:

$$J_A(t) = \frac{2 v_{sat,p} \varepsilon_S V_{A,S}}{W_{SC}^2} - q v_{sat,p} N_D \tag{4.78}$$

The space-charge region punches-through to the N-buffer layer at a time (t_{PT}) when its width becomes equal to the width (W_N) of the lightly doped portion of the N-base region. The corresponding anode current density is:

$$J_{A,PT} = \frac{2 v_{sat,p} \varepsilon_S V_{A,S}}{W_N^2} - q v_{sat,p} N_D \tag{4.79}$$

During the initial portion of the anode current tail, the current decays in accordance with Eq. 4.75, which can be rewritten as:

$$J_A(t) = J_{A,D} e^{-t/\tau_{HL}} \tag{4.80}$$

where

$$J_{A,D} = \frac{qD_n N_{D,BL} K_{av} \sqrt{J_{A,ON}}}{L_{nP+} N_{AA}} \tag{4.81}$$

The time (t_{PT}) at which the space-charge region punches-through to the N-buffer layer is the time taken for the anode current to decay from $J_{A,D}$ to $J_{A,PT}$:

$$t_{PT} = \tau_{HL} \ln\left[\frac{J_{A,D}}{J_{A,PT}}\right] \tag{4.82}$$

Once the space-charge region extends through the entire width of the lightly doped portion of the N-base region, its width cannot increase any further due to the high doping in the N-buffer layer. The remaining stored charge in the N-buffer layer is then removed by recombination under low-level injection conditions at a rate corresponding to the lifetime in the N-buffer layer. The current transient is then described by:

$$J_A(t) = J_{A,PT} e^{-t/\tau_{BL}} \tag{4.83}$$

where τ_{BL} is the low-level lifetime in the N-buffer layer. The lifetime in the buffer layer is smaller than the corresponding value in the lightly doped portion of the N-base region:

$$\tau_{BL} = \frac{\tau_{LL}}{1 + (N_{D,BL}/N_{REF})} \tag{4.84}$$

where τ_{LL} is the low-level lifetime in the lightly doped portion of the N-base region, $N_{D,BL}$ is the doping concentration of the buffer layer, and N_{REF} is a reference doping concentration (typically $5 \times 10^{16}\,\text{cm}^{-3}$). The current fall time (t_i), defined as the time taken for the anode current to reach one tenth of the on-state value, is then given by:

$$t_i = t_{PT} + \tau_{BL} \ln\left[\frac{10 J_{A,PT}}{J_{A,ON}}\right] \tag{4.85}$$

Consider the case of a GTO structure with P-base, N-base, and N-buffer layer widths of 20, 440, and 30 μm, respectively, with a high-level lifetime of 10 μs in the N-base region, N-buffer layer doping concentration of 1×10^{17} cm^{-3}, and an effective anode doping concentration of 5×10^{18} cm^{-3}. The diffusion coefficient for electrons (D_{nP+}) in the anode region for this doping concentration is 3.5 cm^2/V-s. The diffusion length for electrons (L_n) in the anode region is then 0.5 μm if the low-level recombination lifetime in this region is 0.5 ns. Using these values in Eq. 4.81, with an on-state anode current density of 100 A/cm^2, yields an anode current density ($J_{A,D}$) of 28 A/cm^2 at the beginning of the anode current tail. The anode current density, when the space-charge region punches-through to the N-buffer layer, is found to be 18 A/cm^2 based upon Eq. 4.79.

The anode current waveform predicted by the above equation is shown in Fig. 4.23. It can be seen that the anode current undergoes an abrupt reduction followed by a slow decay in magnitude until the space-charge region extends through the entire lightly doped portion of the N-base region. After this, the anode current decays much more rapidly to zero. Using the values given in the previous paragraph, the current turn-off time (t_i) is found to be 4.5 μs. This is a relatively long time interval during which there is a substantial anode current density flowing through the GTO structure while its anode voltage is high. Consequently, the high power dissipation associated with the current tail limits the frequency of operation for the GTO structure.

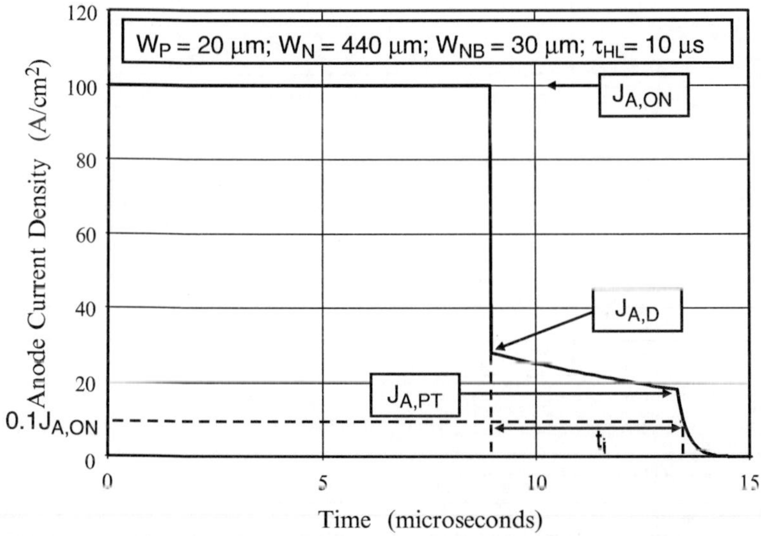

Fig. 4.23 Anode current transient during turn-off for the GTO structure under low-level injection conditions in the buffer-layer

Simulation Example

In order to gain insight into the operation of the asymmetric GTO structure during its turn-off, the results of two-dimensional numerical simulations for a typical structure are discussed here for the case of a ramp gate drive. The device structure used has the cross section shown in Fig. 4.1 with a width of 100 μm. No cathode short was used in the structure. The doping profile for the GTO structure used in the simulations was provided in Fig. 4.8. The widths of the P-base, N-base, and N-buffer layer regions are 20, 440, and 30 μm, respectively.

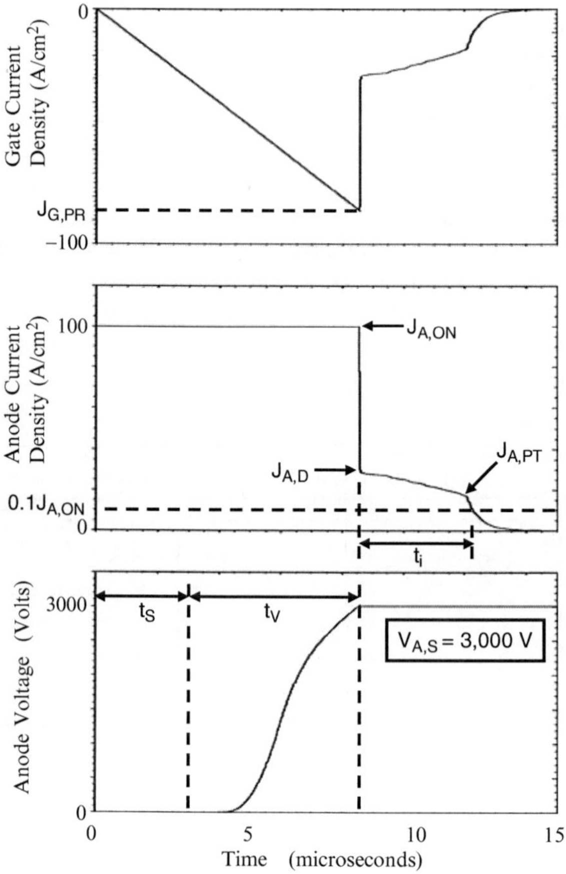

Fig. 4.24 5kV GTO turn-off waveforms

The numerical simulations were performed with a gate ramp rate of −10 A/cm^2-μs starting from an on-state current density of 100 A/cm^2. The resulting waveforms obtained from the numerical simulations for the anode voltage and current, as well as the gate current, are shown in Fig. 4.24 for the case of an anode supply voltage of 3,000 V. The storage time extracted from the simulations was about 3 μs based upon detecting an increase in the anode voltage from 1.744 V in the on-state to 3 V. The simple two-dimensional analysis for the storage time (Eq. 4.44) provides a good estimate for this time interval. After the storage time, the anode voltage increases as the square of the time as predicted by the analytical model until it reaches 2,000 V. It then increases at a slower rate. This is associated with the onset of avalanche multiplication at high anode bias voltages – an effect not included in the analytical model. The anode voltage-rise time for the case of supply voltage of 3,000 obtained in the simulations is 5.5 μs. The one-dimensional analysis for the voltage rise-time (Eq. 4.55) provides a good prediction for this time interval. The anode current waveform exhibits a sharp drop at the end of the voltage rise-time interval and then decays slowly over a period of about 3.5 μs followed by a faster rate of decay and described by the analytical model. The one-dimensional analytical models predict a shape for the anode current waveform (see Figs. 4.22 and 4.23) that is consistent with these results. However, the current fall time observed in the simulations falls between those predicted by the two models (based upon high-level and low-level injection conditions in the buffer layer) consistent with the fact that the carrier concentration in the N-buffer layer varies from high-level injection to low-level injection conditions.

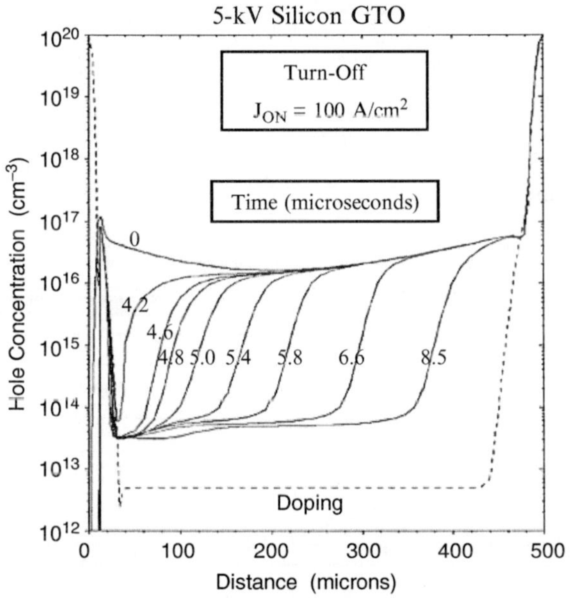

Fig. 4.25 Hole carrier distribution during the storage time and voltage rise-time for the 5-kV GTO turn-off transient

In the textbook [2], three-dimensional views of the carrier distributions were providing to demonstrate that the hole carrier distribution is remarkably one-dimensional in nature even though the turn-off process is two-dimensional in nature. This observation justifies creating one-dimensional analytical models for the turn-off waveforms that provide a good description of the GTO switching behavior. Further insight into the turn-off process, especially during the anode current tail, can be obtained by examination of the carrier density at the center of the N^+ cathode.

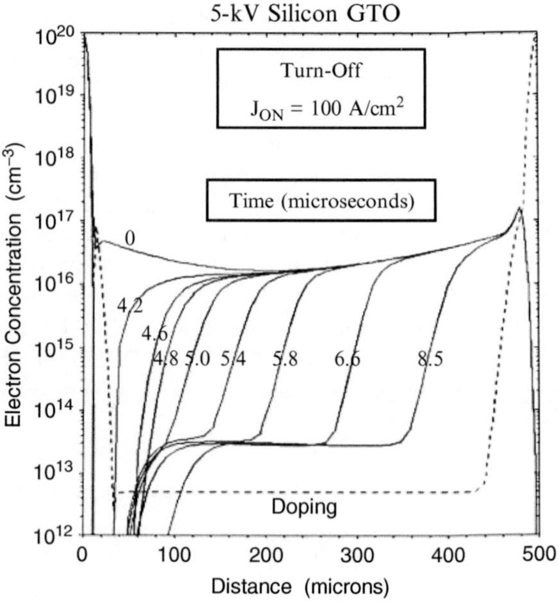

Fig. 4.26 Electron carrier distribution during the storage time and voltage rise-time for the 5-kV GTO turn-off transient

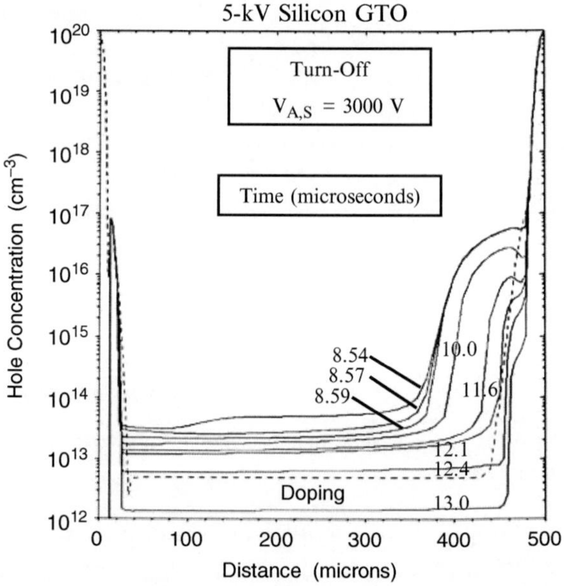

Fig. 4.27 Hole carrier distribution during the current tail-time for the 5-kV GTO turn-off transient

A one-dimensional view of the minority carrier distribution in the 5-kV GTO structure at the center of the cathode is shown in Fig. 4.25 from the initial steady-state operating point ($t = 0$ μs) to the end of the voltage rise-time ($t = 8.5$ μs). The initial carrier distribution has a catenary form within the P-base, N-base, and N-buffer regions because the GTO operates like a P-i-N rectifier in the on-state. The carrier concentration is almost constant through these regions. It can be observed from Fig. 4.25 that the carrier distribution in the N-base region does not change significantly near the anode region during the storage phase and the voltage rise phase. A significant space-charge region begins to form after 4 μs during the turn-off and expands toward the right-hand side. The hole concentration in the space-charge region is about 6–8 × 10^{13} cm^{-3}, which is consistent with the value for p_{SC} obtained using the analytical model with the carriers moving at the saturated drift velocity and an on-state current density of 100 A/cm^2. The width of the space-charge region can be observed to be about 340 μm when the anode voltage reaches 3,000 V, which is consistent with the predictions of the analytical model when the electron concentration in the space-charge region is included (see Eq. 4.47). The electron concentration distribution is provided in Fig. 4.26. It can be observed that the electron concentration in the space-charge region is about 3–4 × 10^{13} cm^{-3} consistent with the value obtained using the analytical model with an electron current density of 50 A/cm^2.

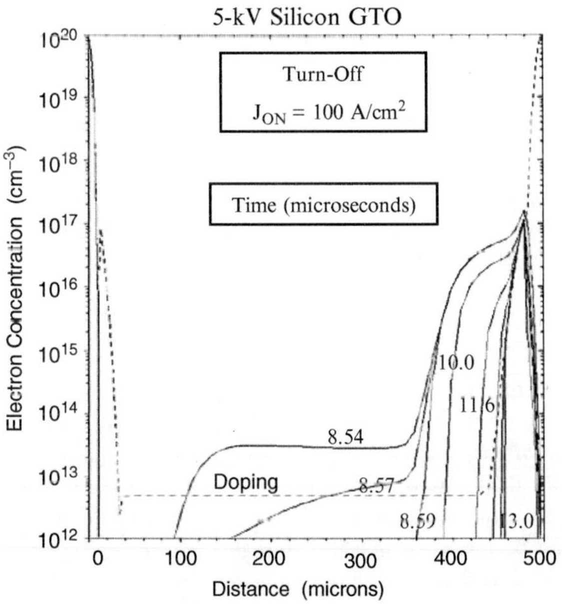

Fig. 4.28 Electron carrier distribution during the current tail-time for the 5-kV GTO turn-off transient

A one-dimensional view of the hole carrier distribution in the GTO structure at the center of the cathode is shown in Fig. 4.27 during the current tail time. The anode voltage is held constant at the anode supply voltage of 3,000 V during this transient. When the anode current reduces abruptly between $t = 8.54$ and $t = 8.59$ μs, the hole concentration decreases in the space-charge region but is not altered within the rest of the N-base region and N-buffer layer. This is consistent with the assumption used in the analytical model that the hole concentration within the stored charge in the N-base region at the beginning of the anode current decay phase is equal to the initial on-state carrier concentration. Subsequently, the hole concentration in the stored-charge region decreases due to the recombination process. At the same time, the space-charge region expands in spite of a constant anode voltage because the hole concentration in the space-charge region is smaller.

The electron concentration profiles during the current tail time are shown in Fig. 4.28. It can be observed that the electron concentration reduces abruptly between $t = 8.54$ and $t = 8.59$ μs because the cathode current ceases at the beginning of the current fall-time. Due to the reduced anode current after the abrupt drop, a smaller net positive charge exists within the space-charge region, which allows it to expand through the entire lightly doped portion of the N-base region as the anode current decays slowly. After the space-charge region extends through the entire lightly doped portion of the N-base region at time $t = 12.1$ μs, further recombination of the stored charge occurs within the N-buffer layer at a more rapid rate due to the lower lifetime for minority carriers in the buffer layer. This produces the faster current tail during the second part of the current transient.

4.2.5 Lifetime Dependence

The optimization of the power losses for the GTO structure requires performing a trade-off between the on-state voltage drop and the switching losses. One approach to achieve this is by adjusting the lifetime in the drift region. A reduction of the lifetime in the drift region also alters the lifetime in the buffer layer. The impact of reducing the lifetime on the on-state voltage drop was previously shown in Sect. 4.2.3. The on-state voltage drop increases when the lifetime is reduced, and at very low lifetime values, the GTO structure is unable to latch up into the regenerative mode.

The two models developed for turn-off of the asymmetric GTO structure presented in the previous section can be used to analyze the impact of changes to the lifetime in the drift region. The anode voltage transients predicted by the model based upon high-level injection conditions in the N-buffer layer are shown in Fig. 4.29. It can be observed that the storage time decreases only slightly when the lifetime is reduced. The voltage rise-time decreases when the lifetime is reduced because of the smaller average concentration for the holes in the N-base region that are being removed during the voltage transient. The anode current transients predicted by the model based upon high-level injection conditions in the N-buffer layer are shown in Fig. 4.30. It can be observed that the current fall-time decreases substantially when the lifetime is reduced. The current fall-time decreases when the lifetime is reduced because of the smaller average concentration for the holes in the N-base region that are being removed during the current transient.

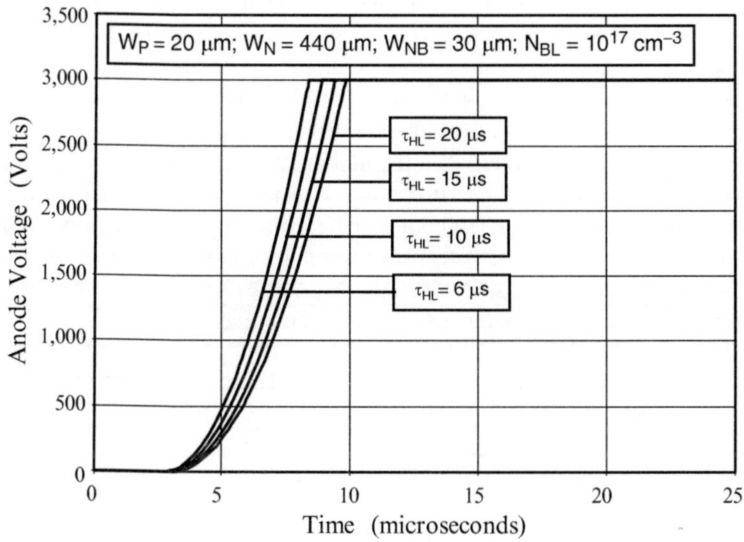

Fig. 4.29 Lifetime dependence of anode voltage transient during turn-off for the 5-kV asymmetric GTO structure assuming high-level injection in the N-buffer layer

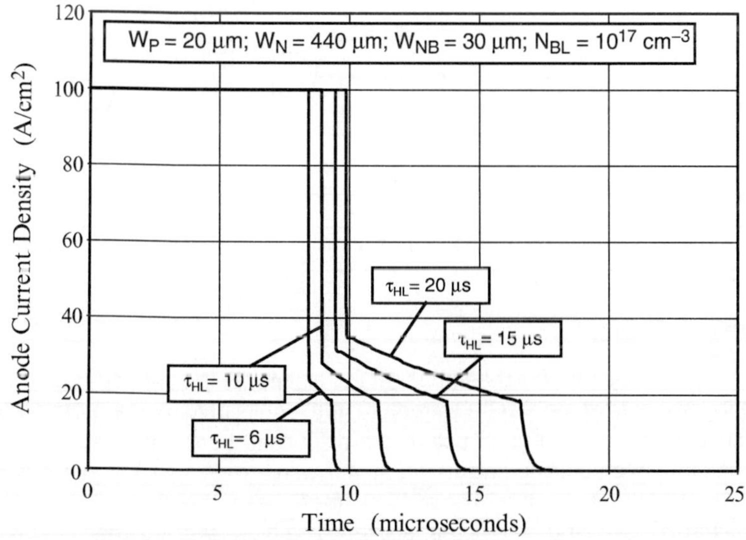

Fig. 4.30 Lifetime dependence of anode current transient during turn-off for the 5-kV asymmetric GTO structure assuming high-level injection in the N-buffer layer

The anode voltage transients predicted by the model based upon low-level injection conditions in the N-buffer layer are the same as those shown in Fig. 4.29. The anode current transients predicted by the model based upon low-level injection conditions in the N-buffer layer are shown in Fig. 4.31. It can be observed that the current fall time decreases substantially when the lifetime is reduced. The current fall-time decreases when the lifetime is reduced because of the smaller average concentration for the holes in the N-base region that are being removed during the current transient. In the case of both analytical models, it can be observed that the initial anode current density $J_{A,D}$ decreases when the lifetime is reduced. In contrast, the anode current density $J_{A,PT}$ at which the space-charge region punches-through to the buffer layer remains the same, i.e., independent of the lifetime.

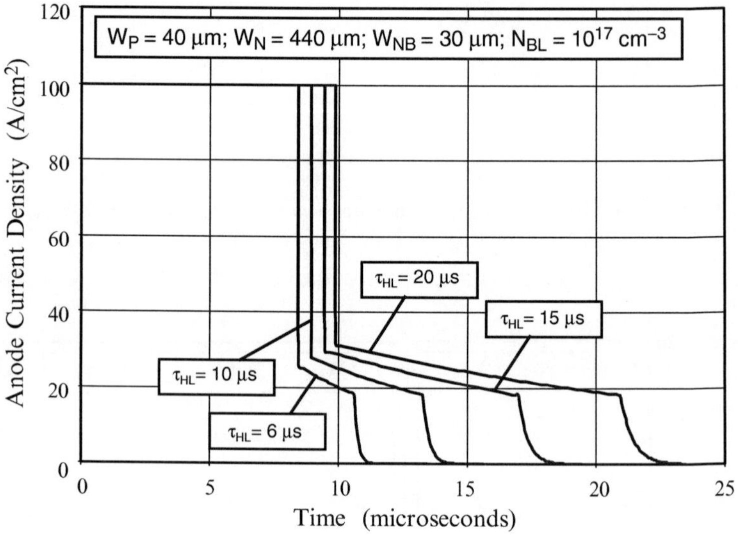

Fig. 4.31 Lifetime dependence of anode current transient during turn-off for the 5-kV asymmetric GTO structure assuming low-level injection in the N-buffer layer

Simulation Example

In order to gain insight into the impact of the lifetime in the N-base region on the operation of the asymmetric GTO structure, the results of two-dimensional numerical simulations for a typical structure are discussed here. The device structure used has the cross section shown in Fig. 4.1 with a width of 100 μm. No cathode short was used in the structure. The widths of the P-base, N-base, and N-buffer layer regions are 20, 440, and 30 μm, respectively. The high-level lifetime in the N-base region was varied between 6 and 20 μs. For turning off the GTO structures, the numerical simulations were performed with a gate ramp rate of -10 A/cm^2-μs starting from an on-state current density of 100 A/cm^2. The resulting waveforms obtained from the numerical simulations for the anode voltage and current, as well as the gate current, are shown in Fig. 4.32 for the case of an anode supply voltage of 3,000 V.

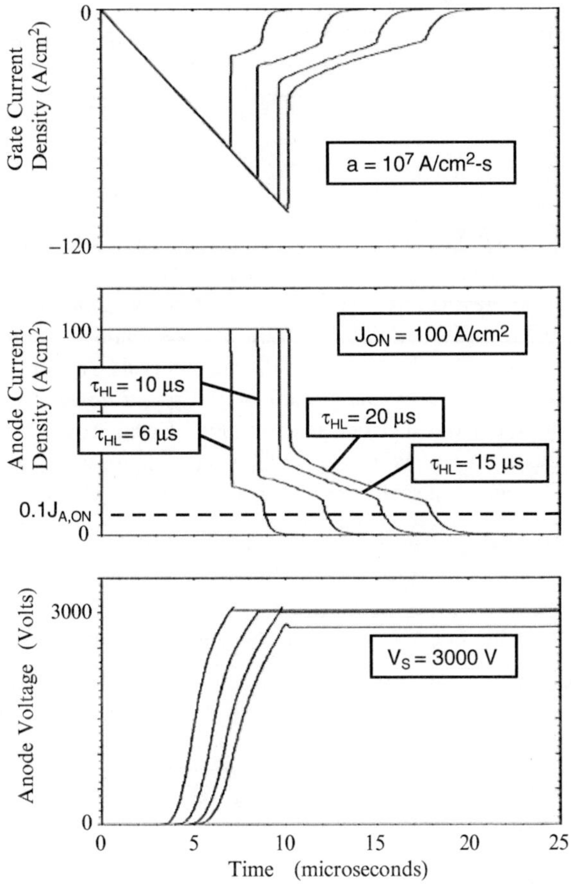

Fig. 4.32 Impact of lifetime on the 5-kV GTO turn-off waveforms

The numerical simulations show a decrease in the voltage rise-time with reduction of the lifetime in the N-base region. The results of the analytical model are consistent with this behavior. The numerical simulations also show a substantial decrease in the anode current fall time. This behavior is also very well modeled by both the analytical models. In addition, the numerical simulations show a reduction of the initial anode current ($J_{A,D}$) while the punch-through anode current ($J_{A,PT}$) is independent of the lifetime in the N-base region. Both analytical models show this behavior as well. The analytical models can therefore also be used to predict the switching energy loss and the turn-off gain.

4.2.6 Switching Energy Loss

The power loss incurred during the switching transient limits the maximum operating frequency for the GTO structure. During the storage time interval, the voltage across the device is equal to its on-state voltage drop. Consequently, although the storage time for the GTO structure is very long, it can be accounted for as a part of the on-time interval. From the point of view of limiting the maximum frequency of operation, consider the case of a device operating at a 50% duty cycle with the on-time being determined by the storage time. In this case, since half the period (T) is equal to the storage time (t_S), the maximum operating frequency is given by:

$$f_{MAX} = \frac{1}{T} = \frac{1}{2t_S} \tag{4.86}$$

A storage time of 10 μs would correspond to a maximum operating frequency of 50 kHz.

In practice, the maximum operating frequency for the GTO structure is limited by the turn-off losses. The turn-off losses are associated with the voltage rise-time interval and the current fall-time interval. The energy loss for each event can be computed by integration of the power loss, as given by the product of the instantaneous current and voltage. During the voltage rise-time interval, the anode current is constant while the voltage increases as the square of the time. The energy loss during the voltage rise-time interval is therefore given by:

$$E_{OFF,V} = \int_0^{t_V} J_{A,ON} V_A(t) \mathrm{d}t \tag{4.87}$$

Using Eq. 4.54 for the anode voltage waveform:

$$E_{OFF,V} = \frac{J_{A,ON}^2 (N_D + p_{SC} - n_{SC})}{6q\varepsilon_S K_{av}^2} t_V^3 \tag{4.88}$$

For the example shown in Fig. 4.21 with an anode supply voltage of 3,000 V, the energy loss per unit area during the voltage rise-time is found to be 0.60 J/cm^2 if the on-state current density is 100 A/cm^2.

During the anode current fall-time interval, the anode voltage is constant while the current decreases in two phases. For the case of high-level injection conditions in the N-buffer layer, the energy loss during the first phase of current fall-time interval is given by:

$$\begin{aligned} E_{OFF,I1} &= \int_0^{t_{PT}} V_{A,S} J_{AD} e^{-2t/\tau_{HL}} \mathrm{d}t \\ &= V_{A,S} J_{AD} \frac{\tau_{HL}}{2} \left(1 - e^{-2t_{PT}/\tau_{HL}}\right) \end{aligned} \tag{4.89}$$

using Eq. 4.63 for the anode current waveform during this time interval. The energy loss during the second phase of current fall-time interval is given by:

$$\begin{aligned} E_{OFF,I2} &= \int_0^{\infty} V_{A,S} J_{A,PT} e^{-2t/\tau_{BL}} \mathrm{d}t \\ &= V_{A,S} J_{A,PT} \frac{\tau_{BL}}{2} \end{aligned} \tag{4.90}$$

For the example shown in Fig. 4.22 with an anode supply voltage of 3,000 V and an on-state current density of 100 A/cm^2, the energy loss per unit area during the first phase of current fall-time is found to be 0.15 J/cm^2 and during the second phase to be 0.0065 J/cm^2. It can be concluded that the turn-off loss during the current transient is mainly due to the first phase. The total energy loss per unit area ($E_{OFF,V}$ + $E_{OFF,I}$) during the turn-off process for the 5-kV GTO structure is found to be 0.76 J/cm^2.

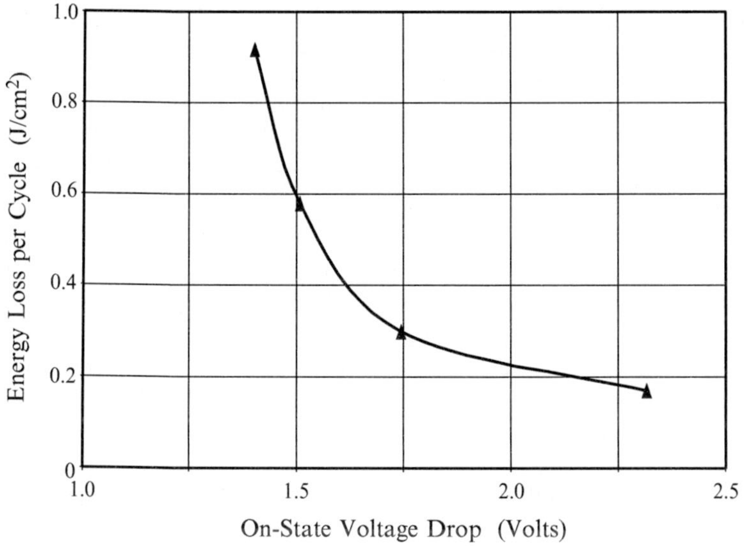

Fig. 4.33 Trade-off curve for the silicon 5-kV asymmetric GTO structure

Using the results obtained from the numerical simulations, the on-state voltage drop and the total energy loss per cycle can be computed. These values are plotted in Fig. 4.33 to create a trade-off curve to optimize the performance of the silicon 5-kV GTO structure by varying the lifetime in the N-base region. Devices used in lower frequency circuits would be chosen from the left-hand side of the trade-off curve, while devices used in higher frequency circuits would be chosen from the right-hand side of the trade-off curve.

4.2.7 Maximum Operating Frequency

An estimate of the maximum operating frequency for operation of the 5-kV GTO structure can be obtained by combining the on-state and switching power losses:

$$P_{D,TOTAL} = \delta P_{D,ON} + E_{OFF} f \tag{4.91}$$

where δ is the duty cycle and f is the switching frequency. In the case of the baseline device structure with a high-level lifetime of 10 μs in the N-base region, the on-state voltage drop is 1.74 V. For the case of a 50% duty cycle, the on-state power dissipation contributes 87 W/cm^2 to the total power loss. Using a total turn-off energy loss of 0.76 J/cm^2 in Eq. 4.91 yields a relatively low maximum operating frequency of about 150 Hz.

High-Level Lifetime (μs)	On-State Voltage Drop (Volts)	On-State Power Dissipation (W/cm^2)	Energy Loss per Cycle (J/cm^2)	Maximum Operating Frequency (Hz)
20	1.403	70.2	1.273	102
15	1.509	75.5	0.994	125
10	1.744	87.2	0.767	147
6	2.315	116	0.590	143

Fig. 4.34 Power loss analysis for the 5-kV asymmetric GTO structure

The maximum operating frequency for the silicon 5-kV GTO structure can be increased by reducing the lifetime in the drift region. Using the results obtained from the numerical simulations, the on-state voltage drop and the energy loss per cycle can be computed. These values are provided in Fig. 4.34 together with the maximum operating frequency as a function of the high-level lifetime in the drift region under the assumption of a 50% duty cycle and a total power dissipation limit of 200 W/cm^2. The maximum operating frequency is plotted in Fig. 4.35 as a function of the high-level lifetime in the drift region. It can be observed that the maximum operating frequency can be increased up to 150 Hz by reducing the high-level lifetime to 10 μs. A further reduction in the lifetime produces a reduction of the maximum operating frequency due to the increase in on-state voltage drop. It can be concluded that the 5-kV silicon GTO is therefore limited to operation at below 150 Hz.

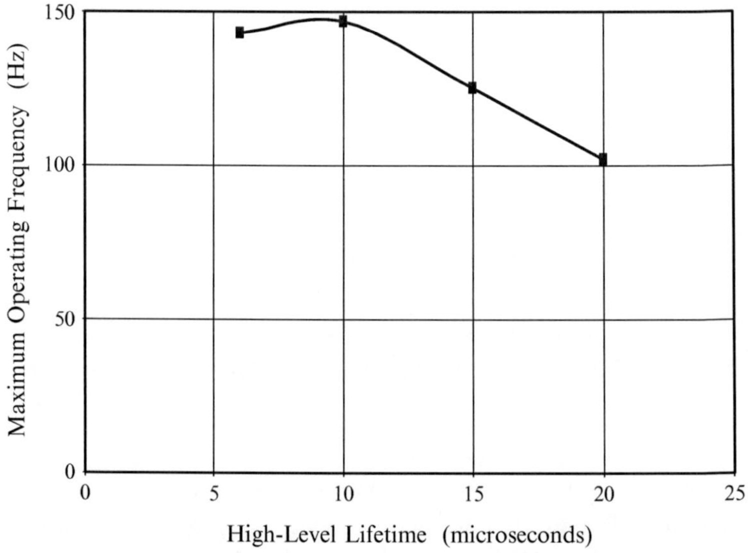

Fig. 4.35 Maximum operating frequency for the 5-kV asymmetric GTO structure

4.2.8 *Turn-Off Gain*

The peak reverse gate current occurs at the end of the voltage rise-time during turn-off of the GTO structure. Consequently:

$$J_{G,PR} = a(t_{S,R} + t_V) \tag{4.92}$$

The turn-off gain is defined as:

$$G_{OFF} = \frac{J_{A,ON}}{J_{G,PR}} \tag{4.93}$$

For the example shown in Figs. 4.21 and 4.22 with an anode supply voltage of 3,000 V and an on-state current density of 100 A/cm^2, the peak reverse gate current is found to be 99 A/cm^2, which is essentially equal to the on-state current density. The turn-off gain under these conditions is therefore close to unity. It can be concluded that the silicon 5-kV GTO requires very large gate drive currents which increases the cost of the gate drive circuit.

From the numerical simulations (see Fig. 4.32), it can be observed that the peak reverse gate drive current becomes smaller when the lifetime in the drift region is reduced. This is mainly due to a reduction of the storage time and a slight reduction in the voltage rise-time. The turn-off gain obtained using the data from the numerical simulations is plotted in Fig. 4.36. It can be observed that reducing the high-level lifetime in the drift region to 6 μs produces an increase in the turn-off gain to 1.4 which is still small. Consequently, the 5-kV silicon GTO structure requires a large reverse gate drive current. This has motivated the development of MOS-gated thyristor structures that are discussed in subsequent chapters.

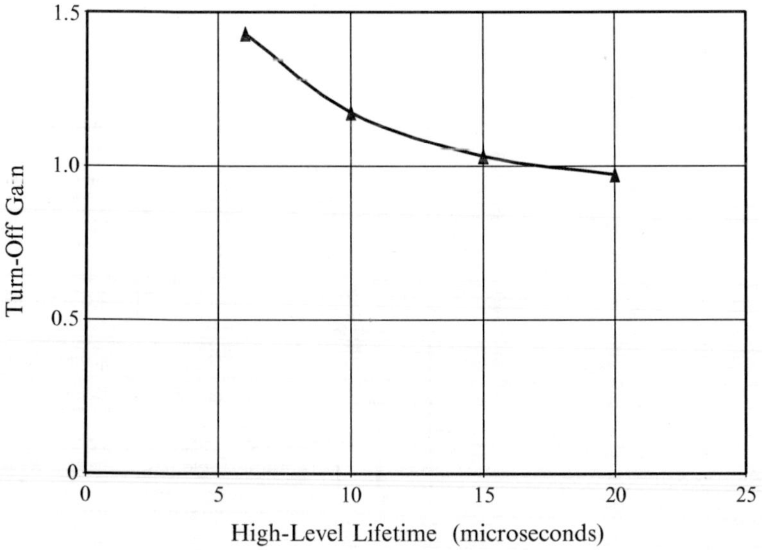

Fig. 4.36 Turn-off gain for the 5-kV asymmetric GTO structure

4.2.9 *Buffer Layer Doping*

As shown in the previous sections, the optimization of the power losses for the GTO structure requires performing a trade-off between the on-state voltage drop and the switching losses. One approach to achieving this is by reducing the lifetime in the drift region as discussed above. An alternative approach to achieve this is by adjusting the doping concentration in the buffer layer [7, 8]. Changes to the buffer layer doping level alter the injection efficiency of the anode junction and the lifetime in the buffer layer. The buffer layer doping concentration can be conveniently altered by changing the surface concentration of the buffer layer diffusion profile. The ability to adjust the device characteristics with the buffer layer doping is constrained at the lower end by the need to suppress reach-through of the depletion region in the forward blocking mode and at the higher end by reduction of the current gain of the P-N-P transistor to a level where regenerative action in the thyristor structure cannot be sustained. The model in Sect. 4.2.4.3 based upon low-level injection in the buffer layer predicts a dependence of the switching behavior on the buffer layer doping concentration.

Simulation Example

In order to gain insight into the impact of the buffer-layer doping concentration on the operation of the asymmetric GTO structure, the results of two-dimensional numerical simulations for a typical structure are discussed here. The device structure used has the cross section shown in Fig. 4.1 with a width of 100 μm. No cathode short was used in the structure. The widths of the P-base, N-base, and N-buffer layer regions are 20, 440, and 30 μm, respectively. The doping concentration of the buffer layer was varied by changing the surface concentration of the diffused region.

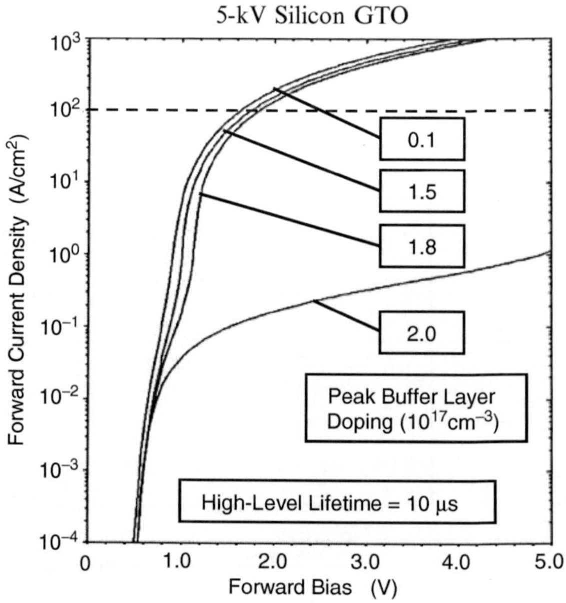

Fig. 4.37 On-state characteristics of the 5-kV asymmetric GTO structure: buffer layer doping

Numerical simulations were first performed to understand the impact of the buffer layer doping on the on-state characteristics while maintaining a high-level lifetime of 10 μs in the drift region. The gate drive current was kept at 0.02 A/cm^2 during these simulations. The resulting *i–v* characteristics are shown in Fig. 4.37. It can be observed that the on-state voltage drop is a weak function of the buffer-layer doping until the concentration is increased to 2×10^{17} cm^{-3}. At this peak buffer-layer doping concentration, the 5-kV silicon GTO structure no longer latches up resulting in a very high on-state voltage drop.

The impact of changing the buffer layer doping concentration on the injected hole concentration in the drift region is shown in Fig. 4.38. The hole concentration at the cathode side does not change with changes in buffer layer doping concentration. As the buffer layer doping increases, the hole concentration on the anode side reduces due to a reduction of the injection efficiency of the anode junction. This smaller hole concentration on the anode side reduces time for the current transient in the 5-kV silicon GTO structure.

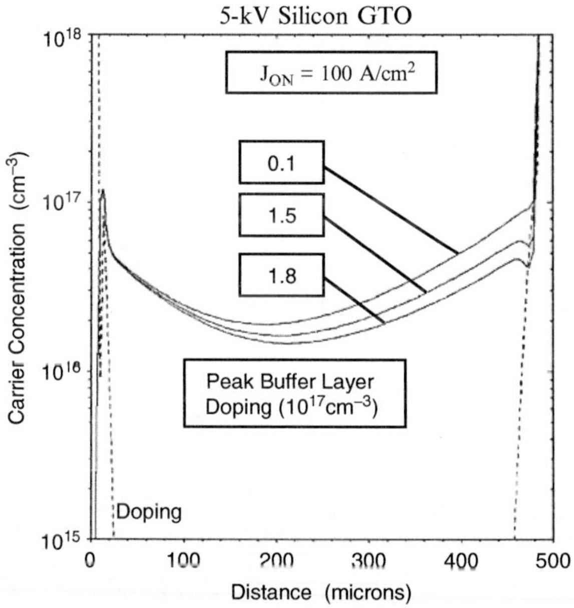

Fig. 4.38 On-state carrier distribution in the 5-kV asymmetric GTO structure as function of the buffer layer doping concentration

The numerical simulations of the turn-off for the 5-kV silicon GTO structure with various buffer layer doping concentrations were performed with a gate ramp rate of −10 A/cm^2-μs starting from an on-state current density of 100 A/cm^2. The resulting waveforms obtained from the numerical simulations for the anode voltage and current, as well as the gate current, are shown in Fig. 4.39 for the case of an anode supply voltage of 3,000 V. It can be observed that all the voltage waveforms begin to increase at the same time indicating that the storage time is independent of the buffer-layer doping concentration. This is consistent with the analytical model for the storage time (see Eq. 4.44). The anode voltage rise time for the various cases of buffer layer doping concentration is almost the same. This is consistent with the analytical model (see Eq. 4.55) for the voltage rise-time. The anode current waveforms exhibit a slightly larger initial abrupt reduction when the buffer layer doping concentration is increased. This is consistent with the analytical model (see Eq. 4.56) because the current gain of the P-N-P transistor reduces when the buffer layer doping concentration is increased because the injection efficiency of the anode junction is smaller.

The current fall-time for the 5-kV silicon GTO structure during the first phase becomes longer with reduced buffer layer doping due to an increase in the initial value ($J_{A,D}$) for the transient. The end point for this transient ($J_{A,PT}$) remains independent of the buffer layer doping concentration as predicted by the analytical model. Consequently, the energy loss during the turn-off event is a strong function of the buffer-layer doping concentration.

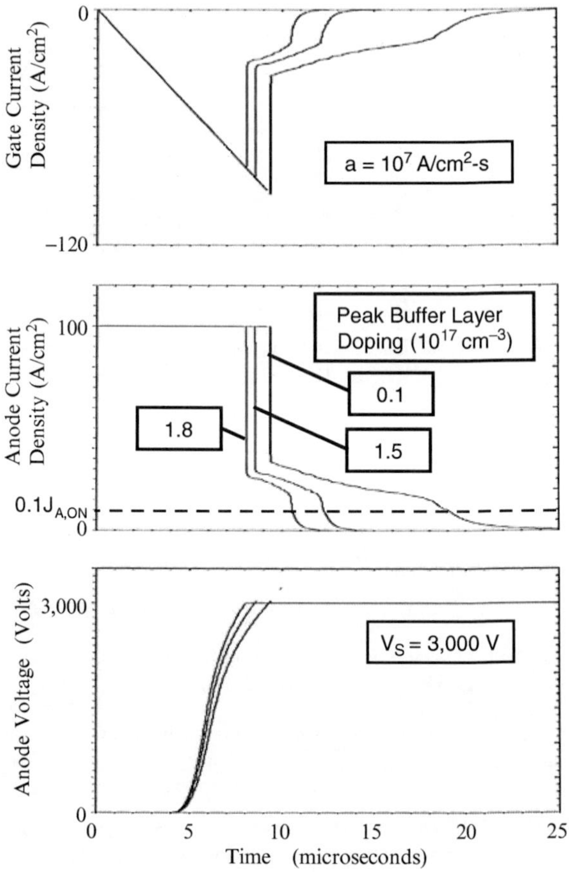

Fig. 4.39 Impact of buffer layer doping concentration on the 5-kV GTO turn-off waveforms

4.2.9.1 Switching Energy Loss

Using the results obtained from the numerical simulations for the silicon 5-kV GTO structures with different buffer layer doping concentrations, the on-state voltage drop and the total energy loss per cycle can be computed. These values are plotted in Fig. 4.40 to create a trade-off curve to optimize the performance of the silicon 5-kV GTO structure by varying the doping concentration in the buffer layer. Devices used in lower frequency circuits would be chosen from the left-hand side of the trade-off curve while devices used in higher frequency circuits would be chosen from the right-hand side of the trade-off curve. The trade-off curve obtained by varying the doping concentration in the buffer layer extends over a much narrower range than when the lifetime in N-base region is varied. It is difficult to precisely control the doping concentration in the buffer layer while the lifetime in the N-base region can be accurately modified by using electron irradiation.

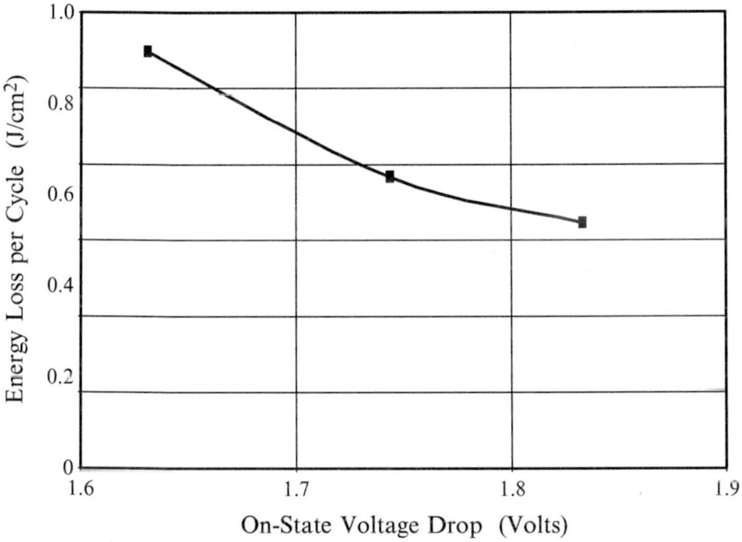

Fig. 4.40 Trade-off curve for the silicon 5 kV asymmetric GTO structure

4.2.9.2 Maximum Operating Frequency

Peak Buffer Layer Doping (cm^{-3})	On-State Voltage Drop (Volts)	On-State Power Dissipation (W/cm^2)	Energy Loss per Cycle (J/cm^2)	Maximum Operating Frequency (Hz)
1 x 10^{16}	1.631	81.5	1.096	108
1.5 x 10^{17}	1.744	87.2	0.767	147
1.8 x 10^{17}	1.833	91.7	0.648	167

Fig. 4.41 Power loss analysis for the 5-kV asymmetric GTO structure assuming high-level injection in the N-buffer layer

As discussed previously, the maximum operating frequency for the GTO structure is limited by the turn-off losses. The turn-off losses are associated with the voltage rise-time interval and the current fall-time interval. The energy loss for each event can be computed by using the equation previously provided in Sect. 4.2.6. Using this information, the maximum operating frequency for the GTO structure can be derived using Eq. 4.91.

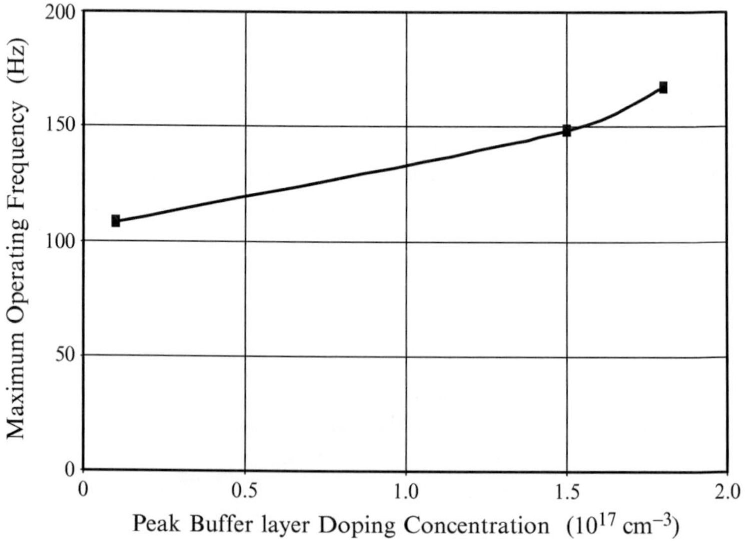

Fig. 4.42 Maximum operating frequency for the 5-kV asymmetric GTO structure

The maximum operating frequency for the silicon 5-kV GTO structure can be increased by increasing the doping concentration in the buffer layer. Using the results obtained from the numerical simulations, the on-state voltage drop and the energy loss per cycle can be computed. These values are provided in Fig. 4.41 together with the maximum operating frequency as a function of the peak doping concentration in the buffer layer under the assumption of a 50% duty cycle and a total power dissipation limit of 200 W/cm^2. The maximum operating frequency is plotted in Fig. 4.42 as a function of the doping concentration in the buffer layer. It can be observed that the maximum operating frequency can be increased only up to 170 Hz. It can be concluded that the 5-kV silicon GTO is therefore limited to operation to about 150 Hz.

4.2.9.3 Turn-Off Gain

From the numerical simulations (see Fig. 4.38), it can be observed that the peak reverse gate drive current becomes smaller when the buffer-layer doping concentration is increased. The turn-off gain obtained using the data from the numerical simulations is plotted in Fig. 4.43. It can be observed that increasing the peak doping concentration of the buffer layer to 1.8 × 10^{17} cm^{-3} produces an increase in the turn-off gain to 1.25 which is still small. Consequently, the 5-kV silicon GTO structure requires a large reverse gate drive current. This has motivated the development of MOS-gated thyristor structures that are discussed in subsequent chapters.

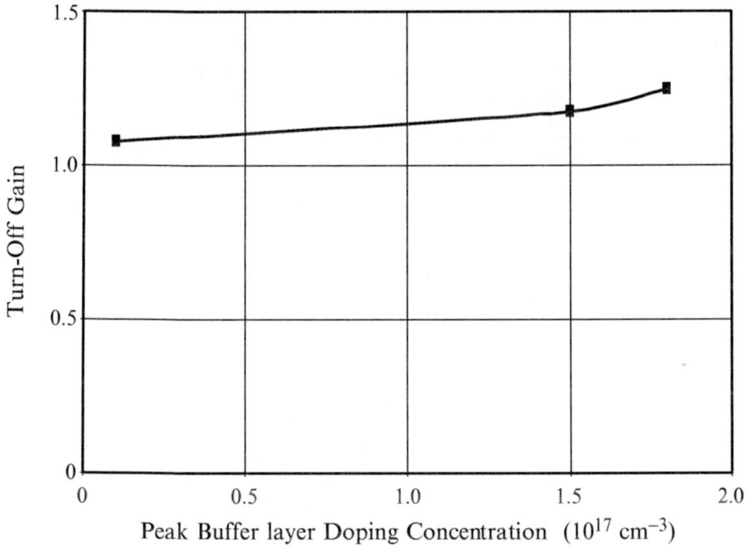

Fig. 4.43 Turn-off gain for the 5-kV asymmetric GTO structure

4.2.10 Transparent Emitter Structure

The optimization of the power losses for the silicon GTO structure can be achieved by performing a trade-off between the on-state voltage drop and the switching losses. In addition to optimization of the lifetime in the drift region and the doping concentration in the buffer layer, this trade-off can be achieved by using a lightly doped anode region with small thickness [9]. This is known as a *transparent emitter structure*. Reducing the doping concentration of the anode region produces a decrease in the injection efficiency of the anode junction. This produces a reduction in the injected hole concentration at the anode side within the N-base region which improves the turn-off time. In addition, holes are injected into the anode region during the turn-off phase, diffuse through its small thickness, and then recombine at the metal contact. This additional process for removal of the holes from the N-base region speeds up the turn-off process during the anode current tail. A high lifetime value is usually utilized in the N-base region for the transparent emitter structure because the injected hole concentration has been reduced using the smaller anode injection efficiency.

Simulation Example

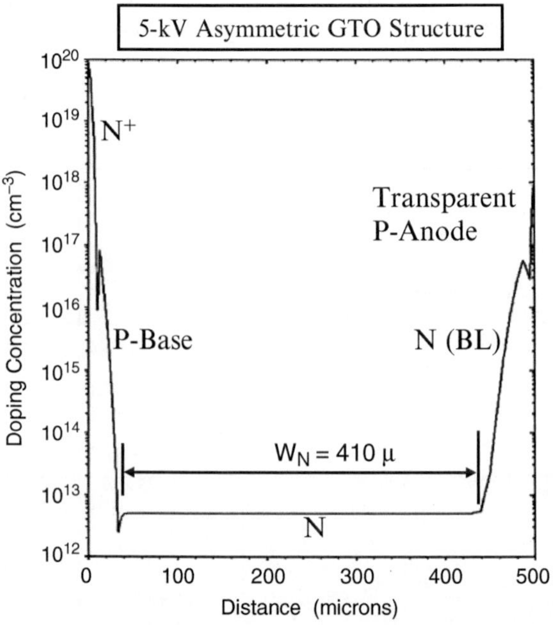

Fig. 4.44 Doping profile for the asymmetric 5-kV GTO with transparent anode region

In order to gain insight into the operation of the asymmetric GTO structure with transparent emitter structure, the results of two-dimensional numerical simulations for a device with anode surface concentration of 1×10^{18} cm^{-3} and thickness of 5 μm are discussed here. The device structure used has the cross section shown in Fig. 4.1 with a width of 100 μm. No cathode short was used in the structure. The widths of the P-base, N-base, and N-buffer layer regions are 20, 440, and 30 μm, respectively. The surface concentration of the buffer-layer diffusion was reduced so that the peak buffer-layer doping concentration was close to the baseline device structure. The doping profile for the silicon 5-kV GTO structure with the transparent emitter region is shown in Fig. 4.44.

Numerical simulations were first performed to understand the impact of the transparent emitter region on the on-state characteristics. The resulting on-state *i–v* characteristics are shown in Fig. 4.45. A high-level lifetime of 100 μs was used in the N-base region for this device. For comparison purposes, the on-state characteristics of the conventional silicon 5-kV GTO structure with highly doped anode region were also obtained for a large high-level lifetime of 100 μs. In addition, the on-state characteristics for baseline conventional silicon 5-kV GTO structure with a lower high-level lifetime of 10 μs were obtained. These device characteristics are also included in Fig. 4.45 for comparison purposes. The gate drive current for all the structures was kept at 0.02 A/cm^2 during these simulations. It can be observed that the on-state voltage drop for the device structure with the transparent emitter region falls between those of the conventional structure with high-level lifetime of 100 and 10 μs. This is due to an alteration of the hole concentration profile in the N-base region.

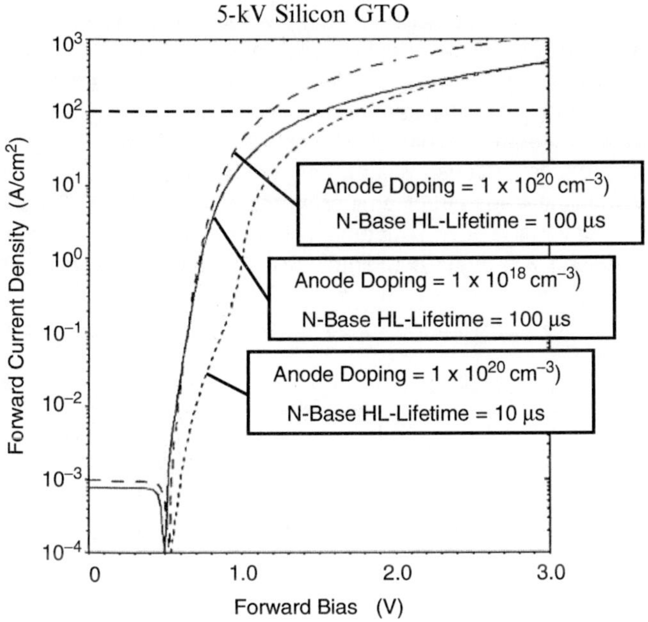

Fig. 4.45 On-state characteristics of the 5-kV asymmetric GTO structure: transparent emitter structure

The injected hole concentration profile in the drift region is shown in Fig. 4.46 for the three silicon 5-kV GTO structures. The hole concentration is the largest throughout the N-base region for the conventional structure with the large high-level lifetime of 100 μs. When the high-level lifetime is reduced to 10 μs in the conventional silicon GTO structure, the hole concentration becomes smaller throughout the N-base region. In contrast, the hole concentration only reduces at the anode side for the transparent emitter structure. The reduced stored charge in the N-base region toward the anode side is beneficial in speeding up the turn-off process.

Numerical simulations of the turn-off for the 5-kV silicon GTO structure with the transparent emitter structure were performed with a gate ramp rate of −10 A/cm^2-μs starting from an on-state current density of 100 A/cm^2. The resulting waveforms obtained from the numerical simulations for the anode voltage and current, as well as the gate current, are shown in Fig. 4.47 for the case of an anode supply voltage of 3,000 V. For comparison purposes, the waveforms for the baseline conventional silicon 5-kV GTO with a high-level lifetime of 10 μs are included in the figure. It can be observed that the storage time for the transparent emitter GTO structure is longer due to the larger lifetime in the N-base region. The anode voltage rise time for the device with the transparent emitter is slightly shorter (0.5 μs) than that of the baseline conventional device structure. The current fall-time for the 5-kV silicon GTO structure with the transparent emitter region during the first phase becomes shorter despite an increase in the initial value ($J_{A,D}$) for the transient while the end point for this transient ($J_{A,PT}$) remains independent of the buffer layer doping concentration as predicted by the analytical model. This occurs because the decay in the current is accelerated by the injection of electrons into the transparent anode region followed by their rapid recombination at the anode contact. The anode current turn-off time is therefore substantially reduced. Consequently, the energy loss per cycle (E_{OFF1}) for the device with the transparent emitter structure is much smaller than for the baseline conventional device structure. This demonstrates that the transparent emitter structure has a better combination of on-state voltage drop and turn-off energy loss.

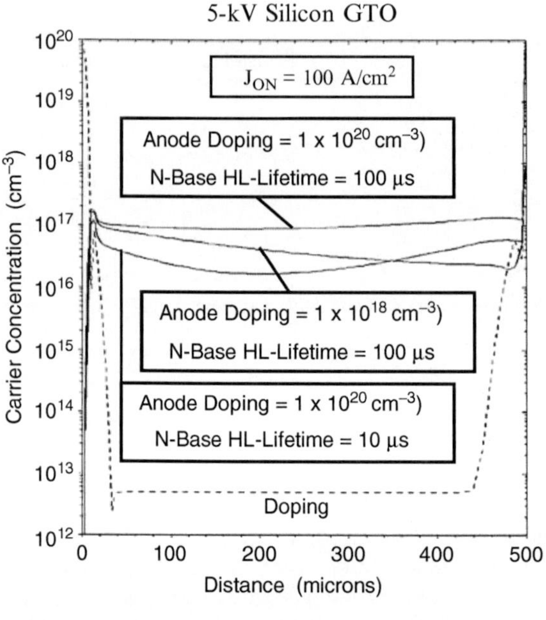

Fig. 4.46 On-state carrier distribution in the 5-kV asymmetric GTO structure: transparent emitter structure

The changes in the carrier distribution within the asymmetric GTO structure with the transparent anode region during the storage time and the voltage rise-time are similar to those for the conventional device structure. However, this does not hold true during the current fall-time. A one-dimensional view of the hole carrier distribution at the center of the cathode in the GTO structure with the transparent anode region is shown in Fig. 4.48 during the current tail time. In this figure, the initial carrier distribution is included as a reference. The anode voltage is held constant at the anode supply voltage of 3,000 V during this transient. When the anode current reduces abruptly between $t = 10.0$ and $t = 10.2$ μs, the hole concentration decreases in the space-charge region but is not altered within the rest of the N-base region and N-buffer layer. This is consistent with the assumption used in the analytical model that the hole concentration within the stored charge in the N-base region at the beginning of the anode current decay phase is equal to the initial on-state carrier concentration. Subsequently, the hole concentration in the stored charge region decreases due to the recombination process and the removal of carriers via the injection into the transparent anode region due to its low doping concentration and small thickness. At the same time, the space-charge region expands in spite of a constant anode voltage because the hole concentration in the space-charge region is smaller.

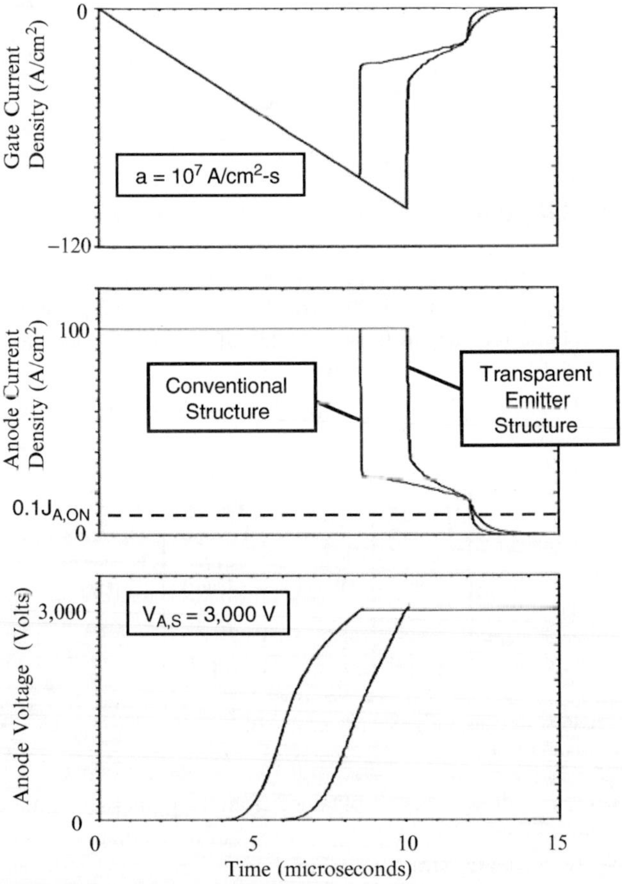

Fig. 4.47 Silicon 5-kV GTO turn-off waveforms: transparent emitter structure

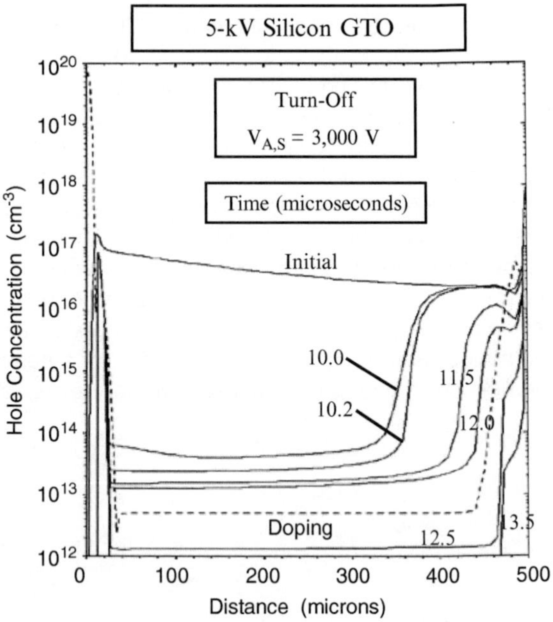

Fig. 4.48 Hole carrier distribution during the current tail-time for the 5-kV GTO with transparent emitter region

4.2.10.1 Switching Energy Loss

As discussed previously, the maximum operating frequency for the GTO structure is limited by the turn-off losses. The turn-off losses are associated with the voltage rise-time interval and the current fall-time interval. The energy loss for each event can be computed by using the equation previously provided in Sect. 4.2.6. Using this information, the maximum operating frequency for the GTO structure can be derived using Eq. 4.91.

Structure	On-State Voltage Drop (Volts)	On-State Power Dissipation (W/cm^2)	Energy Loss per Cycle (J/cm^2)	Maximum Operating Frequency (Hz)
Transparent Emitter	1.513	75.7	0.69	180
Conventional	1.744	87.2	0.767	147

Fig. 4.49 Power loss analysis for the 5-kV asymmertric GTO structure assuming high-level injection in the N-buffer layer

The turn-off energy loss per cycle obtained from the numerical simulations of the transparent emitter silicon 5-kV GTO structure can be compared with that for the baseline conventional device structure in Fig. 4.49. It can be observed that the slightly shorted voltage rise-time and the significantly shorter current fall-time results in smaller energy loss per cycle for the transparent emitter structure.

4.2.10.2 Maximum Operating Frequency

The maximum operating frequency for the silicon 5-kV GTO structure with the transparent emitter structure is larger than that for the conventional GTO structure due to its smaller on-state voltage drop and switching energy loss per cycle. The maximum operating frequency obtained under the assumption of a 50% duty cycle and a total power dissipation limit of 200 W/cm^2 for the transparent emitter structure is 180 Hz when compared with 147 Hz for the baseline conventional device structure.

4.2.10.3 Turn-Off Gain

From the numerical simulations (see Fig. 4.47), it can be observed that the peak reverse gate drive current is larger for the transparent emitter structure when compared with the baseline conventional silicon 5-kV GTO structure. This is due to the larger storage time for the transparent emitter structure. The turn-off gain obtained using the data from the numerical simulations is close to unity for the transparent emitter silicon 5-kV GTO structure.

4.3 10,000-V Silicon GTO

The 10-kV silicon GTO structure can be expected to function just like the 5-kV device. However, its design and operation is constrained by the larger blocking voltage capability. As discussed below, the very low doping concentrations required for the N-base region are challenging to achieve from a fabrication stand point. The lifetime in the N-base region for the 10-kV device must be larger to maintain a reasonable on-state voltage drop and regenerative action for latching-up the thyristor structure. The larger N-base width results in more stored charge within the structure which limits the switching frequency and the maximum turn-off current density. These parameters are discussed in this section.

4.3.1 Blocking Characteristics

The theory for the forward blocking capability of thyristors presented in Sect. 4.2.1 for the 5-kV GTO can also be used for the 10-kV silicon asymmetric GTO device structure. The blocking voltage capability for the silicon asymmetric GTO structure can be increased by reducing the doping concentration of the N-base region and increasing its thickness. The asymmetric silicon GTO structure must have a forward blocking voltage of 11,000 V for a 10-kV rated device. In the case of avalanche breakdown, there is a unique value of 6.45×10^{12} cm^{-3} for the drift region to obtain this blocking voltage. However, in the case of the asymmetric GTO structure, it is advantageous to use a much lower doping concentration for the lightly doped portion of the N-base region in order to reduce its width. The strong conductivity modulation of the N-base region during on-state operation favors a smaller thickness for the N-base region independent of its original doping concentration. A doping concentration of 2×10^{12} cm^{-3} will be assumed for the N-base region.

The doping concentration of the N-buffer layer must be sufficiently large to prevent reach-through of the electric field to the P^+ collector region. Doping concentrations above 1×10^{16} cm^{-3} are sufficient to accomplish this goal. For the baseline device structure, an N-buffer layer doping concentration of 1.2×10^{17} cm^{-3} will be assumed. In this case, the emitter injection efficiency computed using Eq. 4.6 is 0.741. When the device is close to breakdown, the entire N-base region is depleted and the base transport factor computed by using Eq. 4.10 in this case is 0.779. In computing these values, a lifetime of 10 μs was assumed for the N-base region resulting in a lifetime of 2.9 μs in the N-buffer layer due to the scaling according to Eq. 4.11. Based upon Eq. 4.7, open-base transistor breakdown will then occur when the multiplication coefficient becomes equal to 1.73 for the above values for the injection efficiency and base transport factor.

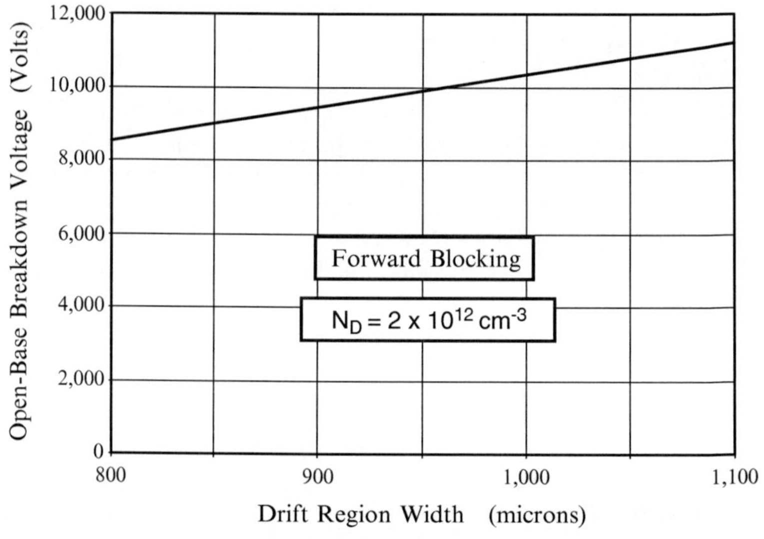

Fig. 4.50 Open-base breakdown voltage for the asymmetric GTO structure in the forward blocking mode

The forward blocking capability for the silicon asymmetric GTO structure can be computed by using Eq. 4.18 for various widths for the N-base region. The analysis requires determination of the voltage V_{NPT} by using Eq. 4.17 for each width of the N-base region. The resulting values for the forward blocking voltage are plotted in Fig. 4.50. From this graph, the N-base region width required to obtain a forward blocking voltage of 11.000-V is 1,100 μm. This width is reduced when taking into account the voltage supported within the P-base region due to its graded doping profile.

Simulation Example

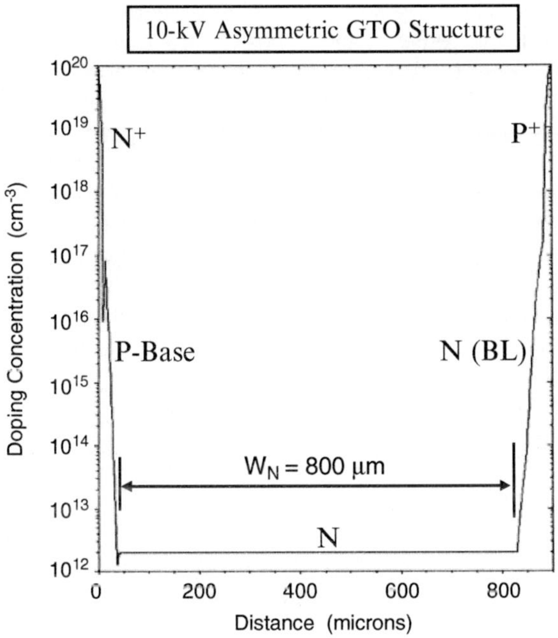

Fig. 4.51 Doping profile for the simulated baseline asymmetric 10-kV GTO structure

In order to gain insight into the physics of operation for the 10-kV asymmetric GTO structure under voltage blocking conditions, the results of two-dimensional numerical simulations are described here for a device with N-base width of 800 μm. The simulations were performed using a cell with the structure shown in Fig. 4.1. This cell has a width of 100 μm (area $= 1.0 \times 10^{-6}$ cm^{-2}) with a cathode width (W_{KS}) of 180 μm. The asymmetric GTO structure used for the simulations was formed by diffusions performed into a uniformly doped N-type drift region with a doping concentration of 2×10^{12} cm^{-3}. A lifetime (τ_{p0}, τ_{n0}) of 10 μs was used for the baseline device. The N-buffer layer was formed by diffusion from the anode side with a depth of 55 μm. For the baseline device structure, the surface concentration of the N-type diffusion was adjusted to achieve a peak doping concentration of 1.2×10^{17} cm^{-3} in the buffer layer. The P-base region was formed with a Gaussian doping profile with a surface concentration of 5×10^{17} cm^{-3} and a depth of 30 μm. The N^{+} cathode region was formed with a Gaussian doping profile with a surface concentration of 1×10^{20} cm^{-3} and a depth of 10 μm. The doping profile in the vertical direction through the N^{+} cathode region is shown in Fig. 4.51 indicating the net width of the lightly doped portion of the N-base region is 800 μm after accounting for the diffusions. The peak doping concentration of the P-base region is 8×10^{16} cm^{-3} and its thickness is 23 μm. The peak doping concentration of the N-buffer layer is 1.2×10^{17} cm^{-3} and its thickness is 50 μm.

The forward blocking capability of the 10-kV silicon GTO structure was obtained by increasing the anode bias while maintaining the gate electrode at zero volts. The characteristics obtained for three lifetime (τ_{p0}) values in the N-base region are provided in Fig. 4.52. In all cases, the leakage current increases rapidly with increasing anode bias voltage until about 1,000 V. This occurs due to the increase in the space-charge generation volume and the increase in the current gain (α_{PNP}) of the open base P-N-P transistor until the anode bias becomes equal to the reach-through voltage of 990 V obtained using the analytical solution given by Eq. 4.2. The leakage current then becomes independent of the collector voltage until close to the breakdown voltage. This behavior is well described by the analytical model. The leakage current becomes larger when the lifetime is reduced consistent with the analytical model but the breakdown voltage changes by only a small amount. A breakdown voltage of 10,500 V is possible with an N-base width of only 800 μm, which is smaller than that predicted by the analytical model (see Fig. 4.50), due to some of the applied anode voltage being supported in the depletion regions formed within the P-base and N-buffer layers.

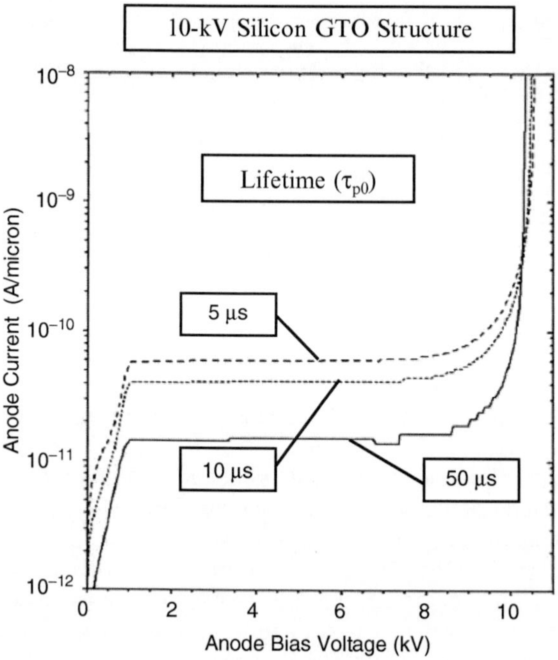

Fig. 4.52 Forward blocking characteristics for the 10-kV asymmetric GTO structure

The voltage is primarily supported within the lightly doped portion of N-base region in the 10-kV asymmetric GTO structure during operation in the forward blocking mode. This is illustrated in Fig. 4.53 where the electric field profiles are shown during operation in the forward blocking mode at several collector voltages. It can be observed that the P-base/N-base junction (J_2) becomes reverse-biased during the forward blocking mode with the depletion region extending toward the right-hand side with increasing (positive) collector bias. The electric field has a triangular shape until the entire lightly doped portion of the N-base region becomes completely depleted. This occurs at a collector bias of 1,000 V in good agreement with the value obtained using the analytical solution (see Eq. 4.2). The electric field profile then takes a trapezoidal shape due to the high doping concentration in the N-buffer layer.

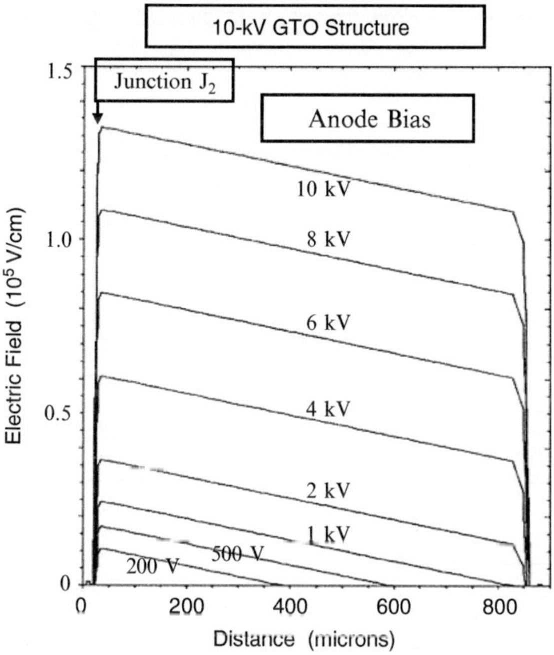

Fig. 4.53 Electric field profiles in the forward blocking mode for the 10-kV asymmetric GTO structure

4.3.2 On-State Voltage Drop

The on-state *i–v* characteristics and on-state voltage drop can be computed using the analytical model discussed in Sect. 4.2.3. It is important to take end-region recombination into account during this analysis. In general, a larger lifetime is required in the N-base region for the 10-kV device when compared with the 5-kV device due to the larger width for the N-base region.

Simulation Results

The results of two-dimensional numerical simulations for the 10-kV asymmetrical silicon GTO structure are described here. The total width of the structure, as shown by the cross section in Fig. 4.1, was 100 μm (area = 1 × 10^{-6} cm^{-2}) with a cathode finger width (W_{KS}) of 180 μm. The doping profile for the device structure was already shown in Fig. 4.51.

The on-state characteristics of the 10-kV silicon asymmetric GTO structure were obtained by using a gate drive current of 2×10^{-8} A/μm using various values for the lifetime in the drift region. The characteristics obtained from the numerical simulations are shown in Fig. 4.54. It can be observed that the on-state voltage drop increases as expected with reduction of the lifetime (τ_{p0}, τ_{n0}) indicated in the figure. Due to the relatively higher on-state voltage drop than for the 5-kV device structure, the on-state current density for the 10-kV silicon GTO must be reduced to 50 A/cm^2 as indicated in the figure by the bold dashed line. A snap-back is observed in the *i–v* characteristics when the lifetime is reduced to 5 μs indicating that this is a lower limit to the lifetime. The on-state voltage drop at a high-level lifetime value of 10 μs is found to be 2.13 V at an on-state current density of 50 A/cm^2.

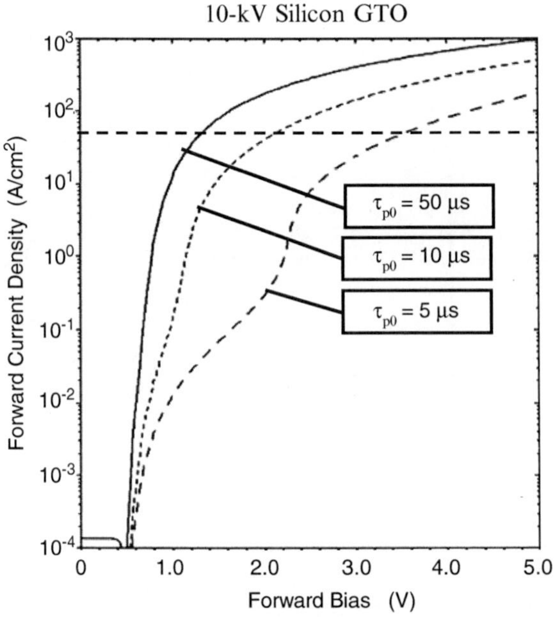

Fig. 4.54 On-state characteristics of the 10-kV asymmetric GTO structure

The good on-state voltage drop for the 10-kV asymmetric GTO structure is due to the large number of carriers injected into the drift region producing a drastic reduction of its resistance. This is illustrated in Fig. 4.55 where the injected carrier density is shown for three cases of the lifetime (τ_{p0}, τ_{n0}) in the drift region of the GTO structure. It can be observed that the injected carrier density is more than three orders of magnitude larger than the doping concentration even for the case of a lifetime of 10 μs. The injected carrier density is reduced by a factor of three times in the middle of the drift region when the lifetime is reduced to 5 μs. The injected carrier density is nearly uniform in the N-base region only when the lifetime is very large (50 μs). There is a significant drop in the injected carrier density in the middle of the drift region when the lifetime is reduced to 10 or 5 μs. This is due to the relatively large width for the N-base region when compared with the 5-kV silicon GTO structure. The buffer layer operates neither at high-level nor low-level injection conditions for all lifetime cases.

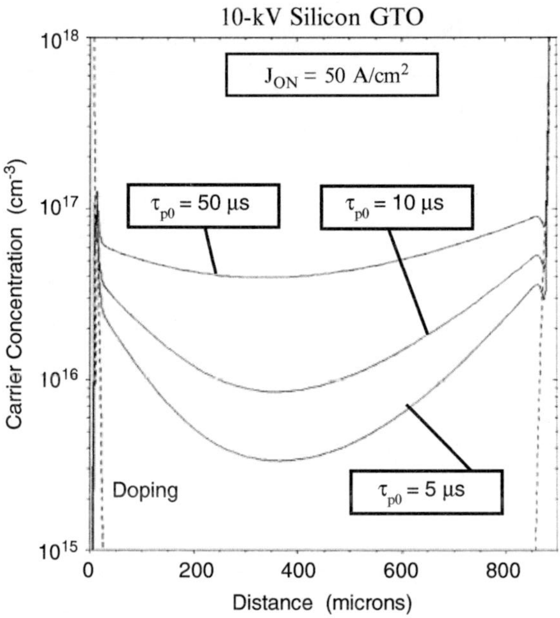

Fig. 4.55 On-state carrier distribution in the 10-kV asymmetric GTO structure

4.3.3 *Turn-Off Characteristics*

The physics for turn-off of the 10-kV silicon asymmetric GTO structure can be expected to be the same as that for the 5-kV device structure. However, it is more difficult to turn off the 10-kV device structure due to the larger amount of stored charge in the N-base region. Consequently, the 10-kV device structure must be operated at a lower on-state current density not only due to its larger on-state voltage drop but from a switching point of view.

Simulation Results

Numerical simulations of the turn-off for the 10-kV silicon GTO structure with a high-level lifetime of 20 μs were performed with a gate ramp rate of −6.25 A/cm^2-μs starting from an on-state current density of 50 A/cm^2. The resulting waveforms obtained from the numerical simulations for the anode voltage and current, as well as the gate current, are shown in Fig. 4.56 for the case of an anode supply voltage of 5,000 V. The 10-kV GTO structure was found to be unable to turn-off at larger anode supply voltages. It can be observed that the storage time for the 10-kV GTO structure is longer than for the 5-kV device structure. The anode voltage rise time for the 10-kV device structure is slightly longer (0.5 μs) than that of the 5-kV baseline device structure. However, the anode current turn-off time is substantially larger due to the longer lifetime required for the 10-kV device structure.

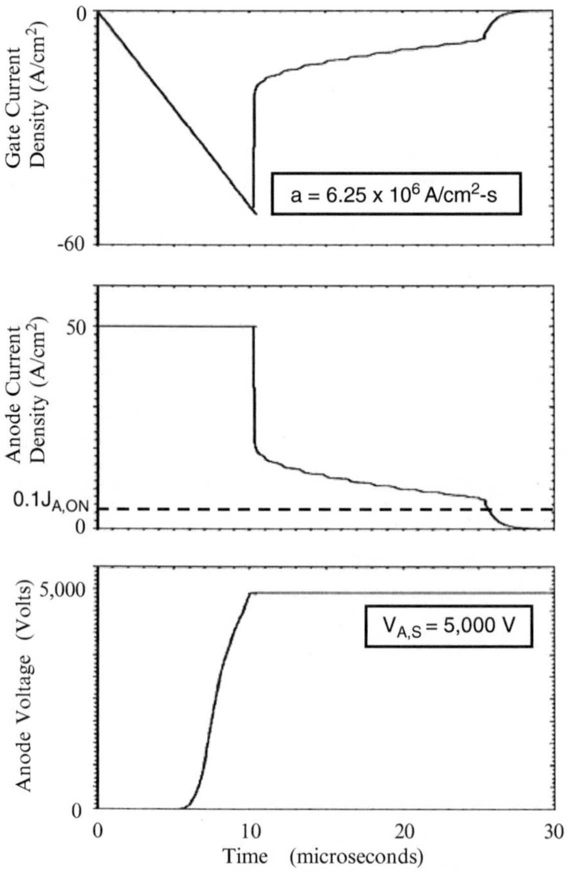

Fig. 4.56 Silicon 10-kV GTO turn-off waveforms

The time at which the space-charge region punches-through to the N-buffer layer (t_{PT}) is the time taken for the anode current to decay from $J_{A,D}$ to $J_{A,PT}$. The value of $J_{A,D}$ and J_{APT} for the 10-kV silicon GTO structure are 22 and 7 A/cm2, resulting in a value of 15.2 μs for t_{PT}. The current fall-time (t_I) for the 10-kV silicon GTO structure is essentially equal to the punch-through time. Since the current fall-time for the 10-kV silicon GTO is much larger than for the 5-kV silicon GTO structure, the energy loss during the turn-off event is very large severely limiting its operating frequency.

4.3.4 *Switching Energy Loss*

As discussed previously, the maximum operating frequency for the GTO structure is limited by the turn-off losses. The turn-off losses are associated with the voltage rise-time interval and the current fall-time interval. The energy loss for each event can be computed by using the equation previously provided in Sect. 4.2.6. Using this information, the maximum operating frequency for the 10-kV silicon asymmetric GTO structure can be derived using Eq. 4.91. The turn-off energy loss per cycle obtained from the numerical simulations of the silicon 10-kV GTO structure can be derived from the waveforms in Fig. 4.56. The energy loss per cycle during the voltage rise-time is 0.50 J/cm^2 while the energy loss per cycle during the current fall time is 0.859 J/cm^2 leading to a total energy loss per cycle of 1.359 J/cm^2 for the 10-kV silicon GTO structure.

4.3.5 *Maximum Operating Frequency*

The maximum operating frequency for the silicon 10-kV GTO structure with the high-level lifetime of 20 μs in the N-base region is smaller than for the baseline 5-kV silicon GTO structure due to its larger on-state voltage drop and switching energy loss per cycle. The maximum operating frequency obtained under the assumption of a 50% duty cycle and a total power dissipation limit of 200 W/cm^2 for the 10-kV silicon GTO structure is 108 Hz when compared with 147 Hz for the 5-kV GTO device structure.

4.3.6 *Turn-Off Gain*

From the numerical simulations (see Fig. 4.56), it can be observed that the peak reverse gate drive current for the silicon 10-kV GTO structure is about 50 A/cm^2. The turn-off gain obtained using the data from the numerical simulations is therefore close to unity for the silicon 10-kV GTO structure.

4.4 Reverse-Biased Safe Operating Area

The ability to turn off the anode current at large anode supply voltages is limited by the onset of avalanche multiplication in the space-charge region due to the high electric field. During turn-off, the electric field in the space-charge region in the GTO structure devices becomes larger than in the blocking mode because of the presence of a high concentration of holes that sustain the anode current flow. The hole concentration in the space-charge region during the voltage transient is related to the on-state anode current density ($J_{A,ON}$) by:

$$p_{SC} = \frac{J_{A,ON}}{q v_{sat,p}} \tag{4.94}$$

where $v_{sat,p}$ is the saturated velocity for holes. The positive charge of the holes adds to the positive charge from the donors in the drift region.

The solution of Poisson's equation taking into account the positive charge due to the holes and the donors in the space-charge region yields an equation for the reverse-biased safe operating area:

$$BV_{RBSOA} = 4.45 \times 10^{13}(N_D + p_{SC})^{-3/4} \tag{4.95}$$

Using Eq. 4.94 for the hole concentration yields:

$$BV_{RBSOA} = 4.45 \times 10^{13}\left(N_D + \frac{J_{A,ON}}{q v_{sat,p}}\right)^{-3/4} \tag{4.96}$$

This expression indicates a reduction in the maximum anode on-state current density with increasing anode supply voltage.

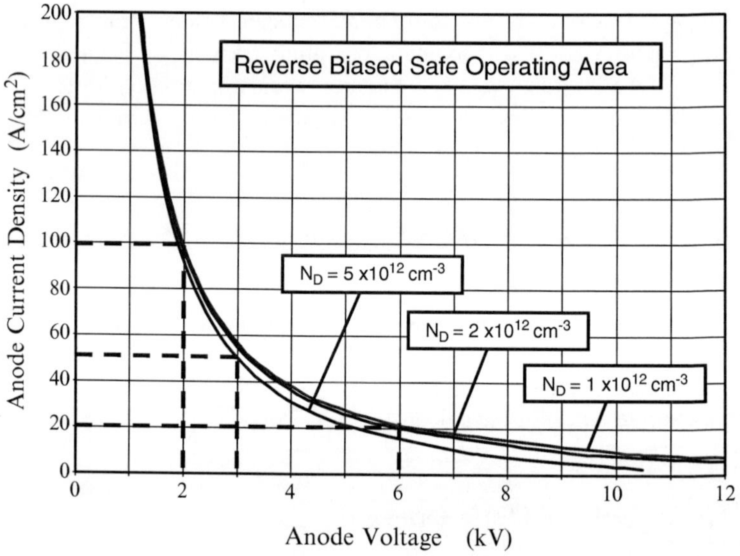

Fig. 4.57 Reverse-biased safe operating area for the GTO structure

The reverse-biased safe operating area (RBSOA) for the silicon GTO structure computed by using the above analysis is provided in Fig. 4.57 for the case of three doping concentrations in the N-drift region. At high anode on-state current densities, the hole concentration in the drift region becomes much larger than

1×10^{13} cm^{-3}. Consequently, the RBSOA boundary becomes independent of the drift region doping concentration. At low anode on-state current densities, the hole concentration in the drift region becomes less than 1×10^{13} cm^{-3} and comparable to the doping concentration in the drift region. Consequently, the RBSOA boundary can be slightly enlarged by reducing the doping concentration in the drift region at low on-state current densities.

Three specific cases for the anode supply voltage are highlighted in the figure. For the case of an anode supply voltage of 2 kV, it is possible to operate the silicon GTO structure up to an anode on-state current density of 100 A/cm^2. For the case of an anode supply voltage of 3 kV, it is possible to operate the silicon GTO structure up to an anode on-state current density of 50 A/cm^2. For the case of an anode supply voltage of 6 kV, it is possible to operate the silicon GTO structure up to an anode on-state current density of 20 A/cm^2. These observations are consistent with the numerical simulations discussed earlier in the chapter.

4.5 Conclusions

The physics of operation and design principles for the silicon GTO structure have been elucidated in this chapter. This device structure was used for high power motor control in traction (electric locomotive) drives from the 1980s until recently. The analysis provided in this chapter demonstrates that the maximum operating frequency of the silicon GTO is limited to 150 Hz and that the device turn-off occurs with approximately unity current gain. Consequently, the large gate drive currents required to operate the device make the drive circuit expensive motivating the replacement of the device with the silicon IGBT structure. In addition, this has motivated the development of alternate MOS-gated thyristor structures that are discussed in this book.

References

1. F.F. Mazda, "Power Electronics Handbook", Chapter 11, pp. 227–245, Butterworths Publishers, 1993.
2. B.J. Baliga, "Fundamentals of Power Semiconductor Devices", Springer-Science, New York, 2008.
3. M. Azuma and M. Kurata, "GTO Thyristors", Proceeding of the IEEE, Vol. 76, pp. 419–427, 1988.
4. B.J. Baliga and M.S. Adler, "Measurement of Carrier Lifetime Profiles in Diffused Layers of Semiconductors", IEEE Transactions on Electron Devices, Vol. ED-25, pp. 472–477, 1978.
5. D.J. Roulston, N.D. Arora, and S.G. Chamberlain, "Modeling and Measurement of Minority Carrier Lifetime versus Doping in Diffused Layers", IEEE Transactions on Electron Devices, Vol. ED–29, pp. 284-291, 1982.
6. Medici User's Manual, Avant! Corporation, Fremont, CA.

7. T. Ogura, et al, "6000-V Gate Turn-Off Thyristors (GTOs) with n-Buffer and New Anode Short Structure", IEEE Transactions on Electron Devices, Vol. ED-38, pp. 1491–1495, 1991.
8. K. Satoh, et al, "6kV/4kA Gate Commutated Turn-Off Thyristor with Operation DC Voltage of 3.6kV", IEEE International Symposium on Power Semiconductor Devices and ICs, Abstract 11.01, pp. 205–208, 1998.
9. S. Eicher, et al, "Design Considerations for a 7 kV/3kA GTO with Transparent Anode and Buffer Layer", IEEE Power Electronics Specialists Conference, Vol. 1, pp. 29–34, 1996.

Chapter 5
Silicon IGBT (Insulated Gate Bipolar Transistor)

The silicon IGBT is arguably the most successful innovation in power semiconductor devices during the past three-decades. By using a combination of bipolar current flow controlled using an MOS-gate structure, the power gain was increased a million fold when compared with existing power bipolar junction transistors and power MOSFET structures with high blocking voltages [1]. The widespread applications for the device in consumer, industrial, transportation, lighting, and even medical applications is a testimonial to its highly desirable characteristics. The IGBT offers a unique combination of ease of control due to its MOS-gate structure, low chip cost due to its relatively high on-state current density, and exception ruggedness. Silicon IGBT modules are now available with blocking voltage capability up to 6.5 kV and current handling capability of 1,000 A. Any new high voltage power device technology must offer significant improvements in performance relative to the silicon IGBT to be considered attractive for applications.

The basic operating principles and characteristics for the silicon IGBT have been described in detail in the textbook [2]. Three types of IGBT devices, namely the symmetric, asymmetric, and transparent emitter structures, were analyzed in detail. The most commonly utilized device is the asymmetric structure because the applications require operation of the device only in the first quadrant. In the case of devices designed for lower blocking voltages (<1,200 V), the buffer layer of the asymmetric structure is uniformly doped using an epitaxial growth process. In contrast, in the case of devices designed for high blocking voltages (>2,000 V), the buffer layer of the asymmetric structure is non-uniformly doped using a diffusion process. In this chapter, the performance of the high voltage (5 and 10 kV) silicon IGBT structure is discussed to serve as a benchmark for purposes of comparison with the other devices covered in the book. Consequently, only the asymmetric device structure is analyzed here with a buffer layer formed by diffusion from the collector side of the device structure.

Historically, the first IGBT structures were fabricated using a planar gate structure with the DMOS process [3]. In order to reduce the on-state voltage drop of the IGBT, the trench gate structure was subsequently introduced [4]. The voltage rating for the trench gate IGBT structure was then scaled upward [5, 6] due to

B.J. Baliga, *Advanced High Voltage Power Device Concepts*,
DOI 10.1007/978-1-4614-0269-5_5, © Springer Science+Business Media, LLC 2011

interest in applications for driving motors in traction (electric street-cars and trains) applications. The forward blocking capability of IGBTs has also been increased to 6.5 kV by using the field stop concept [7, 8], which is basically an asymmetric structure with a diffused buffer layer on the collector side.

5.1 Basic Structure and Operation

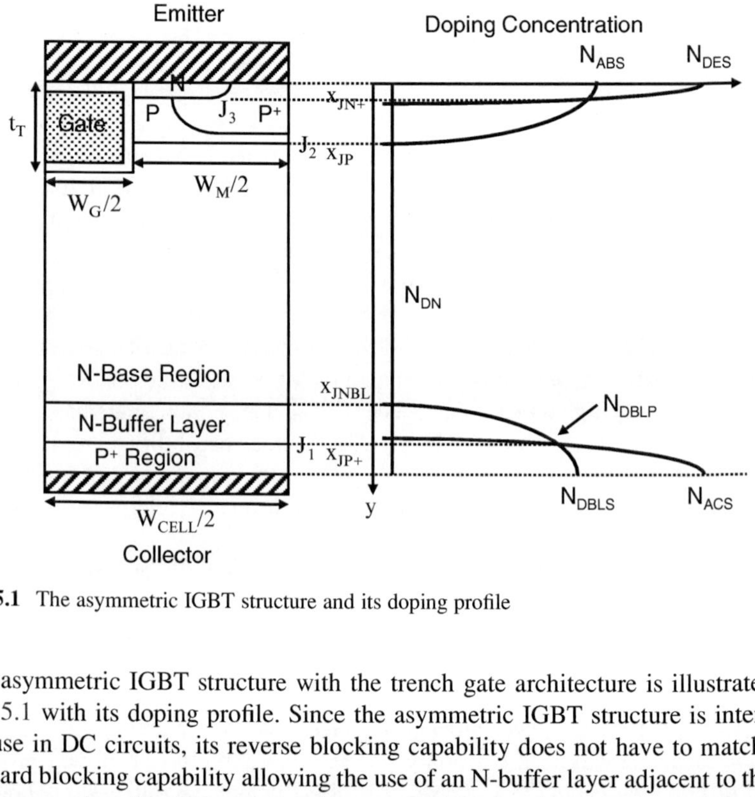

Fig. 5.1 The asymmetric IGBT structure and its doping profile

The asymmetric IGBT structure with the trench gate architecture is illustrated in Fig. 5.1 with its doping profile. Since the asymmetric IGBT structure is intended for use in DC circuits, its reverse blocking capability does not have to match the forward blocking capability allowing the use of an N-buffer layer adjacent to the P^+ collector region. The N-buffer layer has a much larger doping concentration than the lightly doped portion of the N-base region. The electric field in the asymmetric IGBT takes a trapezoidal shape allowing supporting the forward blocking voltage with a thinner N-base region. This allows achieving a lower on-state voltage drop and superior turn-off characteristics. The doping concentration of the buffer layer and the lifetime in the N-base region must be optimized to perform a trade-off between on-state voltage drop and turn-off switching losses. In addition, a thin, lightly doped P^+ diffused region can be utilized on the collector side to create a transparent emitter IGBT structure. These IGBT structures are discussed in this chapter with two blocking voltage ratings for comparison with other device structures in the book.

5.2 5,000-V Silicon Trench-Gate IGBT

The design and characteristics for the 5,000-V asymmetric silicon trench-gate IGBT structure are discussed in this section. The design parameters for the N-base region required to achieve this blocking voltage are first analyzed. Using the optimum N-base width, the blocking characteristics for the device are then obtained as a function of the lifetime in the drift region. The on-state characteristics for the device are obtained for various lifetime values as well. The gate controlled turn-off behavior of the silicon IGBT structure is analyzed including the effect of the lifetime in the drift region, the buffer layer concentration, and the transparent emitter design.

5.2.1 Blocking Characteristics

The physics for blocking voltages in the first and third quadrants by the IGBT structure is discussed in detail in the textbook [2]. When a positive bias is applied to the collector terminal of the asymmetric IGBT structure, the P-base/N-base junction (J_2) becomes reverse biased while the junction (J_1) between the P^+ collector region and the N-base region becomes forward biased. The forward blocking voltage is supported across the P-base/N-base junction (J_2) with a depletion layer extending mostly within the N-base region. The electric field distribution within the asymmetric IGBT structure is essentially the same as that illustrated in Fig. 4.3 for the asymmetric GTO structure. Consequently, the design procedure described in Chap. 4 can be applied to the asymmetric IGBT structure. From Fig. 4.4, the N-base region width required to obtain a forward blocking voltage of 5,500-V is 470 μm. This width can be slightly reduced when taking into account the voltage supported within the P-base region due to its graded doping profile.

5.2.2 Leakage Current

The leakage current in forward blocking mode is produced by space-charge-generation within the depletion region. In the case of the asymmetric IGBT structure in the forward blocking mode, the space-charge-generation current at the reverse biased P-base/N-base junction J_2 is amplified by the gain of the internal P-N-P transistor:

$$J_{\mathrm{L}} = \frac{J_{\mathrm{SCG}}}{(1 - \alpha_{\mathrm{PNP}})} \tag{5.1}$$

The space-charge-generation current density is given by:

$$J_{SCG} = \frac{qW_D n_i}{\tau_{SC}} = \frac{n_i}{\tau_{SC}}\sqrt{\frac{2q\varepsilon_S V_C}{N_D}} \tag{5.2}$$

at low collector bias voltages *before the depletion region in the lightly doped portion of the N-base regions reaches-through to the interface between the N-base region and the N-buffer layer*. The space-charge-generation current increases with increasing collector bias in this regime of operation for the asymmetric IGBT structure. Concurrently, the current gain (α_{PNP}) of the P-N-P transistor is also a function of the collector bias voltage because the base transport factor increases when the collector bias increases. Prior to the complete depletion of the lightly doped portion of the N-base region, the multiplication factor remains close to unity. It is therefore sufficient to account for the increase in the base transport factor with collector bias as given by Eqs. 4.8 and 4.9.

For the case of the silicon asymmetric IGBT structure with a width of 450 μm for the lightly doped portion of the N-base region with a doping concentration of 5×10^{12} cm^{-3}, the entire lightly doped portion of the N-base region is completely depleted at a reach-through voltage of 780 V given by Eq. 4.2. Once the lightly doped portion of the N-base region becomes completely depleted, the electric field becomes truncated at the interface between the lightly doped portion of the N-base region and the N-buffer layer as illustrated at the bottom of Fig. 4.3. The space-charge generation width then becomes independent of the collector bias because the depletion width in the N-buffer layer is small. Under these bias conditions, the base transport factor also becomes independent of the collector bias as given by Eq. 4.10. Consequently, the leakage current becomes independent of the collector bias until the on-set of avalanche multiplication.

The transport of minority carriers through the base region of the P-N-P transistor occurs through the N-buffer layer and the portion of the N-base region that has not been depleted by the applied collector bias. The base transport factor is therefore given by:

$$\alpha_T = \alpha_{T,N-Buffer}\alpha_{T,N-Base} \tag{5.3}$$

The base-transport factor associated with the N-buffer layer can be obtained from the decay of the hole current within the N-buffer layer as given by low-level injection theory [2]:

$$\alpha_{T,N-Buffer} = e^{-W_{NB}/L_{p,NB}} \tag{5.4}$$

where W_{NB} is the width of the N-buffer layer; and $L_{p,NB}$ is the minority carrier diffusion length in the N-buffer layer. The base transport factor for the lightly doped portion of the N-base region under low-level injection conditions appropriate for computation of the leakage current is given by:

$$\alpha_{\mathrm{T,N-Base}} = \frac{1}{\cosh\left[(W_{\mathrm{N}} - W_{\mathrm{D}})/L_{\mathrm{p,N}}\right]} \tag{5.5}$$

where W_{N} is the width of the lightly doped portion of the N-base region and $L_{\mathrm{p,N}}$ is the minority carrier diffusion length in the lightly doped portion of the N-base region. In this expression, the depletion width W_{D} prior to punch-through is given by:

$$W_{\mathrm{D}} = \sqrt{\frac{2\varepsilon_{\mathrm{S}} V_{\mathrm{C}}}{q N_{\mathrm{D}}}} \tag{5.6}$$

As the collector bias voltage increases, the base transport factor for the P-N-P transistor increases until it becomes equal to that for the N-buffer layer.

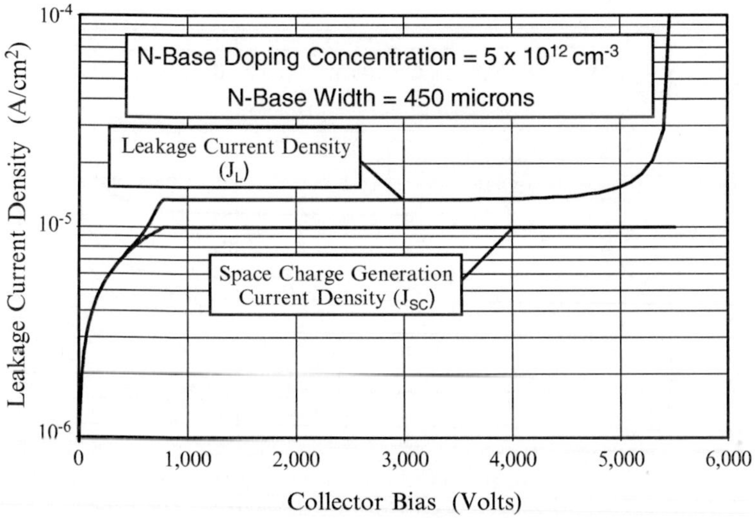

Fig. 5.2 Leakage current for the 5-kV asymmetric IGBT structure

Consider the case of an asymmetric IGBT structure that is designed with a forward blocking capability of 5,000 V. This would be satisfied by using an N-base region with a lightly doped portion having a doping concentration of 5×10^{12} cm^{-3} and width of 450 μm. The N-buffer layer will be assumed to have a doping concentration of 1.2×10^{17} cm^{-3} and thickness of 30 μm (corresponding to the baseline device structure used for the numerical simulations). The leakage current computed by using the above analysis is shown in Fig. 5.2 for the case of a lifetime (τ_{p0}, τ_{n0})

of 5 μs in the lightly doped portion of the N-drift region. This corresponds to a space-charge-generation lifetime (τ_{SC}) of 10 μs if the recombination center is located at mid-gap. In performing the analysis, the reduction of the lifetime with increasing doping concentration in the N-buffer layer was taken into account by using Eq. 4.11. The space-charge-generation current is also included in the figure for comparison purposes. It can be seen that the space-charge-generation leakage current increases with collector bias voltage due to the expansion of the width of the depletion region until 780 V. This corresponds to a reach-through voltage of 780 V obtained by using Eq. 4.2. The space-charge-generation leakage current then becomes independent of the collector bias voltage. The leakage current for the asymmetric IGBT structure is larger than the space-charge-generation current due to the current gain of the P-N-P transistor. For the case of a N-buffer layer doping of 1.2×10^{17} cm^{-3}, the current gain of the P-N-P transistor is 0.261 after the lightly doped portion of the N-base region becomes completely depleted. The leakage current density for this case is 1.35×10^{-5} A/cm^2. When the collector voltage increases above 5,000 V, the multiplication factor starts to increase rapidly producing a breakdown voltage of about 5,500 V.

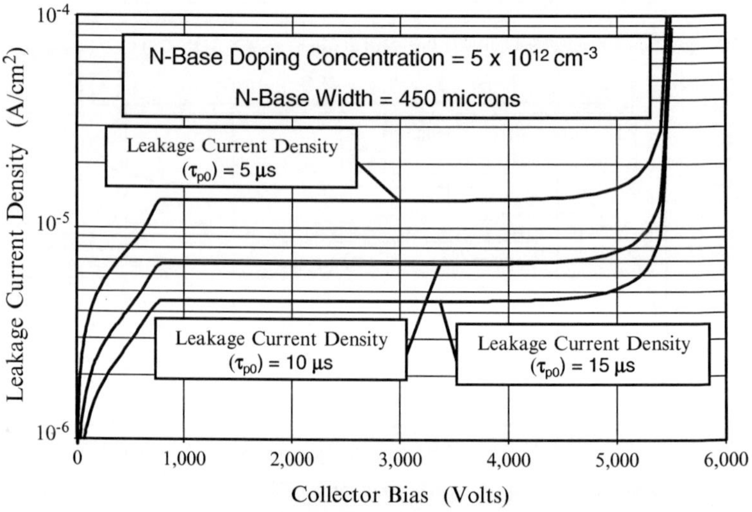

Fig. 5.3 Leakage current for the 5-kV asymmetric IGBT structure: lifetime dependence

The leakage current computed by using the above analysis for the silicon 5-kV IGBT structure is shown in Fig. 5.3 for the case of three lifetime (τ_{p0}, τ_{n0}) values in the lightly doped portion of the N-drift region. This corresponds to space-charge-generation lifetimes (τ_{SC}) of 10, 20, and 30 μs if the recombination center is located at mid-gap. In performing the analysis, the reduction of the lifetime with increasing doping concentration in the N-buffer layer was taken into account by using Eq. 4.11.

In all three cases, the leakage current increases with increasing collector bias voltage until the depletion region reaches through the lightly doped portion of the N-drift region at 780 V, and then becomes independent of the anode bias. The leakage current density is observed to increase when the lifetime is reduced. The breakdown voltage is essentially independent of the lifetime in the N-drift region.

The leakage current computed by using the above analysis for the silicon 5-kV IGBT structure is shown in Fig. 5.4 for the case of three buffer layer doping concentration (N_{BLP}) values. The lifetime (τ_{p0}, τ_{n0}) was assumed to be 5 μs for all three cases. In performing the analysis, the reduction of the lifetime with increasing doping concentration in the N-buffer layer was taken into account by using Eq. 4.11. The leakage current decreases with increasing buffer layer doping concentration due to a reduction of the gain of the P-N-P transistor because of a reduction in the injection efficiency. The breakdown voltage increases slightly with increase in the buffer layer doping concentration.

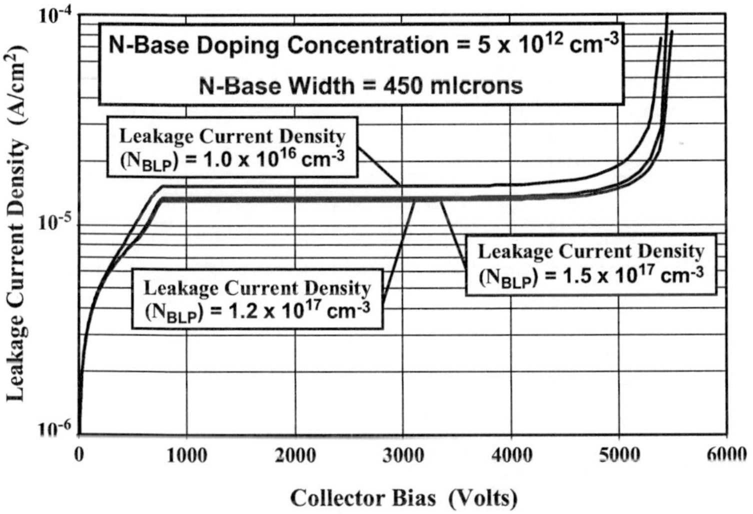

Fig. 5.4 Leakage current for the 5-kV asymmetric IGBT structure: buffer layer doping dependence

Simulation Example

In order to gain insight into the physics of operation for the 5-kV asymmetric IGBT trench gate structure under voltage blocking conditions, the results of two-dimensional numerical simulations for several structures are described here. The simulations were performed using a cell with the structure shown in Fig. 5.1. This half-cell has a width ($W_{CELL}/2$) of 3.5 μm (Area $= 3.5 \times 10^{-8}$ cm^{-2}) with a trench width (W_T) of 1 μm. The asymmetric IGBT structure used for the simulations was formed by diffusions performed into a uniformly doped N-type drift region with a doping concentration of 5×10^{12} cm^{-3}. A lifetime (τ_{p0}, τ_{n0}) of 5 μs was used for the baseline device. The N-buffer layer was formed by diffusion from the collector side with a depth of 55 μm. For the baseline device structure, the surface concentration of the N-type diffusion was adjusted to achieve a peak doping concentration of 1.2×10^{17} cm^{-3} in the buffer layer. The P-base region was formed with a Gaussian doping profile with a surface concentration of 1.5×10^{18} cm^{-3} and a depth of 3 μm. The N$^+$ emitter region was formed with a Gaussian doping profile with a surface concentration of 1×10^{20} cm^{-3} and a depth of 1 μm. The P$^+$ collector region was formed with a Gaussian doping profile with a surface concentration of 1×10^{20} cm^{-3} and a depth of 20 μm. The doping profile in the vertical direction through the N$^+$ emitter region is shown in Fig. 5.5 indicating the net width of the lightly doped portion of the N-base region is 440 μm after accounting for the diffusions. The peak doping concentration of the N-buffer layer is 1.0×10^{17} cm^{-3} and its thickness is 40 μm.

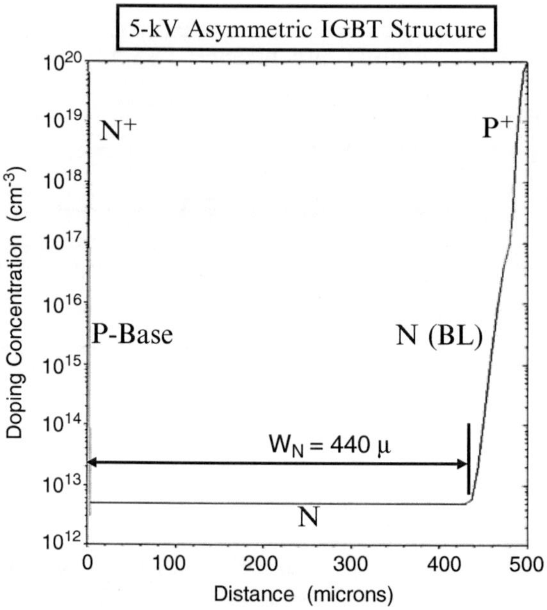

Fig. 5.5 Doping profile for the simulated baseline asymmetric 5-kV IGBT structure

The doping profile of the channel region for the 5-kV asymmetric IGBT structure used for the numerical simulations is provided in Fig. 5.6. This profile was taken along the vertical sidewall of the trench at $x = 3.0$ μm. It can be seen that the peak doping concentration of the P-base region is 1.5×10^{17} cm^{-3} and the channel length is 1.7 μm. This is sufficient to prevent reach-through limited breakdown in the P-base region.

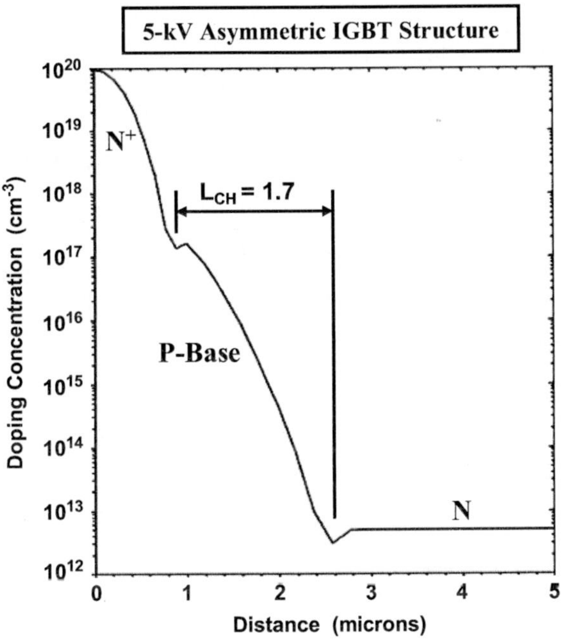

Fig. 5.6 Channel doping profile for the simulated baseline asymmetric 5-kV IGBT structure

The forward blocking capability of the silicon asymmetric IGBT structure was obtained using numerical simulations by increasing the collector bias while maintaining the gate electrode at zero volts. The characteristics obtained for three lifetime (τ_{p0}) values are provided in Fig. 5.7. In all cases, the leakage current increases rapidly with increasing anode bias voltage until about 780 V as predicted by the analytical model (see Fig. 5.2). This occurs due to the increase in the space-charge-generation volume and the increase in the current gain (α_{PNP}) of the open base P-N-P transistor until the collector bias becomes equal to the reach-through voltage obtained using the analytical solution given by Eq. 4.2. The leakage current then becomes independent of the collector voltage until close to the breakdown voltage. This behavior is well described by the analytical model (see Fig. 5.2). The leakage current density obtained using the analytical model is within a factor of 2 of the values derived from the numerical simulations for all cases. The leakage current density increases when the lifetime is reduced. This is due to the increase in the space-charge-generation current even though the gain of the P-N-P transistor is reduced. However, the reduced current gain of the P-N-P transistor produces a small increase in the blocking voltage. The simple analytical theory provides a very good qualitative and quantitative description of the leakage current behavior as a function of both the collector bias voltage and the lifetime in the N-drift region.

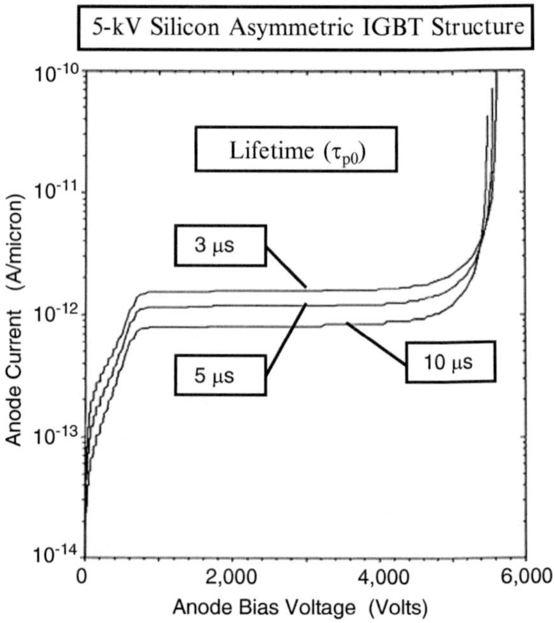

Fig. 5.7 Forward blocking characteristics for the asymmetric IGBT structure: N-base lifetime dependence

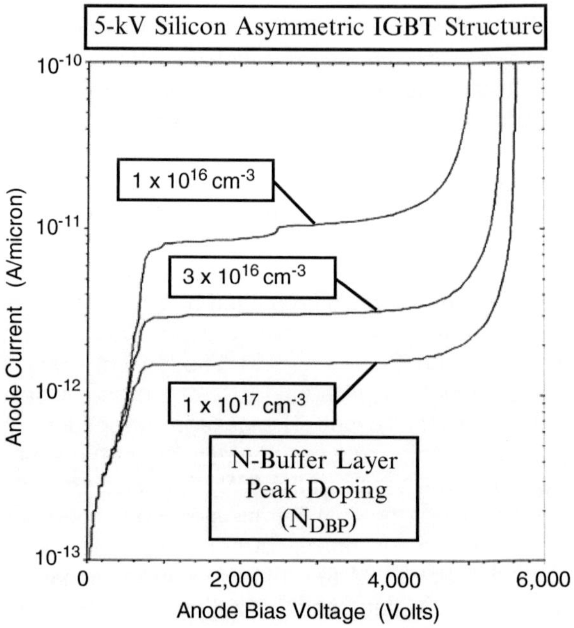

Fig. 5.8 Forward blocking characteristics for the asymmetric IGBT structure: buffer layer doping dependence

The leakage currents obtained for the asymmetric silicon IGBT structure using the numerical simulations are provided in Fig. 5.8 for three doping concentrations in the N-buffer layer. A lifetime of 3 μs (τ_{p0}) was used in the lightly doped portion of the N-base region for all these cases. It can be observed that the leakage current decreases when the N-buffer layer doping concentration is increased due to the reduction of the emitter injection efficiency and base transport factor of the P-N-P transistor. The reduced current gain of the P-N-P transistor also results in an increase in the open-base breakdown voltage. The behavior obtained by using the simple analytical model for the leakage current (see Fig. 5.4) is in good qualitative and quantitative agreement with these simulation results.

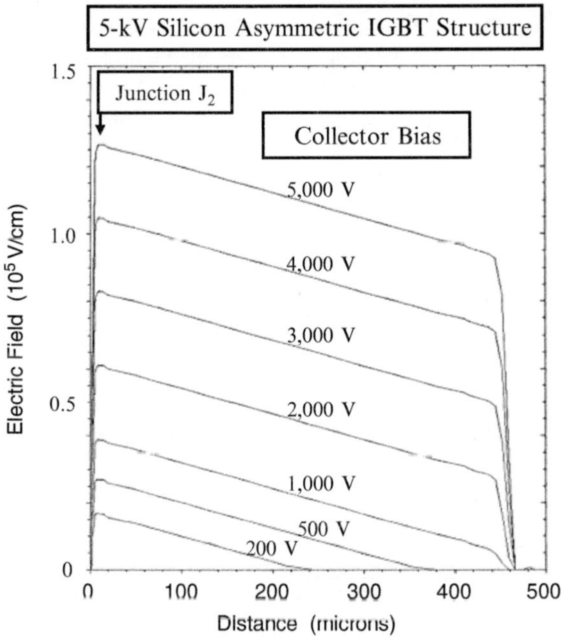

Fig. 5.9 Electric field profiles in the 5-kV asymmetric IGBT structure

The voltage is primarily supported within the lightly doped portion of N-base region in the asymmetric IGBT structure during operation in the forward blocking mode. This is illustrated in Fig. 5.9 where the electric field profiles are shown during operation in the forward blocking mode at several collector bias voltages. It can be observed that the P-Base/N-base junction (J_2) becomes reverse biased during the forward blocking mode with the depletion region extending toward the right-hand-side with increasing (positive) collector bias. The electric field has a triangular shape until the entire lightly doped portion of the N-base region becomes completely depleted. This occurs at a collector bias of about 800 V in agreement with the value obtained using the analytical solution (see Eq. 4.2). The electric field profile then takes a trapezoidal shape due to the high doping concentration in the N-buffer layer.

5.2.3 On-State Voltage Drop

In the textbook [2], two models were developed for the asymmetric IGBT structure based upon the doping concentration of the N-buffer layer relative to the injected carrier concentration. In the first case, the injected hole concentration was assumed to exceed the doping level in the buffer layer. This model is valid for the case of lower buffer layer doping levels. Since the buffer layer and the N-base region operate under high-level injection conditions, the analytical model for the on-state carrier distribution and on-state voltage drop was developed similar to that for the symmetric IGBT structure. However, at high buffer layer doping concentrations, the injected hole concentration in the buffer layer is well below the buffer layer doping concentration. In this case, the analytical model for the on-state carrier distribution and on-state voltage drop was developed with low-level injection in the buffer layer and high-level injection conditions in the N-base region. The analytical model for this case predicts discontinuities in the hole concentration at the boundary between the N-base region and the N-buffer layer.

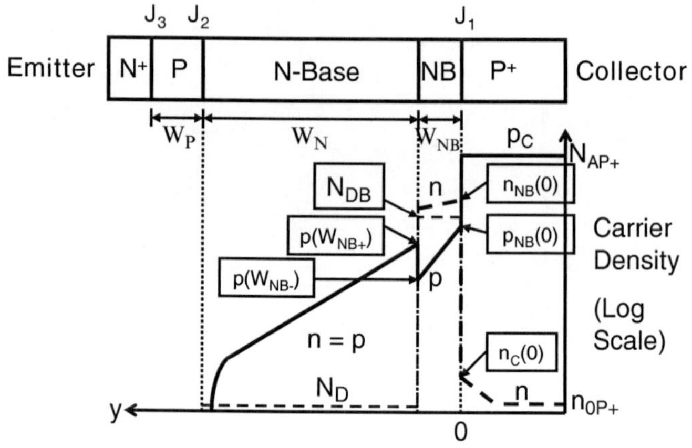

Fig. 5.10 Carrier distribution within the asymmetric IGBT structure in the on-state with mid-level injection conditions in the buffer layer

In this book, a generally applicable analytical model is developed for the asymmetric IGBT structure which is valid for any injection level in the buffer layer. This analytical model is then also applicable for the extreme cases of low-level injection conditions in the buffer layer and high-level injection conditions in the buffer layer. The carrier distribution profiles in the on-state for the asymmetric IGBT structure are shown in Fig. 5.10. The hole and electron concentrations in the N-base region are equal due to charge neutrality and the low doping concentration required for the drift region. The hole concentration in the N-buffer layer is less than the doping concentration (N_{DB}) in the buffer layer. Although the buffer layer doping is shown to be uniform in the figure, the case of a diffused buffer layer can be treated in the analytical model using the peak doping concentration of the buffer layer. Since the hole concentration in the buffer layer is comparable to the doping concentration, the electron concentration in the buffer becomes larger than the doping concentration to preserve charge neutrality. Consequently, the electron concentration ($n_{NB}(0)$) in the buffer layer at the junction (J_1) is given by:

$$n_{NB}(0) = N_{DB} + p_{NB}(0) \tag{5.7}$$

The enhanced electron concentration in the buffer layer produces an increase in the electron injection into the P^+ collector region which reduces the injection efficiency.

Using the Boltzmann quasi-equilibrium boundary condition for the carrier densities on both sides of the junction (J_1):

$$\frac{p_C}{p_{NB}(0)} = \frac{n_{NB}(0)}{n_C(0)} = \frac{N_{DB} + p_{NB}(0)}{n_C(0)} \tag{5.8}$$

by using Eq. 5.7. The hole concentration (p_C) is equal to the doping concentration (N_{AP+}) in the collector region because it is operating under low-level injection conditions. Consequently:

$$n_C(0) = \frac{N_{DB}\,p_{NB}(0) + p_{NB}^2(0)}{N_{AP+}} \tag{5.9}$$

The electron current density at junction (J_1) is given by:

$$J_n(0) = \frac{qD_{nP+}}{L_{nP+}} n_C(0) \tag{5.10}$$

where D_{nP+} and L_{nP+} are the diffusion coefficient and diffusion length for the P^+ collector region, respectively. These parameters must be calculated after taking into account the reduction of the mobility and lifetime in the highly doped collector region. Similarly, the hole current density at junction (J_1) is given by:

$$J_p(0) = \frac{qD_{pNB}}{L_{pNB}} p_{NB}(0) \tag{5.11}$$

where D_{pP+} and L_{pP+} are the diffusion coefficient and diffusion length for holes in the buffer layer, respectively. These parameters must be calculated after taking into account the reduction of the mobility and lifetime in the highly doped buffer layer. The collector current density is obtained by adding these components:

$$J_C = J_n(0) + J_p(0) = \frac{qD_{nP+}}{L_{nP+}} n_C(0) + \frac{qD_{pNB}}{L_{pNB}} p_{NB}(0) \tag{5.12}$$

By using Eq. 5.9, an expression for the hole concentration in the buffer layer at junction (J_1) is obtained:

$$p_{NB}^2(0) + \left(\frac{D_{pNB}N_{AP+}L_{nP+} + D_{nP+}N_{DB}L_{pNB}}{D_{nP+}L_{pNB}}\right)p_{NB}(0) - \frac{N_{AP+}L_{nP+}J_C}{qD_{nP+}} = 0 \quad (5.13)$$

The solution of this quadratic equation for the hole concentration in the buffer layer at junction (J_1) is:

$$p_{NB}(0) = \frac{1}{2}\left(\sqrt{b^2 - 4c} - b\right) \quad (5.14)$$

where

$$b = \frac{D_{pNB}N_{AP+}L_{nP+} + D_{nP+}N_{DB}L_{pNB}}{D_{nP+}L_{pNB}} \quad (5.15)$$

and

$$c = -\frac{N_{AP+}L_{nP+}J_C}{qD_{nP+}} \quad (5.16)$$

Since the unified analytical model presented here is valid for all injection levels in the N-buffer layer, it can be used to predict the variation of the injected hole concentration with lifetime in the N-base region and the doping concentration in the N-buffer layer.

The holes diffuse through the buffer layer producing a concentration ($p(W_{NB-})$) inside the buffer layer at the boundary between the N-buffer layer and the N-base region:

$$p(W_{NB-}) = p_{NB}(0)e^{-\left(W_{NB}/L_{pNB}\right)} \quad (5.17)$$

where W_{NB} is the thickness of the buffer layer. The hole concentration ($p(W_{NB+})$) in the N-base region at the boundary between the N-buffer layer and the N-base region can be obtained by equating the hole current density on the two sides of this boundary [2]:

$$p(W_{NB+}) = \frac{L_a \tanh[(W_N + W_{NB})/L_a]}{2qD_p} J_p(W_{NB-}) \quad (5.18)$$

with

$$J_p(W_{NB-}) = J_p(0)e^{-\left(W_{NB}/L_{pNB}\right)} \quad (5.19)$$

The hole concentration profile in the N-base region as dictated by high-level injection conditions is given by [2]:

$$p(y) = p(W_{NB}+)\frac{\sinh[(W_N + W_{NB} - y)/L_a]}{\sinh[(W_N + W_{NB})/L_a]} \tag{5.20}$$

which is valid for $y > W_{NB}$.

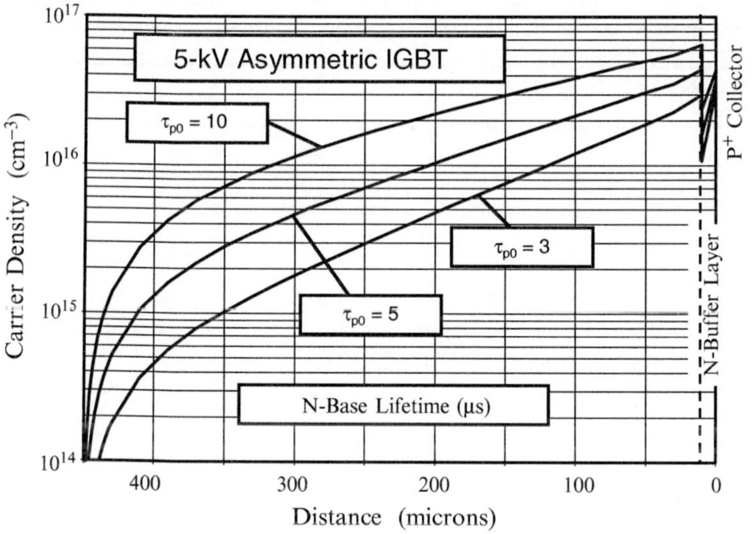

Fig. 5.11 Carrier distribution in the 5-kV asymmetric IGBT structure: lifetime dependence

The free carrier distribution obtained by using the above equations is provided in Fig. 5.11 for the case of a 5-kV asymmetric IGBT structure with an N-base region thickness of 440 μm and an effective buffer layer thickness of 10 μm. The hole lifetime (τ_{p0}) in the N-base region was varied for these plots from 3 to 10 μs. Note that the high-level lifetime (τ_{HL}) in these cases is two-times the hole lifetime (τ_{p0}). It can be observed that the hole concentration ($p_{NB}(0)$) decreases at the collector side of the N-buffer layer (at $y = 0$) from 4.3×10^{16} cm^{-3} to 3.2×10^{16} cm^{-3} when the lifetime in the N-base region is reduced from 10 to 3 μs. In addition, the hole concentration is significantly reduced at the emitter side when the lifetime in the N-base region decreases. The carrier density falls below 1×10^{15} cm^{-3} over a significant portion of the N-base region when the lifetime becomes smaller than 5 μs. This is due to the relatively large width of the N-base region required to support the 5-kV forward blocking capability making the diffusion length (L_a) much shorter than the N-base width (W_N). These results indicate that the on-state voltage drop will increase rapidly when the hole lifetime (τ_{p0}) in the N-base region is reduced below 5 μs. However, the smaller stored charge in the N-base region and buffer layer will reduce the turn-off time and energy loss per cycle.

The on-state voltage drop for the asymmetric IGBT structure can be obtained by using the equations derived in the textbook in Sect. 9.5.5 [2]. The on-state voltage drop for the asymmetric IGBT structure can be obtained by using:

$$V_{ON} = V_{P+NBL} + V_B + V_{MOSFET} \tag{5.21}$$

where V_{P+NBL} is the voltage drop across the P$^+$ collector/N-buffer layer junction (J_1), V_B is the voltage drop across the N-base region after accounting for conductivity modulation due to high-level injection conditions, and V_{MOSFET} is the voltage drop across the MOSFET portion. In the asymmetric IGBT structure, the junction (J_1) between the P$^+$ collector region and the N-buffer layer operates at neither high-level nor low-level injection conditions. Consequently, the voltage drop across the junction (J_1) must be obtained using:

$$V_{P+NB} = \frac{kT}{q} \ln\left(\frac{p_{NB}(0)N_{BL}}{n_i^2}\right) \tag{5.22}$$

The voltage drop across the N-base region can be obtained by integrating the electric field inside the N-base region [2]. The voltage drop is obtained by taking the sum of two parts. The first part is given by:

$$V_{B1} = \frac{2L_a J_C \sinh(W_N/L_a)}{qp(W_{NB+})(\mu_n + \mu_p)} \left\{\tanh^{-1}\left[e^{-(W_{ON}/L_a)}\right] - \tanh^{-1}\left[e^{-(W_N/L_a)}\right]\right\} \tag{5.23}$$

The depletion width (W_{ON}) across the P-base/N-base junction (J_2) in the on-state depends on the on-state voltage drop. The voltage drop associated with the second part is given by:

$$V_{B2} = \frac{kT}{q}\left(\frac{\mu_n - \mu_p}{\mu_n + \mu_p}\right) \ln\left[\frac{\tanh(W_{ON}/L_a)\cosh(W_{ON}/L_a)}{\tanh(W_N/L_a)\cosh(W_N/L_a)}\right] \tag{5.24}$$

For the trench gate IGBT structure considered here, the voltage drop across the MOSFET portion includes only the contribution from channel:

$$V_{CH} = \frac{J_C L_{CH} W_{CELL}}{2\mu_{ni} C_{OX}(V_G - V_{TH})} \tag{5.25}$$

Due to the small cell pitch for the trench-gate IGBT structure, the contribution to the on-state voltage drop from the MOSFET is very small.

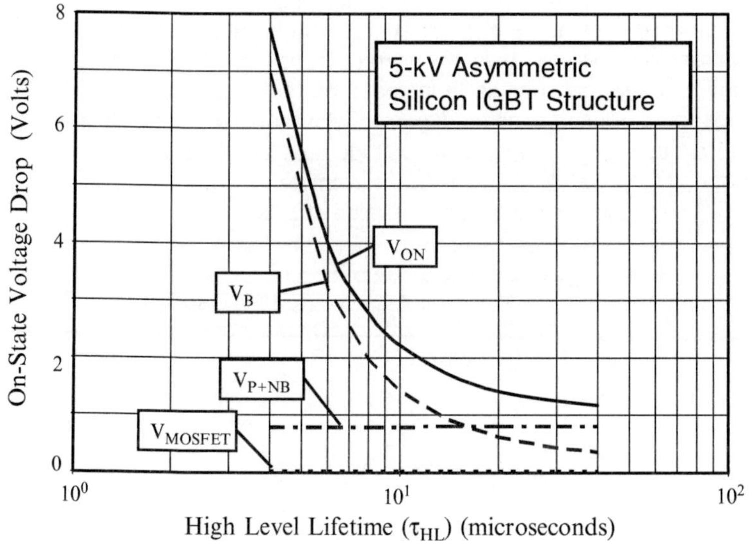

Fig. 5.12 On-state voltage drop for the 5-kV asymmetric IGBT structure: N Base lifetime dependence

The on-state voltage drop (at an on-state current density of 50 A/cm^2) computed for the 5-kV asymmetric silicon IGBT structure by using the above equations is provided in Fig. 5.12 as a function of the high-level lifetime in the N-base region. This asymmetric IGBT structure had the optimized N-base region width of 440 μm and effective N-buffer layer width of 10 μm. The on-state voltage drop is 4.1 V for a high-level lifetime of 6 μs in the N-base region. It can be observed that the on-state voltage drop increases rapidly as the lifetime is reduced due to the increase in the voltage drop (V_B) across the N-base region.

Simulation Results

The results of two-dimensional numerical simulations for the 5-kV asymmetrical silicon IGBT structure are described here. The total width ($W_{CELL}/2$) of the structure, as shown by the cross-section in Fig. 5.1, was 3.5 μm (Area = 3.5 × 10^{-8} cm^{-2}). A trench width (W_G) of 1 μm was used with a gate oxide thickness of 500 Å. The P-base and N$^+$ emitter regions were formed by using Gaussian doping profiles defined from the upper surface. The N-buffer layer and P$^+$ collector regions were formed by using Gaussian doping profiles defined from the lower surface. The doping profiles for the baseline device structure were already shown in Figs. 5.5 and 5.6.

The on-state characteristics of the 5-kV silicon asymmetric IGBT structure were obtained by using a gate bias voltage of 10 V for the case of various values for the lifetime in the drift region. This device structure has a peak buffer layer doping concentration of 1×10^{17} cm^{-3}. The characteristics obtained from the numerical simulations are shown in Fig. 5.13. The current initially increases exponentially with increasing collector bias. At current densities above 0.001 A/cm^2, the non-state voltage drop begins to increase more rapidly. Consequently, the on-state voltage drop increases as expected with reduction of the lifetime (τ_{p0}, τ_{n0}) indicated in the figure. The on-state voltage drop at a hole lifetime (τ_{p0}) value of 5 μs is found to be 3.01 V at an on-state current density of 50 A/cm^2 and increases to 4.1 V at a reduced hole lifetime (τ_{p0}) value of 3 μs.

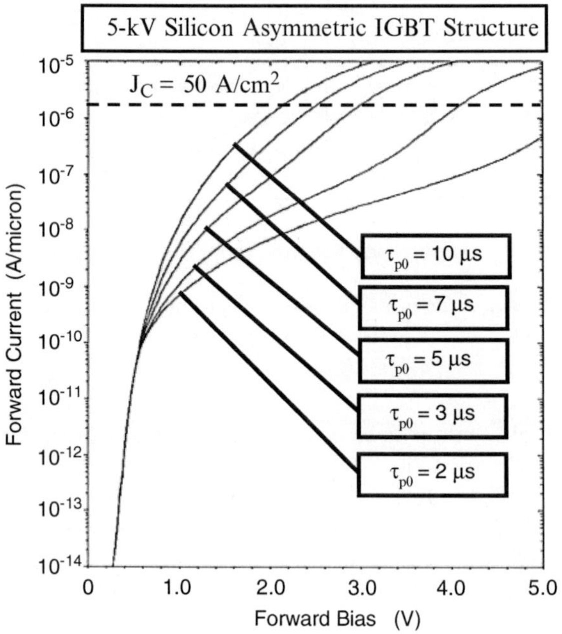

Fig. 5.13 On-state characteristics of the 5-kV asymmetric IGBT structure: lifetime dependence

The variation of the on-state voltage drop as a function of the lifetime in the N-base region predicted by the analytical model is compared with that obtained from the results of the numerical simulation in Fig. 5.14. There is a reasonable agreement between the prediction of the analytical model and the numerical simulations despite the analytical model being based upon a uniform doping concentration in the buffer layer. The results of the numerical simulations indicate a smaller increase in the on-state voltage drop when the high-level lifetime is reduced below 6 μs. This occurs because a larger hole concentration is observed in the N-base region near the emitter side in the results of the numerical simulations than that predicted using the analytical model based upon high-level injection. The values obtained from the numerical simulations will be utilized when developing the power loss trade-off curves for the 5-kV asymmetric IGBT structure later in the chapter.

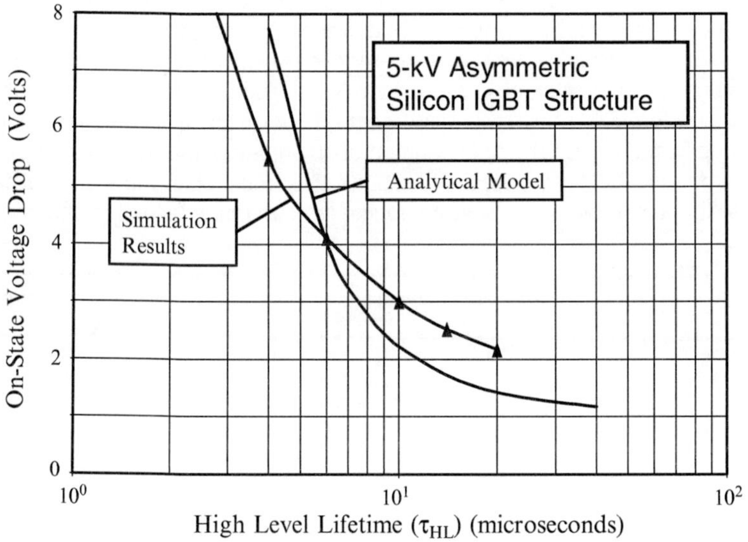

Fig. 5.14 On-state voltage drop for the 5-kV asymmetric IGBT structure: N-base lifetime dependence

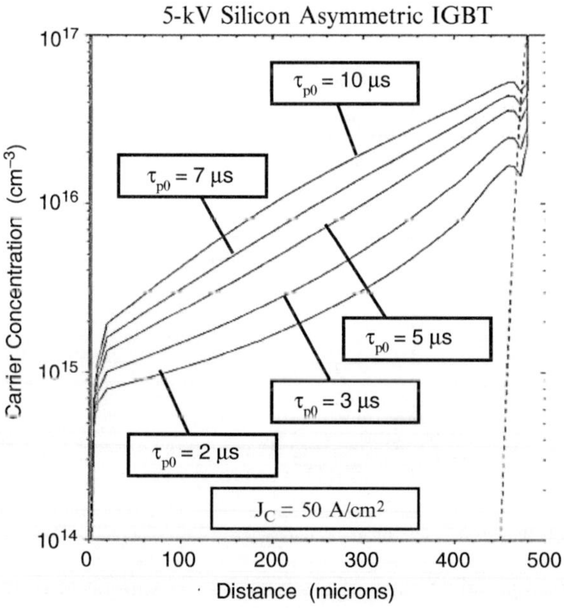

Fig. 5.15 On-state carrier distribution in the 5-kV asymmetric IGBT structure: lifetime dependence

The on-state voltage drop for the 5-kV asymmetric IGBT structure is determined by the distribution of carriers injected into the N-base region producing the desired reduction of its resistance. The hole distribution in the 5-kV asymmetric IGBT structure is provided in Fig. 5.15 for five cases of the lifetime (τ_{p0}, τ_{n0}) in the drift region. It can be observed that the injected carrier density is four orders of magnitude larger than the doping concentration on the collector side but not as large on the emitter side. The injected carrier density is reduced in the middle of the drift region when the lifetime is reduced to 3 ms. The predictions of the analytical model (see Fig. 5.11) have the same general characteristics as observed in the numerical simulations. The hole concentration values at the various interfaces in the asymmetric IGBT structure are also quite well predicted by the analytical model despite the assumption of a uniform doping concentration in the N-buffer layer.

5.2.4 *Turn-Off Characteristics*

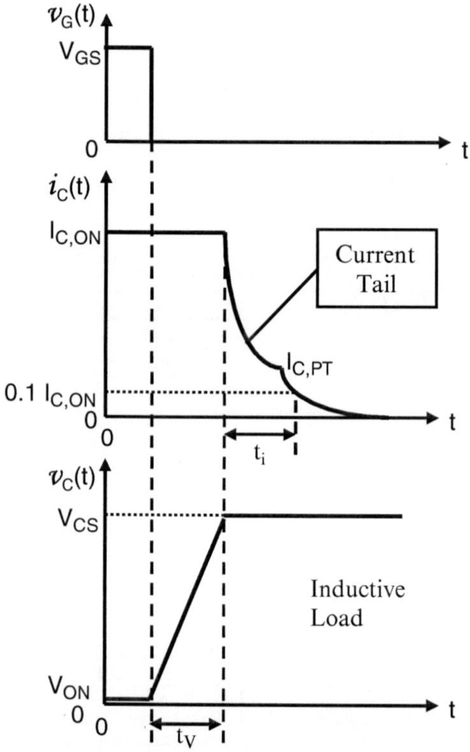

Fig. 5.16 Turn-off waveforms for the asymmetric IGBT structure

One of the important advantages of the IGBT is the simplicity of the gate control circuit due to its MOS-gate structure. In order to turn-off the device, the gate voltage must simply be reduced from the on-state value (V_{GS}, nominally 10 V) to zero as illustrated in Fig. 5.16. The magnitude of the gate current can be limited by using a resistance in series with the gate voltage source. The waveform for the gate voltage shown in the figure is for the case of zero gate resistance. Once the gate voltage falls below the threshold voltage, the electron current from the channel ceases. In the case of an inductive load, the collector current for the IGBT structure is then sustained by the hole current flow due to the presence of stored charge in the N-base region. Unlike the GTO structure, there is no prolonged storage time interval for the IGBT structure during its turn-off because the P-base/N-base junction is reverse biased in the on-state. The collector voltage begins to increase in the IGBT structure immediately after the gate voltage reduces below the threshold voltage.

Once the collector voltage reaches the collector supply voltage, the collector current decreases as shown in the figure. For the asymmetric IGBT structure, the current tail usually occurs in two parts if the collector voltage is insufficient for the space-charge region to extend completely through the N-base region. In this case, there is still some stored charge left in the N-base region near the N-buffer layer after the voltage transient is completed and the collector voltage is equal to the collector supply voltage. During the first part of the collector current decay, the stored charge in the N-base region is removed by recombination, as well the collector current flow. This is a relatively slow decay due to the large high-level lifetime in the N-base region. As the collector current decreases, the hole concentration in the space-charge region decreases allowing the space-charge region to expand even though the collector voltage is constant. Eventually, the space-charge region extends through the entire N-base region when the collector current density becomes equal to the punch-through current density ($I_{C,PT}$). At this point in time, stored charge is present only in the N-buffer layer. The stored charge in the N-buffer layer decreases by recombination at a faster pace due to the smaller lifetime in the N-buffer layer associated with its greater doping concentration than the N-base region. This produces a faster decay of the collector current during the second phase of the current tail as illustrated in the figure.

5.2.4.1 Voltage Rise-Time

The analysis of the turn-off waveform for the collector voltage transient for the asymmetric IGBT structure can be performed by using the charge control principle. In the textbook, it was assumed that recombination in the N-base region can be neglected in the on-state. This results in a linear free carrier (hole) distribution within the lightly doped portion of the N-base region during on-state operation as illustrated in Fig. 5.17. The hole concentration inside the N-base region in the on-state is then given by:

$$p(y) = p(W_{NB+})\left(1 - \frac{y}{W_N}\right) \tag{5.26}$$

if $y = 0$ at the boundary between the N-base region and the N-buffer layer is shown in the figure. In writing this expression, the hole concentration is assumed to be approximately zero at the edge of the space-charge region during the on-state and the space-charge layer width is assumed to be zero due to the small on-state voltage drop of the IGBT structure. The concentration p_{WNB+} in the on-state at the interface between the lightly doped portion of the N-base region and the N-buffer layer was previously derived in Sect. 5.2.3.

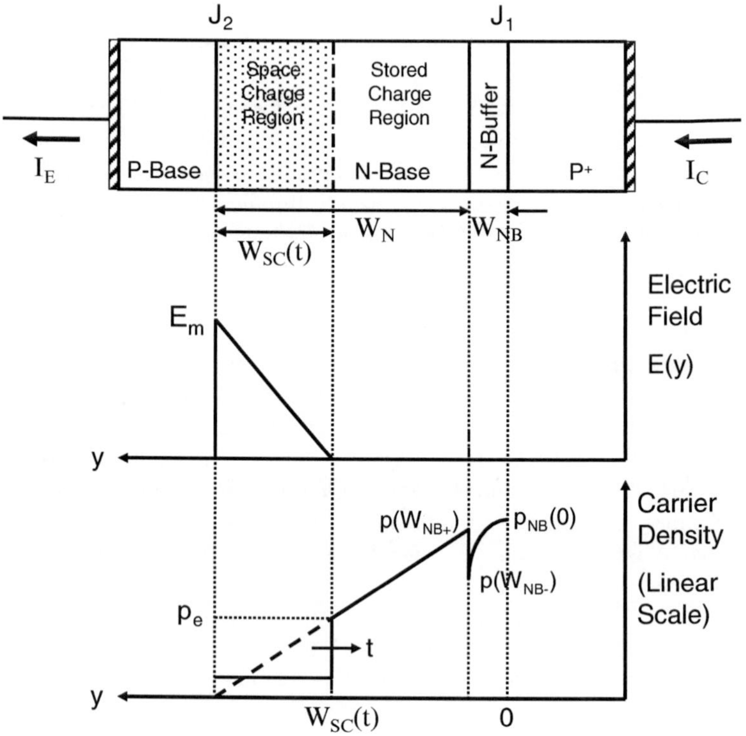

Fig. 5.17 Electric field and free carrier distribution during the voltage rise-time for the asymmetric IGBT structure

To develop the analysis of the collector voltage transient, it will be assumed that the hole concentration profile in the N-base region does not change due to recombination. In this case, the electric field profile in the asymmetric IGBT structure during the collector voltage transient is illustrated in Fig. 5.17. As the space-charge region expands toward the collector side, holes are removed from the stored-charge region at its boundary. The holes then flow through the space-charge region at their saturated drift velocity due to the high electric field in the space-charge region. Due to the high concentration of holes in the space-charge region associated with the collector current flow, the space-charge layer does not reach-through the N-base region during the voltage transient.

For the analysis of the voltage transient, it will be assumed that the hole distribution does not change in the stored-charge region of the N-base region.

Consequently, the concentration of holes at the edge of the space-charge region (p_e) increases during the turn-off process as the space-charge width increases:

$$p_e(t) = p(W_{NB+})\left[\frac{W_{SC}(t)}{W_N}\right] \tag{5.27}$$

According to the charge-control principle, the charge removed by the expansion of the space-charge layer must equal the charge removed due to collector current flow:

$$J_{C,ON} = qp_e(t)\frac{dW_{SC}(t)}{dt} = qp(W_{NB+})\left[\frac{W_{SC}(t)}{W_N}\right]\frac{dW_{SC}(t)}{dt} \tag{5.28}$$

by using Eq. 5.27. Integrating this equation on both sides and applying the boundary condition of zero width for the space-charge layer at time zero provides the solution for the evolution of the space-charge region width with time [2]:

$$W_{SC}(t) = \sqrt{\frac{2W_N J_{C,ON} t}{qp(W_{NB+})}} \tag{5.29}$$

The space-charge layer expands toward the right-hand-side as indicated by the horizontal time arrow in Fig. 5.17 with the hole concentration profile in the stored-charge region remaining unchanged.

The collector voltage supported by the asymmetric IGBT structure is related to the space-charge layer width by:

$$V_C(t) = \frac{q(N_D + p_{SC})W_{SC}^2(t)}{2\varepsilon_S} \tag{5.30}$$

The hole concentration in the space-charge layer can be related to the collector current density under the assumption that the carriers are moving at the saturated drift velocity in the space-charge layer:

$$p_{SC} = \frac{J_{C,ON}}{qv_{sat,p}} \tag{5.31}$$

The hole concentration in the space-charge region remains constant during the voltage rise-time because the collector current density is constant. Consequently, the slope of the electric field profile in the space-charge region also becomes independent of time.

Applying the solution for the evolution of the space-charge layer from Eq. 5.29 in Eq. 5.30:

$$V_C(t) = \frac{W_N(N_D + p_{SC})J_{C,ON}}{\varepsilon_S p(W_{NB+})} t \tag{5.32}$$

The analytical model for turn-off of the asymmetric IGBT structure under inductive load conditions predicts a linear increase in the collector voltage with time. This analytical model does not include the impact of carrier generation due to the impact ionization process at larger collector bias voltages. Impact ionization introduces additional holes and electrons into the space-charge region resulting in a reduction of the rate of rise of the collector voltage prolonging the voltage rise time.

The end of the first phase of the turn-off process, where the collector voltage increases while the collector current remains constant, occurs when the collector voltages reach the collector supply voltage (V_{CS}). This time interval ($t_{V,OFF}$) can be obtained by making the collector voltage equal to the collector supply voltage in Eq. 5.32:

$$t_{V,OFF} = \frac{\varepsilon_S p(W_{NB+})V_{CS}}{W_N(N_D + p_{SC})J_{C,ON}} \tag{5.33}$$

According to the analytical model, the voltage rise-time is proportional to the collector bias supply voltage. However, it is only weakly dependent on the on-state current density (through p_{SC}) because the hole concentration $p(W_{NB+})$ is proportional to the on-state current density.

In this book, the analysis of the collector voltage transient is also provided using the non-linear hole concentration profile given by Eq. 5.20. The concentration $p(W_{NB+})$ in the on-state at the interface between the lightly doped portion of the N-base region and the N-buffer layer was previously derived in Sect. 5.2.3. To develop the analysis of the collector voltage transient, it will be assumed that the hole concentration profile in the N-base region does not change due to recombination. In this case, the electric field profile in the asymmetric IGBT structure during the collector voltage transient is illustrated in Fig. 5.17. As the space-charge region expands toward the collector side, holes are removed from the stored-charge region at its boundary. The holes then flow through the space-charge region at their saturated drift velocity due to the high electric field in the space-charge region. Due to the high concentration of holes in the space-charge region associated with the collector current flow, the space-charge layer does not reach-through the N-base region during the voltage transient.

For the analysis of the voltage transient, it will be assumed that the hole distribution does not change in the stored-charge region of the N-base region. Consequently, the concentration of holes at the edge of the space-charge region (p_e) increases during the turn-off process as the space-charge width increases:

$$p_e(t) = p(W_{NB+}) \frac{\sinh[W_{SC}(t)/L_a]}{\sinh[(W_N + W_{NB})/L_a]} \tag{5.34}$$

According to the charge-control principle, the charge removed by the expansion of the space-charge layer must equal the charge removed due to collector current flow:

$$J_{C,ON} = qp_e(t)\frac{dW_{SC}(t)}{dt} = qp(W_{NB+})\frac{\sinh[W_{SC}(t)/L_a]}{\sinh[(W_N + W_{NB})/L_a]}\frac{dW_{SC}(t)}{dt} \tag{5.35}$$

by using Eq. 5.34. Integrating this equation on both sides and applying the boundary condition of width $W_{SC}(0)$ for the space-charge layer at time zero provides the solution for the evolution of the space-charge region width with time [2]:

$$W_{SC}(t) = L_a\, a\cosh\left\{\frac{J_{C,ON}\sinh[(W_N + W_{NB})/L_a]}{qL_a\, p(W_{NB+})}t + \cosh[W_{SC}(0)/L_a]\right\} \tag{5.36}$$

The space-charge layer expands toward the right-hand-side as indicated by the horizontal time arrow in Fig. 5.17 with the hole concentration profile in the stored-charge region remaining unchanged.

The collector voltage supported by the asymmetric IGBT structure is related to the space-charge layer width by:

$$V_C(t) = \frac{q(N_D + p_{SC})W_{SC}^2(t)}{2\varepsilon_S} \tag{5.37}$$

The hole concentration in the space-charge layer can be related to the collector current density under the assumption that the carriers are moving at the saturated drift velocity in the space-charge layer:

$$p_{SC} = \frac{J_{C,ON}}{qv_{sat,p}} \tag{5.38}$$

The hole concentration in the space-charge region remains constant during the voltage rise-time because the collector current density is constant. Consequently, the slope of the electric field profile in the space-charge region also becomes independent of time. The new analytical model for turn-off of the asymmetric IGBT structure under inductive load conditions predicts a non-linear increase in the collector voltage with time. This analytical model does not include the impact of carrier generation due to the impact ionization process at larger collector bias voltages. Impact ionization introduces additional holes and electrons into the space-charge region resulting in a reduction of the rate of rise of the collector voltage prolonging the voltage rise-time.

The end of the first phase of the turn-off process, where the collector voltage increases while the collector current remains constant, occurs when the collector voltage reaches the collector supply voltage (V_{CS}). This time interval ($t_{V,OFF}$) can be obtained by making the collector voltage equal to the collector supply voltage in Eq. 5.37:

$$t_{V,OFF} = \frac{qL_a p(W_{NB+})}{J_{C,ON}\sinh[(W_N + W_{NB})/L_a]}\left\{\begin{array}{l}\cosh\left[\sqrt{\dfrac{2\varepsilon_S V_{CS}}{qL_a^2(N_D + p_{SC})}}\right] \\ -\cosh[W_{SC}(0)/L_a]\end{array}\right\} \quad (5.39)$$

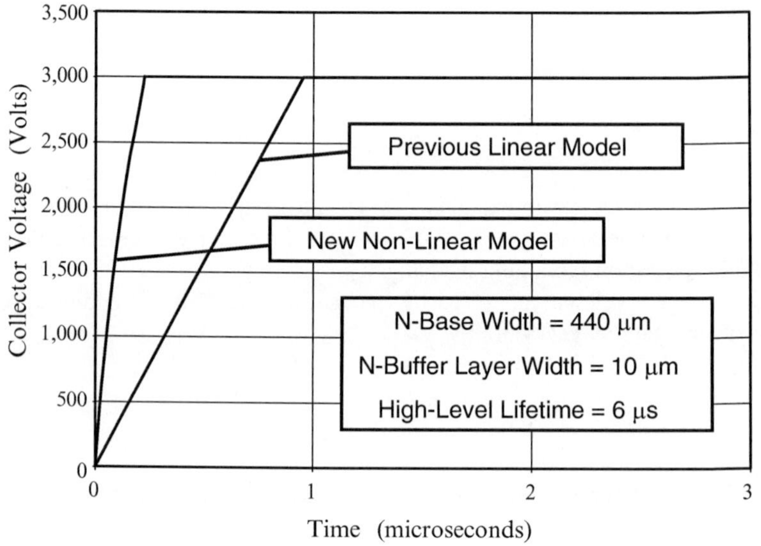

Fig. 5.18 Collector voltage transient during turn-off for the asymmetric IGBT structure

Consider the case of a 5-kV asymmetric IGBT structure with N-base, and N-buffer layer widths of 440 and 10 μm, respectively. The collector voltage transient obtained using the previous and new analytical models for the case of a high-level lifetime of 6 μs in the N-base region are shown in Fig. 5.18. The voltage transient is predicted to occur at a much faster rate with the new model when compared with the previous model. The time interval for the voltage transient ($t_{V,OFF}$) obtained using the new model is only 0.225 μs versus 0.995 μs with the previous model. In comparison with the 5-kV asymmetric GTO structure discussed in the previous chapter, the 5-kV asymmetric IGBT structure does not have a prolonged storage time and the voltage rise-time is substantially smaller for both analytical models. This behavior reduces the switching power losses in the IGBT and allows operation of the IGBT structure at a higher frequency when compared with the GTO.

The width of the space-charge layer at the end of the voltage transient can be obtained by using the collector supply voltage:

$$W_{SC}(t_{V,OFF}) = \sqrt{\frac{2\varepsilon_S V_{CS}}{q(N_D + p_{SC})}} \tag{5.40}$$

It can be concluded that the width of the space-charge layer at the end of the first phase depends upon the collector supply voltage and the initial on-state current density (via p_{SC}). In a typical 5-kV asymmetric IGBT structure, the space-charge layer width obtained by using the above equation with an on-state current density of 50 A/cm^2 and a collector supply voltage of 3,000 V is 327 μm if the doping concentration of the N-base region is 5×10^{12} cm^{-3}. The space-charge region width at the end of the voltage transient is therefore about 100 μm smaller than the width of the N-base region for a typical collector supply voltage of 3,000 V. Consequently, a substantial amount of stored charge remains in the N-base region after the voltage transient.

5.2.4.2 Current Fall-Time

During the second phase of the turn-off process, the collector current decays while the collector voltage remains fixed at the collector supply voltage. The decay of the collector current occurs in two parts for the asymmetric IGBT structure. At the end of the voltage transient, there is a substantial amount of stored charge in the N-base region. Consequently, during the first part, the collector current flow is governed by the recombination of the excess holes that are trapped within the N-base region under high-level injection conditions. This is indicated in Fig. 5.19 by the downward vertical arrow with time indicated near it. At the same time, holes and electrons are also removed from the stored charge region due to the collector current flow. This is indicated by the horizontal arrows with the collector current near them. The arrow on the left-hand-side at the boundary $W_{SC}(t_V)$ indicates the collector current removing holes from the space-charge region toward the left-side. The arrow on the right-hand-side at junction (J_1) indicates the collector current removing electrons from the space-charge region toward the right-side.

As the collector current decreases, the hole concentration in the space-charge region also decreases. Consequently, the space-charge region expands during the first part of the current transient until the space-charge region covers the entire width (W_N) of the N-base region. After this time, the space-charge region width cannot increase any further due to the high doping concentration in the N-buffer layer. Consequently, during the second part of the collector current transient, the collector current flow is governed by the recombination of holes in the N-buffer layer under low-level injection conditions. The second part of the collector current decay occurs at a much faster rate than during the first part due to the smaller lifetime in the N-buffer layer associated with its larger doping concentration.

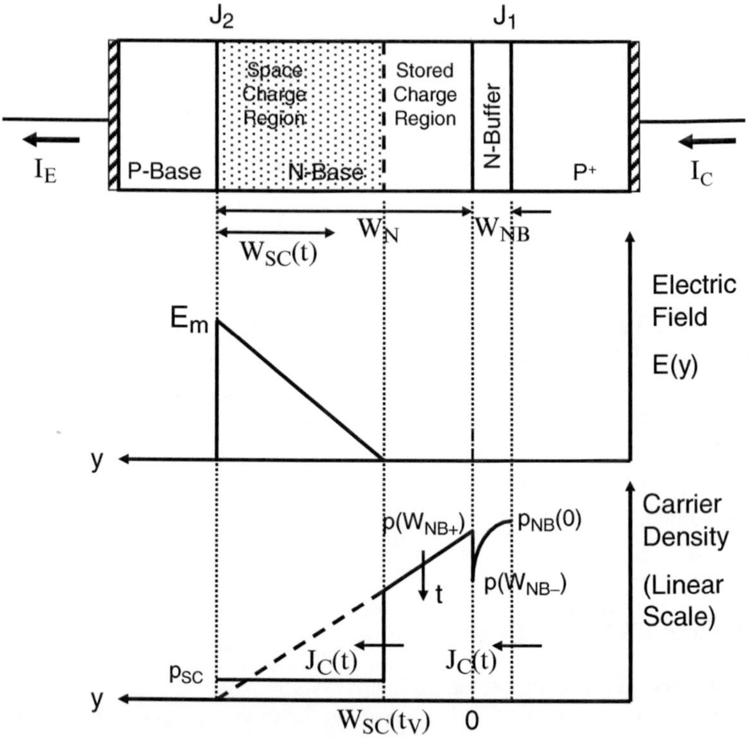

Fig. 5.19 Electric field and free carrier distribution during the current fall-time for the asymmetric IGBT structure

During the first part of the current transient, the continuity equation for holes in the N-base region and the N-buffer layer is given by:

$$\frac{\mathrm{d}\delta p_{\mathrm{N}}}{\mathrm{d}t} = -\frac{\delta p_{\mathrm{N}}}{\tau_{\mathrm{HL}}} - \frac{J_{\mathrm{C}}(t)}{q(W_{\mathrm{N}} - W_{\mathrm{SC}} + W_{\mathrm{NB}})} \tag{5.41}$$

where δp_{N} is the excess hole concentration. The first term on the right-hand-side of this equation is the removal of holes due to recombination while the second term on the right-hand-side of this equation is the removal of holes due to collector current flow. Applying this equation to the hole concentration in the N-buffer layer at junction (J_1):

$$\frac{\mathrm{d}p_{\mathrm{NB}}(0,t)}{\mathrm{d}t} = -\frac{\delta p_{\mathrm{NB}}(0,t)}{\tau_{\mathrm{HL}}} - \frac{J_{\mathrm{C}}(t)}{q(W_{\mathrm{N}} - W_{\mathrm{SC}} + W_{\mathrm{NB}})} \tag{5.42}$$

Using Eq. 5.12 for the collector current density, together with Eqs. 5.9–5.11:

$$J_C(t) = J_n(0,t) + J_p(0,t)$$
$$= \left(\frac{qD_{nP+}}{L_{nP+}} + \frac{qD_{nP+}N_{DB}}{L_{nP+}N_{AP+}}\right)p_{NB}(0,t) + \frac{qD_{nP+}}{L_{nP+}N_{AP+}}p_{NB}^2(0,t) \tag{5.43}$$

Substituting this expression for the collector current density into Eq. 5.42 yields a first order differential equation that the governs the decay of the hole concentration $p_{NB}(0,t)$ as a function of time:

$$\frac{dp_{NB}(0,t)}{dt} = Ap_{NB}(0,t) + Bp_{NB}^2(0,t) \tag{5.44}$$

where the constants are given by:

$$A = -\left\{\frac{1}{(W_N - W_{SC} + W_{NB})}\left[\frac{D_{pNB}}{L_{pNB}} + \frac{D_{nP+}N_{DB}}{L_{nP+}N_{AP+}}\right] + \frac{1}{\tau_{HL}}\right\} \tag{5.45}$$

and

$$B = -\frac{D_{nP+}}{L_{nP+}N_{AP+}(W_N - W_{SC} + W_{NB})} \tag{5.46}$$

The solution for hole concentration $p_{NB}(0, t)$ is:

$$p_{NB}(0,t) = \frac{1}{[(C + B/A)]e^{-At} - (B/A)} \tag{5.47}$$

where C is the reciprocal of the on-state hole concentration in the N-buffer layer at junction (J_1):

$$C = \frac{1}{p_{NB}(0,0)} \tag{5.48}$$

It is worth pointing out that the constant A in the above equation is negative. Consequently, the exponential term in the denominator of Eq. 5.47 increases with time resulting in the expected reduction in the hole concentration with time. The decay of the holes during the first part of the transient is not a simple exponential in nature because holes are removed at a faster rate initially due to the larger collector current levels. As the collector current reduces, the rate of removal of the holes also becomes slower.

The waveform for the collector current decay during the first part can then be obtained by using Eq. 5.43 with the solution for the hole concentration derived from Eq. 5.47. According to this analytical model, the collector current does not reduce exponentially with time. The collector current reduces at a faster rate initially because the hole concentration changes at a faster rate as described above. As the collector current decays during the first part, the concentration of holes in the space-charge region also decreases resulting in an increase in the width of the space-charge region in spite of a constant collector voltage. The collector current decay during the first part occurs until the space-charge region extends completely through the N-base region. The space-charge region punches through to the N-buffer layer at a unique collector current density which is independent of the lifetime in the N-base region. This punch-through collector current density can be derived by equating the space-charge layer width given by Eq. 5.40 to the width of the N-base region (W_N) and using Eq. 5.38 for the hole concentration in the space-charge region:

$$J_{C,PT} = \frac{2v_{sat,p}\varepsilon_S V_{CS}}{W_N^2} - qv_{sat,p}N_D \tag{5.49}$$

During the second part of the collector current decay, the stored charge exists only inside the N-buffer layer. At this point in time, the hole concentration in the N-buffer layer has been substantially reduced during the first part of the collector decay. Consequently, the hole concentration in the N-buffer layer decreases by recombination under low-level injection conditions as determined by the low-level lifetime of the N-buffer layer. The collector current transient is then described by [2]:

$$J_C(t) = J_{C,PT}e^{-t/\tau_{BL}} \tag{5.50}$$

where τ_{BL} is the low-level lifetime in the N-buffer layer. The lifetime in the buffer layer is substantially smaller than the corresponding value in the lightly doped portion of the N-base region:

$$\tau_{BL} = \frac{\tau_{LL}}{1 + (N_{D,BL}/N_{REF})} \tag{5.51}$$

where τ_{LL} is the low-level lifetime in the lightly doped portion of the N-base region, $N_{D,BL}$ is the doping concentration of the buffer layer, and N_{REF} is a reference doping concentration (typically 5×10^{16} cm^{-3}). Consequently, the collector current transient during the second part occurs exponentially at a faster rate than during the first part.

Consider the case of a 5-kV asymmetric IGBT structure with N-base and N-buffer layer widths of 440 and 10 μm, respectively, and an N-buffer layer doping concentration of 1×10^{17} cm^{-3}. The collector current waveform predicted by the new analytical model presented in this section is shown in Fig. 5.20 for a high-level lifetime of 6 μs in the N-base region. It can be observed that the collector current decays in two stages. During the first part, the decay is much faster than predicted by a simple exponential variation with recombination occurring at the high-level lifetime in the N-base region. When the collector reaches the punch-through current density ($J_{C,PT}$) of 19.3 A/cm^2 predicted by Eq. 5.49, the second part of the collector current transient begins to occur. The current fall time ($t_{I,OFF}$) is usually defined as the time taken for the anode current to reach one-tenth of the on-state value as shown in Fig. 5.20. For comparison purposes, the collector current transient for the previous exponential decay model is also shown in the figure using a high-level lifetime of 6 μs for recombination of holes in the N-base region. It can be observed that the new model predicts a much faster decay of the collector current during the first part because it takes into account the removal of holes from the stored-charge region by the collector current flow.

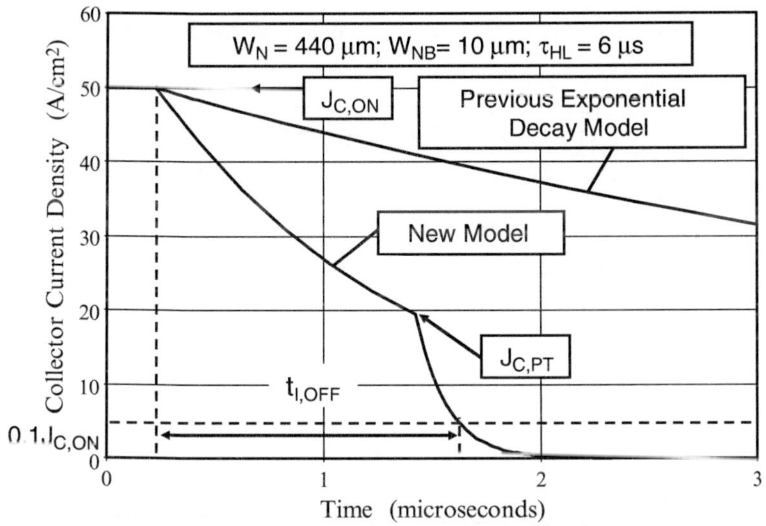

Fig. 5.20 Collector current transient during turn-off for the 5-kV asymmetric IGBT structure

Simulation Example

In order to gain insight into the operation of the asymmetric 5-kV IGBT structure during its turn-off, the results of two-dimensional numerical simulations for a typical structure are discussed here. The device structure used has the cross-section shown in Fig. 5.1 with a cell half-width of 3.5 μm. The doping profile for the IGBT structure used in the numerical simulations was provided in Fig. 5.5. The widths of the uniformly doped N-base region and the diffused N-buffer layer are 440 and 30 μm, respectively. For the typical case discussed here, a high-level lifetime of 6 μs was used in the N-base region.

The numerical simulations were performed with an abrupt reduction of the gate voltage from 10 to 0 V in 10 ns starting from an on-state current density of 50 A/cm^2. The resulting waveforms obtained from the numerical simulations for the anode voltage and current are shown in Fig. 5.21 for the case of a collector supply voltage of 3,000 V. Unlike the GTO structure discussed in the previous chapter, there is no storage time associated with the turn-off of the IGBT structure. The collector voltage begins to increase immediately at the end of the gate voltage transient because the P-base/N-base junction (J_2) is already reverse biased in the on-state. Initially, the collector voltage increases rapidly with the time as predicted by the analytical model until it reaches about 1,000 V. It then increases at a slower rate. This is associated with the on-set of avalanche multiplication at high collector bias voltages – an effect not included in the analytical model. The dotted line in the figure provides an extrapolation of the collector voltage transient without the effect of impact ionization. The collector voltage rise-time obtained using this extrapolated line is 0.25 μs in agreement with the analytical model (see Fig. 5.18). The anode voltage rise-time obtained in the numerical simulations for the case of supply voltage of 3,000 V including the effect of impact ionization is much larger (1.37 μs) than from the analytical model.

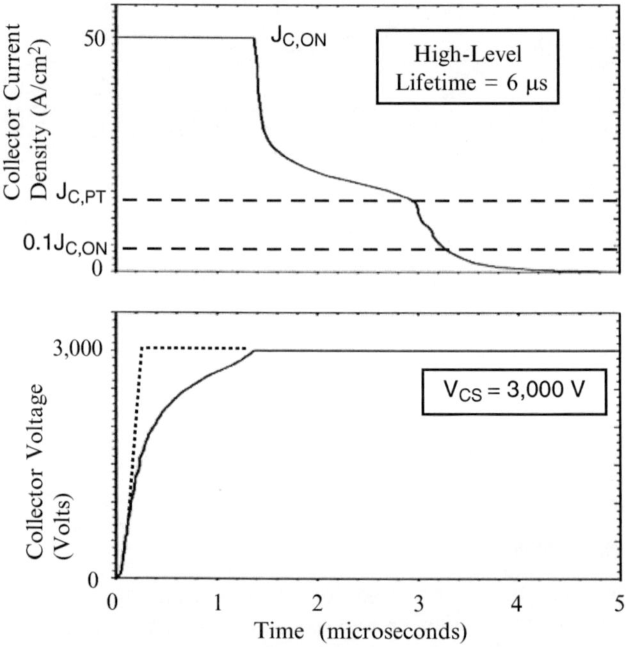

Fig. 5.21 Typical turn-off waveforms for the asymmetric 5-kV IGBT structure

After the completion of the collector voltage transient, the collector current waveform decays from the initial on-state current density at a rate that decreases with time. The current decays to the punch-through current density (indicated in the figure) in 1.6 μs. This time interval is well predicted by the new analytical model (see Fig. 5.20). The punch-through current density of about 20 A/cm^2 observed in the numerical simulations is also very well predicted by the analytical model. After reaching the punch-through current density, the collector current is observed to decay at a faster rate as described by the analytical model. It can be concluded that the new one-dimensional analytical model predict a shape for the collector current waveform that is consistent with the numerical simulation results.

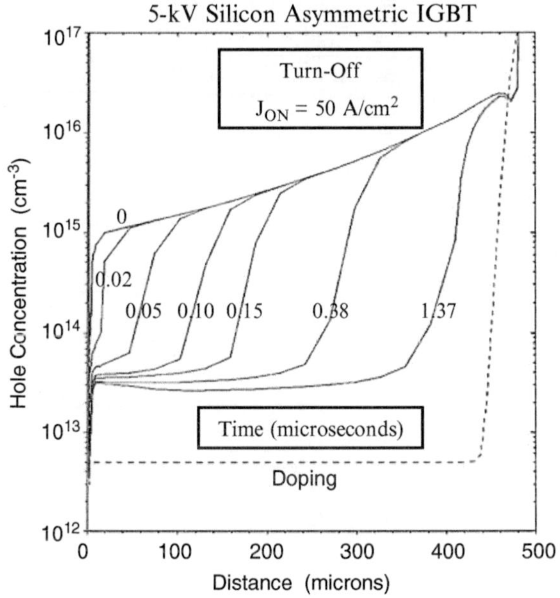

Fig. 5.22 Hole carrier distribution for the 5-kV IGBT turn-off transient during the voltage rise-time

A one-dimensional view of the minority carrier distribution in the 5 kV asymmetric IGBT structure is shown in Fig. 5.22 from the initial steady-state operating point ($t = 0$ μs) to the end of the voltage rise-time ($t = 1.37$ μs). These carrier profiles were taken at $x = 1$ μm through the P-base region. The initial carrier distribution has the distribution predicted by the analytical model (see Eq. 5.20) for the IGBT structure. It can be observed from Fig. 5.22 that the carrier distribution in the N-base region near the collector does not change during the collector voltage rise phase. A significant space-charge region begins to form immediately during the turn-off and expands toward the right-hand side demonstrating that there is no storage phase for the IGBT structure. At larger collector voltages, the hole concentration in the space-charge region is about 3×10^{13} cm^{-3}, which is consistent with the value for p_{SC} obtained using the analytical model with the carriers moving at the saturated drift velocity and a on-state current density of 50 A/cm^2. The width of the space-charge region can be observed to be about 360 μm when the collector voltage reaches 3,000 V, which is slightly larger than that predicted by the analytical model.

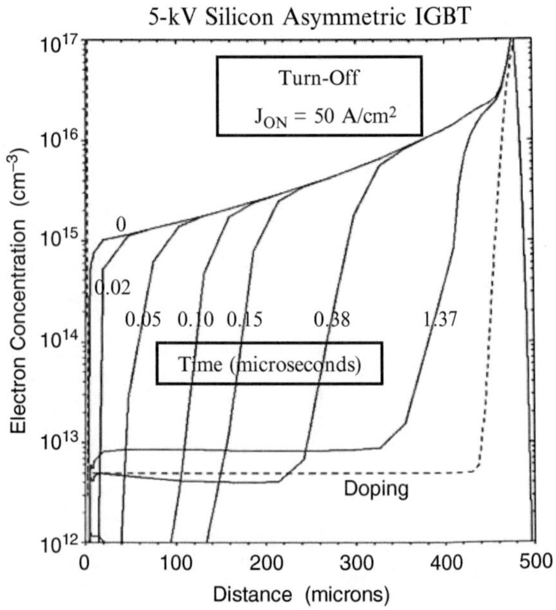

Fig. 5.23 Electron carrier distribution for the 5-kV IGBT during the voltage rise-time

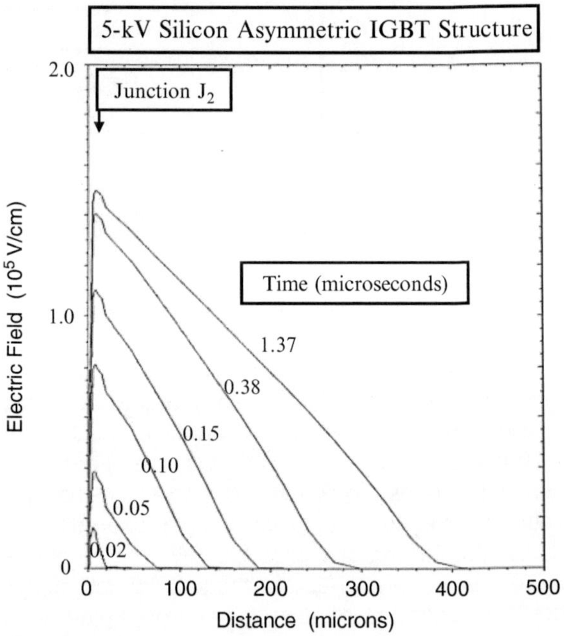

Fig. 5.24 Electric field distribution for the 5-kV asymmetric IGBT during the voltage rise-time

The electron concentration distribution during the voltage rise phase is provided in Fig. 5.23. The electron concentration is equal to the hole concentration in the stored-charge region (i.e., outside the space-charge region) due to charge neutrality. No electrons are observed in the space-charge region at lower collector bias voltages because the electric field is too low for significant impact ionization. However, when the collector voltage reaches 2,000 V at time $t = 0.38$ μs, electrons are observed on the space-charge region due to the on-set of impact ionization. It can be observed that the electron concentration in the space-charge region increases from 5 to 9×10^{12} cm^{-3} when the collector voltage increases from 2,000 to 3,000 V due to enhanced impact ionization at the larger electric field. The collector current created by impact ionization reduces the collector current available for removal of carriers from the stored-charge region. Consequently, the rate of rise of the collector voltage is greatly reduced as the collector voltage increases.

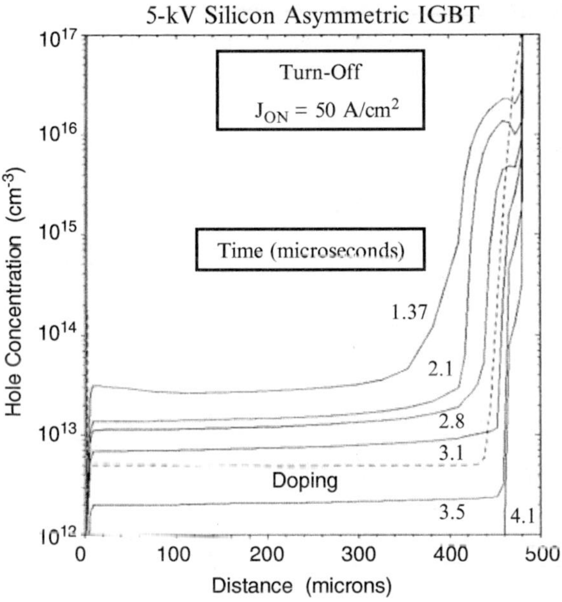

Fig. 5.25 Hole carrier distribution for the 5-kV asymmetric IGBT turn off transient during the current tail-time

The electric field profiles in the 5-kV asymmetric IGBT structure obtained from the numerical simulations are shown in Fig. 5.24 for various time instances during the voltage rise-time. It can be observed that the peak electric field occurs at the P-base/N-base junction (J_2) as expected. The peak electric field increases with time due to supporting a larger collector voltage. The electric field is triangular in shape, unlike during the blocking mode (see Fig. 5.9), even at high collector bias voltages due to the large hole charge in the space-charge region. It can also be observed that the slope of the electric field profile decreases when the time exceeds 0.38 μs because of the addition of electrons in the space-charge region caused by the impact of ionization. The negative charge of the electrons counteracts the positive charge of the holes producing a larger space-charge region width at the end of the voltage transient than predicted by the analytical model based on just the hole charge.

A one-dimensional view of the hole carrier distribution in the 5-kV asymmetric IGBT structure is shown in Fig. 5.25 during the current tail time. The collector voltage was held constant at the collector supply voltage of 3,000-V during this transient. The hole concentration in the stored-charge region begins to decrease immediately after the end of the voltage transient due to the recombination process and the removal of holes and electrons by the collector current flow. At the same time, the space-charge region expands in spite of a constant collector voltage because the hole concentration in the space-charge region reduces. From Fig. 5.25, it can be observed that all the holes in the N-base region have been removed at time $t = 3\ \mu s$ corresponding to the end of the first phase of the collector current transient (see Fig. 5.21). Subsequently, the holes remaining in the N-buffer layer are at concentrations well below its doping concentration. Consequently, the recombination of holes in the N-buffer layer during the second part of the collector current transient occurs under low-level injection conditions as assumed in the analytical model.

5.2.5 *Lifetime Dependence*

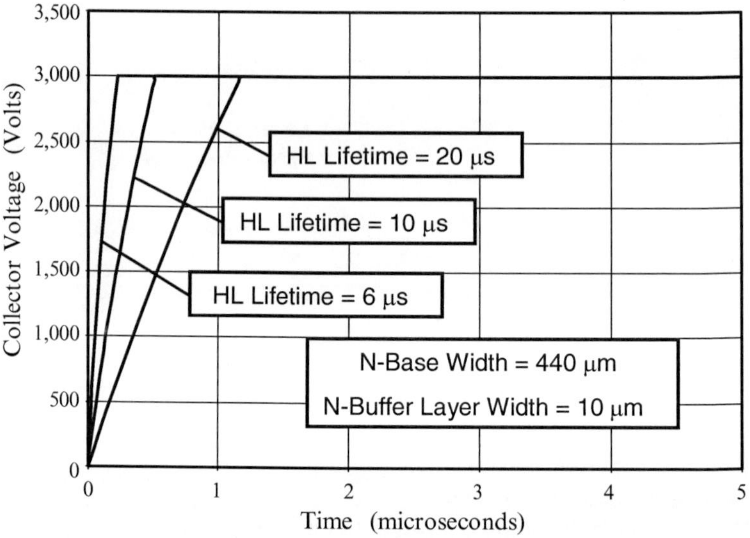

Fig. 5.26 Lifetime dependence of collector voltage transient during turn-off for the 5-kV asymmetric IGBT structure

The optimization of the power losses for the IGBT structure requires performing a trade-off between the on-state voltage drop and the switching losses. One approach to achieve this is by adjusting the lifetime in the drift (N-base) region. A reduction of the lifetime in the drift region also alters the lifetime in the N-buffer layer. The impact of reducing the lifetime on the on-state voltage drop was previously shown in Sect. 5.2.3. The on-state voltage drop increases when the lifetime is reduced.

The new analytical model developed for turn-off of the asymmetric IGBT structure presented in the previous section can be used to analyze the impact of changes to the lifetime in the drift region. The collector voltage transients predicted by the new analytical model are shown in Fig. 5.26. The voltage rise-time increases when the lifetime is increased because of the larger concentration for the holes in the N-base region that are being removed during the collector voltage transient. The voltage rise-time obtained by using the analytical model is 0.255, 0.51, and 1.17 μs for high-level lifetime values of 6, 10, and 20 μs, respectively. Using these values with a collector supply voltage of 3,000 V yields an average [d*V*/d*t*] for the collector voltage transient of 1.18×10^{10} V/s, 5.88×10^{9} V/s, and 2.56×10^{9} V/s for high-level lifetime values of 6, 10, and 20 μs, respectively.

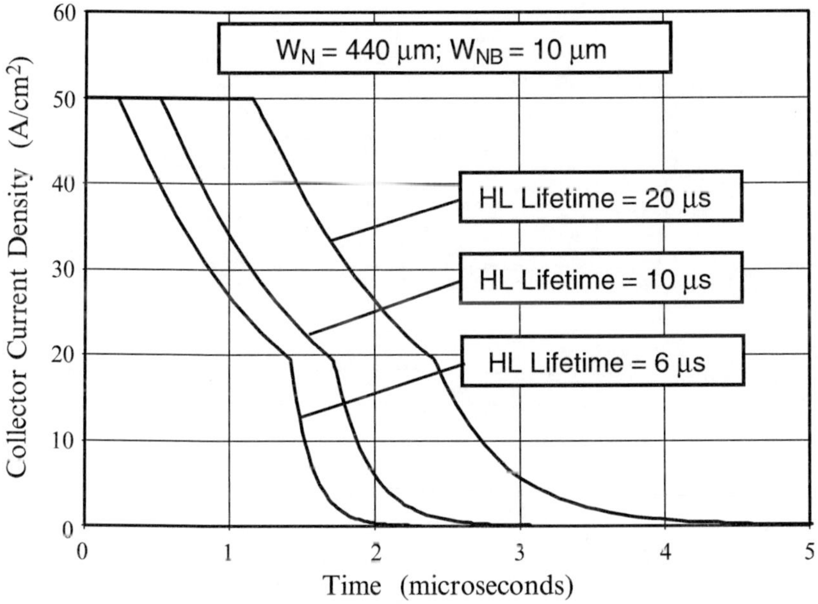

Fig. 5.27 Lifetime dependence of collector current transient during turn-off for the 5-kV asymmetric IGBT structure

The collector current transients predicted by the new analytical model are shown in Fig. 5.27. It can be observed that the current transient becomes longer when the lifetime in the N-base region increases. The current fall-time increases when the lifetime is increased because of the larger concentration for the holes in the N-base region that are being removed during the current transient. The new analytical model for the collector current transient does not take into account the influence of the impact ionization generated current. The addition of holes and electrons due to impact ionization can be expected to prolong the collector current transient during the first part.

Simulation Example

In order to gain insight into the impact of the lifetime in the N-base region on the operation of the 5-kV asymmetric IGBT structure, the results of two-dimensional numerical simulations for a typical structure are discussed here. The device structure used has the cross-section shown in Fig. 5.1 with a half-cell width of 3.5 μm. The widths of the N-base and N-buffer layer regions are 440 and 30 μm, respectively. The high-level lifetime in the N-base region was varied between 4 and 20 μs. For turning-off the IGBT structures, the numerical simulations were performed with gate voltage rapidly ramped down from 10 to 0 V in 10 ns starting from an on-state current density of 50 A/cm^2. The resulting waveforms obtained from the numerical simulations for the collector voltage and current are shown in Fig. 5.28 for the case of a collector supply voltage of 3,000 V.

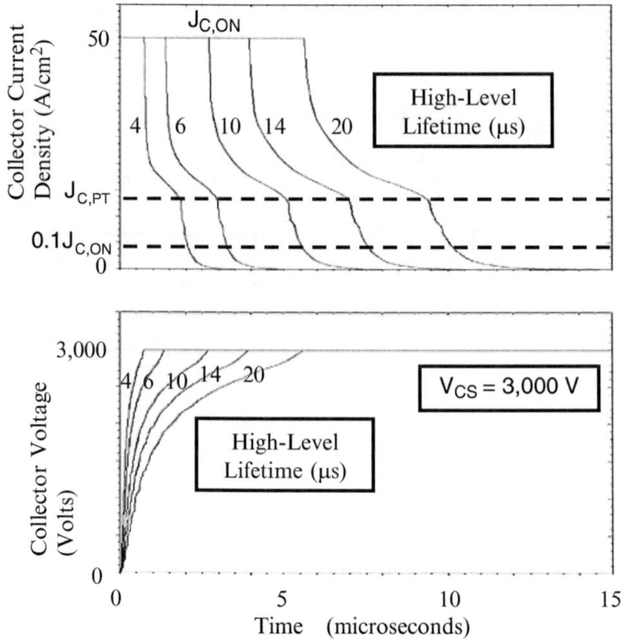

Fig. 5.28 Impact of lifetime on the 5-kV asymmetric IGBT turn-off waveforms

The numerical simulations show a decrease in the voltage rise-time with reduction of the lifetime in the N-base region. The [dV/dt] values for the collector voltage transients, at voltages below 100 V where impact ionization is negligible, are close to those obtained with the new analytical model. However, the on-set of impact ionization at larger collector voltages greatly prolongs the collector voltage transient. This produces a substantial increase in the power dissipation within the IGBT structure during the voltage rise-time phase. In practice, the device temperature will rise during the transient due to the power dissipation resulting in suppressing the impact ionization.

The numerical simulations of the 5-kV asymmetrical IGBT structure also show a substantial increase in the collector current fall-time when the lifetime increases. The numerical simulations show a much larger time for the first part of the collector current decay when compared with the analytical model due to the influence of impact ionization generated carriers. The numerical simulations show a reduction of the collector current during the first part of the decay to the punch-through anode current ($J_{C,PT}$) which is independent of the lifetime in the N-base region as predicted by the analytical model.

5.2.6 Switching Energy Loss

The power loss incurred during the switching transients limit the maximum operating frequency for the IGBT structure. Power losses during the turn-on of the IGBT structure are significant but strongly dependent on the reverse recovery behavior of the fly-back rectifiers in circuits. Consequently, it is common practice to use only the turn-off energy loss per cycle during characterization of IGBT devices. The turn-off losses are associated with the voltage rise-time interval and the current fall-time interval. The energy loss for each event can be computed by integration of the power loss, as given by the product of the instantaneous current and voltage. During the voltage rise-time interval, the anode current is constant while the voltage increases in a non-linear manner as a function of time. In order to simplify the analysis, the energy loss during the voltage rise-time interval will be computed using:

$$E_{\mathrm{OFF,V}} = \frac{1}{2} J_{\mathrm{C,ON}} V_{\mathrm{CS}} t_{\mathrm{V,OFF}} \tag{5.52}$$

For the typical switching waveforms for the 5-kV asymmetric IGBT structure shown in Fig. 5.21 with a collector supply voltage of 3,000 V, the energy loss per unit area during the collector voltage rise-time is found to be 0.10 J/cm^2 if the on-state current density is 50 A/cm^2.

During the collector current fall-time interval, the collector voltage is constant while the current decreases in two phases. In order to simplify the analysis, the energy loss during the collector current fall-time interval will be computed using:

$$E_{\mathrm{OFF,I}} = \frac{1}{2} J_{\mathrm{C,ON}} V_{\mathrm{CS}} t_{\mathrm{I,OFF}} \tag{5.53}$$

For the typical switching waveforms for the 5-kV asymmetric IGBT structure shown in Fig. 5.21 with a collector supply voltage of 3,000 V, the energy loss per unit area during the collector current fall-time is found to be 0.14 J/cm^2 if the on-state current density is 50 A/cm^2. The total energy loss per unit area ($E_{\mathrm{OFF,V}} + E_{\mathrm{OFF,I}}$) during the turn-off process for the 5-kV asymmetric IGBT structure is found to be 0.24 J/cm^2.

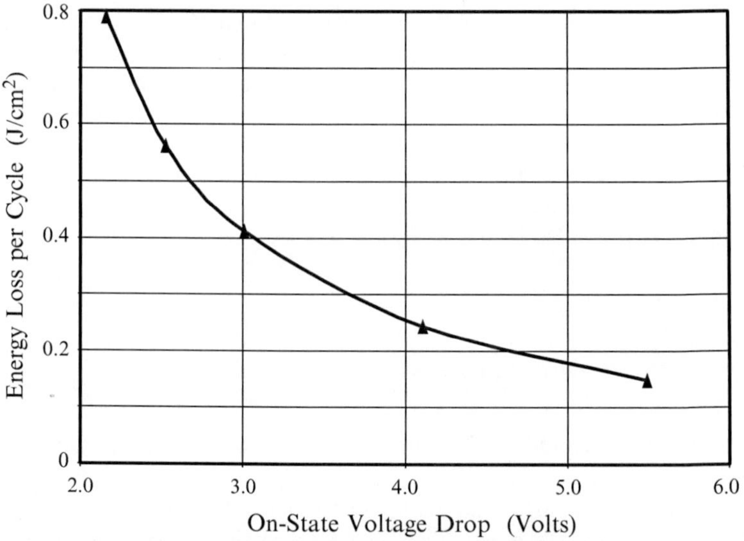

Fig. 5.29 Trade-off curve for the silicon 5-kV asymmetric IGBT structure: lifetime in N-base region

Using the results obtained from the numerical simulations, the on-state voltage drop and the total energy loss per cycle can be computed. These values are plotted in Fig. 5.29 to create a trade-off curve to optimize the performance of the silicon 5-kV asymmetric IGBT structure by varying the lifetime in the N-base region. Devices used in lower frequency circuits would be chosen from the left-hand-side of the trade-off curve while devices used in higher frequency circuits would be chosen from the right-hand-side of the trade-off curve.

5.2.7 *Maximum Operating Frequency*

The maximum operating frequency for operation of the 5-kV asymmetric IGBT structure can be obtained by combining the on-state and switching power losses:

$$P_{\mathrm{D,TOTAL}} = \delta\, P_{\mathrm{D,ON}} + E_{\mathrm{OFF}} f \tag{5.54}$$

where δ is the duty cycle and f is the switching frequency. In the case of the baseline asymmetric IGBT device structure with a high-level lifetime of 6 μs in the N-base region, the on-state voltage drop is 4.11 V at an on-state current density of 50 A/cm^2. For the case of a 50% duty cycle, the on-state power dissipation contributes

103 W/cm^2 to the total power loss. Using a total turn-off energy loss per cycle of 0.24 J/cm^2 in Eq. 5.54 yields a maximum operating frequency of about 400 Hz.

High-Level Lifetime (μs)	On-State Voltage Drop (Volts)	On-State Power Dissipation (W/cm^2)	Energy Loss per Cycle (J/cm^2)	Maximum Operating Frequency (Hz)
20	2.16	54.0	0.791	185
14	2.53	63.2	0.563	243
10	3.01	75.2	0.413	302
6	4.11	103	0.244	399
4	5.49	137	0.150	419

Fig. 5.30 Power loss analysis for the 5-kV asymmetric IGBT structure

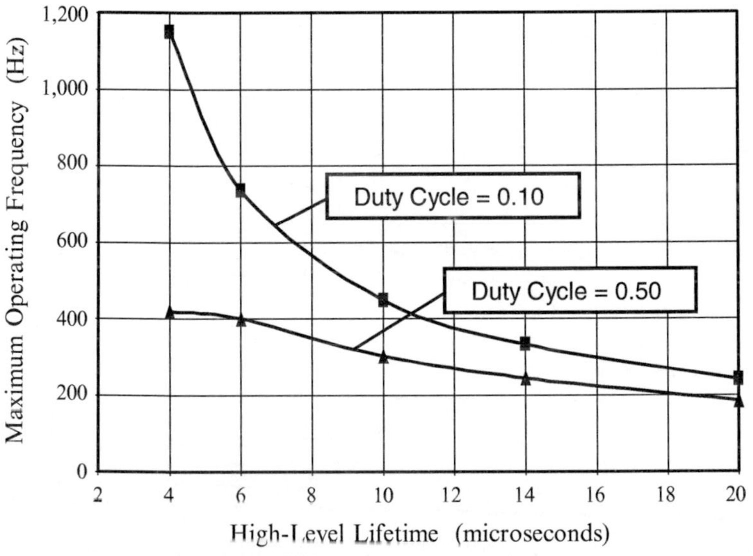

Fig. 5.31 Maximum operating frequency for the 5-kV asymmetric IGBT structure

The maximum operating frequency for the silicon 5-kV asymmetric IGBT structure can be increased by reducing the lifetime in the N-base region. Using the results obtained from the numerical simulations, the on-state voltage drop and the energy loss per cycle can be computed. These values are provided in Fig. 5.30 together with the maximum operating frequency as a function of the high-level lifetime in the N-base region under the assumption of a 50% duty cycle and a total power dissipation limit of 200 W/cm^2. The maximum operating frequency is plotted in Fig. 5.31 as a function of the high-level lifetime in the N-base region.

It can be observed that the maximum operating frequency can be increased up to 400 Hz by reducing the high-level lifetime to 4 μs. This is much superior to the maximum operating frequency of 150 Hz for the 5-kV GTO structure. The IGBT is often operated with pulse-width-modulation to synthesize variable frequency output power for motor control. In these applications, the duty cycle can be much shorter than 50%. In this case, the maximum operating frequency for the 5-kV asymmetric IGBT structure can be increased. As an example, the maximum operating frequency for the 5-kV asymmetric IGBT structure operated at a 10% duty cycle is included in Fig. 5.31. It can be seen that the maximum operating frequency can now exceed 1,000 Hz.

5.2.8 *Buffer Layer Doping*

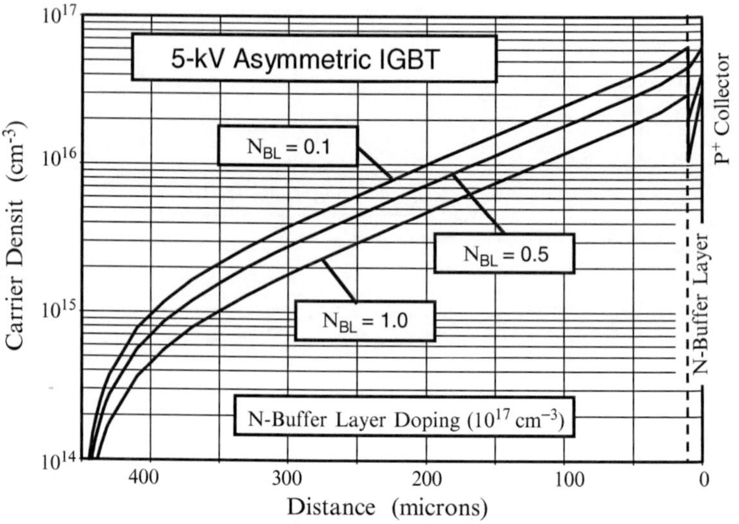

Fig. 5.32 Carrier distribution in the 5-kV asymmetric IGBT structure: buffer layer doping dependence

The free carrier distribution obtained by using the analytical model developed in Sect. 5.2.3 is provided in Fig. 5.32 for the case of a 5-kV asymmetric IGBT structure with a N-base region thickness of 440 μm and an effective N-buffer layer thickness of 10 μm. The buffer layer doping concentration was varied from 1×10^{16} cm^{-3} to 1×10^{17} cm^{-3} while using a hole lifetime (τ_{p0}) in the N-base region of 3 μs. It can be observed that the hole concentration ($p_{NB}(0)$) increases at the collector side of the N-buffer layer (at $y = 0$) from 3.2 to 6.1×10^{16} cm^{-3}

when the buffer layer doping concentration is reduced. In addition, the hole concentration is increased at the emitter side when the N-buffer layer doping concentration decreases. These results also indicate that the on-state voltage drop will increase rapidly if the doping concentration of the buffer layer is increased beyond 1×10^{17} cm^{-3}. However, the smaller stored charge in the N-base region and buffer layer will reduce the turn-off time and energy loss per cycle.

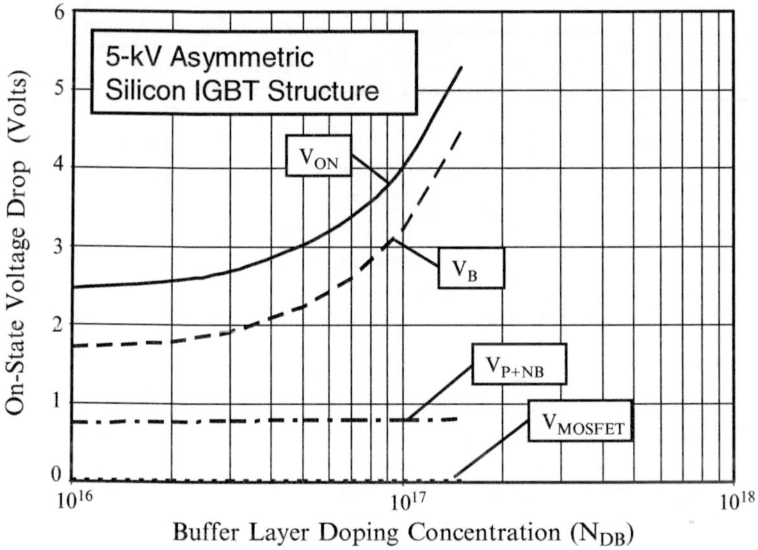

Fig. 5.33 On-state voltage drop for the 5-kV asymmetric IGBT structure: N-buffer layer doping dependence

The characteristics of the 5-kV asymmetric silicon IGBT structure can also be optimized by changing the doping concentration in the N-buffer layer. Consider the case of the 5-kV asymmetric IGBT structure with the optimized N-base region width of 440 μm and high level lifetime of 6 μs, and an effective N-buffer layer width of 10 μm. The on-state voltage drop for this device structure obtained by using the analytical model is provided in Fig. 5.33 as a function of the doping concentration in the N-buffer layer. The on-state voltage drop is 4.0 V for an N-buffer layer doping concentration of 1×10^{17} cm^{-3}. From the figure, it can be concluded that the changes in the on-state voltage drop are determined by the variation of the voltage drop in the N-base region. It can be seen that the on-state voltage drop can be reduced by decreasing the doping concentration in the N-buffer layer. However, this will increase the hole concentration in the N-base region (see Fig. 5.32) resulting in longer switching times. These results indicate that the doping concentration of the N-buffer layer must typically [7] range from 1×10^{16} cm^{-3} to 1×10^{17} cm^{-3}.

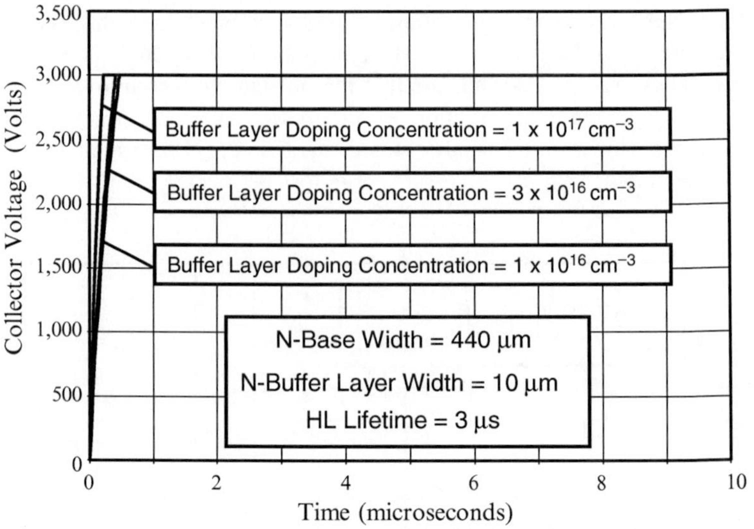

Fig. 5.34 Collector voltage transient during turn-off for the asymmetric IGBT structure

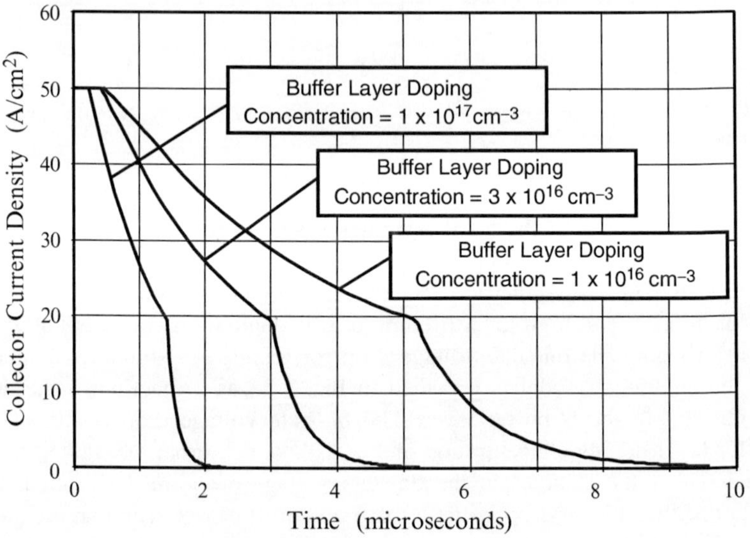

Fig. 5.35 Collector Current Transient during Turn-off for the Asymmetric IGBT Structure

The turn-off waveforms predicted by using the analytical model described in this chapter for the collector voltage and collector current transients are shown in Figs. 5.34 and 5.35 for various doping concentrations in the buffer layer. The analytical model predicts only a small change in the rate of rise of the collector voltage with buffer layer doping concentration. However, the analytical model predicts a strong impact of the buffer layer doping concentration of the collector current waveform.

Simulation Example

The on-state i–v characteristics for the 5-kV asymmetric IGBT structure were obtained with various doping concentrations for the diffused doping profile of the N-buffer layer by using a gate bias voltage of 10 V. A high-level lifetime of 6 μs was used for all of these cases. The on-state characteristics obtained from the numerical simulations are shown in Fig. 5.36. It can be seen from this figure that the on-state characteristics degrade rapidly when the peak N-buffer layer doping concentration is increased above 1×10^{17} cm^{-3}.

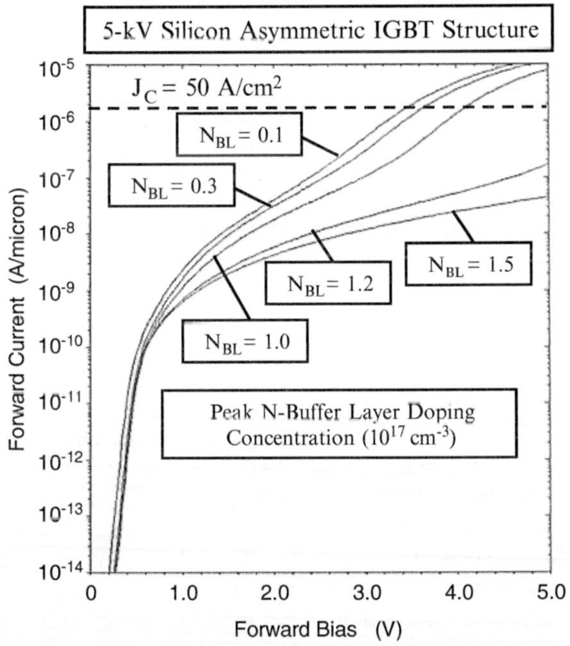

Fig. 5.36 On-state characteristics of the 5-kV asymmetric IGBT structure: buffer layer doping dependence

The on-state voltage drop predicted by the analytical model when the N-buffer layer doping is altered is compared with that obtained from the numerical simulation results in Fig. 5.37. At low N-buffer layer doping levels, the analytical model predicts a lower on-state voltage drop than observed with the numerical simulations by about 1 V. A very rapid increase in the on-state voltage drop is observed in the numerical simulations when the peak N-buffer layer doping concentration is increased beyond 1×10^{17} cm^{-3} for the case of a high-level lifetime of 6 μs in the N-base region. For this reason, it is common practice to use a peak N-buffer doping concentration of between 1 and 5×10^{16} cm^{-3} in high voltage asymmetric IGBT devices.

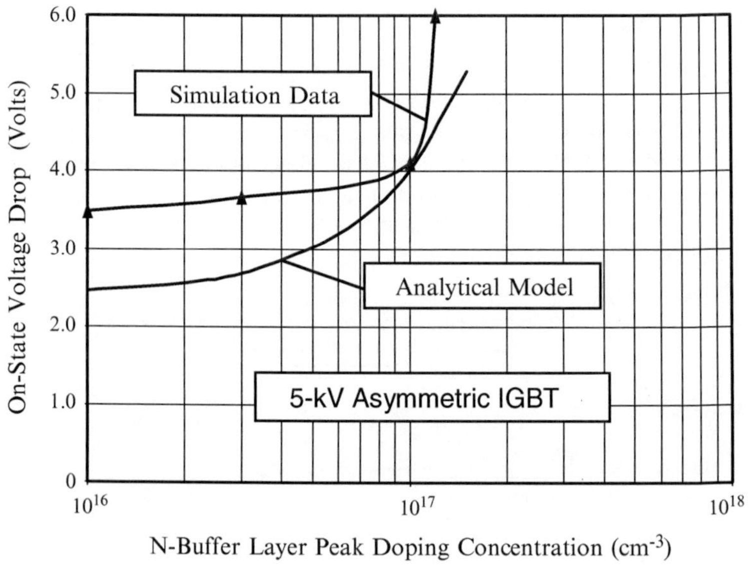

Fig. 5.37 On-state voltage drop for the 5-kV asymmetric IGBT structure: N-buffer layer doping dependence

The hole concentration in the on-state at a current density of 50 A/cm^2 obtained using the numerical simulations of the 5-kV asymmetric IGBT structure is shown in Fig. 5.38. It can be observed that the hole concentration throughout the N-base region is reduced when the doping concentration of the N-buffer layer is increased. This occurs due to a reduction of the injection efficiency of the N-buffer/P$^+$ collector junction (J_1) when the doping concentration of the N-buffer layer is increased. For the range of buffer layer doping concentrations shown in this figure, the hole concentration is not significantly changed throughout the N-base region. Consequently, it can be expected that the voltage drop across the N-base region will also not increase significantly when the doping concentration of the N-buffer layer remains below 1×10^{17} cm^{-3}. This is consistent with the relatively small increase in the on-state voltage drop for the 5-kV asymmetric IGBT structure observed in the numerical simulations as shown in Fig. 5.37 for buffer layer doping concentrations below 1×10^{17} cm^{-3}.

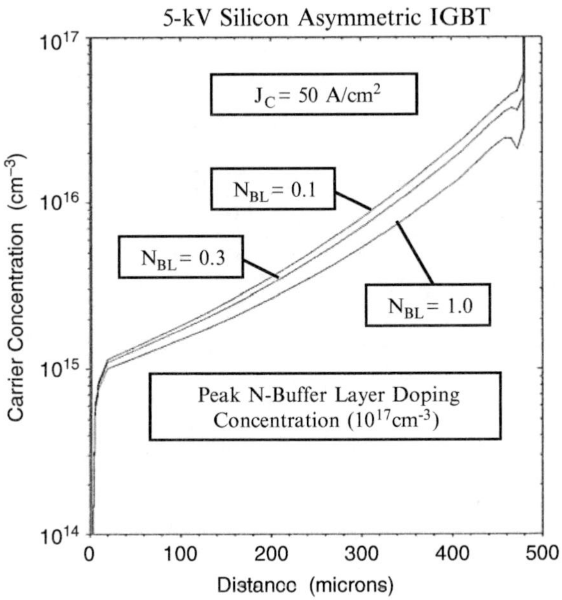

Fig. 5.38 On-state carrier distribution in the 5-kV asymmetric IGBT structure: buffer layer doping dependence

Numerical simulations of the turn-off for the 5-kV silicon asymmetric IGBT structure with various buffer layer doping concentrations were performed by stepping the gate voltage down from 10 V to 0 V in 10 ns starting from an on-state current density of 50 A/cm^2. The resulting waveforms obtained from the numerical simulations for the collector voltage and current are shown in Fig. 5.39 for the case of a collector supply voltage of 3,000 V. It can be observed that all the voltage waveforms begin to increase immediately indicating that there is no storage time for the IGBT structure. This is because the P-base/N-base junction (J_2) is reverse biased for the IGBT structure operating in the on-state. The rate of increase in the collector voltage for the various cases of buffer layer doping concentration is almost the same at lower collector voltages. This is consistent with the predictions of the analytical model (see Fig. 5.34) for the voltage waveform.

The collector current waveforms for the 5-kV asymmetric IGBT structure exhibit a significant change when the buffer layer doping concentration is increased. This is consistent with the predictions of the analytical model that were shown in Fig. 5.35. In all cases, the collector current decays at a faster rate during the initial part of the transient and occurs at a faster rate than obtained using an exponential decay with the lifetime in the N-base region. In all cases, the end of the first part of the collector current transient occurs when it reaches the same current density (about 20 A/cm^2) which is equal to the punch-through current density predicted by the analytical model. Subsequently, the collector current transient occurs at a faster rate when the buffer layer doping concentration is increased as predicted by the analytical model. By comparing Figs. 5.35 and 5.39, it can be concluded that the analytical model provides a good qualitative and quantitative prediction of the collector current transient.

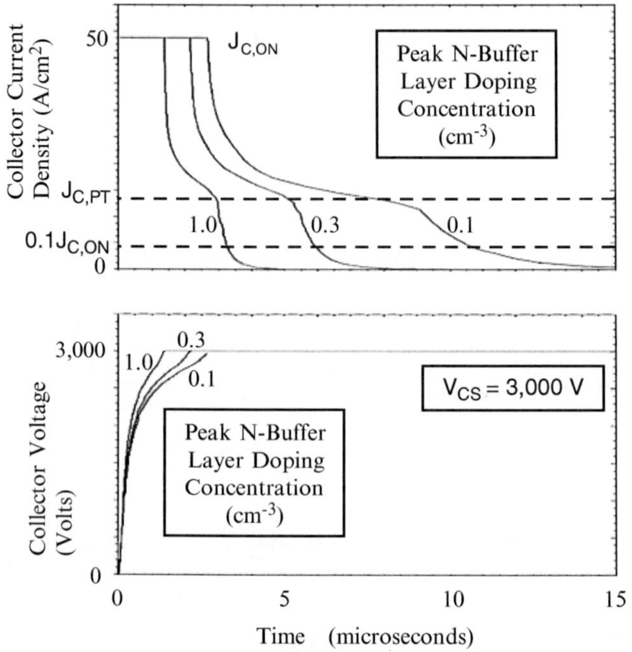

Fig. 5.39 Impact of buffer layer doping concentration on the 5 kV IGBT turn-off waveforms

5.2.8.1 Switching Energy Loss

As shown in the previous sections, the optimization of the power losses for the IGBT structure requires performing a trade-off between the on-state voltage drop and the switching losses. One approach to achieving this is by reducing the lifetime in the drift region as discussed in Sect. 5.2.7. An alternative approach to achieve this is by adjusting the doping concentration in the buffer layer [9]. Changes to the buffer layer doping level alter the injection efficiency of the collector junction and the lifetime in the buffer layer. The buffer layer doping concentration can be conveniently altered by changing the surface concentration of the buffer layer diffusion profile. The ability to adjust the device characteristics with the buffer layer doping is constrained at the lower end by the need to suppress reach-through of the depletion region in the forward blocking mode and at the higher end by reduction of the current gain of the P-N-P transistor to a level where the on-state voltage drop of the IGBT becomes very large. The analytical model presented in Sect. 5.2.4.3 predicts a dependence of the switching behavior of the asymmetric IGBT structure on the buffer layer doping concentration.

Using the results obtained from the numerical simulations for the silicon 5-kV asymmetric IGBT structures with different buffer layer doping concentrations, the on-state voltage drop and the total energy loss per cycle can be computed. These values are plotted in Fig. 5.40 to create a trade-off curve to optimize the performance of the silicon 5-kV asymmetric IGBT structure by varying the doping concentration in the buffer layer. Devices used in lower frequency circuits would be chosen from the left-hand-side of the trade-off curve while devices used in higher frequency circuits would be chosen from the right-hand-side of the trade-off curve. The trade-off curve obtained by varying the doping concentration in the buffer layer extends over a much narrower range than when the lifetime in N-base region is varied. It is difficult to precisely control the doping concentration in the buffer layer while the lifetime in the N-base region can be accurately modified by using electron irradiation.

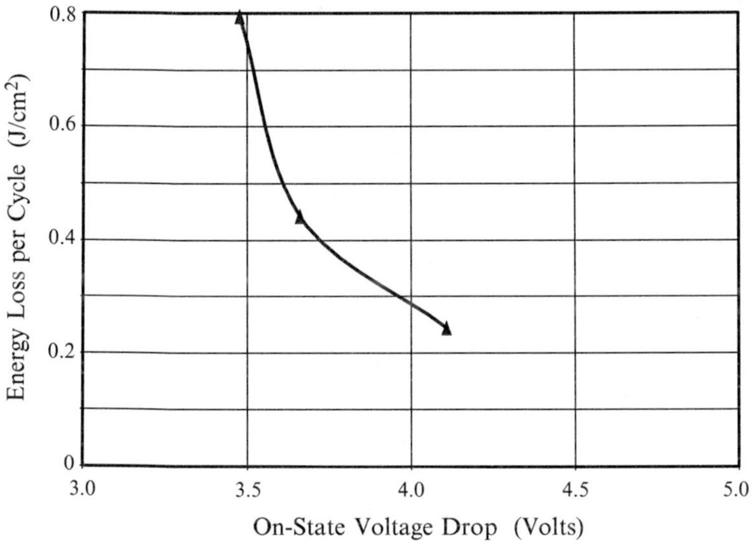

Fig. 5.40 Trade-off curve for the silicon 5-kV asymmetric IGBT structure: buffer layer doping concentration

5.2.8.2 Maximum Operating Frequency

As discussed previously, the maximum operating frequency for the IGBT structure is limited by the turn-off losses. The turn-off losses are associated with the voltage rise-time interval and the current fall-time interval. The energy loss for each event can be computed by using results of the numerical simulations previously provided because they include the adverse influence of impact ionization during the transients. Using this information, the maximum operating frequency for the IGBT structure can be derived using Eq. 5.54.

Peak Buffer Layer Doping (cm^{-3})	On-State Voltage Drop (Volts)	On-State Power Dissipation (W/cm^2)	Energy Loss per Cycle (J/cm^2)	Maximum Operating Frequency (Hz)
1×10^{16}	3.48	87.0	0.795	142
3×10^{16}	3.66	91.5	0.443	245
1×10^{17}	4.11	103	0.244	399

Fig. 5.41 Power loss analysis for the 5-kV asymmetric IGBT structure

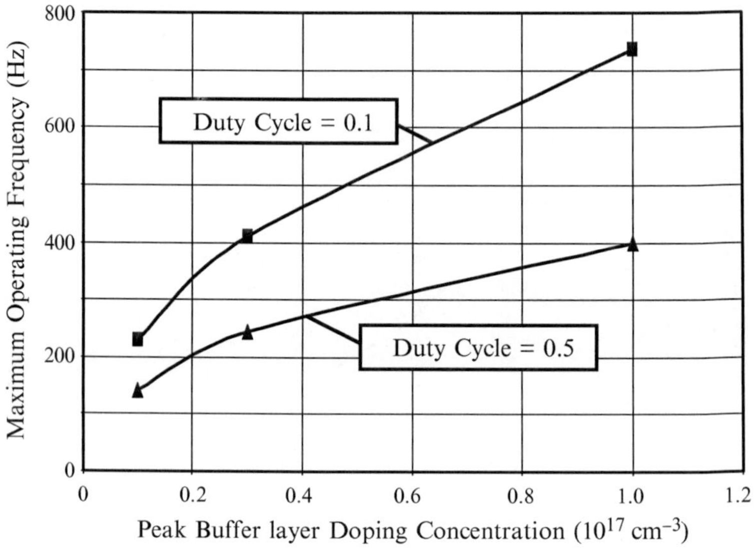

Fig. 5.42 Maximum operating frequency for the 5-kV asymmetric IGBT structure: buffer layer doping concentration

The maximum operating frequency for the silicon 5-kV IGBT structure can be increased by increasing the doping concentration in the buffer layer. Using the results obtained from the numerical simulations, the on-state voltage drop and the energy loss per cycle can be computed. These values are provided in Fig. 5.41 together with the maximum operating frequency as a function of the peak doping concentration in the N-buffer layer under the assumption of a 50% duty cycle and a total power dissipation limit of 200 W/cm^2. The maximum operating frequency is plotted in Fig. 5.42 as a function of the doping concentration in the buffer layer. It can be observed that the maximum operating frequency can be increased only up to 400 Hz by using an N-buffer layer doping concentration of 1×10^{17} cm^{-3}. The figure also provides the maximum operating frequency for the 5-kV asymmetric IGBT structure for the case of operation with a duty cycle of

10%. In this case, the maximum operating frequency for the 5-kV IGBT structure can be increased to over 700 Hz.

5.2.9 Transparent Emitter Structure

The optimization of the power losses for the silicon IGBT structure can be achieved by performing a trade-off between the on-state voltage drop and the switching losses. In addition to optimization of the lifetime in the drift region and the doping concentration in the buffer layer, this trade-off can be achieved by using a lightly doped collector region with small thickness [10]. This is known as a *transparent emitter structure* because the collector region serves as the emitter of the internal P-N-P transistor in the IGBT structure. Reducing the doping concentration of the collector region of the IGBT produces a decrease in the injection efficiency of the collector junction. This produces a reduction in the injected hole concentration at the collector side within the N-base region which improves the turn-off time. In addition, holes are injected into the collector region during the turn-off phase, diffused through its small thickness, and then recombine at the metal contact. This additional process for removal of the holes from the N-base region speeds up the turn-off process during the collector current tail. A large lifetime value is usually utilized in the N-base region for the transparent emitter structure because the injected hole concentration has been reduced using the smaller collector injection efficiency.

Simulation Example

In order to gain insight into the operation of the 5-kV asymmetric IGBT structure with transparent emitter structure, the results of two-dimensional numerical simulations for a device with collector surface concentration of 1×10^{18} cm^{-3} and thickness of 5 μm are discussed here. The device structure used has the cross-section shown in Fig. 5.1 with a half-cell width of 3.5 μm. The widths of the N-base and N-buffer layer regions were 440 and 30 μm, respectively. The surface concentration of the buffer-layer diffusion was varied to optimize the device characteristics. The doping profile for the silicon 5 kV asymmetric IGBT structure with the transparent emitter region is shown in Fig. 5.43 for the case of a peak buffer layer doping concentration of 6×10^{16} cm^{-3}. The peak N-buffer layer doping concentration was varied from 1.5 to 6×10^{16} cm^{-3}. The on-state *i–v* characteristics obtained using the numerical simulations are shown in Fig. 5.44 by the solid lines. A high-level lifetime of 20 μs was used in the N-base region for these devices. It can be observed that the on-state characteristics are sensitive to the doping concentration of the N-buffer layer. The on-state voltage drop increases significantly when the peak buffer layer doping concentration is made larger than 3×10^{16} cm^{-3}.

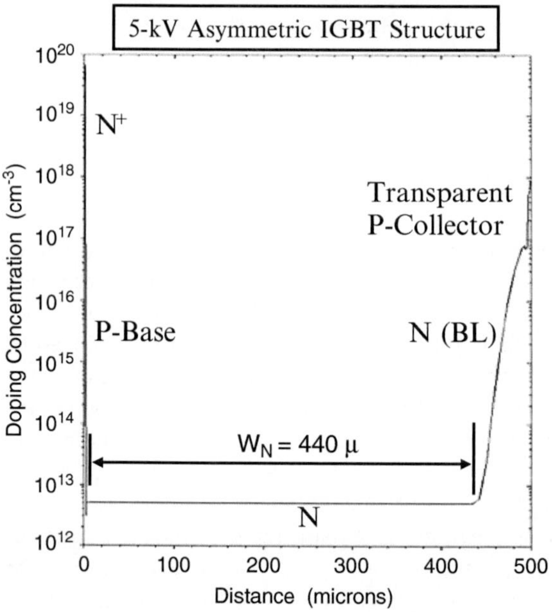

Fig. 5.43 Doping profile for the asymmetric 5-kV IGBT with transparent collector

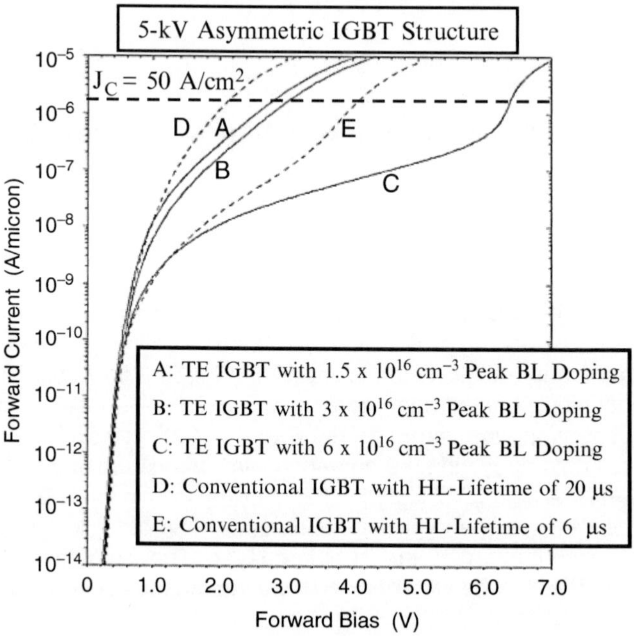

Fig. 5.44 On-state characteristics of the 5-kV asymmetric IGBT structures

For comparison purposes, the on-state characteristics of the conventional silicon 5 kV asymmetric IGBT structure with highly doped collector region are also provided in the figure (see dashed lines) for the case of a high-level lifetime of 3 and 10 μs. It can be observed that the on-state voltage drop for the 5-kV asymmetric IGBT structure with the transparent emitter region falls between those of the conventional structure with high-level lifetime of 6 and 20 μs when the buffer layer doping concentration falls between 1 and 3 × 10^{16} cm^{-3}. For an equal high-level lifetime value of 20 μs, the transparent emitter IGBT structure has a larger on-state voltage drop than the conventional IGBT structure due to the smaller injection efficiency of its collector junction.

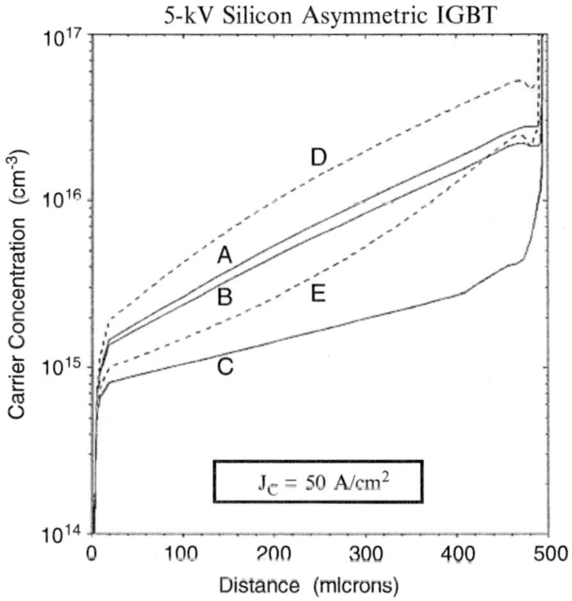

Fig. 5.45 On-state carrier distribution in the 5-kV asymmetric IGBT structure: see Fig. 5.44 for legend

The injected hole concentration profile in the N-base region is shown in Fig. 5.45 for the three silicon 5-kV asymmetric IGBT structures with transparent emitter regions. For comparison purposes, the hole profiles for the conventional asymmetric IGBT structure is also provided in the figure with dashed lines for two lifetime values. The hole concentration in the N-base region at the collector side for the transparent emitter structures (*A*, *B*) with buffer layer doping concentration of 1.5 and 3 × 10^{16} cm^{-3} are close to that for the conventional structure with high-level lifetime of 6 μs but the hole concentration for these devices is larger at the emitter side of the N-base region due to the larger high-level lifetime in the transparent emitter structures. The improved hole distribution for these transparent emitter IGBT structures (*A*, *B*) results in a lower on-state voltage drop than for the conventional IGBT structure (*E*) with a high-level lifetime of 6 μs. When the buffer layer doping concentration of the transparent emitter IGBT structure (*C*) is increased to 6×10^{16} cm^{-3}, the hole concentration becomes smaller throughout the N-base region. This produces the relatively large on-state voltage drop observed in Fig. 5.44 for structure *C*. However, the reduced stored charge in the N-base region is beneficial in speeding up the turn-off process.

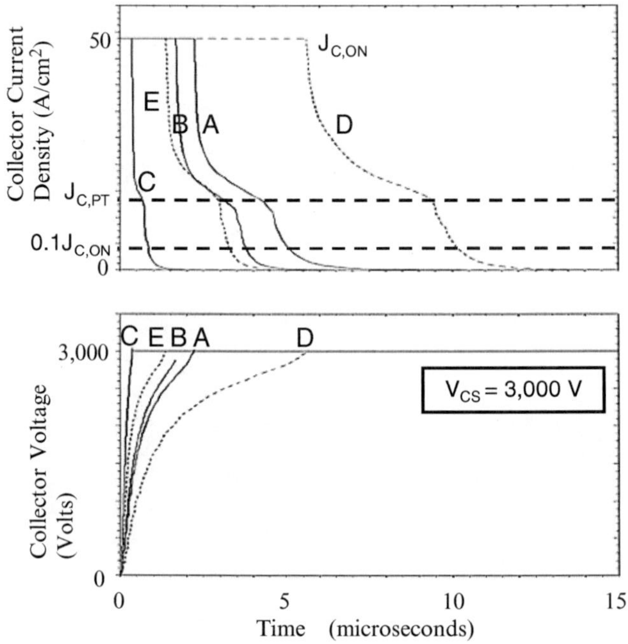

Fig. 5.46 Impact of buffer layer doping concentration on the 5 kV IGBT turn-off waveforms: see Fig. 5.44 for legend

Numerical simulations of the turn-off for the 5-kV silicon asymmetric IGBT structure with the transparent emitter structure were performed with the gate voltage stepped down from 10 to 0 V in 10 ns from an on-state current density of 50 A/cm^2. The resulting waveforms obtained from the numerical simulations for the collector voltage and current are shown in Fig. 5.46 for the case of a collector supply voltage of 3,000 V. As in the case of the two previous figures, the transients are shown for three values of the peak buffer layer doping concentrations for the transparent emitter structure. For comparison purposes, the collector voltage and current transients for the conventional asymmetric IGBT structure are also provided in the figure with dashed lines for two lifetime values. The collector voltage rise-time for the device (*C*) with the transparent emitter is very small for the highest buffer layer doping concentration because of the reduced hole concentration in the N-base region (see Fig. 5.45). The current fall-time for the 5-kV silicon asymmetric IGBT structure with the transparent emitter region during the first phase becomes longer when the buffer layer doping concentration is reduced. However, the end point for this transient ($J_{C,PT}$) is the same as that for the conventional IGBT structure and remains independent of the buffer layer doping concentration as predicted by the analytical model.

The collector voltage and current transients for the transparent emitter IGBT structure (*B*) with buffer layer doping concentration of 3×10^{16} cm^{-3} is very similar to those observed for the conventional IGBT structure (*E*) with a high-level lifetime of 6 μs. Consequently, the energy loss per cycle ($E_{OFF,T}$) for these device structures are nearly the same. However, the on-state voltage drop for the IGBT device with the transparent emitter structure is 3.06 V at an on-state current density of 50 A/cm^2 compared with 4.11 V for the conventional IGBT device structure. This demonstrates that the transparent emitter structure has a better combination of on-state voltage drop and turn-off energy loss.

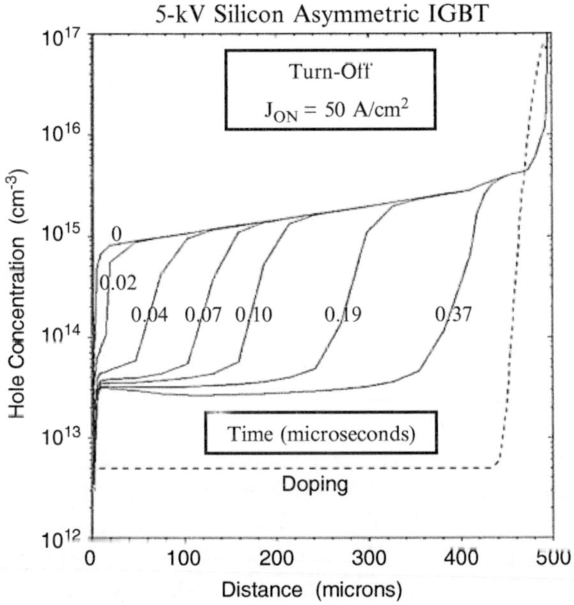

Fig. 5.47 Hole carrier distribution during the current fall-time for the 5-kV asymmetric IGBT structure (C) with transparent emitter region

The changes in the carrier distribution within the 5-kV asymmetric IGBT structure with the transparent emitter region during the voltage rise-time are similar to those for the conventional device structure as shown in Fig. 5.47 for the case of the device (*C*) with buffer layer doping concentration of 6×10^{16} cm^{-3}. This also holds true during the current fall-time as shown in Fig. 5.48. The faster transients for the voltage and current observed for the transparent emitter IGBT structure are associated with the reduced stored charge in the N-base region.

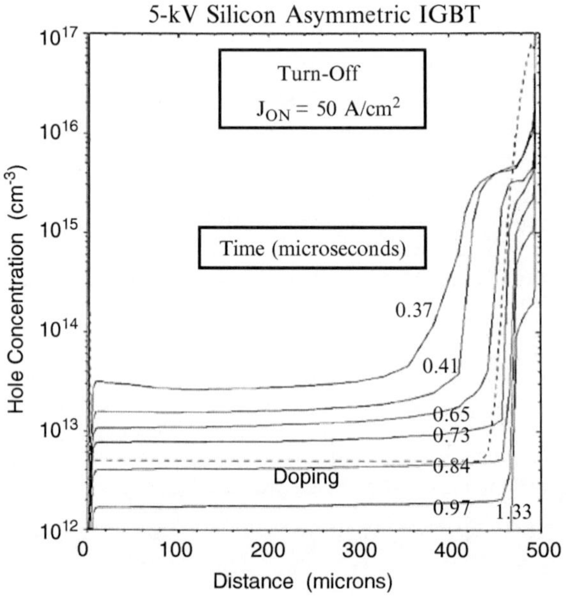

Fig. 5.48 Hole carrier distribution during the current fall-time for the 5-kV asymmetric IGBT structure (C) with transparent emitter region

5.2.9.1 Switching Energy Loss

Using the results obtained from the numerical simulations for the silicon 5-kV asymmetric IGBT structures with transparent emitter region and different buffer layer doping concentrations, the on-state voltage drop and the total energy loss per cycle can be computed. These values are plotted in Fig. 5.49 to create a trade-off curve to optimize the performance of the silicon 5-kV asymmetric IGBT structure with transparent emitter region by varying the doping concentration in the buffer layer. Devices used in lower frequency circuits would be chosen from the left-hand-side of the trade-off curve while devices used in higher frequency circuits would be chosen from the right-hand-side of the trade-off curve. The trade-off curve obtained by varying the doping concentration in the buffer layer extends over a smaller range of energy loss per cycle than when the lifetime in the N-base region is altered. However, the transparent emitter structure provides a better trade-off between the on-state voltage drop and the energy loss per cycle.

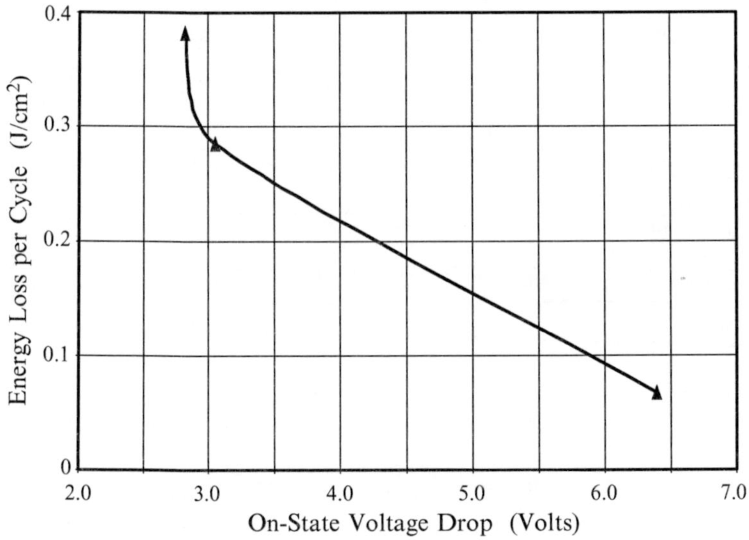

Fig. 5.49 Trade-off curve for the silicon 5-kV asymmetric IGBT structure with transparent emitter structure

5.2.9.2 Maximum Operating Frequency

Structure	On-State Voltage Drop (Volts)	On-State Power Dissipation (W/cm^2)	Energy Loss per Cycle (J/cm^2)	Maximum Operating Frequency (Hz)
Transparent Emitter	3.056	76.4	0.285	434
Conventional	4.109	103	0.244	399

Fig. 5.50 Power loss analysis for the 5-kV asymmetric IGBT structure with transparent emitter region

The maximum operating frequency for the 5-kV asymmetric IGBT structure is limited by the turn-off losses. The turn-off losses are associated with the voltage rise-time interval and the current fall-time interval. The energy loss for each event can be computed by using the equation previously provided in Sect. 5.2.6. Using this information, the maximum operating frequency for the IGBT structure can be derived using Eq. 5.54.

The maximum operating frequency for the silicon 5-kV asymmetric IGBT structure with the transparent emitter structure is larger than that for the conventional 5-kV Asymmetric IGBT structure due to its smaller combination of on-state voltage drop and switching energy loss per cycle. The performance of these devices is compared in Fig. 5.50. The maximum operating frequency obtained under the assumption of a 50% duty cycle and a total power dissipation limit of 200 W/cm^2 for the transparent emitter structure is 434 Hz when compared with 399 Hz for the conventional device structure.

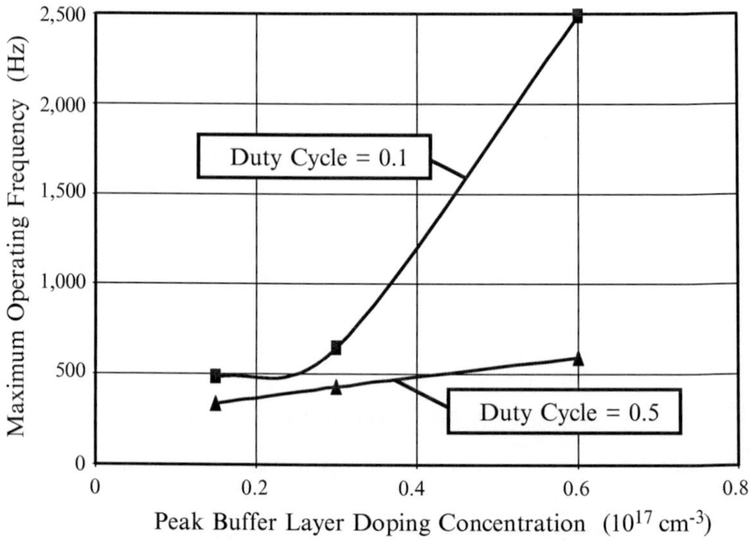

Fig. 5.51 Maximum operating frequency for the 5-kV asymmetric IGBT structure: Fig. 5.52 leakage current for the asymmetric 5-kV IGBT with transparent emitter: see Fig. 5.44 for legend. Buffer layer doping concentration

The maximum operating frequency for the silicon 5-kV IGBT structure can be increased by increasing the doping concentration in the buffer layer. Using the results obtained from the numerical simulations, the on-state voltage drop and the energy loss per cycle can be computed. The maximum operating frequency is plotted in Fig. 5.51 as a function of the doping concentration in the buffer layer. It can be observed that, for a duty cycle of 0.5, the maximum operating frequency can be increased to 500 Hz by using an N-buffer layer doping concentration of 6×10^{16} cm^{-3}. The figure also provides the maximum operating frequency for the 5-kV asymmetric IGBT structure for the case of operation with a duty cycle of 10%. In this case, the maximum operating frequency for the 5-kV asymmetric IGBT structure with transparent emitter region can be increased to 2,500 Hz.

5.2.9.3 Leakage Current

For completeness, the blocking characteristics for the 5-kV asymmetric IGBT structure with transparent emitter region are discussed in this section. The reduced doping concentration of the collector region reduces the injection efficiency of the N-buffer layer/P$^+$ collector junction. This results in a smaller current gain for the wide-base P-N-P transistor in the IGBT structure. Since the space-charge-generation current in the IGBT device is amplified by the P-N-P transistor, the reduced current gain of the P-N-P transistor in the transparent emitter structure will result in a smaller leakage current.

Simulation Example

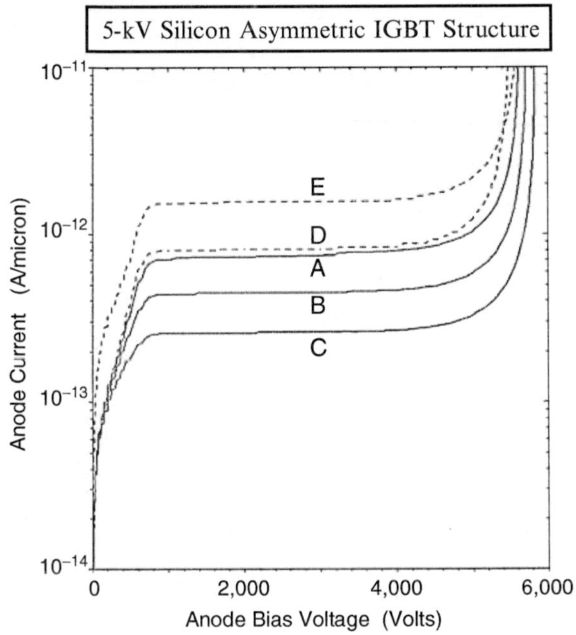

Fig. 5.52 Leakage current for the asymmetric 5-kV IGBT with transparent emitter: see Fig. 5.44 for legend

The forward blocking characteristics of the 5-kV asymmetric IGBT structure was obtained at 300 K by increasing the collector bias voltage while using zero gate bias. The characteristics of three transparent emitter devices with different buffer layer doping concentrations are provided in Fig. 5.52. It can be observed that the leakage current for the device reduces when the peak doping concentration of the N-buffer layer is increased. This is due a reduction of the injection efficiency of the N-buffer layer/P-collector junction which makes the gain of the P-N-P transistor smaller. For comparison purposes, the leakage currents for the conventional 5-kV asymmetric IGBT structure are also provided in the figure with dashed lines for two lifetime values. It is found that the leakage current of the conventional 5-kV asymmetric IGBT structure with high-level lifetime of 6 μs is four-times larger than the leakage current for the transparent emitter device with peak buffer layer doping of 3×10^{16} cm^{-3}. These results indicate that the transparent emitter IGBT structure offers superior forward blocking characteristics as well.

5.3 5,000-V Silicon Planar-Gate IGBT

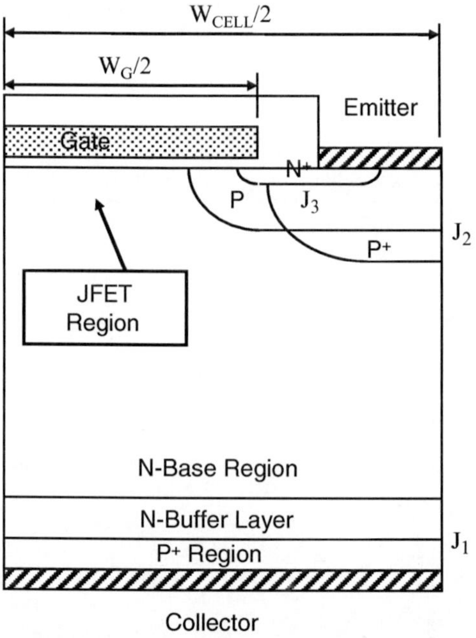

Fig. 5.53 The silicon asymmetric planar-gate IGBT structure

The MOS-gated thyristor structures discussed in subsequent chapters of this book are based up on the planar gate architecture. The characteristics for the 5,000-V asymmetric silicon planar-gate IGBT structure are therefore discussed in this section for purposes of comparison. The design parameters for the N-base region required to achieve this blocking voltage are the same as those for the trench-gate structure. All the analytical models described in the previous section for the trench-gate structure are also valid for the planar gate IGBT structure as well.

The planar-gate IGBT structure is illustrated in Fig. 5.53. In this device, the width of the JFET (Junction Field-Effect Transistor) region and its doping must be carefully selected. On the one hand, if the JFET doping concentration is too low, a snap-back in the on-state characteristics is observed. On the other hand, if the JFET doping concentration is too high, the blocking characteristics will be degraded.

5.3.1 Blocking Characteristics

The physics for blocking voltages in the first and third quadrants by the trench-gate IGBT structure are the same as those for the trench-gate structure. Consequently, only the results of numerical simulations on a typical planar-gate IGBT structure are discussed in this section.

Simulation Example

In order to gain insight into the physics of operation for the 5-kV asymmetric planar-gate IGBT structure under voltage blocking conditions, the results of two-dimensional numerical simulations are described here. The simulations were performed using a cell with the structure shown in Fig. 5.53. This half-cell has a width ($W_{CELL}/2$) of 15 μm (Area $= 1.5 \times 10^{-7}$ cm^{-2}) with a gate width (W_G) of 20 μm. The asymmetric IGBT structure used for the simulations was formed by diffusions performed into a uniformly doped N-type drift region with a doping concentration of 5×10^{12} cm^{-3}. The N-buffer layer was formed by diffusion from the collector side with a depth of 55 μm. For the baseline device structure, the surface concentration of the N-type diffusion was adjusted to achieve a peak doping concentration of 1.2×10^{17} cm^{-3} in the buffer layer. The P-base region was formed with a Gaussian doping profile with a surface concentration of 4.5×10^{17} cm^{-3} and a depth of 3 μm. The N$^+$ emitter region was formed with a Gaussian doping profile with a surface concentration of 1×10^{20} cm^{-3} and a depth of 1 μm. The vertical doping profile for the planar structure is similar to that shown for the trench-gate structure in Fig. 5.5.

The doping profile across the silicon surface for the 5-kV asymmetric planar-gate IGBT structure used for the numerical simulations is provided in Fig. 5.54. It can be seen that the peak doping concentration of the P-base region is 1.3×10^{17} cm^{-3} and the channel length is 1.2 μm. This is sufficient to prevent reach-through limited breakdown in the P-base region.

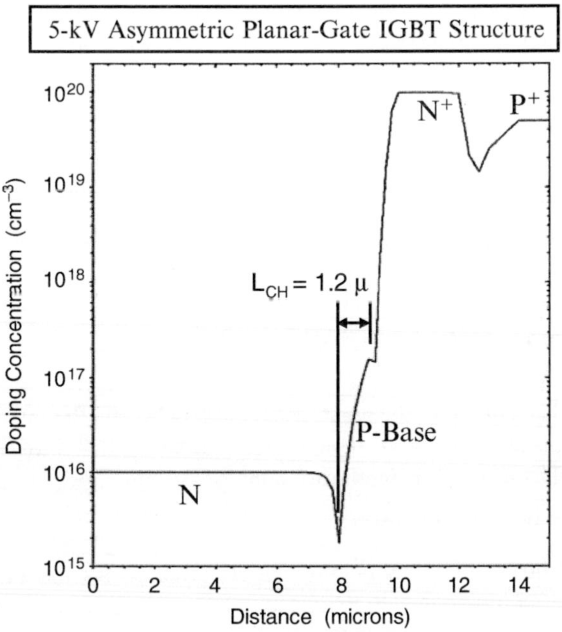

Fig. 5.54 Channel doping profile for the asymmetric 5-kV planar-gate IGBT structure

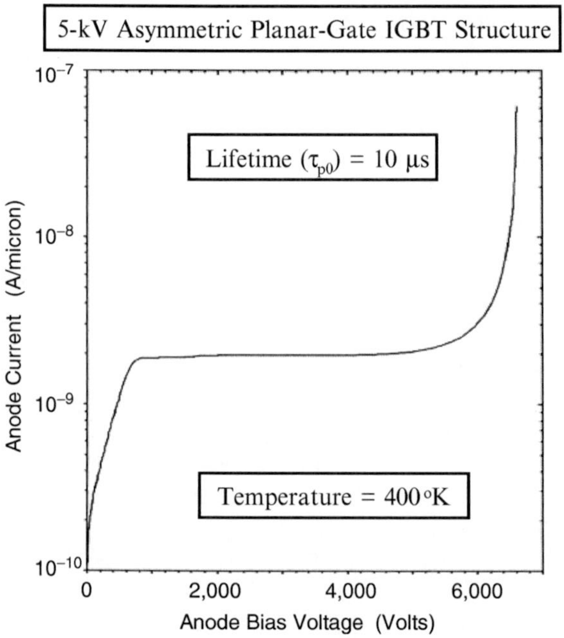

Fig. 5.55 Forward blocking characteristics for the asymmetric planar-gate IGBT structure

The forward blocking capability of the silicon asymmetric planar-gate IGBT structure was obtained using numerical simulations by increasing the collector bias while maintaining the gate electrode at zero volts. The characteristics obtained for a lifetime (τ_{p0}) of 10 μs is provided in Fig. 5.55 at 400 K. As in the case of the trench-gate structure, the leakage current increases rapidly with increasing anode bias voltage until about 780 V as predicted by the analytical model (see Fig. 5.2). The voltage is primarily supported within the lightly doped portion of N-base region in the asymmetric planar-gate IGBT structure during operation in the forward blocking mode. The electric field profiles are similar to those previously shown in Fig. 5.9.

5.3.2 *On-State Voltage Drop*

The model for the on-state characteristics and the injected carrier distribution in the planar-gate IGBT structure is similar to that provided in the previous section. The results obtained from numerical simulations are provided in this section.

Simulation Results

The results of two-dimensional numerical simulations for the 5-kV asymmetrical silicon planar-gate IGBT structure are described here. The total width ($W_{CELL}/2$)

of the structure, as shown by the cross-section in Fig. 5.53, was 15 μm (Area = 1.5×10^{-7} cm^{-2}). A gate width (W_G) of 20 μm was used with a gate oxide thickness of 500 Å. The P-base and N^+ emitter regions were formed by using Gaussian doping profiles defined from the upper surface. The N-buffer layer and P^+ collector regions were formed by using Gaussian doping profiles defined from the lower surface. The doping profiles for the baseline device structure were already shown in Figs. 5.5 and 5.54.

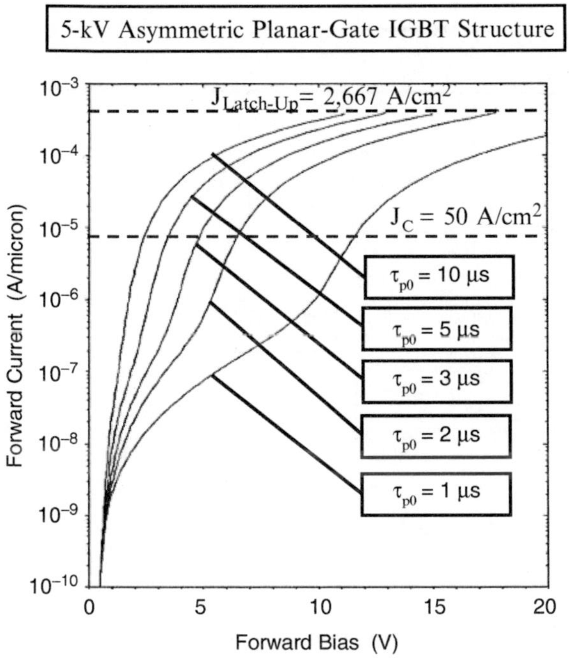

Fig. 5.56 On-state carrier distribution in the 5-kV asymmetric planar gate IGBT structure: lifetime dependence

The on-state characteristics of the 5-kV silicon asymmetric planar-gate IGBT structure were obtained by using a gate-bias voltage of 10 V for the case of various values for the lifetime in the drift region. The characteristics obtained from the numerical simulations are shown in Fig. 5.56. The current initially increases exponentially with increasing collector bias. At current densities above 0.005 A/cm^2, the on-state voltage drop begins to increase more rapidly. Consequently, the on-state voltage drop increases as expected with reduction of the lifetime (τ_{p0}, τ_{n0}) indicated in the figure. The on-state voltage drop at a hole lifetime (τ_{p0}) value of 10 μs is found to be 2.43 V at an on-state current density of 50 A/cm^2 and increases to 11.5 V when the hole lifetime (τ_{p0}) is reduced to 1 μs.

The on-state voltage drop for the 5-kV asymmetric planar-gate IGBT structure is determined by the distribution of carriers injected into the N-base region producing the desired reduction of its resistance. The hole distribution in the device is provided in Fig. 5.57 at two locations. It can be observed that the injected carrier density is four orders of magnitude larger than the doping concentration on the collector side but not as large on the emitter side. The hole concentration goes to zero at the junction between the P^+ region and the N-drift region ($x = 15$ μm) but is enhanced under the gate electrode ($x = 0$ μm).

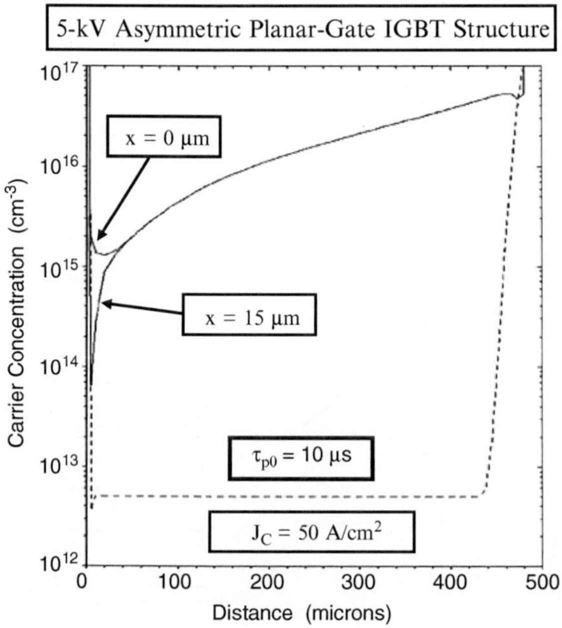

Fig. 5.57 On-state carrier distribution in the 5-kV asymmetric planar-gate IGBT structure: position dependence

The hole concentration in the drift region of the 5-kV asymmetric planar-gate IGBT structure is shown in Fig 5.58 for various values of the lifetime (τ_{p0}, τ_{n0}). It can be observed that the injected carrier density is four orders of magnitude larger than the doping concentration on the collector side but not as large on the emitter side. The injected carrier density is significantly reduced in the middle of the drift region when the lifetime is reduced to 1 μs. The predictions of the analytical model (see Fig. 5.11) have the same general characteristics as observed in the numerical simulations. The hole concentration values at the buffer-layer interfaces in the asymmetric IGBT structure are also quite well predicted by the analytical model despite the assumption of a uniform doping concentration in the N-buffer layer. The hole concentration in the planar gate IGBT are slightly smaller than those observed in the trench-gate IGBT resulting in a larger on-state voltage drop.

The variation of the on-state voltage drop for the 5-kV asymmetric planar-gate IGBT structure as a function of the lifetime in the N-base region is compared with that obtained for the trench-gate structure in Fig. 5.59. It can be observed that the on-state voltage drop of the planar-gate structure is only slightly (~1 V) greater than that for the trench-gate device at this voltage rating.

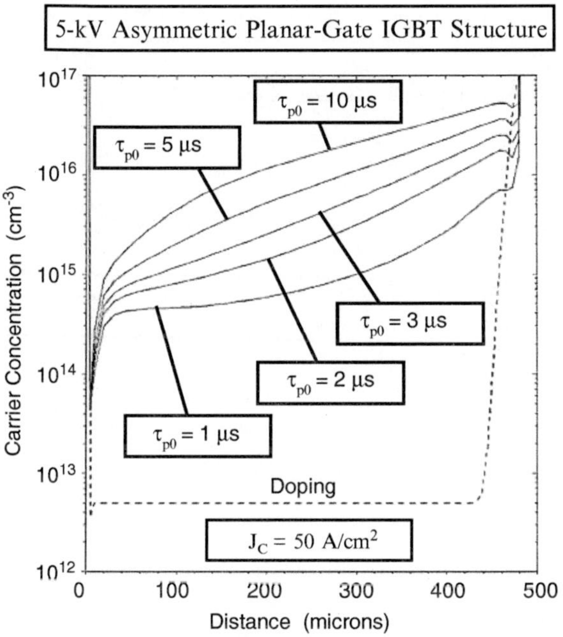

Fig. 5.58 On-state carrier distribution in the 5-kV asymmetric planar-gate IGBT structure: lifetime dependence

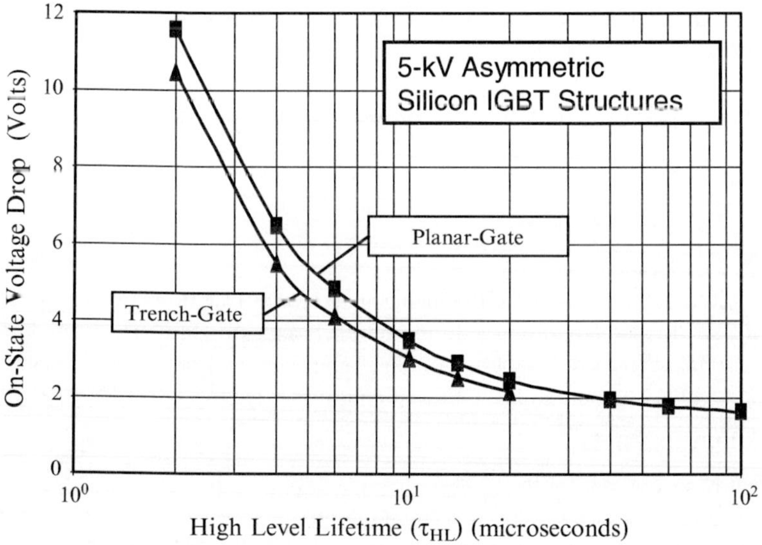

Fig. 5.59 On-state voltage drop for the 5-kV asymmetric IGBT structure: planar vs trench gate

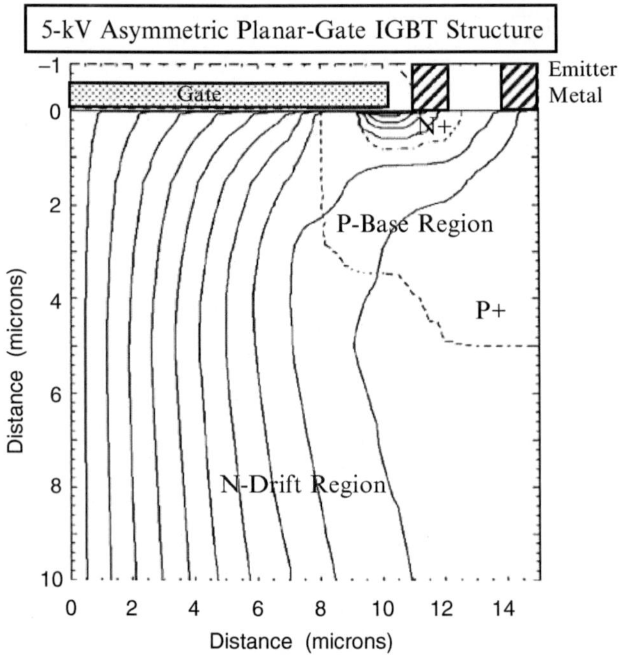

Fig. 5.60 On-state current distribution in the 5-kV asymmetric planar-gate IGBT structure

The current flow-lines in the on-state for the 5-kV asymmetric planar-gate IGBT structure are provided in Fig. 5.60 for the case of a high-level lifetime of 2 μs in the drift region. It can be observed that most of the current flows via the channel due to the low gain of the P-N-P transistor when the lifetime is small.

5.3.3 Turn-Off Characteristics

The physics for the turn-off for the planar-gate IGBT structure can be expected to be similar to that already discussed for the trench-gate structure. The same analytical models can therefore be applied to the planar-gate device. The results obtained by using numerical simulations are discussed in this section.

Simulation Example

In order to gain insight into the operation of the asymmetric 5-kV planar-gate IGBT structure during its turn-off, the results of two-dimensional numerical simulations for a typical structure are discussed here. The device structure used has the cross-section shown in Fig. 5.53 with a cell half-width of 15 μm. The doping profile for the IGBT structure used in the numerical simulations was provided in

Figs. 5.5 and 5.54. The widths of the uniformly doped N-base region and the diffused N-buffer layer are 440 and 30 μm, respectively. For the typical case discussed here, a high-level lifetime of 4 μs was used in the N-base region.

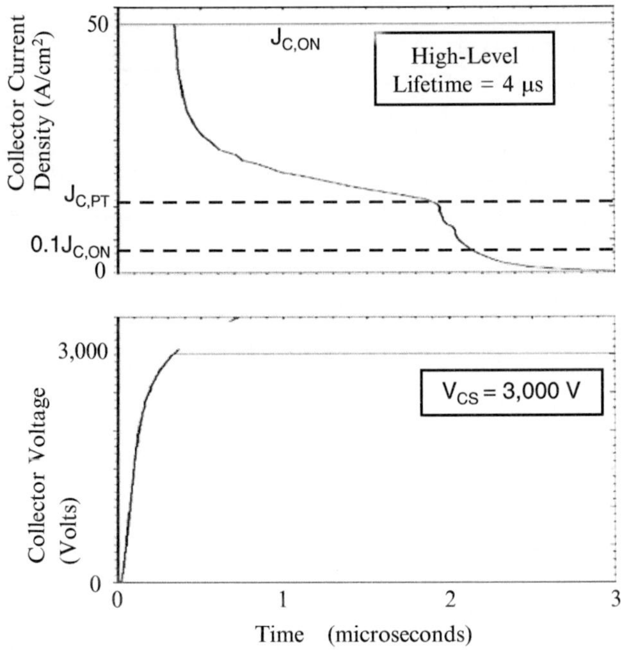

Fig. 5.61 Typical turn-off waveforms for the asymmetric 5-kV planar-gate IGBT structure

The numerical simulations of the 5-kV asymmetric planar-gate IGBT structure were performed with an abrupt reduction of the gate voltage from 10 V to 0 V in 10 ns starting from an on-state current density of 50 A/cm^2. The resulting waveforms obtained from the numerical simulations for the anode voltage and current are shown in Fig. 5.61 for the case of a collector supply voltage of 3,000 V. The turn-off waveforms are similar to, those obtained for the trench-gate structure. The hole carrier and electric field distributions during turn-off for the planar-gate structure are also similar to those previously shown for the trench-gate structure.

5.3.4 Lifetime Dependence

The optimization of the power losses for the IGBT structure requires performing a trade-off between the on-state voltage drop and the switching losses. One approach to achieve this is by adjusting the lifetime in the drift (N-base) region. The results obtained by using numerical simulations are described in this section.

Simulation Example

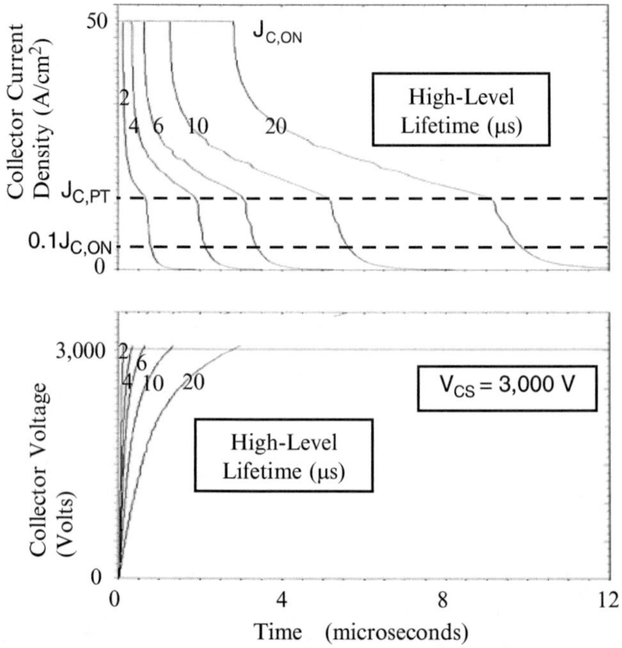

Fig. 5.62 Impact of lifetime on the 5 kV asymmetric planar-gate IGBT turn-off waveforms

In order to gain insight into the impact of the lifetime in the N-base region on the operation of the 5-kV asymmetric planar-gate IGBT structure, the results of two-dimensional numerical simulations for a typical structure are discussed here. The device structure used has the cross-section shown in Fig. 5.53 with a half-cell width of 15 μm. The widths of the N-base and N-buffer layer regions are 440 and 30 μm, respectively. The high-level lifetime in the N-base region was varied between 2 and 20 μs. For turning-off the IGBT structures, the numerical simulations were performed with gate voltage rapidly ramped down from 10 to 0 V in 10 ns starting from an on-state current density of 50 A/cm^2. The resulting waveforms obtained from the numerical simulations for the collector voltage and current are shown in Fig. 5.62 for the case of a collector supply voltage of 3,000 V. In comparison with the trench-gate structure (see Fig. 5.28), the voltage rise-time for the planar gate structure is much shorter while the current fall-time is much longer. The total energy loss per cycle for both devices is similar.

5.3.5 *Switching Energy Loss*

The power loss incurred during the switching transients limits the maximum operating frequency for the IGBT structure. The turn-off energy loss per cycle for the planar-gate IGBT can be computed by the same equations previously developed for the trench-gate structure. Using the results obtained from the numerical simulations, the on-state voltage drop and the total energy loss per cycle can be computed. These values are plotted in Fig. 5.63 to create a trade-off curve to optimize the performance of the silicon 5-kV asymmetric IGBT structure by varying the lifetime in the N-base region. Devices used in lower frequency circuits would be chosen from the left-hand-side of the trade-off curve while devices used in higher frequency circuits would be chosen from the right-hand-side of the trade-off curve.

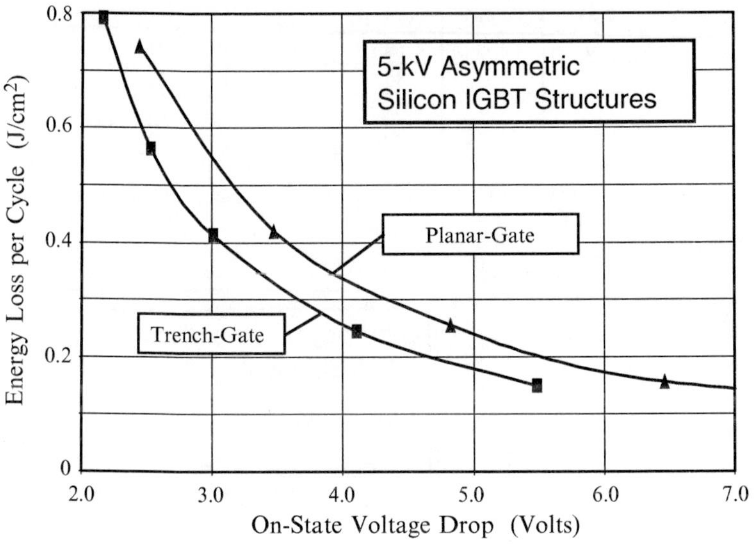

Fig. 5.63 Trade-off curve for the silicon 5-kV asymmetric planar-gate IGBT structure: lifetime in N-base region

For purposes of comparison, the trade-off curve for the 5-kV trench-gate asymmetric IGBT structure has been included in Fig. 5.63. It can be observed that the trade-off curve for the trench-gate structure is superior to that for the planar-gate structure. This is mainly due to the smaller on-state voltage drop observed for the trench-gate structure.

5.3.6 Maximum Operating Frequency

The maximum operating frequency for operation of the 5-kV asymmetric planar-gate IGBT structure can be obtained by combining the on-state and switching power losses as discussed earlier for the trench-gate structure with the aid of

Eq. 5.54. In the case of the asymmetric planar-gate IGBT device structure with a high-level lifetime of 6 μs in the N-base region, the on-state voltage drop is 4.82 V at an on-state current density of 50 A/cm^2. For the case of a 50% duty cycle, the on-state power dissipation contributes 120.5 W/cm^2 to the total power loss. Using a total turn-off energy loss per cycle of 0.255 J/cm^2 in Eq. 5.54 yields a maximum operating frequency of about 312 Hz.

High-Level Lifetime (μs)	On-State Voltage Drop (Volts)	On-State Power Dissipation (W/cm^2)	Energy Loss per Cycle (J/cm^2)	Maximum Operating Frequency (Hz)
20	2.43	60.8	0.743	187
10	3.47	86.8	0.420	270
6	4.82	120.5	0.255	312
4	6.46	161.5	0.097	399

Fig. 5.64 Power loss analysis for the 5-kV asymmetric planar-gate IGBT structure

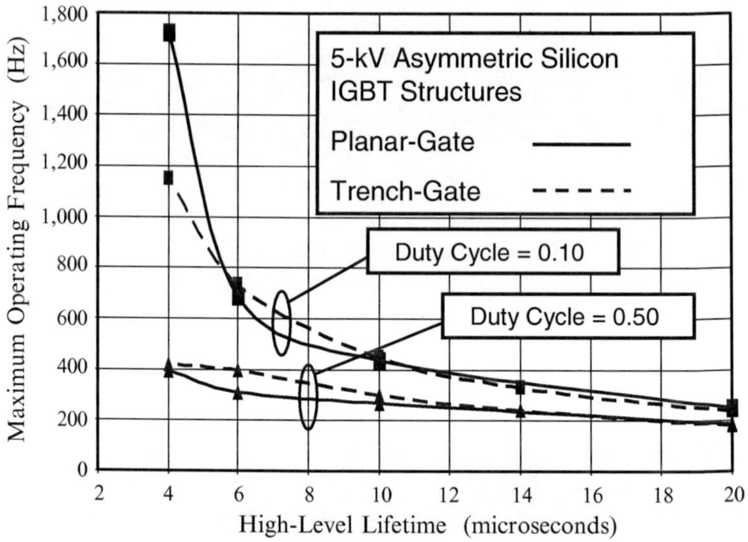

Fig. 5.65 Maximum operating frequency for the 5-kV asymmetric IGBT structures

The maximum operating frequency for the silicon 5-kV asymmetric planar-gate IGBT structure can be increased by reducing the lifetime in the N-base region. Using the results obtained from the numerical simulations, the on-state voltage drop and the energy loss per cycle can be computed. These values are provided in Fig. 5.64 together with the maximum operating frequency as a function of the high level lifetime in the N-base region under the assumption of a 50% duty cycle and a total power dissipation limit of 200 W/cm^2. The maximum operating frequency is plotted in Fig. 5.65 as a function of the high-level lifetime in the N-base region. It can be observed that the maximum operating frequency can be increased up to 400 Hz by reducing the high-level lifetime to 4 μs. In comparison with the trench-gate IGBT structure, it can be observed that the planar-gate structure has a similar maximum frequency of operation.

5.4 10,000-V Silicon IGBT

The 10-kV silicon asymmetric IGBT structure can be expected to function just like the 5-kV device. However, its design and operation is constrained by the larger blocking voltage capability. As discussed below, the very low doping concentrations required for the N-base region are challenging to achieve from a fabrication stand point. The lifetime in the N-base region for the 10-kV device must be larger to maintain a reasonable on-state voltage drop. The larger N-base width results in more stored charge within the structure which limits the switching frequency.

In the previous chapter, it was demonstrated that the GTO structure has a limited reverse biased safe operating area due to influence of the holes in the space-charge region due to current flow. The analysis of the reverse biased safe operating area for the IGBT structure is identical to that provided in Sect. 4.4. Using the results shown in Fig. 5.57, it can be concluded that in order to turn-off the 10-kV asymmetric IGBT structure with a collector supply voltage of 6-kV, it is necessary to reduce the collector current density to only 20 A/cm^2. This value will therefore be utilized when determining the on-state voltage drop and switching transients for the 10-kV asymmetric IGBT structures.

5.4.1 Blocking Characteristics

The electric field distribution within the asymmetric IGBT structure is essentially the same as that illustrated in Fig. 4.3 for the asymmetric GTO structure. Consequently, the design procedure described in Chap. 4 can be applied to the asymmetric IGBT structure. From Fig. 5.50, the N-base region width required to obtain a forward blocking voltage of 11,000 V is 1,100 μm. However, the results of the numerical simulation shown in Chap. 4 for the 10-kV GTO structure demonstrate that an N-base width of 800 μm is sufficient.

Simulation Example

In order to gain insight into the physics of operation for the 10-kV asymmetric trench-gate IGBT structure under voltage blocking conditions, the results of two-dimensional numerical simulations are described here for a device with N-base width of 850 μm. The simulations were performed using a cell with the structure shown in Fig. 5.1. This half-cell has a width of 3.5 μm (Area = 3.5×10^{-8} cm^{-2}). The asymmetric IGBT structure used for the simulations was formed by diffusions performed into a uniformly doped N-type drift region with a doping concentration of 2×10^{12} cm^{-3}. The N-buffer layer was formed by diffusion from the collector side with a depth of 50 μm. The surface concentration of the N-type diffusion was adjusted to achieve a peak doping concentration of 7×10^{16} cm^{-3} in the buffer layer based upon the results previously described for the 5-kV structure. The P-base region was formed with a Gaussian doping profile with a surface concentration of 1.5×10^{18} cm^{-3} and a depth of 3 μm. The N$^+$ emitter region was formed with a Gaussian doping profile with a surface concentration of 1×10^{20} cm^{-3} and a depth of 1 μm. The doping profile in the vertical direction through the N$^+$ emitter region is shown in Fig. 5.66 indicating that the net width of the lightly doped portion of the N-base region is 850 μm after accounting for the diffusions. The P-base and N$^+$ source regions are too shallow to be observed in this figure. Their doping profiles are the same as those for the 5-kV asymmetric IGBT structure previously shown in Fig. 5.6.

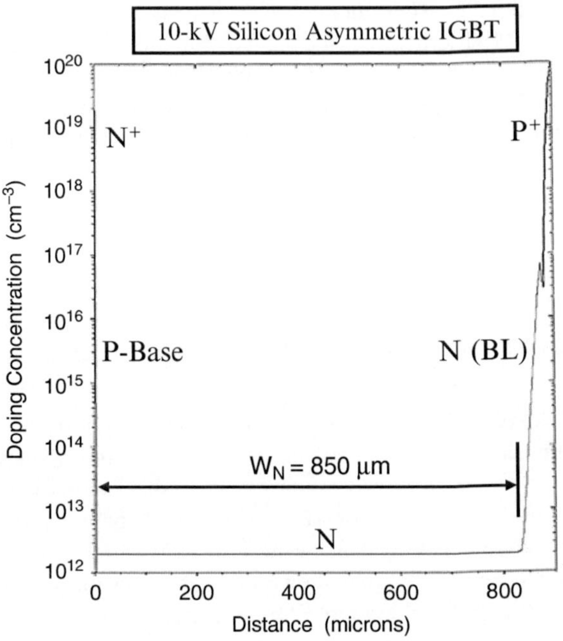

Fig. 5.66 Doping profile for the simulated asymmetric 10-kV IGBT structure

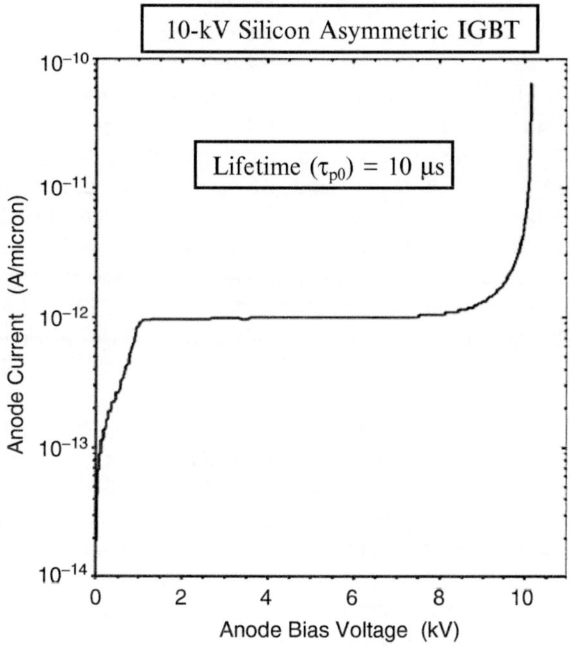

Fig. 5.67 Forward blocking characteristics for the 10-kV asymmetric IGBT structure

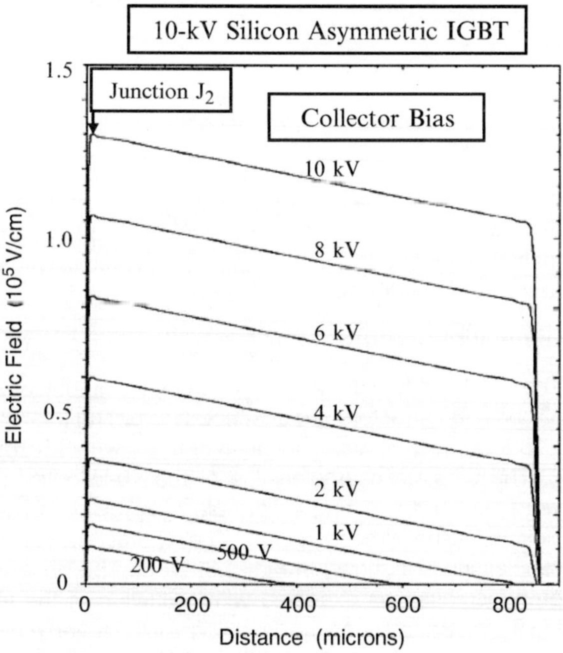

Fig. 5.68 Electric field profiles in the forward blocking mode for the 10-kV asymmetric IGBT structure

The forward blocking capability of the 10-kV silicon asymmetric IGBT structure was obtained by increasing the collector bias while maintaining the gate electrode at zero volts. The characteristics obtained for a lifetime (τ_{p0}) of 10 μs is shown in Fig. 5.67. The leakage current increases rapidly with increasing collector bias voltage until about 1,000 V. This occurs due to the increase in the space-charge-generation volume and the increase in the current gain (α_{PNP}) of the open base P-N-P transistor until the collector bias becomes equal to the reach-through voltage of 1,115 V obtained using the analytical solution given by Eq. 4.2. The leakage current then becomes independent of the collector voltage until close to the breakdown voltage. This behavior is well described by the analytical model. The numerical simulations indicate that a breakdown voltage of 10,500 V is possible with an N-base width of only 850 μm. This blocking voltage is a little lower than that for the GTO structure due to the shallower junctions in the IGBT structures.

The voltage is primarily supported within the lightly doped portion of N-base region in the 10-kV asymmetric IGBT structure during operation in the forward blocking mode. This is illustrated in Fig. 5.68 where the electric field profiles are shown during operation in the forward blocking mode at several collector voltages. It can be observed that the P-Base/N-base junction (J_2) becomes reverse biased during the forward blocking mode with the depletion region extending toward the right-hand-side with increasing (positive) collector bias. The electric field has a triangular shape until the entire lightly doped portion of the N-base region becomes completely depleted. This occurs at a collector bias just above 1,000 V in good agreement with the reach-through voltage of 1,115 V obtained using the analytical solution (see Eq. 4.2). The electric field profile then takes a trapezoidal shape due to the high doping concentration in the N-buffer layer.

5.4.2 *On-State Voltage Drop*

The on-state i–v characteristics and on-state voltage drop can be computed using the analytical model discussed in Sect. 5.2.3. In general, a larger lifetime is required in the N-base region for the 10-kV device when compared with the 5-kV device due to the larger width for the N-base region.

Simulation Results

The results of two-dimensional numerical simulations for the 10-kV asymmetrical silicon IGBT structure are described here. The total half-cell width of the structure, as shown by the cross-section in Fig. 5.1, was 3.5 μm (Area $= 3.5 \times 10^{-8}$ cm^{-2}).

The on-state characteristics of the 10-kV silicon asymmetric IGBT structure were obtained by using a gate bias voltage of 10-V using various values for the lifetime in the N-base region. The characteristics obtained from the numerical simulations are shown in Fig. 5.69. It can be observed that the on-state voltage drop increases as expected with reduction of the lifetime (τ_{p0}, τ_{n0}) indicated in the figure. Based upon the limitations of the reverse biased safe operating area, the on-state current density for the 10-kV silicon asymmetric IGBT structure must be reduced to 20 A/cm^2 as indicated in the figure by the bold dashed line.

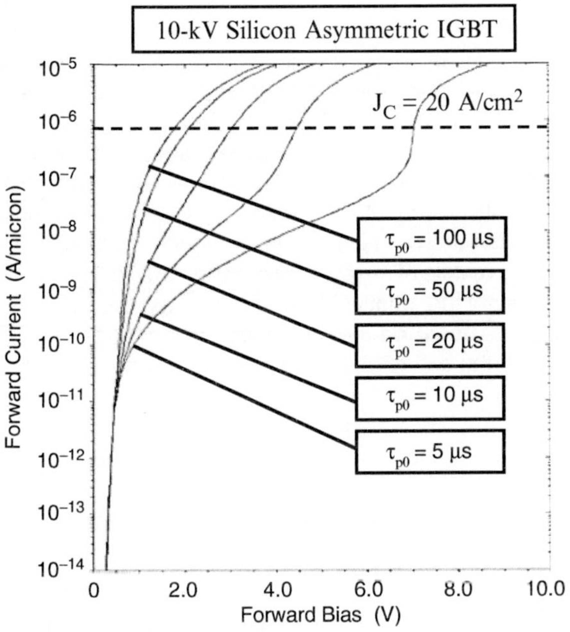

Fig. 5.69 On-state carrier distribution in the 10-kV asymmetric IGBT structure

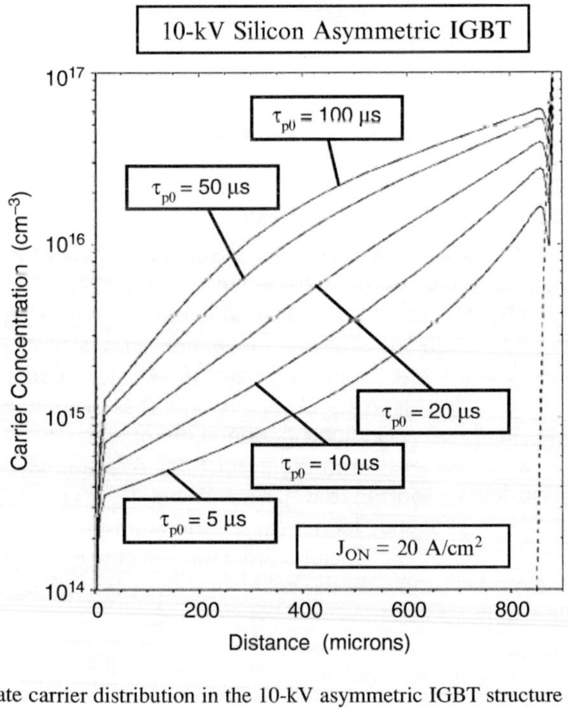

Fig. 5.70 On-state carrier distribution in the 10-kV asymmetric IGBT structure

The good on-state voltage drop for the 10-kV asymmetric IGBT structure for larger values of the lifetime in the N-base region is due to the large number of carriers injected into the drift region producing a drastic reduction of its resistance. This is illustrated in Fig. 5.70 where the injected carrier density is shown for five cases of the lifetime (τ_{p0}, τ_{n0}) in the N-base region of the IGBT structure. It can be observed that the injected carrier density is more than three orders of magnitude larger than the doping concentration for the case of a lifetime of 100 μs. The injected carrier density is reduced by a factor of 3-times near the collector junction when the lifetime is reduced to 5 μs. There is a significant reduction in the injected carrier density in the middle of the drift region when the lifetime is reduced to 10 or 5 ms. This is due to the relatively large width for the N-base region when compared with the 5-kV silicon IGBT structure. The reduced hole concentration in the drift region on the emitter side produces the observed increase in on-state voltage drop.

5.4.3 Turn-Off Characteristics

The physics for turn-off of the 10-kV silicon asymmetric IGBT structure can be expected to be the same as that for the 5-kV device structure. However, it is more difficult to turn-off the 10-kV device structure due to the larger amount of stored charge in the N-base region and its limited reverse biased safe operating area. Consequently, the 10-kV device structure must be operated at a lower on-state current density not only due to its larger on-state voltage drop but from a switching point of view.

Simulation Results

Numerical simulations of the turn-off for the 10-kV silicon IGBT structure with a high-level lifetime of 20 μs were performed by stepping the gate voltage down from 10 to 0 V in 10 ns using an on-state current density of 20 A/cm^2. The resulting waveforms obtained from the numerical simulations for the collector voltage and current are shown in Fig. 5.71 for the case of a collector supply voltage of 5,000 V. It can be observed that there is no storage time for the 10-kV asymmetric IGBT structure because the P-base/N-base junction is reverse biased in the on-state. The collector voltage initially increases approximately linearly as described by the analytical model. The collector voltage increases at a lower rate once impact ionization sets-in at collector voltages above 2,000 V. The collector voltage almost saturates at 5,000 V indicating operation of the device close to its RBSOA limit. This is consistent with the predictions of the analytical model for the RBSOA of the IGBT (see Fig. 5.57). The collector voltage rise-time for the 10-kV device structure is slightly longer (by 1 μs) than that of the 5-kV baseline device structure. However, the collector current turn-off time is substantially larger due to the extra stored charge near the collector side of the drift region at the end of the voltage transient for the 10-kV device structure.

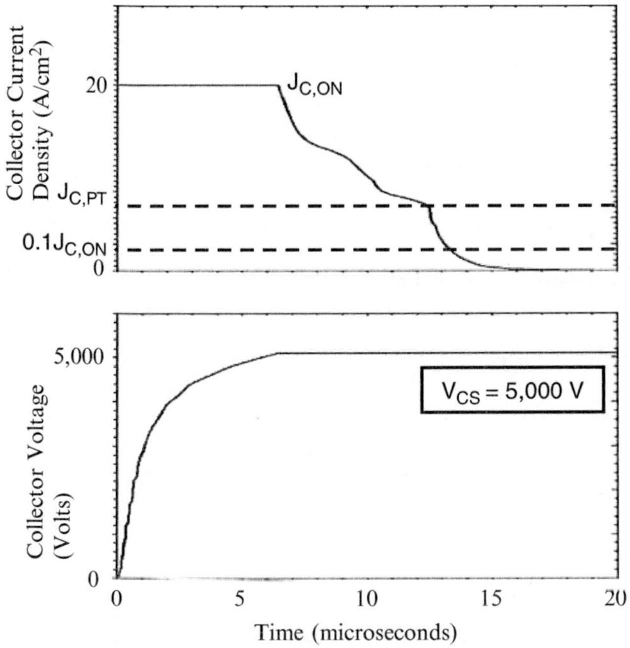

Fig. 5.71 10-kV asymmetric IGBT turn-off waveforms

The time at which the space-charge region punches-through to the N-buffer layer (t_{PT}) is the time taken for the collector current to decay from $J_{C,ON}$ to $J_{C,PT}$. The value of $J_{C,PT}$ for the 10-kV silicon asymmetric IGBT structure observed in the results of the numerical simulations is 7 A/cm^2 which is close to 9 A/cm^2 obtained by using the analytical model (Eq. 5.49). The current fall-time ($t_{I,OFF}$) for the 10-kV silicon asymmetric IGBT structure is essentially equal to the punch-through time. Since the current fall-time for the 10-kV silicon asymmetric IGBT is much larger than for the 5-kV silicon asymmetric IGBT structure, the energy loss during the turn-off event is very large severely limiting its operating frequency.

5.4.4 Switching Energy Loss

As discussed previously, the maximum operating frequency for the IGBT structure is limited by the turn-off losses. The turn-off losses are associated with the voltage rise-time interval and the current fall-time interval. The energy loss for each event can be computed by using the equations previously provided in Sect. 5.2.6. Using this information, the maximum operating frequency for the 10-kV silicon asymmetric IGBT structure can be derived using Eq. 5.54. The turn-off energy loss per cycle obtained from the numerical simulations of the silicon 10-kV asymmetric

IGBT structure can be derived from the waveforms in Fig. 5.71. For case of a high-level lifetime of 20 μs in the N-base region, the energy loss per cycle during the voltage rise-time is 0.33 J/cm^2 while the energy loss per cycle during the current fall time is 0.345 J/cm^2 in the case of an on-state current density of 20 A/cm^2 and a collector supply voltage of 5,000 V. The total energy loss per cycle is 0.675 J/cm^2 for the 10-kV silicon asymmetric IGBT structure.

5.4.5 *Maximum Operating Frequency*

The maximum operating frequency for the 10-kV asymmetric IGBT structure is limited by the turn-off losses. The turn-off losses are associated with the voltage rise-time interval and the current fall-time interval. The energy loss for each event can be computed by using the equation previously provided in Sect. 5.2.6. Using this information, the maximum operating frequency for the IGBT structure can be derived using Eq. 5.54. The data acquired from the numerical simulations of the 10-kV asymmetric IGBT structure is provided in Fig. 5.72.

High-Level Lifetime (μs)	On-State Voltage Drop (Volts)	On-State Power Dissipation (W/cm^2)	Energy Loss per Cycle (J/cm^2)	Maximum Operating Frequency (Hz)
20	4.456	44.6	0.675	230

Fig. 5.72 Power loss analysis for the 10-kV asymmetric IGBT structure

The maximum operating frequency obtained under the assumption of a 50% duty cycle and a total power dissipation limit of 200 W/cm^2 for the 10-kV asymmetric IGBT structure is found to be 230 Hz. The maximum operating frequency for the silicon 10-kV asymmetric IGBT structure is less than that for the 5-kV asymmetric IGBT structure due to its larger combination of on-state voltage drop and switching energy loss per cycle.

5.5 Forward Biased Safe Operation Area

One of unique strengths of the IGBT structure from an application standpoint has been its excellent forward-biased-safe-operating-area (FBSOA). The IGBT is commonly used in an H-bridge circuit for creating the variable frequency drive in motor control applications as illustrated in Fig. 1.11. The H-bridge circuit contains fly-back rectifiers across each of the IGBT switches. A wide FBSOA is beneficial when the power switch turns-on during each PWM cycle because it allows controlling the rate of rise of the current. In the case of thyristor structures, the device has an inherent fast rate of rise of current once the regenerative action commences. A high rate of rise of the current in the power switches produces very large reverse recovery currents in the P-i-N rectifiers [11] commonly used in the circuits. In the case of thyristor-based switches, the rate of rise of the current must be limited by using a snubber circuit to prevent destructive failure of the rectifier and the switch [12].

In the case of the IGBT structure, it is possible to control the rate of rise of the current by adjusting the gate drive resistance. This approach is possible because of the wide FBSOA boundary for the IGBT structure. The FBSOA boundary is limited by avalanche breakdown in the presence of a large density of holes and electrons in the space-charge region as discussed in the textbook [2]. The analytical model for the FBSOA of the IGBT structure can be found in the textbook. In this chapter, the FBSOA of the 5-kV asymmetric silicon IGBT structure is described by using the results of numerical simulations.

Simulation Results

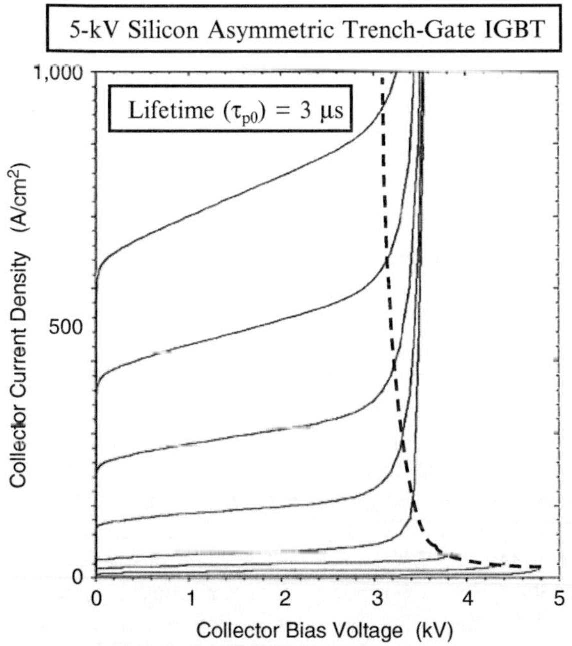

Fig. 5.73 5-kV asymmetric trench-gate IGBT FBSOA boundary

Numerical simulations of the 5-kV silicon asymmetric trench-gate IGBT structure were performed for the case of a high-level lifetime of 6 μs with various values for the gate-bias voltage while sweeping the collector voltage. The resulting output characteristics are shown in Fig. 5.73. At low collector bias voltages, the output characteristics are relatively flat. However, at large collector bias voltages, the collector current begins to increase rapidly delineating the FBSOA boundary.

The dashed line in the figure provides a demarcation of the FBSOA boundary. From this line, it can be concluded that the 5-kV asymmetric trench-gate IGBT structure has an excellent FBSOA.

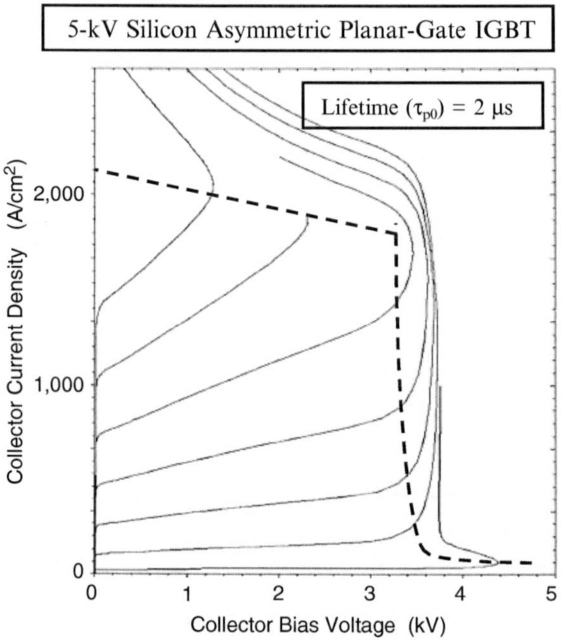

Fig. 5.74 5-kV asymmetric planar-gate IGBT FBSOA boundary

Numerical simulations of the 5-kV silicon asymmetric planar-gate IGBT structure were performed for the case of a high-level lifetime of 4 μs with various values for the gate-bias voltage while sweeping the collector voltage. The resulting output characteristics are shown in Fig. 5.74. At smaller gate-bias voltages and low collector bias voltages, the output characteristics are relatively flat. In these cases, at large collector bias voltages, the collector current begins to increase rapidly delineating the FBSOA boundary. The dashed line in the figure provides a demarcation of the FBSOA boundary. This boundary is wider than that for the trench gate structure due to the high electric field generated at the corner of the trenches. For the planar-gate structure, the simulations were also performed at larger gate bias voltages to find the upper current limit for the FBSOA boundary. It can be observed from the figure that there is an upper collector current density limit of about 2,000 A/cm^2 for the 5-kV planar-gate asymmetric IGBT structure. This boundary is associated with latch-up of the parasitic thyristor in the structure. From these simulation results, it can be concluded that the 5-kV asymmetric planar-gate IGBT structure also has an excellent FBSOA.

5.6 Reverse Biased Safe Operation Area

The analytical solution for the reverse biased safe operating area (RBSOA) for the IGBT structure can be obtained by using Eq. 4.97 provided for the GTO structure because the physics of operation is similar. However, the GTO structure suffers from current crowding during the turn-off process. This problem does not occur in the IGBT structure. The RBSOA boundaries for the 5-kV asymmetric IGBT structures obtained by using numerical simulations are provided in this section.

Simulation Results

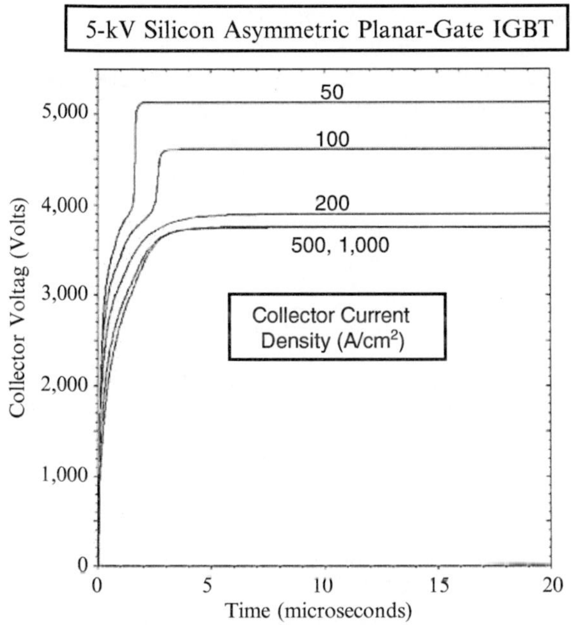

Fig. 5.75 5-kV asymmetric planar-gate IGBT RBSOA turn-off waveforms

The RBSOA boundary for the IGBT structure can be obtained by turning-off the structure starting with various on-state current densities. The presence of holes in the space-charge region enhances the electric field at the junction between the P-base region and the drift region. The electric field becomes larger for larger initial on-state current densities. Consequently, the collector voltage at which the on-state current density can be sustained by the impact ionization process becomes smaller. During turn-off, the collector voltage becomes limited as a function of time providing the RBSOA limit for each corresponding on-state current density.

Numerical simulations of the 5-kV silicon asymmetric planar-gate IGBT structure were performed for the case of a high-level lifetime of 2 μs with various values for the initial on-state current density. The resulting collector voltage waveforms are provided in Fig. 5.75. In each case, the collector voltage increases and becomes limited by the on-set of avalanche breakdown. Using the collector turn-off waveforms, the RBSOA boundary can be determined as shown in Fig. 5.76. The planar-gate IGBT exhibits an excellent RBSOA boundary making it a very popular device for switching inductive loads such as in motor control applications.

Numerical simulations of the 5-kV silicon asymmetric trench-gate IGBT structure were performed for the case of a high-level lifetime of 2 μs with various values for the initial on-state current density. The resulting collector voltage waveforms are provided in Fig. 5.77. In each case, the collector voltage increases and becomes limited by the on-site of avalanche breakdown. Using the collector turn-off waveforms, the RBSOA boundary can be determined as shown in Fig. 5.76. The trench-gate IGBT also exhibits an excellent RBSOA boundary, although slightly inferior to that for the planar-gate structure due to enhanced electric field at the trench corner, making it a very popular device for switching inductive loads such as in motor control applications.

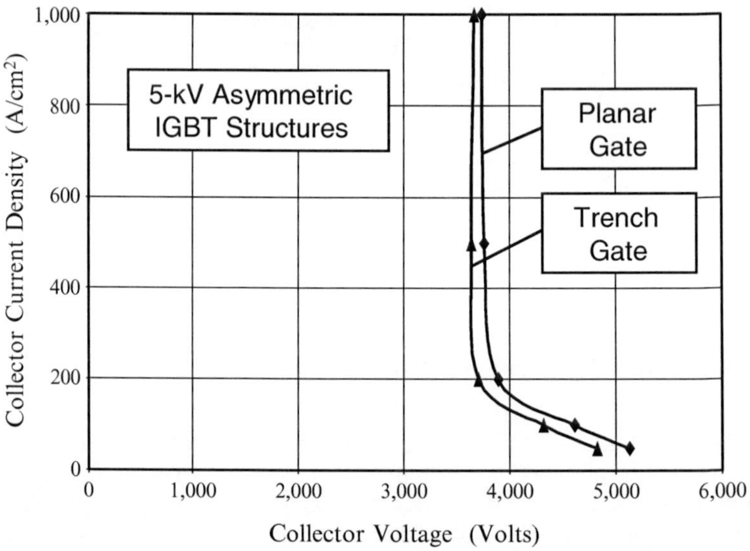

Fig. 5.76 RBSOA boundary for the 5-kV asymmetric IGBT structures

Fig. 5.77 RBSOA turn-off waveforms for the 5-kV asymmetric trench-gate IGBT structures

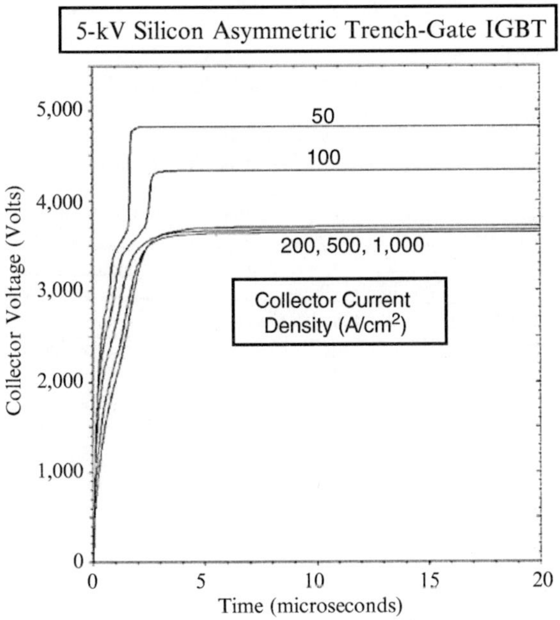

5.7 Conclusions

The physics of operation and design principles for the silicon IGBT structure have been elucidated in this chapter. This device structure is extensively used for high power motor control in traction (electric locomotive) drives since the later part of the 1990s when high voltage (>4 kV) devices were developed with excellent ruggedness. The analysis provided in this chapter demonstrates that the silicon 5-kV asymmetric IGBT structure offers excellent characteristics for motor control applications. However, the on-state characteristics of the IGBT degrade considerably when the switching speed is reduced as well as when the forward blocking voltage is extended to 10 kV. This has motivated the development of alternate MOS-gated thyristor structures that are discussed in this book.

References

1. B.J. Baliga, "How the Super-Transistor Works", Scientific American Magazine, Special Issue on 'The Solid-State Century', pp. 34–41, January 22, 1998.
2. B.J. Baliga, "Fundamentals of Power Semiconductor Devices", Springer-Science, New York, 2008.
3. B.J. Baliga, et al, "The Insulated Gate Rectifier: A New Power Switching Device", IEEE International Electron Devices Meeting, Abstract 10.6, pp. 264–267, 1982.

4. H.R. Chang, et al, "Insulated Gate Bipolar Transistors (IGBT) with a Trench Gate Structure", IEEE International Electron Devices Meeting, Abstract 29.5, pp. 674–677, 1987.
5. M. Pfaffenlehner, et al, "New 3300 V Chip Generation with a Trench IGBT and an Optimized Field Stop Concept with a Smooth Switching Behavior", IEEE International Symposium on Power Semiconductor Devices and ICs, Abstract 2.2, pp. 107–110, 2004.
6. T. Ogura, et al, "4.5 kV Injection Enhanced Insulated Gate Bipolar Transistors with High Turn-Off Ruggedness", IEEE Transactions on Electron Devices, Vol. ED-51, pp. 636–641, 2004.
7. J.G. Bauer, et al, "6.5 kV Modules using IGBTs with Field Stop Technology", IEEE International Symposium on Power Semiconductor Devices and ICs, Abstract 6.2, pp. 121–124, 2001.
8. J.G. Bauer, et al, "Investigations on the Ruggedness Limit of 6.5 kV IGBTs", IEEE International Symposium on Power Semiconductor Devices and ICs, pp. 71–74, 2005.
9. H. Dettmer, et al, "Punch-Through IGBTs with Homogeneous N-Base Operating at 4-kV Line Voltage", IEEE International Symposium on Power Semiconductor Devices and ICs, Abstract 12.4, pp. 492–496, 1995.
10. F. Bauer, et al, "Design Considerations and Characteristics of Rugged Punch-through (PT) IGBTs with 4.5 kV Blocking Capability", IEEE International Symposium on Power Semiconductor Devices and ICs, Abstract 14.1, pp. 327–330, 1996.
11. B.J. Baliga, "Advanced Power Rectifier Concepts", Springer-Science, New York, 2009.
12. N. Mohan, T.M. Undeland, and W.P Robbins, "Power Electronics", John Wiley and Sons, Inc., New York, 1995.

Chapter 6
SiC Planar MOSFET Structures

In the previous chapters, it was demonstrated that the maximum operating frequency of high voltage bipolar silicon power devices is limited by the power dissipation due to their slow switching transients. The rate of rise of the voltage and rate of fall of the current during the turn-off process in these devices is slowed down by the presence of the large amount of stored charge in the drift region. Consequently, high voltage silicon carbide unipolar power MOSFET devices are a very attractive alternative to silicon bipolar power devices [1, 2]. Silicon carbide power device structures have been discussed in detail in a previous book [3]. In that book, it was shown that the conventional planar power D-MOSFET structure, developed and widely utilized for silicon, is not suitable for the development of silicon carbide devices. Two problems are encountered when utilizing the conventional power D-MOSFET structure for silicon carbide. The first problem is the much larger threshold voltage required to create an inversion layer in silicon carbide due to its much greater band gap. The doping concentration required in the P-base region to achieve a typical threshold voltage of 2 V is so low that the device cannot sustain a high blocking voltage due to reach through of the depletion layer in the base region. The second problem is the very high electric field generated in the gate oxide because the electric field in the silicon carbide drift region under the gate is an order to magnitude larger than for silicon devices. This leads to rupture of the gate oxide at large blocking voltages.

For a planar power MOSFET structure, these issues can be addressed in a satisfactory manner by shielding the channel from the high electric field developed in the drift region. The concept of shielding of the channel region in a planar power MOSFET structure was first proposed [4] at PSRC in the early 1990s with a US patent issued in 1996. The shielding was accomplished by formation of either a P^+ region under the channel or by creating a high resistivity conduction barrier region under the channel. The shielding approach also allowed the creation of a new power MOSFET structure called the ACCUFET where an accumulation layer is utilized to create the channel. The accumulation mode of operation allows achieving the desired typical threshold voltage and also provides a much larger channel mobility to reduce the channel resistance contribution.

B.J. Baliga, *Advanced High Voltage Power Device Concepts*,
DOI 10.1007/978-1-4614-0269-5_6,

In this chapter, the characteristics of the shielded planar inversion-mode high voltage (5, 10, and 20 kV) silicon carbide power MOSFET structures are analyzed using analytical modeling and numerical simulations. The impact of the shielding concept on ameliorating the reach-through breakdown in silicon carbide is described. The results of the analysis of the shielded planar silicon carbide MOSFET structures by using two-dimensional numerical simulations are described in this chapter as in the case of all the silicon devices. It is shown that the shielding concept also enables reduction of the electric field developed in the gate oxide leading to the possibility of fully utilizing the breakdown field strength of the underlying semiconductor drift region. Only the inversion-mode silicon carbide power MOSFET structure is included in this book because most of the research activity has been relegated to this structure.

6.1 Shielded Planar Inversion-Mode MOSFET Structure

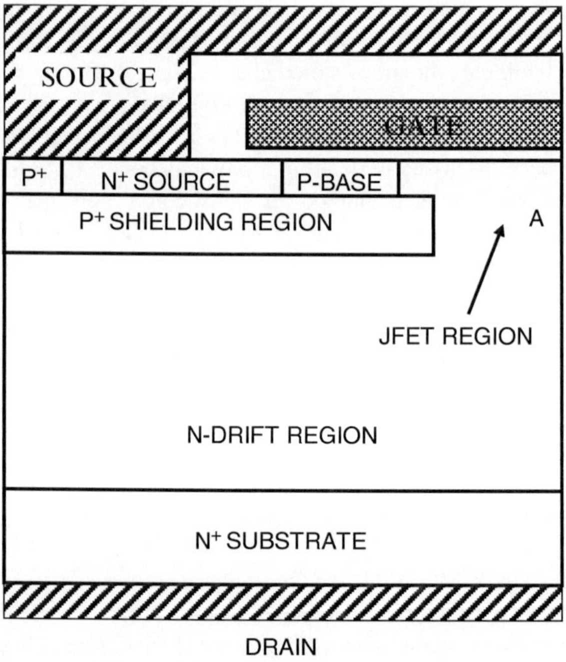

Fig. 6.1 Shielded planar inversion-mode power MOSFET structure

The basic structure of the shielded planar inversion-mode power MOSFET structure is shown in Fig. 6.1. The structure contains a sub-surface P^+ shielding region which extends under both the N^+ source region and the P-base region. The P^+ shielding region is shown to extend beyond the edge of the P-base region in the figure. However, the P-base and the P^+ shielding region can also be formed by using a self-aligned ion-implantation process. The space between the P^+ shielding regions, indicated in the figure as the JFET region, is optimized to obtain a low specific on-resistance while simultaneously shielding the gate oxide interface and the P-base region from the high electric field in the drift region. A potential barrier is formed at location A after the JFET region becomes depleted by the applied drain bias in the blocking mode. This barrier prevents the electric field from becoming large at the gate oxide interface and at the P-base/N-drift junction. When a positive bias is applied to the gate electrode, an inversion layer channel is formed at the surface of the P-base region in the structure enabling the conduction of drain current with a low specific on-resistance. The specific on-resistance of the silicon carbide inversion-mode power MOSFET structures with lower breakdown voltages (up to 1,000 V) was shown to be limited by the channel resistance [5] due to the very low specific resistance of the drift region. In the case of the high voltage devices that are of interest in this book, the drift region resistance increases by many orders of magnitude. Despite this, it is shown in this chapter that the channel resistance is still a limiting factor for the higher voltage inversion-mode silicon carbide power MOSFETs due to the poor channel mobility. Inversion layer mobility in high voltage silicon carbide power MOSFETs has been reported [6] to be in the range of 15–20 cm^2/V-s although larger values (up to 165 cm^2/V-s) have been observed in lateral MOSFETs [7].

6.2 Blocking Mode

In the forward blocking mode of the silicon carbide shielded planar inversion-mode power MOSFET structure, the voltage is supported by a depletion region formed on both sides of the P^+ region/N-drift junction. The maximum blocking voltage is determined by the electric field at this junction becoming equal to the critical electric field for breakdown if the parasitic N^+/P/N bipolar transistor is completely suppressed. This suppression is accomplished by short-circuiting the N^+ source and P^+ regions using the source metal as shown on the upper left-hand-side of the cross-section. If the doping concentration of the P^+ region is large, the reach-through breakdown problem is completely eliminated. In addition, the high doping concentration in the P^+ region promotes the depletion of the JFET region at lower drain voltages providing enhanced shielding of the channel and gate oxide.

In the conventional silicon carbide power D-MOSFET structure, the minimum P-base thickness and doping concentration are constrained by the reach-through limitation [8]. This does not occur in the silicon carbide shielded planar inversion-mode power MOSFET structure due to shielding of the P-base region from the drain potential by the P^+ shielding region. This allows reducing the channel length to less than 1 μm. In addition, the doping concentration of the P-base region can be reduced to achieve a desired threshold voltage without reach-through-induced breakdown. The smaller channel length and threshold voltage reduce the channel resistance contribution.

As in the case of the planar silicon power D-MOSFET structure, the maximum blocking voltage capability of the silicon carbide shielded inversion-mode planar MOSFET structure is determined by the drift region doping concentration and thickness. However, to fully utilize the high breakdown electric field strength available in silicon carbide, it is necessary to screen the gate oxide from the high field within the semiconductor. In the shielded planar MOSFET structure, this is achieved by the formation of a potential barrier at location A by the depletion of the JFET region at a low drain bias voltage. The maximum electric field in the gate oxide can be made not only below its rupture strength but also lower than values required for reliable operation over long time durations.

6.3 Threshold Voltage

The threshold voltage of the power MOSFET devices is an important design parameter from an application stand-point. The threshold voltage must be maintained at above 1 V for most system applications to provide immunity against inadvertent turn-on due to voltage spikes arising from noise. At the same time, a high threshold voltage is not desirable because the voltage available for creating the charge in the channel is determined by (V_G–V_{TH}) where V_G is the applied gate bias voltage and V_{TH} is the threshold voltage. Most power electronic systems designed for high voltage operation provide a gate drive voltage of only up to 10 V. Based upon this criterion, the threshold voltage should be kept at about 2 V in order to obtain a low channel resistance contribution.

For the inversion-mode shielded planar MOSFET structure, the threshold voltage can be modeled by defining it as the gate bias at which on-set of *strong inversion* begins to occur in the channel. This voltage can be determined using [8]:

$$V_{TH} = \frac{\sqrt{4\varepsilon_S kTN_A \ln(N_A/n_i)}}{C_{ox}} + \frac{2kT}{q}\ln\left(\frac{N_A}{n_i}\right) \tag{6.1}$$

where N_A is the doping concentration of the P-base region, k is Boltzmann's constant, and T is the absolute temperature. The presence of positive fixed oxide charge shifts the threshold voltage in the negative direction by:

$$\Delta V_{TH} = \frac{Q_F}{C_{ox}} \tag{6.2}$$

The analytically calculated threshold voltages for 4H-SiC inversion-mode MOSFET structures are provided in Fig. 6.2 for the case of a gate oxide thickness of 0.05 μm as a function of the P-base doping concentration with the inclusion of a metal-semiconductor work-function difference of 1 V. It can be observed that the analytical model predicts a threshold voltage of about 5 V for a P-base doping concentration of 2×10^{16} cm^{-3}. Such a low P-base doping concentration is only feasible for the shielded silicon carbide power MOSFET structure. The threshold voltage model is the same for all the shielded silicon carbide power MOSFET structures discussed in this chapter irrespective of their blocking voltage capability.

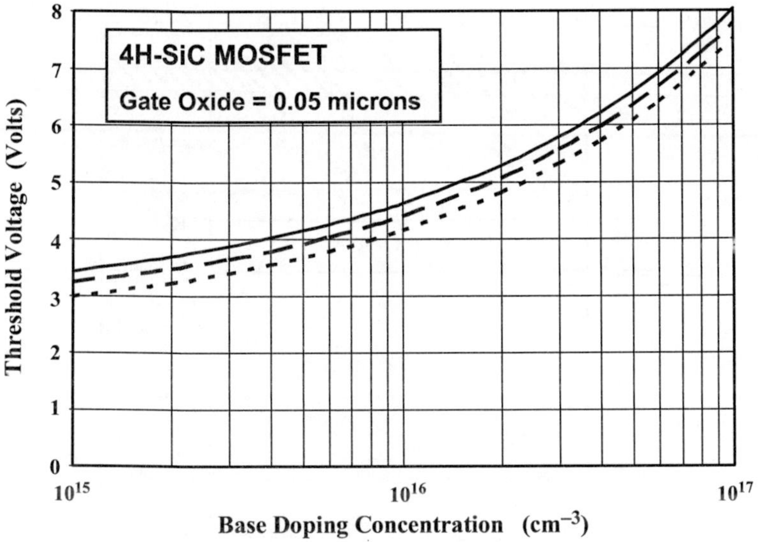

Fig. 6.2 Threshold voltage of 4H-SiC inversion-mode MOSFET structures (*Solid line* 300 K, *dash line* 400 K, *dotted line* 500 K)

6.4 On-State Resistance

In the silicon carbide shielded inversion-mode planar MOSFET structure, current flow between the drain and source can be induced by creating an inversion layer channel on the surface of the P-base region. The current flows through the channel formed due to the applied gate bias into the JFET region via the accumulation layer formed above it under the gate oxide. It then spreads into the N-drift region at a 45° angle and becomes uniform through the rest of the structure. The total on-resistance for the silicon carbide shielded inversion-mode planar SiC MOSFET structure is determined by the resistance of the components in the current path:

$$R_{\mathrm{on,sp}} = R_{\mathrm{CH}} + R_{\mathrm{A}} + R_{\mathrm{JFET}} + R_{\mathrm{D}} \tag{6.3}$$

where R_{CH} is the channel resistance, R_{A} is the accumulation region resistance, R_{JFET} is the resistance of the JFET region, R_{D} is the resistance of the drift region after taking into account current spreading from the JFET region. The resistance of the N^+ substrate has been omitted in the above analysis even though the substrate contribution for 4H-SiC can be very large unless its thickness is reduced to below 50 μm. The resistances can be analytically modeled by using the current flow pattern indicated by the shaded regions in Fig. 6.3.

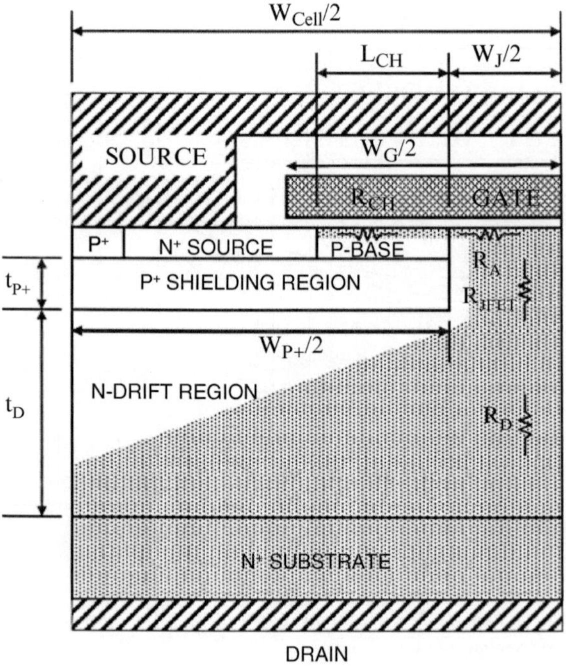

Fig. 6.3 Current path and resistances in the shielded planar SiC inversion-mode power MOSFET structure

6.4.1 *Channel-Resistance*

For the shielded planar SiC MOSFET structure with the P-base region, the specific channel resistance is given by:

$$R_{CH} = \frac{(L_{CH} W_{Cell})}{2\mu_{inv} C_{ox}(V_G - V_{TH})} \quad (6.4)$$

where L_{CH} is the channel length as shown in Fig. 6.3, μ_{inv} is the mobility for electrons in the inversion layer channel, C_{ox} is the specific capacitance of the gate oxide, V_G is the applied gate bias, and V_{TH} is the threshold voltage. As mentioned earlier, although an inversion layer mobility of 165 cm^2/V-s has been observed in lateral MOSFET structures [7], the inversion layer mobility reported for high voltage 4H-SiC power MOSFET structures [6] is usually only 15–20 cm^2/V-s.

The relatively low inversion layer mobility can make the channel resistance component dominant in the shielded planar SiC MOSFET structure. The channel resistance can be reduced by achieving submicron channel length dimensions without resorting to E-beam lithography with special processing tricks [9]. However, it is more convenient and practical to employ staggered ion implants for the P-base and N^+ source regions to control the channel length [10]. This process allows achieving a channel length of 1 μm.

6.4.2 Accumulation-Resistance

In the shielded planar SiC MOSFET structure, the current flowing through the inversion channel enters the JFET region at the edge of the P-base junction. The current spreads downwards from the edge of the P-base junction into the JFET region. The current spreading phenomenon is aided by the formation of an accumulation layer in the semiconductor below the gate oxide due to the positive gate bias applied to turn-on the device. The specific on-resistance contributed by the accumulation layer in the shielded planar SiC MOSFET structure is given by:

$$R_{\mathrm{A,SP}} = K_{\mathrm{A}} \frac{W_{\mathrm{J}} W_{\mathrm{Cell}}}{4 \mu_{\mathrm{nA}} C_{\mathrm{OX}} (V_{\mathrm{G}} - V_{\mathrm{TH}})} \tag{6.5}$$

In writing this expression, a coefficient K_{A} has been introduced to account for the current spreading from the accumulation layer into the JFET region. A typical value for this coefficient is 0.6 based upon the current flow observed from numerical simulations of shielded planar SiC MOSFET structures. The threshold voltage in the expression is for the on-set of formation of the accumulation layer. A zero threshold voltage will be assumed here when performing the analytical computations. Note that the width of the JFET region defines the length of the accumulation region.

6.4.3 JFET-Resistance

The electrons entering from the channel into the drift region are distributed into the JFET region via the accumulation layer formed under the gate electrode. The spreading of current in this region was accounted for by using a constant K_{A} of 0.6 for the accumulation layer resistance. Consequently, the current flow through the JFET region can be treated with a uniform current density. In the shielded planar SiC

MOSFET structure, the cross-sectional area for the JFET region is uniform with the width given by:

$$a = (W_J - 2W_0) \tag{6.6}$$

where W_0 is the zero-bias depletion width for the JFET region. The zero-bias depletion width (W_0) in the JFET region can be computed by using its doping concentration, which is usually increased above that for the drift region to shorten the JFET width and increase the channel density:

$$W_0 = \sqrt{\frac{2\varepsilon_S V_{bi}}{qN_{DJ}}} \tag{6.7}$$

where N_{DJ} is the doping concentration in the JFET region. The built-in potential is also related to the doping concentrations on both sides of the junction:

$$V_{bi} = \frac{kT}{q} \ln\left(\frac{N_A N_{DJ}}{n_i^2}\right) \tag{6.8}$$

where N_A is the doping concentration in the P^+ shielding region. Compared with silicon devices, the built-in potential for 4H-SiC is about three-times larger.

The specific on-resistance contributed by the JFET region in the shielded planar SiC MOSFET structure can be obtained by using:

$$R_{JFET,SP} = \frac{\rho_{JFET} t_{P+} W_{Cell}}{(W_J - 2W_0)} \tag{6.9}$$

where ρ_{JFET} is the resistivity of the JFET region given by:

$$\rho_{JFET} = \frac{1}{q\mu_n N_{DJ}} \tag{6.10}$$

where μ_n is the bulk mobility appropriate to the doping level of the JFET region.

6.4.4 Drift-Resistance

The resistance contributed by the drift region in the shielded planar SiC MOSFET structure is enhanced above that for the ideal drift region due to current spreading from the JFET region. The cross-sectional area for the current flow in the drift region increases from the width “a” of the JFET region at a 45° angle as illustrated in Fig. 6.3 by the shaded area. For the high voltage shielded planar SiC MOSFET structure discussed in this book, the current paths in the drift region overlap at a depth of $W_{P+}/2$ from the bottom of the P$^+$ shielding region. The specific on-resistance contributed by the drift region with this model is given by:

$$R_{D,SP} = \frac{\rho_D W_{Cell}}{2} \ln\left[\frac{a+W_{P+}}{a}\right] + \rho_D[t_D - (W_{P+}/2)] \tag{6.11}$$

6.4.5 *Total On-Resistance*

The total specific on-resistance for the shielded silicon carbide inversion-mode MOSFET structure can be obtained by adding the above components. The specific on-resistance can be minimized to adjusting the width of the JFET region in the structure. The channel resistance will increase when the JFET region width is increased due to an increase in the cell pitch. In addition, the accumulation resistance will also increase due to the longer accumulation layer path and the larger cell pitch. At the same time, the JFET and drift region resistances will become smaller due to the larger width “a” for current flow from the upper surface into the drift region. Consequently, a minimum specific on-resistance is observed at an optimum JFET width (and cell pitch). The minimum specific on-resistance value and the optimum cell pitch size are also a function of the doping concentration of the JFET region. A JFET doping concentration of 1×10^{16} cm^{-3} will be used for all the shielded silicon carbide inversion-mode MOSFET structures in this chapter. This JFET doping concentration is sufficiently high to reduce the zero-bias depletion width of the JFET region while being sufficiently low to avoid a big increase in the electric field at the junction between the P$^+$ buried layer and the N-drift region which would degrade the breakdown voltage.

6.5 Capacitances

The capacitances within the shielded planar 4H-SiC MOSFET structure can be analytically modeled using the same approach as described in the textbook [8] and the companion book on power MOSFETs [5]. The specific input (or gate) capacitance for the shielded planar 4H-SiC MOSFET structure is given by:

$$C_{IN,SP} = C_{N+} + C_P + C_{SM} = \frac{(W_G - W_J)}{W_{Cell}}\left(\frac{\varepsilon_{OX}}{t_{OX}}\right) + \frac{W_G}{W_{Cell}}\left(\frac{\varepsilon_{OX}}{t_{IEOX}}\right) \tag{6.12}$$

where t_{OX} and t_{IEOX} are the thicknesses of the gate and inter-electrode oxides, respectively.

The capacitance between the gate and drain electrodes (also called the reverse transfer capacitance) is determined by the width of the JFET region where the gate electrode overlaps the N-drift region. The MOS structure in this portion of the shielded planar 4H-SiC power MOSFET structure operates under deep depletion conditions when a positive voltage is applied to the drain. The gate-drain capacitance for the shielded planar 4H-SiC MOSFET power structure is given by:

$$C_{GD,SP} = \frac{W_J}{W_{Cell}}\left(\frac{C_{OX}C_{S,M}}{C_{OX} + C_{S,M}}\right) \tag{6.13}$$

where $C_{S,M}$ is the semiconductor capacitance under the gate oxide, which decreases with increasing drain bias voltage. The specific capacitance of the semiconductor depletion region can be obtained by computation of the depletion layer width. The depletion layer width in the semiconductor under the gate oxide can be obtained using:

$$W_{D,MOS} = \frac{\varepsilon_S}{C_{OX}}\left\{\sqrt{1 + \frac{2V_D C_{OX}^2}{q\varepsilon_S N_{DJ}}} - 1\right\} \tag{6.14}$$

where N_{DJ} is the doping concentration of the JFET region. The specific capacitance for the semiconductor is then obtained using:

$$C_{S,M} = \frac{\varepsilon_S}{W_{D,MOS}} \tag{6.15}$$

The gate-drain (or reverse transfer) capacitance can be computed by using Eq. 6.13 with the above equations to determine the semiconductor capacitance as a function of the drain bias voltage.

However, the above equation is only valid until the depletion region from the P^+ shielding regions pinches-off the JFET region in the shielded planar 4H-SiC power MOSFET structure. The gate-drain capacitance then decreases at a different rate because the gate is screened from the drain. The drain voltage at which the JFET region is pinched-off is given by:

$$V_{P,JFET} = \frac{qN_{DJ}}{8\varepsilon_S}W_J^2 - V_{bi} \tag{6.16}$$

For the shielded planar 4H-SiC power MOSFET structure with JFET region doping concentration (N_{DJ}) of 1×10^{16} cm^{-3} and JFET width (W_J) of 4 μm, the

JFET region pinch-off voltage is 37.2 V. After the JFET region is pinched-off, the gate-drain capacitance is determined by the edge of the depletion region located below the P^+ shielding region. This distance below the gate oxide is given by:

$$W_S = t_{P+} + \sqrt{\frac{2\varepsilon_S V_D}{qN_D}} \tag{6.17}$$

The specific capacitance for the semiconductor below the gate oxide can then be obtained using Eq. 6.15 with the width W_S.

The output capacitance for the shielded planar 4H-SiC power MOSFET structure is associated with the capacitance of the junction between the P^+ shielding region and the N-drift region. Due to pinch-off of the JFET region with increasing drain bias voltage, it is necessary to examine the change in the depletion region boundary with applied voltage. The depletion region has a vertical boundary inside the JFET region and a horizontal boundary below the P^+ shielding region [5]. The specific junction capacitance associated with the JFET region is given by:

$$C_{S1,SP} = \frac{\varepsilon_S}{W_{DJ}}\left(\frac{2t_{P+}}{W_{Cell}}\right) \tag{6.18}$$

where the depletion region thickness (W_{DJ}) in the JFET region is related to the drain bias voltage:

$$W_{DJ} = \sqrt{\frac{2\varepsilon_S(V_D + V_{bi})}{qN_{DJ}}} \tag{6.19}$$

The specific junction capacitance associated with the bottom of the P^+ shielding region is given by [6]:

$$C_{S2,SP} = \frac{\varepsilon_S}{W_{DD}}\left(\frac{W_{P+}}{W_{Cell}}\right) \tag{6.20}$$

where the depletion region thickness (W_{DD}) in the drift region is related to the drain bias voltage:

$$W_{DD} = \sqrt{\frac{2\varepsilon_S(V_D + V_{bi})}{qN_D}} \tag{6.21}$$

where N_D is the doping concentration of the drift region. The specific output capacitance for the shielded planar 4H-SiC power MOSFET structure can then be obtained by combining the above values:

$$C_{O,SP} = C_{S1,SP} + C_{S2,SP} \tag{6.22}$$

Once the JFET region is pinched-off, the output capacitance is determined by only the second term in the above equation.

6.6 Inductive Load Turn-Off Characteristics

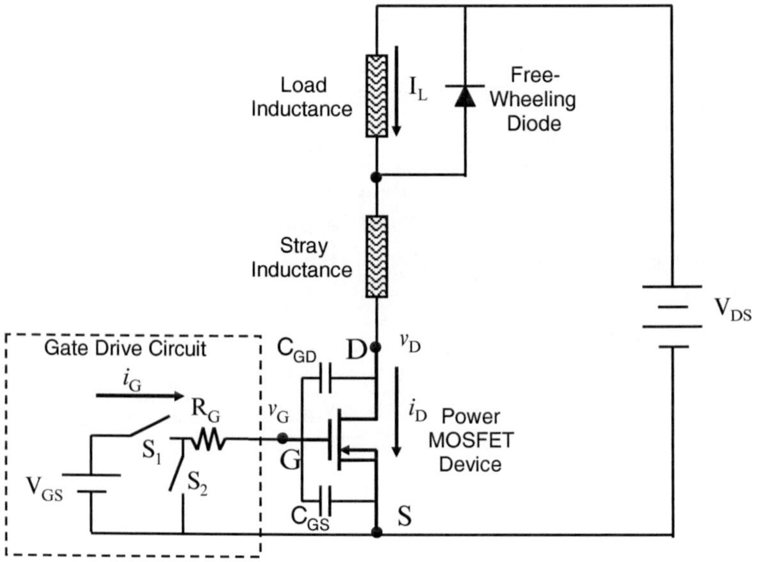

Fig. 6.4 Power MOSFET device operating in an inductive load circuit

The operation of the shielded planar 4H-SiC power MOSFET structure in an inductive load circuit can be analyzed using the same approach as used for the power MOSFET structure in the textbook [8] and the companion book on power MOSFETs [5]. The operation of a power MOSFET device in an inductive load circuit is illustrated in Fig. 6.4. The textbook provides a description of the basic operation of this circuit. After carrying the load current during its duty cycle, the power MOSFET device is switched off to transfer the current back to the freewheeling diode. Prior to the turn-off transient, the device is operating in its on-state because switch S_1 is closed and switch S_2 is open. These initial conditions are defined by: $v_G = V_{GS}$; $i_D = I_L$; and $v_D = V_{ON}(V_{GS})$.

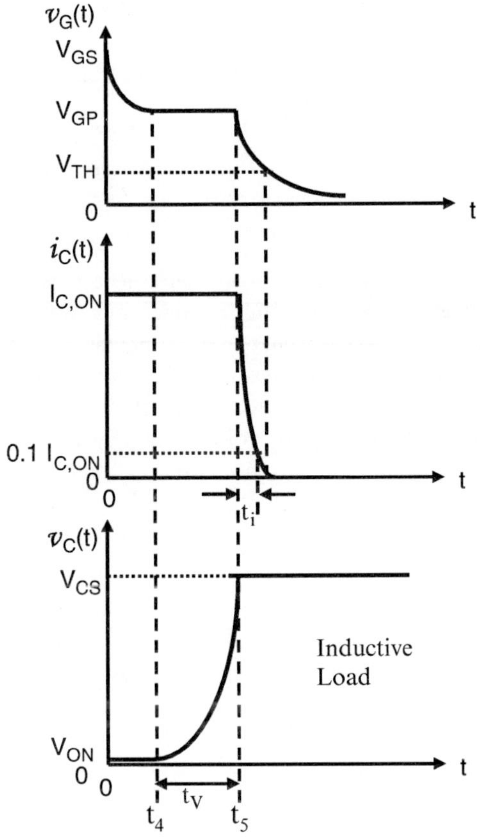

Fig. 6.5 Waveforms during turn-off for the power MOSFET device operating in an inductive load circuit

The basic turn-off waveforms for the power MOSFET structure are illustrated in Fig. 6.5. In order to initiate the turn-off process, switch S_1 is opened and switch S_2 is subsequently closed by the control circuit. The gate electrode of the power MOSFET device is then connected to the source via the gate resistance to discharge its capacitances. However, no changes in the drain current or voltage can occur until the gate voltage reaches the magnitude required to operate the power MOSFET device at a saturated drain current equal to the load current. (The small increase in the drain voltage, due to the increase in on-resistance resulting from the reduction of the gate bias voltage, has been neglected here).

The gate plateau voltage for the shielded planar 4H-SiC power MOSFET structure is given by:

$$V_{GP} = V_{TH} + \sqrt{\frac{J_{D,ON} W_{Cell} L_{CH}}{\mu_{ni} C_{GOX}}} \tag{6.23}$$

where C_{GOX} is the gate oxide capacitance. Since the time constant for discharging the gate of the shielded planar 4H-SiC power MOSFET structure is $R_{G,SP}$*[C_{GS} + $C_{GD}(V_{ON})$], the gate voltage at first decreases exponentially with time as given by:

$$v_G(t) = V_{GS}e^{-t/R_{G,SP}[C_{GS}+C_{GD}(V_{ON})]} \tag{6.24}$$

The time t_4 (using the notation from the textbook) for reaching the gate plateau voltage can be obtained by using this equation with Eq. 6.23 for the plateau voltage:

$$t_4 = R_{G,SP}[C_{GS} + C_{GD}(V_{ON})]\ln\left[\frac{V_{GS}}{V_{GP}}\right] \tag{6.25}$$

This time can be considered to a *turn-off delay time* before the drain voltage begins to increase after the turn-off is initiated by the control circuit.

The drain voltage begins to increase at time t_4 as shown in Fig. 6.5, but the drain current remains constant at the load current I_L because the current cannot be transferred to the diode until the voltage at the drain of the MOSFET device exceeds the supply voltage V_{DS} by one diode drop to forward bias the diode. Since the drain current density is constant, the gate voltage also remains constant at the gate plateau voltage. Consequently:

$$J_{GP} = \frac{V_{GP}}{R_{G,SP}} \tag{6.26}$$

where $R_{G,SP}$ is the specific gate resistance. Since the entire gate current is used to discharge the gate-drain capacitance during the plateau phase (because there is no change in the voltage across the gate-source capacitance):

$$J_{GP} = C_{GD,SP}\frac{dv_D}{dt} \tag{6.27}$$

where $C_{GD,SP}$ is the specific gate transfer capacitance of the power MOSFET structure which is a function of the drain voltage. This voltage dependence of the gate transfer capacitance was not taken into account in the derivation provided in the textbook but is important to include here to allow comparison of the behavior of various power device structures.

For simplicity of analysis, it will be assumed that any screening effect can be ignored. The gate transfer capacitance for the shielded planar 4H-SiC power MOSFET structure is then given by:

$$C_{GD,SP} = \frac{W_J}{W_{Cell}}\left(\frac{C_{OX}C_{S,M}}{C_{OX} + C_{S,M}}\right) \tag{6.28}$$

Using this expression in Eq. 6.27 yields the following differential equation for the voltage increase phase of the turn-off transient:

$$\mathrm{d}t = \left(\frac{W_\mathrm{J}}{W_\mathrm{Cell}}\right)\frac{1}{J_\mathrm{GP}}\left[\frac{C_\mathrm{GOX}}{\sqrt{1+\frac{2v_\mathrm{D}(t)C_\mathrm{GOX}^2}{q\varepsilon_\mathrm{S}N_\mathrm{D}}}}\right]\mathrm{d}v_\mathrm{D} \tag{6.29}$$

Integration of this equation yields:

$$(t-t_4) = \left(\frac{W_\mathrm{J}}{W_\mathrm{Cell}}\right)\frac{q\varepsilon_\mathrm{S}N_\mathrm{D}}{J_\mathrm{GP}C_\mathrm{GOX}}\left[\sqrt{1+\frac{2v_\mathrm{D}(t)C_\mathrm{GOX}^2}{q\varepsilon_\mathrm{S}N_\mathrm{D}}}-\sqrt{1+\frac{2V_\mathrm{ON}C_\mathrm{GOX}^2}{q\varepsilon_\mathrm{S}N_\mathrm{D}}}\right] \tag{6.30}$$

In the case of the shielded planar 4H-SiC power MOSFET structure, the drain voltage increases from the on-state voltage drop (V_ON) until it reaches the drain-supply voltage (V_DS). The voltage rise-time, i.e., the time taken for the voltage to increase from the on-state voltage drop (V_ON) to the drain supply voltage (V_DS), is:

$$\begin{aligned} t_\mathrm{V,OFF} &= (t_5 - t_4) \\ &= \left(\frac{W_\mathrm{J}}{W_\mathrm{Cell}}\right)\frac{q\varepsilon_\mathrm{S}N_\mathrm{D}}{J_\mathrm{GP}C_\mathrm{GOX}}\left[\sqrt{1+\frac{2V_\mathrm{DS}C_\mathrm{GOX}^2}{q\varepsilon_\mathrm{S}N_\mathrm{D}}}-\sqrt{1+\frac{2V_\mathrm{ON}C_\mathrm{GOX}^2}{q\varepsilon_\mathrm{S}N_\mathrm{D}}}\right] \end{aligned} \tag{6.31}$$

A closed form solution for the rise in the drain voltage can be obtained from Eq. 6.30:

$$\begin{aligned} v_\mathrm{D}(t) &= \frac{q\varepsilon_\mathrm{S}N_\mathrm{D}}{2C_\mathrm{GOX}^2} \\ &\times\left\{\left[\frac{J_\mathrm{GP}C_\mathrm{GOX}}{q\varepsilon_\mathrm{S}N_\mathrm{D}}\left(\frac{W_\mathrm{Cell}}{W_\mathrm{J}}\right)(t-t_4)+\sqrt{1+\frac{2V_\mathrm{ON}C_\mathrm{GOX}^2}{q\varepsilon_\mathrm{S}N_\mathrm{D}}}\right]^2-1\right\} \end{aligned} \tag{6.32}$$

This equation describes the increase in the drain voltage from the on-state voltage drop until it reaches the drain supply voltage. The drain voltage has an approximately quadratic shape as a function of the time after t_4 as illustrated in Fig. 6.5.

At the end of the plateau phase (at time t_5), the load current begins to transfer from the power MOSFET device to the freewheeling diode. Since the drain voltage remains constant, the gate-drain capacitance can also be assumed to remain constant during this phase. The current flowing through the gate resistance (R_G) discharges both the gate-drain capacitance (which is negligibly small at high drain bias voltages) and gate-source capacitance leading to an exponential fall in gate voltage from the plateau voltage:

$$v_G(t) = V_{GP}e^{-(t-t_5)/R_{G,SP}C_{GS}} \tag{6.33}$$

The drain current follows the gate voltage as given by:

$$J_D(t) = g_m[v_G(t) - V_{TH}] = \frac{\mu_{ni}C_{OX}}{L_{CH}W_{Cell}}[v_G(t) - V_{TH}]^2 \tag{6.34}$$

The drain current decreases rapidly with time as shown in Fig. 6.5 due to the exponential reduction of the gate voltage, as given by Eq. 6.33, during the current fall phase. The drain current becomes equal to zero when the gate voltage reaches the threshold voltage.

The current fall time is usually defined as the time taken for the drain current to reach 10% of the on-state current value. Due to the rapid drain current transient, the current fall time can be obtained from Eq. 6.33 using a gate voltage equal to the threshold voltage:

$$t_{I,OFF} = R_{G,SP}C_{GS}\ln\left(\frac{V_{GP}}{V_{TH}}\right) \tag{6.35}$$

Specific capacitances should be used in this expression for computation of the current fall time. Beyond this point in time, the gate voltage decreases exponentially until it reaches zero. The time constant for this exponential decay is different from the initial phase due to the smaller gate-drain capacitance.

The turn-off energy loss per cycle can be obtained using:

$$E_{OFF} = \frac{1}{2}J_{ON}V_{DS}(t_{V,OFF} + t_{I,OFF}) \tag{6.36}$$

under the assumption that the drain current and voltage excursions are approximately linear with time. The energy loss during the voltage rise-time interval is comparable to the energy loss during the current fall-time interval for the shielded planar 4H-SiC power MOSFET structure.

6.7 5-kV Inversion-Mode MOSFET

The specific on-resistance of the drift region for silicon carbide power devices is many orders of magnitude smaller than that for silicon devices with the same voltage rating [8]. This stems from the much greater critical electric field for breakdown in silicon carbide which in turn allows higher doping concentrations and smaller widths for the drift region. However, these properties can only be exploited if the electric field in the gate oxide for silicon carbide devices can be suppressed below the breakdown field strength. It is possible to reduce the drift region resistance by using the shielded silicon carbide inversion-mode power MOSFET structure while suppressing the electric field in the gate oxide.

This section discusses the performance of the shielded silicon carbide inversion-mode power MOSFET structure with the 5-kV blocking capability. The results of numerical simulations are used to describe the internal electric field and potential distribution in the blocking mode. The analytical model is then used to predict the specific on-resistance for this structure. The turn-off characteristics of the device are next analyzed using the analytical model with results of numerical simulations provided for gaining further insight into the device physics.

6.7.1 Blocking Characteristics

The blocking voltage capability for the shielded silicon carbide inversion mode power MOSFET structure is determined by the doping concentration and thickness of the drift region if the JFET region width is sufficiently small to prevent rupture of the gate oxide. Under the assumption of a typical edge termination with 80% of the parallel-plane breakdown voltage, the drift region for the 5-kV device structure must have the properties required for a parallel-plane breakdown voltage of 6,250 V. The analytical model for breakdown voltage in 4H-SiC predicts the following relationship [8]:

$$\mathrm{BV_{PP}(4H - SiC)} = 3.0 \times 10^{15} N_{\mathrm{D}}^{-3/4} \tag{6.37}$$

Using this equation, a breakdown voltage of 6,250 V can be obtained in 4H-SiC by using a doping concentration of 3.5×10^{15} cm^{-3}. In the case of the power MOSFET structure, the thickness of the drift region is the depletion layer width at this doping concentration when the drain bias reaches 5,000 V. By using Eq. 6.21, the drift region thickness for the 5-kV shielded silicon carbide inversion-mode power MOSFET structure is found to be 38 μm. This combination of drift region doping concentration and thickness will be used when modeling the specific on-resistance, capacitance, and turn-off behavior of the 5-kV shielded silicon carbide inversion-mode power MOSFET structure.

Simulation Results

The results of two-dimensional numerical simulations on the 5-kV shielded 4H-SiC inversion-mode power MOSFET structure are described here to provide a more detailed understanding of the underlying device physics and operation during the blocking mode. For the numerical simulations, the half-cell structure with a width ($W_{\mathrm{Cell}}/2$) of 5 μm as illustrated in Fig. 6.1 was utilized as representative of the structure. The device used for the numerical simulations had a drift region thickness of 38 μm below the P$^+$ shielding region with a doping concentration of 3.5×10^{15} cm^{-3}. The P$^+$ region extended from a depth of 0.2 to 1.0 μm with a doping concentration of 1×10^{19} cm^{-3}. The P-base and N$^+$ source regions were formed within the 0.2 μm of the N-drift region located above the P$^+$ region. The doping concentration of the P-base region was 2×10^{16} cm^{-3}. Due to the low doping concentration in the drift region for the 5-kV 4H-SiC devices, the uniform doping concentration in the JFET region was enhanced to 1×10^{16} cm^{-3} as is usually required for silicon devices. The enhanced doping concentration was extended to 0.5 μm below the P$^+$ shielding region. This is similar to the current enhancement layer (CEL) utilized in high voltage silicon carbide IGBT structures [11]. The charge in this layer must be sufficiently small to prevent degradation of the breakdown voltage at the junction between the P$^+$ shielding region and the drift region.

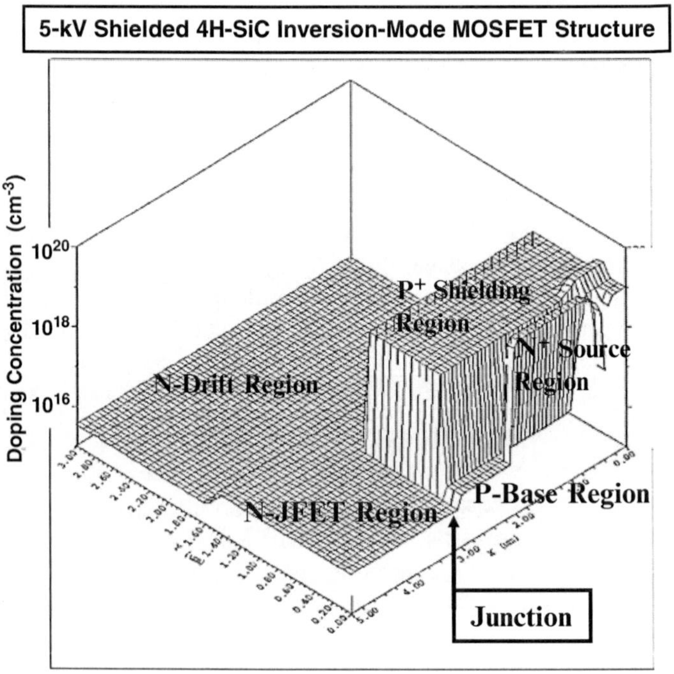

Fig. 6.6 Doping distribution in the shielded 4H-SiC planar power MOSFET structure

A three dimensional view of the doping distribution in the 5-kV shielded 4H-SiC inversion-mode power MOSFET structure is shown in Fig. 6.6 with the upper surface of the structure located on the right-hand-side in order to display the doping concentration in the vicinity of the channel. The highly doped P$^+$ shielding region with doping concentration of 1×10^{19} cm^{-3} is prominently located just below the surface. The P-base region can be observed to have a much lower doping concentration of 2×10^{16} cm^{-3}. The JFET region can be observed to have a lower doping concentration of 1×10^{16} cm^{-3}. The junction between the P-base region and N-JFET region is indicated in the figure. The doping concentration of the N-drift region is less than that of the JFET region as expected to achieve the desired 5-kV breakdown voltage. It can be seen that the enhanced JFET doping is extended below the P$^+$ shielding region to a depth of 1.5 μm from the surface.

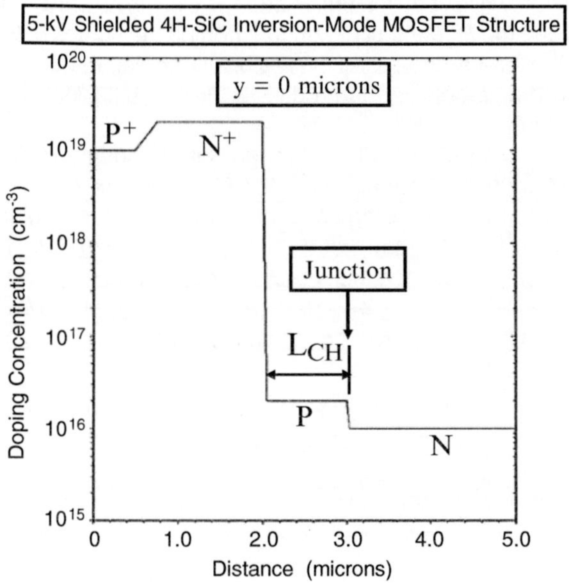

Fig. 6.7 Lateral doping profile for the shielded 4H-SiC inversion-mode MOSFET

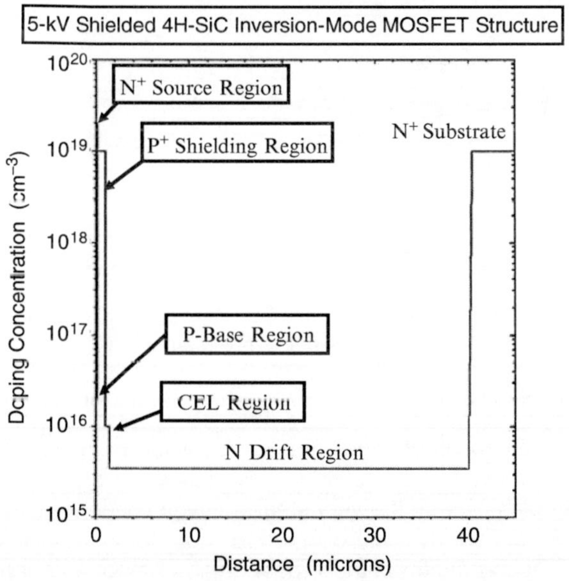

Fig. 6.8 Vertical doping profile in the shielded 4H-SiC inversion-mode MOSFET

The lateral doping profile taken along the surface of the 5-kV shielded 4H-SiC inversion-mode power MOSFET structure is shown in Fig. 6.7. From the profile, it can be observed that the channel extends from 2 to 3 μm creating a channel length of 1 μm in the P-base region. The doping concentration of the JFET region is 1×10^{16} cm^{-3} while that for the P-base region is 2×10^{16} cm^{-3}. The N^{+} source region and the P^{+} contact region for shorting the source to the base region are visible on the left-hand-side. All the regions were defined with uniform doping with abrupt interfaces between them due to the low diffusion rates for dopants in 4H-SiC material.

The vertical doping profile taken through the N^{+} source region of the 5-kV shielded 4H-SiC inversion-mode power MOSFET structure is provided in Fig. 6.8. It can be observed that the doping concentration of the P^{+} shielding region has a maximum value of 1×10^{19} cm^{-3} at a depth ranging from 0.2 to 1.0 μm. The P-base region is located between the N^{+} source region and the P^{+} shielding region with a doping concentration of 2×10^{16} cm^{-3}. The N-drift region has a doping concentration of 3.5×10^{15} cm^{-3} and thickness of 40 μm.

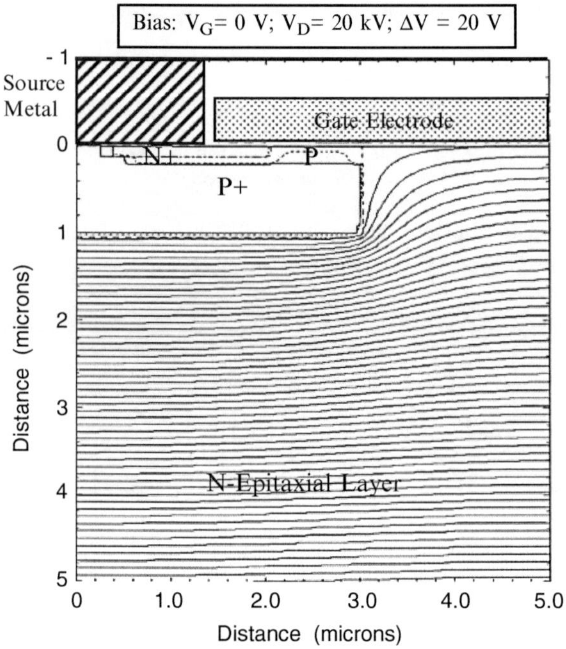

Fig. 6.9 Potential contours in the shielded 4H-SiC inversion-mode MOSFET

The blocking characteristics for the 5-kV shielded 4H-SiC inversion-mode power MOSFET structure were obtained by increasing the drain voltage while using zero gate bias. Due to the very small intrinsic concentration in 4H-SiC, no substantial leakage current was observed at room temperature. This also confirmed that the reach-through of the P-base region has been suppressed by the P^{+} shielding region. The potential contours within the 5-kV shielded 4H-SiC inversion-mode power MOSFET structure at a drain bias of 5,000 V are provided in Fig. 6.9 for the upper part of the device structure. It can be observed that the drain voltage is supported below the P^{+} shielding region. The potential contours do not extend into the P-base region indicating that it is shielded from the drain potential by the P^+ shielding region. The potential contours are crowding at the edge of the P^+ shielding region indicating an enhanced electric field. This can be clearly observed in Fig. 6.10 which provides a three-dimensional view of the electric field distribution. In this figure, it can also be observed that the electric field in the JFET region, and most importantly at the surface under the gate oxide, has been greatly reduced by the presence of the P^+ shielding region.

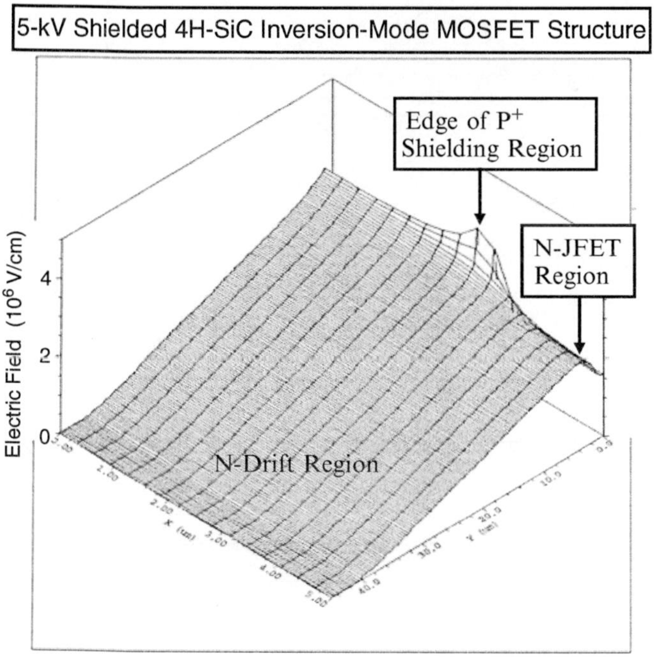

Fig. 6.10 Electric field distribution in the shielded 4H-SiC inversion-mode MOSFET

The electric field distribution near the surface of the 5-kV shielded 4H-SiC inversion-mode power MOSFET structure is shown in Fig. 6.11 to allow examination of the electric field in the JFET region and the gate oxide. It can be observed that the suppression of the electric field in the JFET region by the P^+ shielding region greatly reduces the electric field in the semiconductor at its surface when compared with the maximum electric field at the junction between the P^+ shielding region and the drift region. This produces a reduced electric field in the gate oxide as well. The electric field in the gate oxide has its highest value at the middle point of the JFET region. It is worth pointing out that the value of this oxide field is comparable to the maximum electric field observed in the semiconductor at the edge of the P^+ shielding region in spite of the 2.5x difference in the relative permittivity between the oxide and the semiconductor. These results clearly demonstrate the importance of using the shielding concept to achieve a practical device structure in silicon carbide.

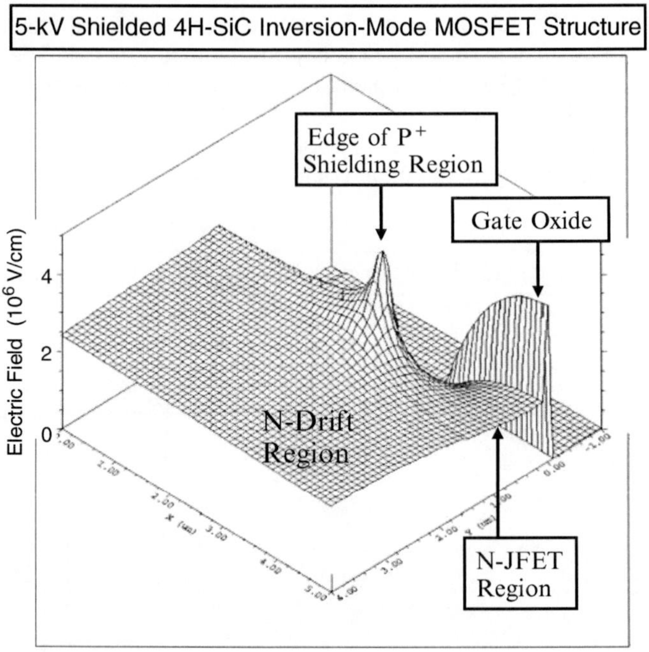

Fig. 6.11 Electric field distribution in the shielded 4H-SiC inversion-mode MOSFET

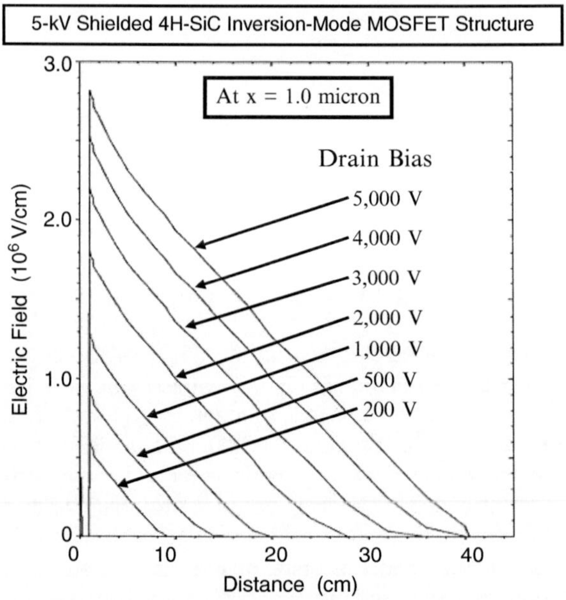

Fig. 6.12 Electric field distribution in the shielded 4H-SiC inversion-mode MOSFET

It is insightful to examine the electric field profile within the 5-kV shielded 4H-SiC inversion-mode power MOSFET structure when it is operating in the blocking mode. The electric field profile through the junction between the P^+ shielding region and the N-drift region is provided in Fig. 6.12. The peak of the electric field occurs at the junction as expected and is essentially triangular in shape in accordance with the predictions of Poisson's equation with a uniform doping profile. There is a discontinuity in the electric field near the junction due to the higher doping concentration of the CEL layer when compared with the drift region. However, the enhancement in the electric field is small and occurs over a very small distance which results in minimal degradation of the breakdown voltage. The maximum electric field at the junction at a drain bias of 5,000 V is 2.8×10^6 V/cm which is close to the critical electric for breakdown for 4H-SiC at the doping concentration of the drift region [8].

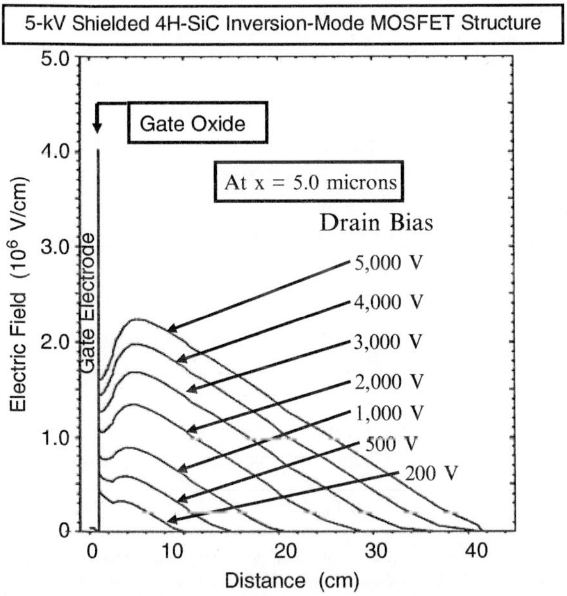

Fig. 6.13 Electric field distribution in the shielded 4H-SiC inversion-mode MOSFET

The electric field profiles obtained through the middle of the JFET region of the 5-kV shielded 4H-SiC inversion-mode power MOSFET structure are shown in Fig. 6.13 at various drain bias voltages. It can be observed that the maximum electric field occurs at a depth of 5 μm from the surface. The shielding effect produced by the P^+ shielding region reduces the electric field at the surface under the gate oxide to only 1.6×10^6 V/cm which is about one-half of the electric field in the bulk below the P^+ shielding region. Consequently, the electric field in the gate oxide is reduced to about 4×10^6 V/cm even when a drain bias of 5,000 V is applied. The low electric field in the gate oxide prevents gate oxide rupture and allows stable device performance over long periods of time. An even further reduction of electric field in the gate oxide can be achieved by reducing the width of the JFET region to 4 μm.

6.7.2 *On-Resistance*

The analytical model for the specific on-resistance for the shielded silicon carbide inversion-mode power MOSFET structure was provided in Sect. 6.4. In the case of the 5-kV shielded silicon carbide inversion-mode power MOSFET structure, the baseline cell width (W_{Cell}) will be assumed to be 10 μm with a channel length of 1 μm. A threshold voltage for the inversion mode structure of 3.7 V can be achieved in the shielded silicon carbide inversion-mode power MOSFET structure by using a low doping concentration of 2×10^{16} cm^{-3} for the P-base region and a gate oxide thickness is 500 Å. Based upon reports in the literature [6], a low inversion layer mobility of 20 cm^2/V-s will be used in the analytical model. For the baseline device structure, the JFET region will be assumed to have a width (W_{JFET}) of 4 μm and its doping concentration will be assumed to be 1×10^{16} cm^{-3}. The thickness (t_{P+}) of the P$^+$ shielding region will be assumed to be 1 μm.

Using the above device structural parameters in Eq. 6.4, the specific resistance contributed by the channel at a gate bias of 10 V is found to be 5.64 mΩ-cm^2. This value is comparable to the ideal specific on-resistance of 3.95 mΩ-cm^2 for the drift region of a 5-kV 4H-SiC unipolar device structure.

In the case of n-channel 4H-SiC MOSFET structures, accumulation layer mobility values of 100–200 cm^2/V-s have been experimentally observed [12]. Using an accumulation layer mobility of 100 cm^2/V-s for the above 5-kV shielded silicon carbide inversion-mode power MOSFET structure, the specific resistance contributed by the accumulation layer at a gate bias of 10 V is found to be 0.88 mΩ-cm^2 by using Eq. 6.5. This value is also comparable to the ideal specific on-resistance of 3.95 mΩ-cm^2 for the drift region of a 5-kV 4H-SiC unipolar device structure.

In the case of the above 5-kV shielded silicon carbide inversion-mode power MOSFET structure, the resistivity for the JFET region is found to be 0.625 Ω-cm. The zero-bias depletion width in the JFET region for the JFET doping concentration of 1×10^{16} cm^{-3} is 0.57 μm based up on a built-in potential of 3 V. Using these values, the specific resistance contributed by the JFET region is found to be 0.22 mΩ-cm^2 by using Eq. 6.9. This value is small when compared to the ideal specific on-resistance of 3.95 mΩ-cm^2 for the drift region of a 5-kV 4H-SiC unipolar device structure.

For the parameters given above for the 5-kV shielded silicon carbide inversion-mode power MOSFET structure, the dimension "a" in Eq. 6.11 is found to be 2.86 μm. The specific resistance contributed by the drift region is then found to be 6.60 mΩ-cm^2 by using Eq. 6.11 with a resistivity of the drift region of 1.62 Ω-cm (based upon a doping concentration of 3.5×10^{15} cm^{-3}) and a drift region thickness of 38 μm below the P$^+$ shielding region.

The total specific on-resistance for the 5-kV shielded silicon carbide inversion-mode power MOSFET structure with cell width of 10 μm can be obtained by adding the above components of the resistances within the device structure. For a gate bias of 10 V, the total specific on-resistance is found to be 13.33 mΩ-cm^2. The channel resistance constitutes 42% of the total specific on-resistance due to the poor mobility for the electrons in the inversion layer. The specific on-resistance for the 5-kV shielded planar 4H-SiC MOSFET structure is about three orders of magnitude smaller than that for the silicon 5-kV power D-MOSFET and U-MOSFET structures.

Note that the substrate resistance has not been included in the above analysis of the total specific on-resistance of the 5-kV shielded silicon carbide inversion-mode power MOSFET structure. A typical N$^+$ 4H-SiC substrate with resistivity of 0.02 Ω-cm and thickness of 200 μm will contribute an additional 4 mΩ-cm^2 to the specific on-resistance. Since this contribution is significant when compared with the 13.33 mΩ-cm^2 specific on-resistance obtained in the analytical model, it is important to reduce the substrate thickness to maximize the performance of the 5-kV shielded silicon carbide inversion-mode power MOSFET structure.

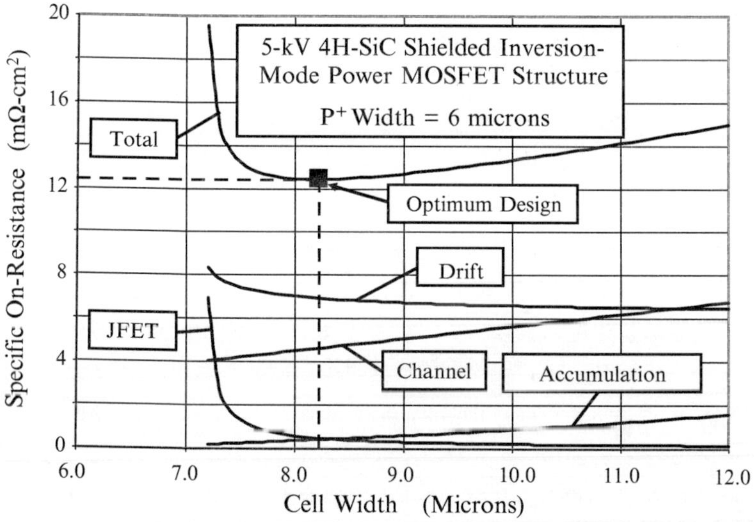

Fig. 6.14 On-resistance for the 5-kV 4H-SiC shielded inversion-mode MOSFET structures

The impact of changing the width of the JFET region on the specific on-resistance of the 5-kV shielded silicon carbide inversion-mode power MOSFET structure can be determined by using the analytical model. The results obtained for the case of a P$^+$

shielding region width (W_{P+}) of 6 μm are provided in Fig. 6.14. For this analysis, an inversion layer mobility of 20 cm^2/V-s was used. The cell width (W_{Cell}) is equal to the width of the JFET region (W_{JFET}) plus the width of the P$^+$ shielding region. It can be observed that the specific on-resistance goes through a minimum as the cell width is increased. At very small cell widths, the resistance from the JFET region and the drift region becomes comparable to the channel contribution producing an increase in the total specific on-resistance. When the cell width is increased beyond 8 μm, the contributions from the channel and accumulation resistances produce a monotonic increase in the specific on-resistance. The smallest specific on-resistance of 12.44 mΩ-cm^2 is observed at an optimum cell width of 8.2 μm.

Simulation Results

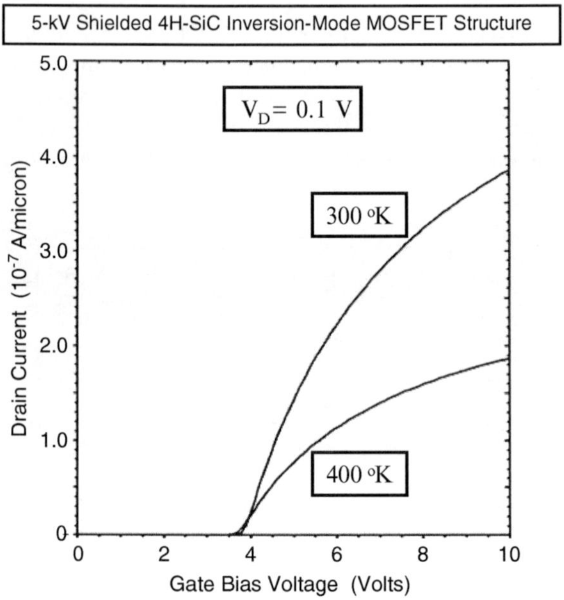

Fig. 6.15 Transfer characteristic of the 5-kV shielded 4H-SiC inversion-mode MOSFET

The transfer characteristics for the 5-kV shielded 4H-SiC inversion-mode power MOSFET structure were obtained using numerical simulations with a drain bias of 0.1 V. The device parameters for the structure used for the numerical simulations were provided in the previous section. The channel mobility was degraded during the simulations to about 20 cm^2/V-s. The resulting transfer characteristic is shown in Fig. 6.15. From this graph, a threshold voltage of 3.7 can be extracted at 300 K and does not change significantly when the temperature is increased to 400 K. This demonstrates that an adequate threshold voltage can be achieved for the inversion-mode 4H-SiC power MOSFET structure by using a low (2 × 10^{16} cm^{-3}) doping concentration for the P-base region. For the case of a gate bias of 10 V and 300 K, the specific on-resistance is found to be 13 mΩ-cm^2 providing validation of the analytical model. The monotonic increase in drain current with gate voltage indicates that the channel resistance is dominant in the 5-kV shielded 4H-SiC inversion-mode power MOSFET structure.

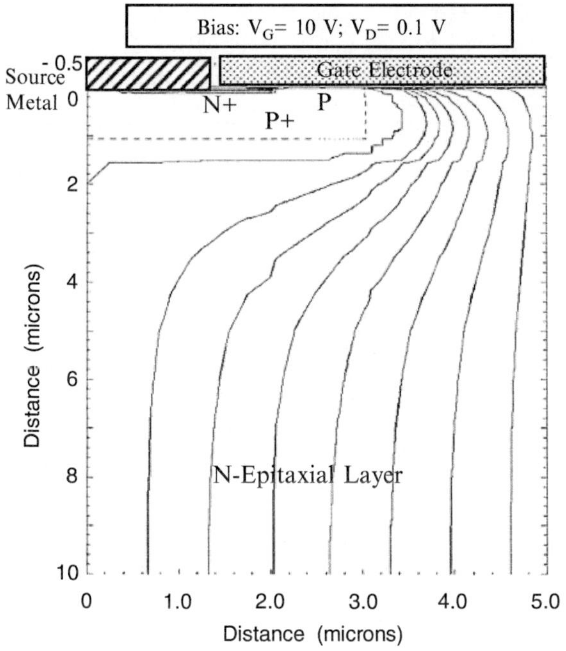

Fig. 6.16 Current distribution in the shielded 4H-SiC inversion-mode MOSFET structure

The on-state current flow pattern within the 5-kV shielded 4H-SiC inversion-mode power MOSFET structure, at a small drain bias of 0.1 V and a gate bias of 10 V, is shown in Fig. 6.16 for the upper 10 μm of the device structure. In the figure, the junction boundary is delineated by the dashed line. The P^+ region extends to $x = 3$ μm leaving a JFET region with half-width of 2 μm. The current flows via the inversion layer formed on the surface of the P-base region. It can be observed that the current flows from the channel and distributes into the JFET region via the accumulation layer. Within the JFET region, the cross-sectional area is approximately constant with a width (a/2) of 1.6 μm in agreement with value obtained using the analytical model. From the figure, it can be seen that the current spreads from the JFET region to the drift region at a 45° angle as assumed in the model and becomes uniform beyond a depth of 4 μm of the drift region. These results provide justification for the assumptions used in formulating the analytical model.

6.7.3 Device Capacitances

Analytical models of the capacitances for the 5-kV shielded silicon carbide inversion-mode power MOSFET structure were derived in Sect. 6.5. The input capacitance can be computed by using Eq. 6.12. The baseline 5-kV shielded silicon carbide inversion-mode power MOSFET structure has a cell width of 10 μm with the other parameters as defined in the previous section. Its JFET region width is 4 μm and its gate width is 6 μm. Using these parameters, the input capacitance for the 5-kV shielded silicon carbide inversion-mode power MOSFET structure is found to be 25.3 nF/cm^2 when an inter-electrode oxide thickness of 5,000 Å is assumed.

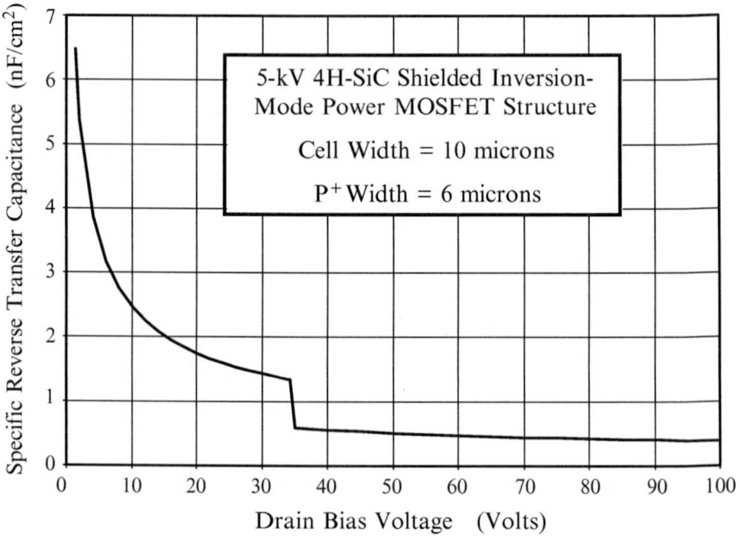

Fig. 6.17 Gate-drain capacitance for the 5-kV 4H-SiC shielded inversion-mode MOSFET

The gate-drain capacitance (or reverse transfer capacitance) for the baseline 5-kV shielded silicon carbide inversion-mode power MOSFET structure can be computed by using Eq. 6.13. The pinch-off voltage for the JFET region is found to be 34 V by using Eq. 6.16. The gate-drain capacitance reduces rapidly as the drain voltage is increased from the on-state voltage drop to the JFET pinch-off voltage. The on-state current density is determined by the power dissipation and the specific on-resistance of the power MOSFET:

$$J_{ON} = \sqrt{\frac{P_{D,ON}}{R_{ON,sp}}} \tag{6.38}$$

The on-state current density is found to be close to 100 A/cm^2 for the 5-kV shielded silicon carbide inversion-mode power MOSFET structure based upon its specific on-resistance of 13.3 mΩ-cm^2 and an on-state power dissipation of 100 W/cm^2. The on-state voltage drop for the device at an on-state current density of 100 A/cm^2 is 1.33 V. At this drain bias voltage, the gate-drain capacitance determined by using the analytical model is 6.47 nF/cm^2. The gate-drain capacitance reduces rapidly to 0.51 nF/cm^2 at a drain bias of 50 V. The variation of the gate-drain capacitance with drain bias voltage predicted by the analytical model is shown in Fig. 6.17 for the 5-kV shielded silicon carbide inversion-mode power MOSFET structure.

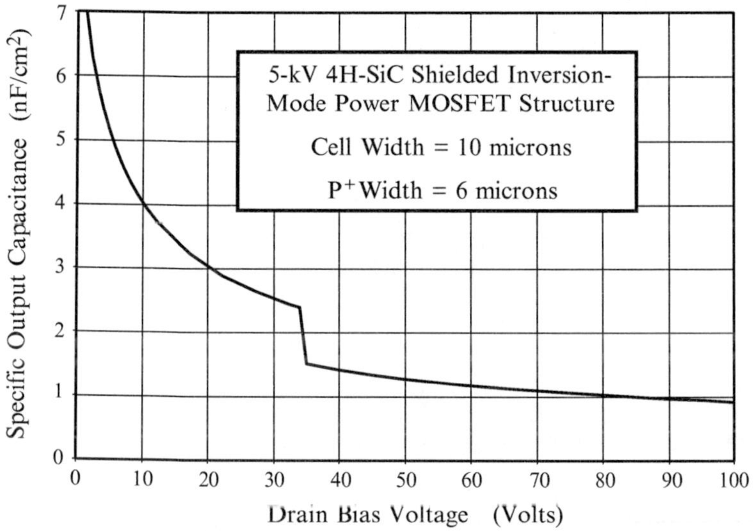

Fig. 6.18 Drain-source capacitance for the 5-kV 4H-SiC shielded inversion-mode MOSFET structure

The drain-source capacitance (or output capacitance) for the baseline 5 kV shielded silicon carbide inversion-mode power MOSFET structure can be computed by using Eq. 6.22. The pinch-off voltage for the JFET region is found to be 34 V by using Eq. 6.16. The drain-source capacitance reduces rapidly as the drain voltage is increased from the on-state voltage drop to the JFET pinch-off voltage. At the on-state voltage, the drain-source capacitance determined by using the analytical model is 6.99 nF/cm^2. The drain-source capacitance reduces rapidly to 1.28 nF/cm^2 at a drain bias of 50 V. The variation of the drain-source capacitance with drain-bias voltage predicted by the analytical model is shown in Fig. 6.18 for the 5-kV shielded silicon carbide inversion-mode power MOSFET structure.

Simulation Results

The input capacitance of the 5-kV shielded 4H-SiC inversion-mode power MOSFET structure was extracted by superposing a small 1-MHz ac signal on the gate electrode while sweeping the gate voltage from 0 to 10 V. The input capacitance was found to be 22 nF/cm^2 in good agreement with the analytical model.

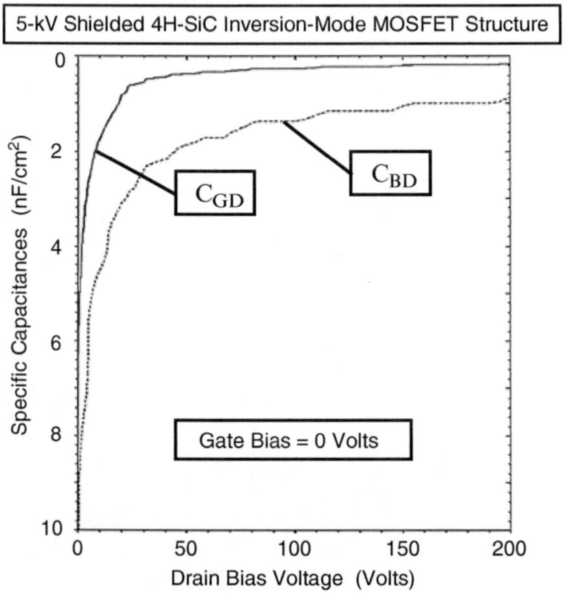

Fig. 6.19 Output and gate-drain capacitance of the shielded 4H-SiC inversion-mode MOSFET structure

The gate-drain (or reverse transfer capacitance) and the output capacitances for the 5-kV shielded 4H-SiC inversion-mode power MOSFET structure was extracted by superposing a small 1-MHz ac signal on the drain electrode while sweeping the drain voltage from 0 to 200 V. The results obtained from the numerical simulations at zero gate bias are shown in Fig. 6.19. In the case of the 5-kV shielded 4H-SiC inversion-mode power MOSFET structure, the source electrode is shielded from the drain by the intervening P$^+$ region. Consequently, the output capacitance is obtained by examining the capacitance (C_{BD}) between the drain electrode and the electrode connected to the P$^+$ region. It can be observed that both of the gate-drain and the output capacitance decrease rapidly until the drain bias reaches close to 30 V and then decreases at a less rapid rate. This trend was predicted by the analytical model by taking into account the pinch-off of the JFET region. The gate-drain capacitance obtained from the numerical simulations is 6.2 nF/cm^2 at the on-state voltage drop of 1.3 V and decreases to 0.4 nF/cm^2 at a drain bias of 50 V. The analytical model provides a very good prediction of these values (see Fig. 6.17). The output capacitance obtained from the numerical simulations is 7 nF/cm^2 at the on-state voltage drop of 1.3 V and decreases to 1.9 nF/cm^2 at a drain bias of 50 V. The analytical model provides a very good prediction of these values (see Fig. 6.18).

6.7.4 Inductive Load Turn-Off Characteristics

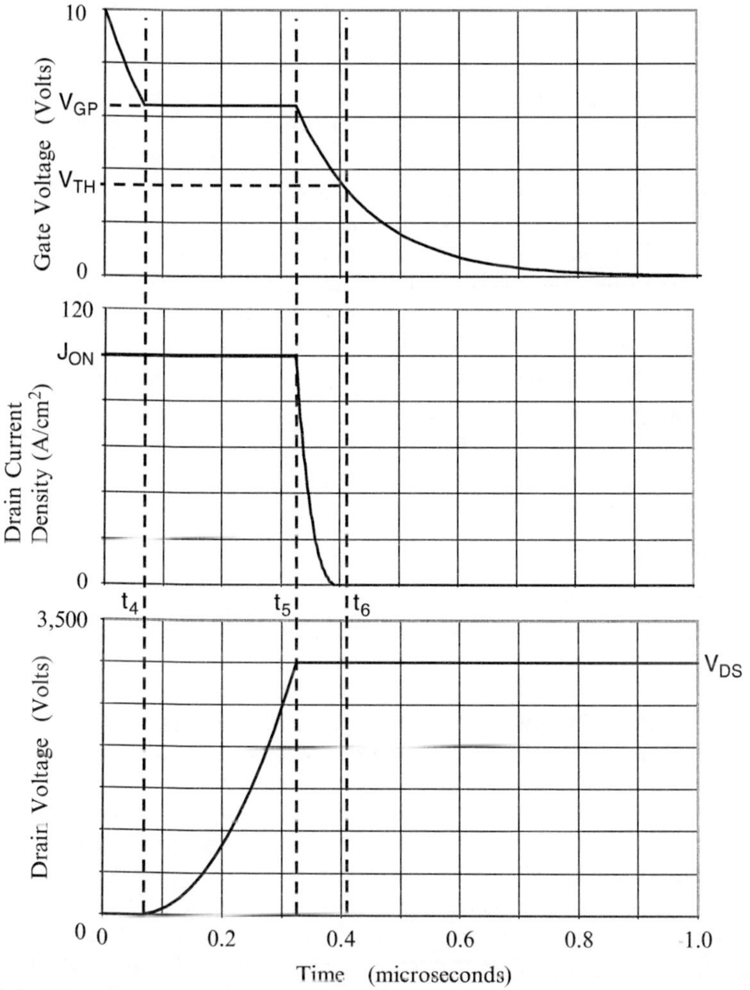

Fig. 6.20 Inductive load turn-off waveforms for the 5-kV 4H-SiC shielded inversion-mode MOSFET structure

The analytical model for the turn-off behavior of the shielded silicon carbide inversion-mode power MOSFET structure was derived in Sect. 6.5. It was demonstrated that the charging of the device capacitances dictates the rate of rise of the drain voltage. Consider the baseline 5-kV shielded silicon carbide inversion-mode power MOSFET structure with a cell width of 10 μm and JFET width of 4 μm. As discussed in the previous sections, this structure has a specific on-resistance of 13.3 mΩ-cm^2 at a gate bias of 10 V. Its specific input capacitance is 25 nF/cm^2 while its gate-drain capacitance varies from 6.47 nF/cm^2 at an on-state voltage drop of 1.33 V to 0.082 nF/cm^2 at a drain bias of 3,000 V. The gate plateau voltage for this device calculated by using Eq. 6.23 is 6.41 V for a threshold voltage of 3.7 V and an inversion layer mobility of 20 cm^2/V-s. The drain voltage transient is dependent on the gate-drain capacitance at the beginning of the gate plateau phase. The gate-drain capacitance at the gate plateau voltage (6.41 V) is found to be 5.3 nF/cm^2 by using the analytical model. This value is used for computation of the drain voltage waveform using the analytical model.

The gate voltage, drain current, and drain voltage transients for the baseline 5-kV shielded silicon carbide inversion-mode power MOSFET structure obtained by using the analytical model with a specific gate resistance of 5 Ω-cm^2 are shown in Fig. 6.20. The gate voltage initially decays exponentially to the gate plateau voltage in 0.067 μs. After the gate voltage reaches the plateau voltage, the drain voltage increases approximately with a square law behavior in accordance with Eq. 6.32 until it reaches 3,000 V at 0.326 μs. The voltage rise-time (t_V) obtained by using the analytical model is 0.259 μs. The gate voltage then reduces exponentially from the gate plateau voltage to zero. The drain current follows this exponential decay in accordance with Eq. 6.34 and reaches zero at time 0.409 μs when the gate voltage becomes equal to the threshold voltage. The current fall-time (t_I) obtained by using the analytical model is 0.083 μs.

Simulation Results

The results of two-dimensional numerical simulations of the inductive load turn-off for the 5-kV shielded 4H-SiC inversion-mode MOSFET structure are described here. The device parameters for the structure used for the numerical simulations were provided in the previous section. The drain supply voltage was chosen as 3,000 V for the turn-off analysis with an initial on-state current density of 100 A/cm^2. During the turn-off simulations, the gate voltage was reduced to zero with a gate resistance of 1×10^8 Ω-μm for the 5 μm half-cell structure, which is equivalent to a specific gate resistance of 5 Ω-cm^2. The current density was initially held constant at an on-state current density of 100 A/cm^2 allowing the drain voltage to rise to the drain supply voltage during the plateau phase. The drain supply voltage was then held constant allowing the drain current density to reduce to zero.

The turn-off waveforms obtained for the 5-kV shielded 4H-SiC inversion-mode MOSFET structure by using the numerical simulations are shown in Fig. 6.21. The gate voltage initially reduces to the gate plateau voltage corresponding to the on-state current density. The gate plateau voltage observed in the numerical simulation is 6.5 V which is close to that obtained with the analytical model. The drain voltage then increases quadratically from the on-state voltage drop to the drain supply voltage ($V_{D,S}$) as predicted by the analytical model to the drain supply voltage. After this, the drain current reduces exponentially. The drain voltage rise-time ($t_5 - t_4$) is much larger than the drain current fall time ($t_6 - t_5$) for the shielded 4H-SiC inversion-mode MOSFET structure. The drain voltage rise-time obtained from the simulations of the 5-kV shielded 4H-SiC inversion-mode MOSFET structure is 0.28 μs and the drain current fall-time obtained from the simulations is 0.09 μs. The values predicted by the analytical model are close to those extracted from the numerical simulations.

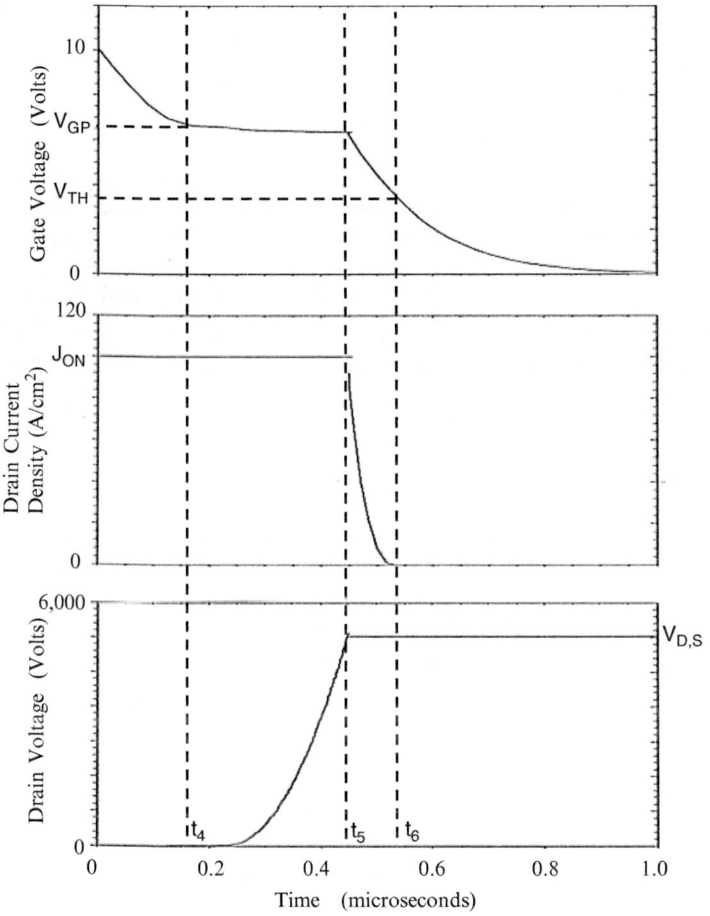

Fig. 6.21 Turn-off waveforms for the 5,000-V shielded 4H-SiC inversion-mode MOSFET structure

6.7.5 Switching Energy Loss

The power loss incurred during the switching transients limit the maximum operating frequency for the 5-kV shielded silicon carbide inversion-mode power MOSFET structure. Power losses during the turn-on of the MOSFET structure are significant but strongly dependent on the reverse recovery behavior of the fly-back rectifiers in circuits. Consequently, it is common practice to use only the turn-off energy loss per cycle during characterization of devices. As shown in the previous section, the turn-off losses are associated with the voltage rise-time interval and the current fall-time interval. The energy loss for each event can be computed by integration of the power loss, as given by the product of the instantaneous current and voltage. During the voltage rise-time interval, the anode current is constant while the voltage increases in a non-linear manner as a function of time. In order to simplify the analysis, the energy loss during the voltage rise-time interval will be computed using:

$$E_{\mathrm{OFF,V}} = \frac{1}{2} J_{\mathrm{C,ON}} V_{\mathrm{CS}} t_{\mathrm{V,OFF}} \tag{6.39}$$

For the typical switching waveforms of the 5-kV shielded silicon carbide inversion-mode power MOSFET structure shown in Fig. 6.20 with a collector supply voltage of 3,000 V, the energy loss per unit area during the collector voltage rise-time is found to be 0.039 J/cm^2 if the on-state current density is 100 A/cm^2.

During the collector current fall-time interval, the collector voltage is constant while the current decreases rapidly. In order to simplify the analysis, the energy loss during the collector current fall-time interval will be computed using:

$$E_{\mathrm{OFF,V}} = \frac{1}{2} J_{\mathrm{C,ON}} V_{\mathrm{CS}} t_{\mathrm{I,OFF}} \tag{6.40}$$

For the typical switching waveforms of the 5-kV shielded silicon carbide inversion-mode power MOSFET structure shown in Fig. 6.20 with a collector supply voltage of 3,000 V, the energy loss per unit area during the collector current fall-time is found to be 0.012 J/cm^2 if the on-state current density is 100 A/cm^2. The total energy loss per unit area ($E_{\mathrm{OFF,V}}$ + $E_{\mathrm{OFF,I}}$) during the turn-off process for the 5-kV asymmetric IGBT structure is found to be 0.051 J/cm^2. This switching energy loss per cycle is much smaller than that for the 5-kV silicon IGBT structure allowing the 5-kV shielded silicon carbide inversion-mode power MOSFET structure to operate at higher frequencies.

6.7.6 Maximum Operating Frequency

The maximum operating frequency for operation of the 5-kV shielded silicon carbide inversion-mode power MOSFET structure can be obtained by combining the on-state and switching power losses:

$$P_{\mathrm{D,TOTAL}} = \delta P_{\mathrm{D,ON}} + E_{\mathrm{OFF}} f \tag{6.41}$$

where δ is the duty cycle and f is the switching frequency. In the case of the 5-kV shielded silicon carbide inversion-mode power MOSFET structure, the on-state voltage drop is 1.33 V at an on-state current density of 100 A/cm^2 because the specific on-resistance for this structure at 300 K is 13.33 mΩ-cm^2. For the case of a 50% duty cycle, the on-state power dissipation contributes 67 W/cm^2 to the total power loss. Using a total turn-off energy loss per cycle of 0.051 J/cm^2 in Eq. 6.41 yields a maximum operating frequency of 2,600 Hz if the total power dissipation is 200 W/cm^2. In the case of operation at a reduced duty cycle of 10%, the maximum operating frequency for the 5-kV shielded silicon carbide inversion-mode power MOSFET structure increases to 3,640 Hz if the total power dissipation is 200 W/cm^2.

6.8 10-kV Inversion-Mode MOSFET

This section discusses the performance of the shielded silicon carbide inversion-mode power MOSFET structure with the 10-kV blocking capability. The results of numerical simulations are used to describe the internal electric field and potential distribution in the blocking mode. The analytical model is then used to predict the specific on-resistance for this structure. The turn-off characteristics of the device are analyzed using the results of numerical simulations to determine the maximum operating frequency.

6.8.1 Blocking Characteristics

The blocking voltage capability for the shielded silicon carbide inversion mode power MOSFET structure is determined by the doping concentration and thickness of the drift region if the JFET region width is sufficiently small to prevent rupture of the gate oxide. Under the assumption of a typical edge termination with 80% of the parallel-plane breakdown voltage, the drift region for the 10-kV device structure must have the properties required for a parallel-plane breakdown voltage of 12,500 V. Using Eq. 6.37, a breakdown voltage of 12,500 V can be obtained in 4H-SiC by using a doping concentration of 1.5×10^{15} cm^{-3}. In the case of the power MOSFET structure, the thickness of the drift region is the depletion layer width at this doping concentration when the drain bias reaches 10,000 V. By using Eq. 6.21, the drift region thickness for the 10-kV shielded silicon carbide inversion-mode power MOSFET structure is found to be 85 μm. This combination of drift region doping concentration and thickness will be used when modeling the behavior of the 10-kV shielded silicon carbide inversion-mode power MOSFET structure.

Simulation Results

The results of two-dimensional numerical simulations on the 10-kV shielded 4H-SiC inversion-mode power MOSFET structure are described here. For the numerical simulations, the half-cell structure with a width ($W_{Cell}/2$) of 5 μm as illustrated in Fig. 6.1 was utilized as representative of the structure. The device used for the numerical simulations had a drift region thickness of 85 μm below the P$^+$ shielding region with a doping concentration of 1.5×10^{15} cm^{-3}. The P$^+$ region extended from a depth of 0.2 to 1.0 μm with a doping concentration of 1×10^{19} cm^{-3}. The P-base and N$^+$ source regions were formed within the 0.2 μm of the N-drift region located above the P$^+$ region. The doping concentration of the P-base region was 2×10^{16} cm^{-3}. Due to the low doping concentration in the drift region for the

10-kV 4H-SiC devices, the uniform doping concentration in the JFET region was enhanced to 1×10^{16} cm^{-3}. The enhanced doping concentration was extended to 0.5 μm below the P^+ shielding region.

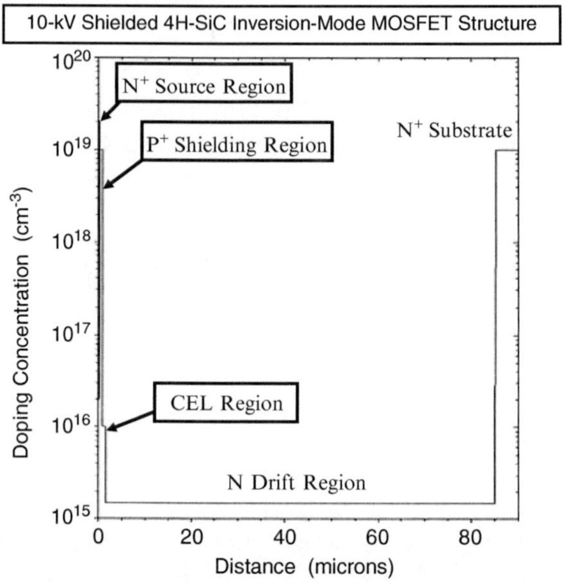

Fig. 6.22 Vertical doping profile in the 10-kV shielded 4H-SiC inversion-mode MOSFET

The lateral doping profile taken along the surface of the 10-kV shielded 4H-SiC inversion-mode power MOSFET structure is identical to that previously shown for the 5-kV device structure in Fig. 6.7. From the profile, it can be observed that the channel extends from 2 to 3 μm creating a channel length of 1 μm in the P-base region. The doping concentration of the JFET region is 1×10^{16} cm^{-3} while that for the P-base region is 2×10^{16} cm^{-3}. The N^+ source region and the P^+ contact region for shorting the source to the base region are visible on the left-hand-side. All the regions were defined with uniform doping with abrupt interfaces between them due to the low diffusion rates for dopants in 4H-SiC material.

The vertical doping profile taken through the N^+ source region of the 10-kV shielded 4H-SiC inversion-mode power MOSFET structure is provided in Fig. 6.22. It can be observed that the doping concentration of the P^+ shielding region has a maximum value of 1×10^{19} cm^{-3} at a depth ranging from 0.2 to 1.0 μm. The P-base region is located between the N^+ source region and the P^+ shielding region with a doping concentration of 2×10^{16} cm^{-3}. The N-drift region has a doping concentration of 1.5×10^{15} cm^{-3} and thickness of 85 μm.

The blocking characteristics for the 10-kV shielded 4H-SiC inversion-mode power MOSFET structure were obtained by increasing the drain voltage while using zero gate bias. Due to the very small intrinsic concentration in 4H-SiC, no substantial leakage current was observed at room temperature.

This also confirmed that the reach-through of the P-base region has been suppressed by the P^+ shielding region. The potential contours within the 10-kV shielded 4H-SiC inversion-mode power MOSFET structure at a drain bias of 10-kV volts are similar to those shown in Fig. 6.9 for the 5-kV device structure. The potential contours crowd at the edge of the P^+ shielding region indicating an enhanced electric field. The electric field distribution in the 10-kV shielded 4H-SiC inversion-mode power MOSFET structure is also similar to those shown in Figs. 6.10 and 6.11 for the 5-kV device structure. There is an enhanced electric field at the edge of the P^+ shielding region. More importantly, there is substantial suppression of the electric field under the gate oxide.

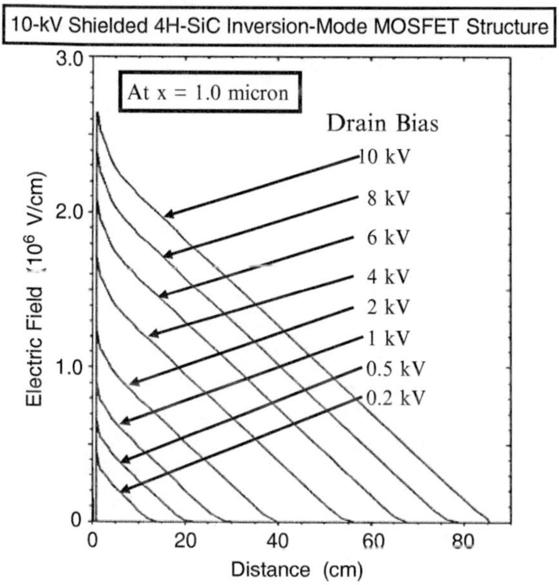

Fig. 6.23 Electric field distribution in the 10 kV shielded 4H-SiC inversion-mode MOSFET structure

It is insightful to examine the electric field profile within the 10-kV shielded 4H-SiC inversion-mode power MOSFET structure when it is operating in the blocking mode. The electric field profile through the junction between the P^+ shielding region and the N-drift region is provided in Fig. 6.23. The peak of the electric field occurs at the junction as expected and is essentially triangular in shape in accordance with the predictions of Poisson's equation with a uniform doping profile. There is a slight increase in the electric field near the junction due to the higher doping concentration of the CEL layer when compared with the drift region. However, the enhancement in the electric field is small and occurs over a very small distance which results in minimal degradation of the breakdown voltage. The maximum electric field at the junction at a drain bias of 10-kV is 2.6×10^6 V/cm which is close to the critical electric for breakdown for 4H-SiC at the doping concentration of the drift region [8].

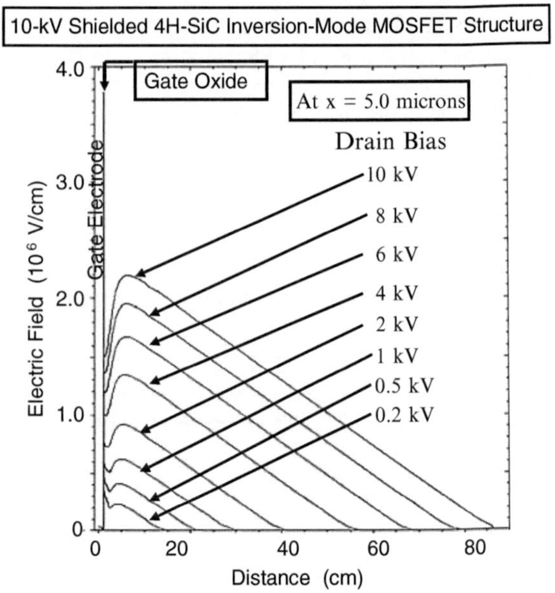

Fig. 6.24 Electric field distribution in the 10-kV shielded 4H-SiC inversion-mode MOSFET structure

The electric field profiles obtained through the middle of the JFET region of the 10-kV shielded 4H-SiC inversion-mode power MOSFET structure are shown in Fig. 6.24 at various drain-bias voltages. It can be observed that the maximum electric field occurs at a depth of 5 μm from the surface. The shielding effect produced by the P^+ shielding region reduces the electric field at the surface under the gate oxide to only 1.5×10^6 V/cm which is about one-half of the electric field in the bulk below the P^+ shielding region. Consequently, the electric field in the gate oxide is reduced to about 3.7×10^6 V/cm even when a drain bias of 10-kV is applied. The low electric field in the gate oxide prevents gate oxide rupture and allows stable device performance over long periods of time. An even further reduction of electric field in the gate oxide can be achieved by reducing the width of the JFET region to 4 μm.

6.8.2 On-Resistance

The analytical model for the specific on-resistance for the shielded silicon carbide inversion-mode power MOSFET structure was provided in Sect. 6.4. In the case of the 10-kV shielded silicon carbide inversion-mode power MOSFET structure, the baseline cell width (W_{Cell}) will be assumed to be 10 μm with a channel length of 1 μm. A threshold voltage for the inversion mode structure of 3.7 V can be achieved in the shielded silicon carbide inversion-mode power MOSFET structure by using a low doping concentration of 2×10^{16} cm^{-3} for the P-base region and a gate oxide thickness is 500 Å. Based upon reports in the literature [6], a low inversion layer mobility of 20 cm^2/V−s will be used in the analytical model. For the baseline device structure, the JFET region will be assumed to have a width (W_{JFET}) of 4 μm and its doping concentration will be assumed to be 1×10^{16} cm^{-3}. The thickness (t_{P+}) of the P$^+$ shielding region will be assumed to 1 μm.

Using the above device structural parameters in Eq. 6.4, the specific resistance contributed by the channel at a gate bias of 10 V is found to be 5.64 mΩ-cm^2. This value is smaller than the ideal specific on-resistance of 21.4 mΩ-cm^2 for the drift region of a 10-kV 4H-SiC unipolar device structure. Using an accumulation layer mobility of 100 cm^2/V-s for the above 10-kV shielded silicon carbide inversion-mode power MOSFET structure, the specific resistance contributed by the accumulation layer at a gate bias of 10 V is found to be 0.88 mΩ-cm^2 by using Eq. 6.5. This value is small when compared with the ideal specific on-resistance of 21.4 mΩ-cm^2 for the drift region of a 10-kV 4H-SiC unipolar device structure. In the case of the above 10-kV shielded silicon carbide inversion-mode power MOSFET structure, the resistivity for the JFET region is found to be 0.625 Ω-cm. The zero-bias depletion width in the JFET region for the JFET doping concentration of 1×10^{16} cm^{-3} is 0.57 μm based up on a built-in potential of 3 V. Using these values, the specific resistance contributed by the JFET region is found to be 0.22 mΩ-cm^2 by using Eq. 6.9. This value is small when compared to the ideal specific on-resistance of 21.4 mΩ-cm^2 for the drift region of a 10-kV 4H-SiC unipolar device structure.

For the parameters given above for the 10-kV shielded silicon carbide inversion-mode power MOSFET structure, the dimension "a" in Eq. 6.11 is found to be 2.86 μm. The specific resistance contributed by the drift region is then found to be 33.2 mΩ-cm^2 by using Eq. 6.11 with a resistivity of the drift region of 3.79 Ω-cm (based upon a doping concentration of 1.5×10^{15} cm^{-3}) and a drift region thickness of 85 μm below the P$^+$ shielding region.

The total specific on-resistance for the 10-kV shielded silicon carbide inversion-mode power MOSFET structure with cell width of 10 μm can be obtained by adding the above components of the resistances within the device structure. For a gate bias of 10 V, the total specific on-resistance is found to be 39.93 mΩ-cm^2. The channel resistance constitutes 14% of the total specific on-resistance despite the poor mobility for the electrons in the inversion layer. The specific on-resistance for the 10-kV shielded planar 4H-SiC MOSFET structure is about three orders of magnitude smaller than that for the silicon 10-kV power D-MOSFET and U-MOSFET structures.

Note that the substrate resistance has not been included in the above analysis of the total specific on-resistance of the 10-kV shielded silicon carbide inversion-mode power MOSFET structure. A typical N$^+$ 4H-SiC substrate with resistivity of 0.02 Ω-cm and thickness of 200 μm will contribute an additional 4 mΩ-cm^2 to the specific on-resistance. This contribution is small when compared with the 39.93 mΩ-cm^2 specific on-resistance obtained in the analytical model.

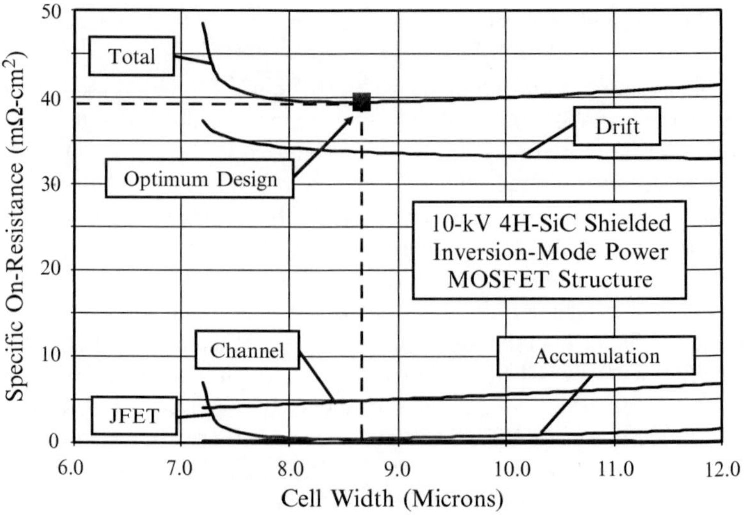

Fig. 6.25 On-resistance for the 10-kV 4H-SiC shielded inversion-mode MOSFET structures

The impact of changing the width of the JFET region on the specific on-resistance of the 10-kV shielded silicon carbide inversion-mode power MOSFET structure can be determined by using the above analytical model. The results obtained for the case of a P^+ shielding region width (W_{P+}) of 6 μm are provided in Fig. 6.25. For this analysis, an inversion layer mobility of 20 cm^2/V-s was used. The cell width (W_{Cell}) is equal to the width of the JFET region (W_{JFET}) plus the width of the P^+ shielding region. It can be observed that the specific on-resistance goes through a minimum as the cell width is increased. The smallest specific on-resistance of 39.4 mΩ-cm^2 is observed at an optimum cell width of 8.6 μm. The drift region contribution can be observed to be a dominant portion of the specific on-resistance for the 10-kV shielded silicon carbide inversion-mode power MOSFET structure.

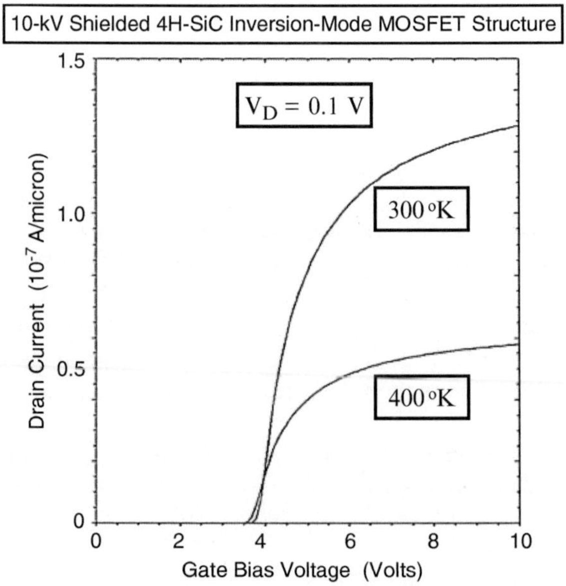

Fig. 6.26 Transfer characteristic of the 10-kV shielded 4H-SiC inversion-mode MOSFET

Simulation Results

The transfer characteristic for the 10-kV shielded 4H-SiC inversion-mode power MOSFET structure was obtained using numerical simulations with a drain bias of 0.1 V. The device parameters for the structure used for the numerical simulations were provided in the previous section. The channel mobility was degraded during the simulations to about 20 cm^2/V-s. The resulting transfer characteristic is shown in Fig. 6.26. From this graph, a threshold voltage of 3.7 can be extracted at 300 K and does not change significantly when the temperature is increased to 400 K. This demonstrates that an adequate threshold voltage can be achieved for the inversion-mode 4H-SiC power MOSFET structure by using a low (2×10^{16} cm^{-3}) doping concentration for the P-base region. For the case of a gate bias of 10 V and 300 K, the specific on-resistance is found to be 39.1 mΩ-cm^2 providing validation of the analytical model. The drain current begins to saturate when the gate voltage reaches 10 V indicating that the channel resistance is not dominant in the 10-kV shielded 4H-SiC inversion-mode power MOSFET structure.

6.8.3 Inductive Load Turn-Off Characteristics

The analytical model for the turn-off behavior of the shielded silicon carbide inversion-mode power MOSFET structure was derived in Sect. 6.5. It was demonstrated that the charging of the device capacitances dictates the rate of rise of the drain voltage. This model was validated for the 5-kV device structure. This model can therefore also be utilized to generate the inductive load turn-off waveforms for the 10 kV shielded silicon carbide inversion-mode power MOSFET structure.

Simulation Results

The results of two-dimensional numerical simulations of the inductive load turn-off for the 10-kV shielded 4H-SiC inversion-mode MOSFET structure are described here. The device parameters for the structure used for the numerical simulations were provided in the previous section. The drain supply voltage was chosen as 6,000 V for the turn-off analysis with an initial on-state current density of 50 A/cm^2. During the turn-off simulations, the gate voltage was reduced to zero with a gate resistance of 1×10^8 Ω-μm for the 5 μm half-cell structure, which is equivalent to a specific gate resistance of 5 Ω-cm^2. The current density was initially held constant at an on-state current density of 50 A/cm^2 allowing the drain voltage to rise to the drain supply voltage ($V_{D,S}$) during the plateau phase. The drain supply voltage was then held constant allowing the drain current density to reduce to zero.

The turn-off waveforms obtained for the 10-kV shielded 4H-SiC inversion-mode MOSFET structure by using the numerical simulations are shown in Fig. 6.27. The gate voltage initially reduces to the gate plateau voltage corresponding to the on-state current density. The gate plateau voltage observed in the numerical

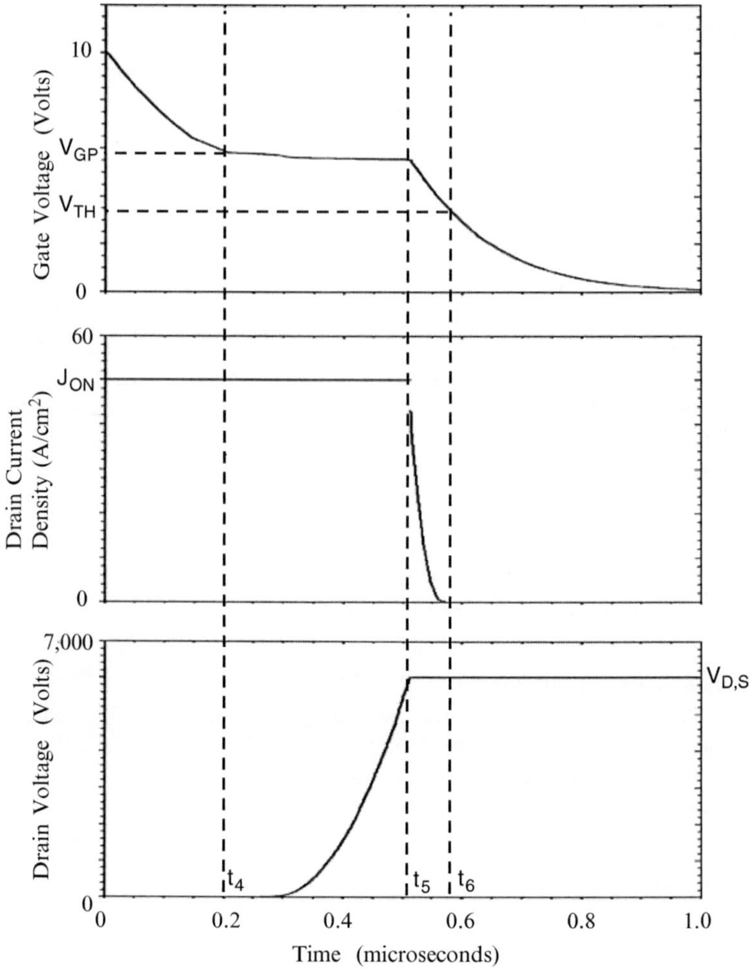

Fig. 6.27 Turn-off waveforms for the 10-kV shielded 4H-SiC inversion-mode MOSFET structure

simulation is 5.56 V. The drain voltage then increases quadratically from the on-state voltage drop to the drain supply voltage. After this, the drain current reduces exponentially. The drain voltage rise-time (t_5-t_4) is much larger than the drain current fall time (t_6-t_5) for the shielded 4H-SiC inversion-mode MOSFET structure. The drain voltage rise-time obtained from the simulations of the 10-kV shielded 4H-SiC inversion-mode MOSFET structure is 0.31 μs and the drain current fall-time obtained from the simulations is 0.08 μs. These values are slightly larger than those observed in the numerical simulations for the 5-kV device structure.

6.8.4 Switching Energy Loss

The power loss incurred during the switching transients limit the maximum operating frequency for the 10-kV shielded silicon carbide inversion-mode power MOSFET structure. Power losses during the turn-on of the MOSFET structure are significant but strongly dependent on the reverse recovery behavior of the fly-back rectifiers in circuits. Consequently, it is common practice to use only the turn-off energy loss per cycle during characterization of devices. As shown in the previous section, the turn-off losses are associated with the voltage rise-time interval and the current fall-time interval. The energy loss for each event can be computed by integration of the power loss, as given by the product of the instantaneous current and voltage. During the voltage rise-time interval, the anode current is constant while the voltage increases in a non-linear manner as a function of time. In order to simplify the analysis, the energy loss during the voltage rise-time interval will be computed using Eq. 6.39. For the typical switching waveforms of the 10-kV shielded silicon carbide inversion-mode power MOSFET structure shown in Fig. 6.27 with a collector supply voltage of 6,000 V, the energy loss per unit area during the collector voltage rise-time is found to be 0.047 J/cm^2 if the on-state current density is 50 A/cm^2.

During the collector current fall-time interval, the collector voltage is constant while the current decreases rapidly. In order to simplify the analysis, the energy loss during the collector current fall-time interval will be computed using Eq. 6.40. For the typical switching waveforms of the 10-kV shielded silicon carbide inversion-mode power MOSFET structure shown in Fig. 6.27 with a collector supply voltage of 6,000 V, the energy loss per unit area during the collector current fall-time is found to be 0.012 J/cm^2 if the on-state current density is 50 A/cm^2. The total energy loss per unit area ($E_{OFF,V} + E_{OFF,I}$) during the turn-off process for the 10-kV asymmetric IGBT structure is found to be 0.059 J/cm^2. This switching energy loss per cycle is much smaller than that for the 10-kV silicon IGBT structure allowing the 10-kV shielded silicon carbide inversion-mode power MOSFET structure to operate at higher frequencies.

6.8.5 Maximum Operating Frequency

The maximum operating frequency for operation of the 10-kV shielded silicon carbide inversion-mode power MOSFET structure can be obtained by combining the on-state and switching power losses and using Eq. 6.41. In the case of the 10-kV shielded silicon carbide inversion-mode power MOSFET structure, the on-state voltage drop is 1.95 V at an on-state current density of 50 A/cm^2 because the specific on-resistance for this structure at 300 K is 39.1 mΩ-cm^2. For the case of a 50% duty cycle, the on-state power dissipation contributes 48.9 W/cm^2 to the total power loss. Using a total turn-off energy loss per cycle of 0.059 J/cm^2 in Eq. 6.41

yields a maximum operating frequency of 2,600 Hz if the total power dissipation is 200 W/cm^2. In the case of operation at a reduced duty cycle of 10%, the maximum operating frequency for the 10-kV shielded silicon carbide inversion-mode power MOSFET structure increases to 3,250 Hz if the total power dissipation is 200 W/cm^2.

6.9 20-kV Inversion-Mode MOSFET

This section discusses the performance of the shielded silicon carbide inversion-mode power MOSFET structure with the 20-kV blocking capability. The results of numerical simulations are used to describe the internal electric field and potential distribution in the blocking mode. The analytical model is then used to predict the specific on-resistance for this structure. The turn-off characteristics of the device are the analyzed using the results of numerical simulations to determine the maximum operating frequency.

6.9.1 Blocking Characteristics

The blocking voltage capability for the shielded silicon carbide inversion mode power MOSFET structure is determined by the doping concentration and thickness of the drift region if the JFET region width is sufficiently small to prevent rupture of the gate oxide. Under the assumption of a typical edge termination with 80% of the parallel-plane breakdown voltage, the drift region for the 20-kV device structure must have the properties required for a parallel-plane breakdown voltage of 25,000 V. Using Eq. 6.37, a breakdown voltage of 25,000 V can be obtained in 4H-SiC by using a doping concentration of 6.0×10^{14} cm^{-3}. In the case of the power MOSFET structure, the thickness of the drift region is the depletion layer width at this doping concentration when the drain bias reaches 20,000 V. By using Eq. 6.21, the drift region thickness for the 20-kV shielded silicon carbide inversion-mode power MOSFET structure is found to be 190 μm. This combination of drift region doping concentration and thickness will be used when modeling the behavior of the 20-kV shielded silicon carbide inversion-mode power MOSFET structure.

Simulation Results

The results of two-dimensional numerical simulations on the 20-kV shielded 4H-SiC inversion-mode power MOSFET structure are described here. For the numerical simulations, the half-cell structure with a width ($W_{Cell}/2$) of 5 μm as illustrated in Fig. 6.1 was utilized as representative of the structure. The device used for the numerical simulations had a drift region thickness of 190 μm below

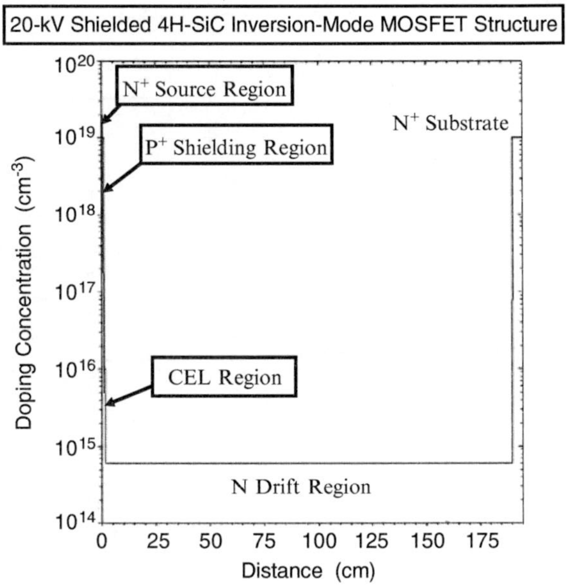

Fig. 6.28 Vertical doping profile in the 20-kV shielded 4H-SiC inversion-mode MOSFET

the P^+ shielding region with a doping concentration of 6.0×10^{14} cm^{-3}. The P^+ region extended from a depth of 0.2 to 1.0 μm with a doping concentration of 1×10^{19} cm^{-3}. The P-base and N^+ source regions were formed within the 0.2 μm of the N-drift region located above the P^+ region. The doping concentration of the P-base region was 2×10^{16} cm^{-3}. Due to the low doping concentration in the drift region for the 20-kV 4H-SiC devices, the uniform doping concentration in the JFET region was enhanced to 1×10^{16} cm^{-3}. The enhanced doping concentration was extended to 0.5 μm below the P^+ shielding region.

The lateral doping profile taken along the surface of the 20-kV shielded 4H-SiC inversion-mode power MOSFET structure is identical to that previously shown for the 5-kV device structure in Fig. 6.7. From the profile, it can be observed that the channel extends from 2 to 3 μm creating a channel length of 1 μm in the P-base region. The doping concentration of the JFET region is 1×10^{16} cm^{-3} while that for the P-base region is 2×10^{16} cm^{-3}. The N^+ source region and the P^+ contact region for shorting the source to the base region are visible on the left-hand-side. All the regions were defined with uniform doping with abrupt interfaces between them due to the low diffusion rates for dopants in 4H-SiC material.

The vertical doping profile taken through the N^+ source region of the 20-kV shielded 4H-SiC inversion-mode power MOSFET structure is provided in Fig. 6.28. It can be observed that the doping concentration of the P^+ shielding region has a maximum value of 1×10^{19} cm^{-3} at a depth ranging from 0.2 to 1.0 μm. The P-base region is located between the N^+ source region and the P^+ shielding region with a doping concentration of 2×10^{16} cm^{-3}. The N-drift region has a doping concentration of 6.0×10^{14} cm^{-3} and thickness of 190 μm.

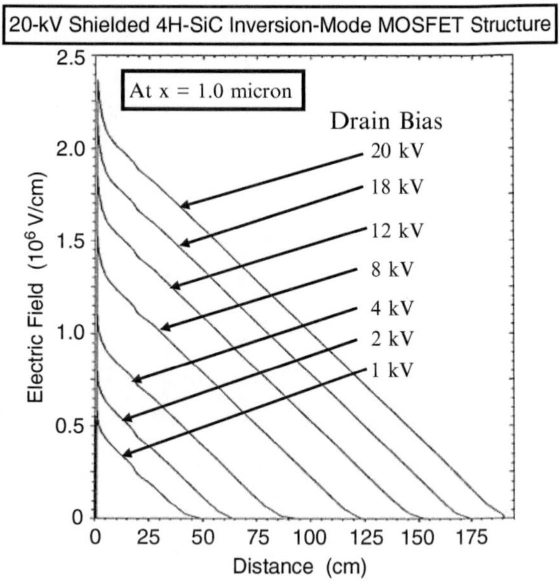

Fig. 6.29 Electric field distribution in the 20-kV shielded 4H-SiC inversion-mode MOSFET structure

The blocking characteristics for the 20-kV shielded 4H-SiC inversion-mode power MOSFET structure were obtained by increasing the drain voltage while using zero gate bias. Due to the very small intrinsic concentration in 4H-SiC, no substantial leakage current was observed at room temperature. This also confirmed that the reach-through of the P-base region has been suppressed by the P^+ shielding region. The potential contours within the 20-kV shielded 4H-SiC inversion-mode power MOSFET structure at a drain bias of 20-kV volts are similar to those shown in Fig. 6.9 for the 5-kV device structure. The potential contours crowd at the edge of the P^+ shielding region indicating an enhanced electric field. The electric field distribution in the 20-kV shielded 4H-SiC inversion-mode power MOSFET structure is also similar to those shown in Figs. 6.10 and 6.11 for the 5-kV device structure. There is an enhanced electric field at the edge of the P^+ shielding region. More importantly, there is substantial suppression of the electric field under the gate oxide.

It is insightful to examine the electric field profile within the 20-kV shielded 4H-SiC inversion-mode power MOSFET structure when it is operating in the blocking mode. The electric field profile through the junction between the P^+ shielding region and the N-drift region is provided in Fig. 6.29. The peak of the electric field occurs at the junction as expected and is essentially triangular in shape in accordance with the predictions of Poisson's equation with a uniform doping profile. There is a slight increase in the electric field near the junction due to the higher doping concentration of the CEL layer when compared with the drift region. However, the enhancement in the electric field is small and occurs over a

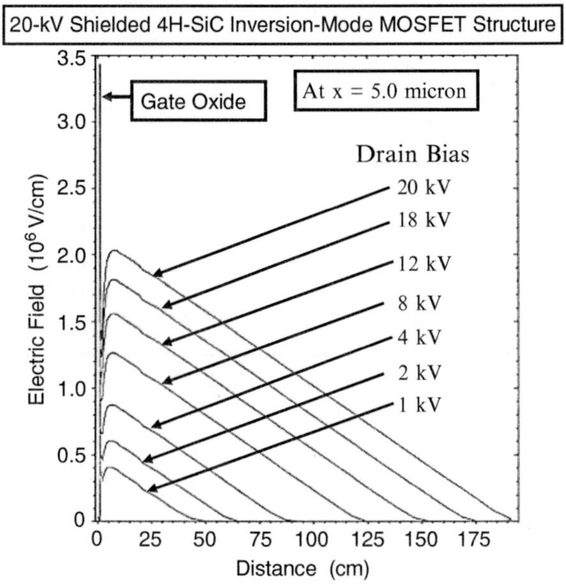

Fig. 6.30 Electric field distribution in the 20-kV shielded 4H-SiC inversion-mode MOSFET structure

very small distance which results in minimal degradation of the breakdown voltage. The maximum electric field at the junction at a drain bias of 20-kV is 2.3×10^6 V/cm which is close to the critical electric for breakdown for 4H-SiC at the doping concentration of the drift region [8].

The electric field profiles obtained through the middle of the JFET region of the 20-kV shielded 4H-SiC inversion-mode power MOSFET structure are shown in Fig. 6.30 at various drain-bias voltages. It can be observed that the maximum electric field occurs at a depth of 5 μm from the surface. The shielding effect produced by the P^+ shielding region reduces the electric field at the surface under the gate oxide to only 1.4×10^6 V/cm which is about one-half of the electric field in the bulk below the P^+ shielding region. Consequently, the electric field in the gate oxide is reduced to about 3.4×10^6 V/cm even when a drain bias of 20-kV is applied. The low electric field in the gate oxide prevents gate oxide rupture and allows stable device performance over long periods of time. An even further reduction of electric field in the gate oxide can be achieved by reducing the width of the JFET region to 4 μm.

6.9.2 On-Resistance

The analytical model for the specific on-resistance for the shielded silicon carbide inversion-mode power MOSFET structure was provided in Sect. 6.4. In the case of the 20-kV shielded silicon carbide inversion-mode power MOSFET structure, the baseline cell width (W_{Cell}) will be assumed to be 10 μm with a channel length of 1 μm. A threshold voltage for the inversion mode structure of 3.7 V can be achieved in the shielded silicon carbide inversion-mode power MOSFET structure by using a low doping concentration of 2×10^{16} cm^{-3} for the P-base region and a gate oxide thickness is 500 Å. Based upon reports in the literature [6], a low inversion layer mobility of 20 cm^2/V-s will be used in the analytical model. For the baseline device structure, the JFET region will be assumed to have a width (W_{JFET}) of 4 μm and its doping concentration will be assumed to be 1×10^{16} cm^{-3}. The thickness (t_{P+}) of the P$^+$ shielding region will be assumed to 1 μm.

Using the above device structural parameters in Eq. 6.4, the specific resistance contributed by the channel at a gate bias of 10 V is found to be 5.64 mΩ-cm^2. This value is much smaller than the ideal specific on-resistance of 118 mΩ-cm^2 for the drift region of a 20-kV 4H-SiC unipolar device structure. Using an accumulation layer mobility of 100 cm^2/V-s for the above 20-kV shielded silicon carbide inversion-mode power MOSFET structure, the specific resistance contributed by the accumulation layer at a gate bias of 10 V is found to be 0.88 mΩ-cm^2 by using Eq. 6.5. This value is small when compared with the ideal specific on-resistance of 118 mΩ-cm^2 for the drift region of a 20-kV 4H-SiC unipolar device structure. In the case of the above 5-kV shielded silicon carbide inversion-mode power MOSFET structure, the resistivity for the JFET region is found to be 0.625 Ω-cm. The zero-bias depletion width in the JFET region for the JFET doping concentration of 1×10^{16} cm^{-3} is 0.57 μm based up on a built-in potential of 3 V. Using these values, the specific resistance contributed by the JFET region is found to be 0.22 mΩ-cm^2 by using Eq. 6.9. This value is small when compared to the ideal specific on-resistance of 118 mΩ-cm^2 for the drift region of a 20-kV 4H-SiC unipolar device structure.

For the parameters given above for the 20-kV shielded silicon carbide inversion-mode power MOSFET structure, the dimension "a" in Eq. 6.11 is found to be 2.86 μm. The specific resistance contributed by the drift region is then found to be 182 mΩ-cm^2 by using Eq. 6.11 with a resistivity of the drift region of 9.47 Ω-cm (based upon a doping concentration of 6.0×10^{14} cm^{-3}) and a drift region thickness of 190 μm below the P$^+$ shielding region.

The total specific on-resistance for the 20-kV shielded silicon carbide inversion-mode power MOSFET structure with cell width of 10 μm can be obtained by adding the above components of the resistances within the device structure. For a gate bias of 10 V, the total specific on-resistance is found to be 189.2 mΩ-cm^2. The channel resistance constitutes 3% of the total specific on-resistance despite the poor mobility for the electrons in the inversion layer. The specific on-resistance for the 20-kV shielded planar 4H-SiC MOSFET structure is about three orders of magnitude smaller than that for the silicon 20-kV power D-MOSFET and U-MOSFET structures.

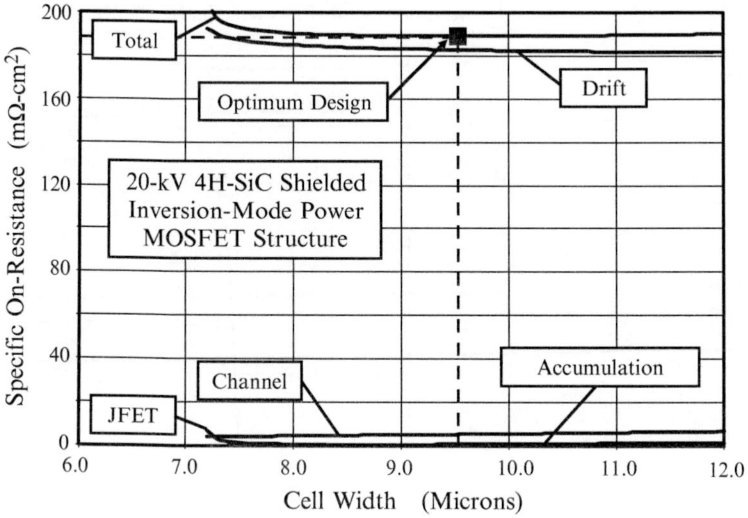

Fig. 6.31 On-resistance for the 20-kV 4H-SiC shielded inversion-mode MOSFET structures

Note that the substrate resistance has not been included in the above analysis of the total specific on-resistance of the 20-kV shielded silicon carbide inversion-mode power MOSFET structure. A typical N^+ 4H-SiC substrate with resistivity of 0.02 Ω-cm and thickness of 200 μm will contribute an additional 4 mΩ-cm^2 to the specific on-resistance. This contribution is small when compared with the 189 mΩ-cm^2 specific on-resistance obtained in the analytical model.

The impact of changing the width of the JFET region on the specific on-resistance of the 20-kV shielded silicon carbide inversion-mode power MOSFET structure can be determined by using the analytical model. The results obtained for the case of a P^+ shielding region width (W_{P+}) of 6 μm are provided in Fig. 6.31. For this analysis, an inversion layer mobility of 20 cm^2/V-s was used. The cell width (W_{Cell}) is equal to the width of the JFET region (W_{JFET}) plus the width of the P^+ shielding region. It can be observed that the specific on-resistance goes through a minimum as the cell width is increased. The smallest specific on-resistance of 189.1 mΩ-cm^2 is observed at an optimum cell width of 9.5 μm. The drift region contribution can be observed to be a dominant portion of the specific on-resistance for the 20-kV shielded silicon carbide inversion-mode power MOSFET structure.

Simulation Results

The transfer characteristic for the 20-kV shielded 4H-SiC inversion-mode power MOSFET structure was obtained using numerical simulations with a drain bias of 0.1 V. The device parameters for the structure used for the numerical simulations were provided in the previous section. The channel mobility was degraded during the simulations to about 20 cm^2/V-s. The resulting transfer characteristic is shown in Fig. 6.32. From this graph, a threshold voltage of 3.7 can be extracted at 300 K

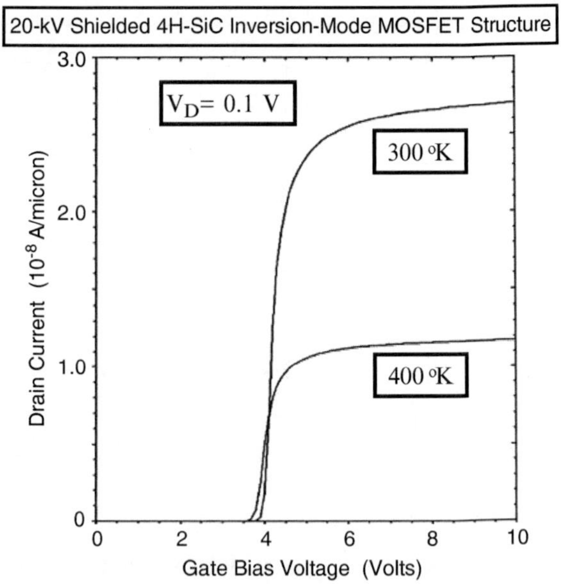

Fig. 6.32 Transfer characteristic of the 20-kV shielded 4H-SiC inversion-mode MOSFET

and does not change significantly when the temperature is increased to 400 K. This demonstrates that an adequate threshold voltage can be achieved for the inversion-mode 4H-SiC power MOSFET structure by using a low (2×10^{16} cm^{-3}) doping concentration for the P-base region. For the case of a gate bias of 10 V and 300 K, the specific on-resistance is found to be 185 mΩ-cm^2 providing validation of the analytical model. The drain current is saturated when the gate voltage reaches 10 V indicating that the channel resistance is not dominant in the 20-kV shielded 4H-SiC inversion-mode power MOSFET structure.

6.9.3 Inductive Load Turn-Off Characteristics

The analytical model for the turn-off behavior of the shielded silicon carbide inversion-mode power MOSFET structure was derived in Sect. 6.5. It was demonstrated that the charging of the device capacitances dictates the rate of rise of the drain voltage. This model was validated for the 5-kV device structure. This model can therefore also be utilized to generate the inductive load turn-off waveforms for the 20-kV shielded silicon carbide inversion-mode power MOSFET structure.

Simulation Results

The results of two-dimensional numerical simulations of the inductive load turn-off for the 20-kV shielded 4H-SiC inversion-mode MOSFET structure are described here. The device parameters for the structure used for the numerical simulations were provided in the previous section. The drain supply voltage was chosen as 12,000 V for the turn-off analysis with an initial on-state current density of 25 A/cm^2. During the turn-off simulations, the gate voltage was reduced to zero with a gate resistance of 1×10^8 Ω-μm for the 5 μm half-cell structure, which is equivalent to a specific gate resistance of 5 Ω-cm^2. The current density was initially held constant at an on-state current density of 25 A/cm^2 allowing the drain voltage to rise to the drain supply voltage during the plateau phase. The drain supply voltage was then held constant allowing the drain current density to reduce to zero.

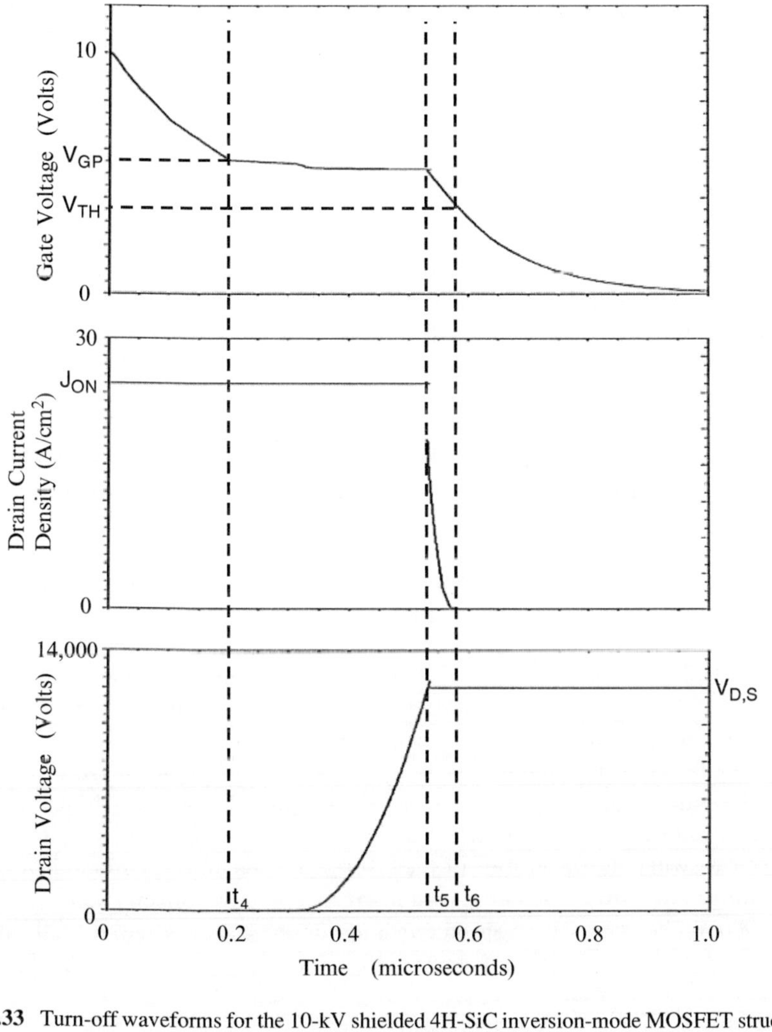

Fig. 6.33 Turn-off waveforms for the 10-kV shielded 4H-SiC inversion-mode MOSFET structure

The turn-off waveforms obtained for the 20-kV shielded 4H-SiC inversion-mode MOSFET structure by using the numerical simulations are shown in Fig. 6.33. The gate voltage initially reduces to the gate plateau voltage corresponding to the on-state current density. The gate plateau voltage observed in the numerical simulation is 5.56 V. The drain voltage then increases quadratically from the on-state voltage drop to the drain supply voltage ($V_{D,S}$). After this, the drain current reduces exponentially. The drain voltage rise-time (t_5-t_4) is much larger than the drain current fall time (t_6-t_5) for the shielded 4H-SiC inversion-mode MOSFET structure. The drain voltage rise-time obtained from the simulations of the 10-kV shielded 4H-SiC inversion-mode MOSFET structure is 0.31 μs and the drain current fall-time obtained from the simulations is 0.08 μs. These values are slightly larger than those observed in the numerical simulations of the 5-kV device structure.

6.9.4 Switching Energy Loss

The power loss incurred during the switching transients limit the maximum operating frequency for the 20-kV shielded silicon carbide inversion-mode power MOSFET structure. Power losses during the turn-on of the MOSFET structure are significant but strongly dependent on the reverse recovery behavior of the fly-back rectifiers in circuits. Consequently, it is common practice to use only the turn-off energy loss per cycle during characterization of devices. As shown in the previous section, the turn-off losses are associated with the voltage rise-time interval and the current fall-time interval. The energy loss for each event can be computed by integration of the power loss, as given by the product of the instantaneous current and voltage. During the voltage rise-time interval, the anode current is constant while the voltage increases in a non-linear manner as a function of time. In order to simplify the analysis, the energy loss during the voltage rise-time interval will be computed using Eq. 6.39. For the typical switching waveforms of the 20-kV shielded silicon carbide inversion-mode power MOSFET structure shown in Fig. 6.33 with a collector supply voltage of 12,000 V, the energy loss per unit area during the collector voltage rise-time is found to be 0.057 J/cm^2 if the on-state current density is 25 A/cm^2.

During the collector current fall-time interval, the collector voltage is constant while the current decreases rapidly. In order to simplify the analysis, the energy loss during the collector current fall-time interval will be computed using Eq. 6.40. For the typical switching waveforms of the 20-kV shielded silicon carbide inversion-mode power MOSFET structure shown in Fig. 6.33 with a collector supply voltage of 12,000 V, the energy loss per unit area during the collector current fall-time is found to be 0.008 J/cm^2 if the on-state current density is 25 A/cm^2. The total energy loss per unit area ($E_{OFF,V} + E_{OFF,I}$) during the turn-off process for the 20-kV asymmetric IGBT structure is found to be 0.057 J/cm^2.

6.9.5 Maximum Operating Frequency

The maximum operating frequency for operation of the 20-kV shielded silicon carbide inversion-mode power MOSFET structure can be obtained by combining the on-state and switching power losses and using Eq. 6.41. In the case of the 20-kV shielded silicon carbide inversion-mode power MOSFET structure, the on-state voltage drop is 4.625 V at an on-state current density of 25 A/cm^2 because the specific on-resistance for this structure at 300 K is 185 mΩ-cm^2. For the case of a 50% duty cycle, the on-state power dissipation contributes 57.8 W/cm^2 to the total power loss. Using a total turn-off energy loss per cycle of 0.057 J/cm^2 in Eq. 6.41 yields a maximum operating frequency of 2,500 Hz if the total power dissipation is 200 W/cm^2. In the case of operation at a reduced duty cycle of 10%, the maximum operating frequency for the 20-kV shielded silicon carbide inversion-mode power MOSFET structure increases to 3,300 Hz if the total power dissipation is 200 W/cm^2.

6.10 Conclusions

The characteristics of the shielded planar 4H-SiC inversion-mode MOSFET structure have been reviewed in this chapter. It has been demonstrated that the specific on-resistance of the 5-kV device is three orders of magnitude smaller than that for the 5-kV silicon MOSFET structure. The channel resistance for the case of a typical channel inversion layer mobility of 20 cm^2/V-s is found to constitute 42.5% of the total specific on-resistance. The drift resistance is found to constitute 49.5% of the total specific on-resistance. The total specific on-resistance for the 5-kV shielded planar 4H-SiC inversion-mode MOSFET structure is found to be 13.33 mΩ-cm^2. Consequently, the on-state voltage drop for the 5-kV shielded planar 4H-SiC inversion-mode MOSFET structure is 1.33 V at an on-state current density of 100 A/cm^2.

The on-state *i–v* characteristics for the 5-kV shielded planar 4H-SiC inversion-mode MOSFET structure (solid-lines) are compared with those for the 5-kV asymmetric IGBT structure (dashed-lines) in Fig. 6.34. The gate-bias voltage is 10-volts for both devices. The 5-kV IGBT structure has a buffer layer doping concentration of 1×10^{17} cm^{-3} and the high-level lifetime in the N-base region is 10 μs. As discussed in Chap. 5, the on-state current density for the 5-kV asymmetric IGBT structure is 50 A/cm^2. The on-state voltage drop for the 5-kV asymmetric IGBT structure at this current density is 3.01 V.

The on-state operating points on the *i–v* curves at 300 K are shown in Fig. 6.34 with a square point for the 5-kV shielded planar 4H-SiC inversion-mode MOSFET structure and a circular point for the 5-kV asymmetric IGBT structure. From Fig. 6.34, it can be concluded that the on-state *i–v* characteristic for the 5-kV shielded planar 4H-SiC inversion-mode MOSFET structure at 300 K is much superior to that for the 5-kV asymmetric IGBT structure at 300 K. However, the on-state characteristics for the 5-kV shielded planar 4H-SiC inversion-mode

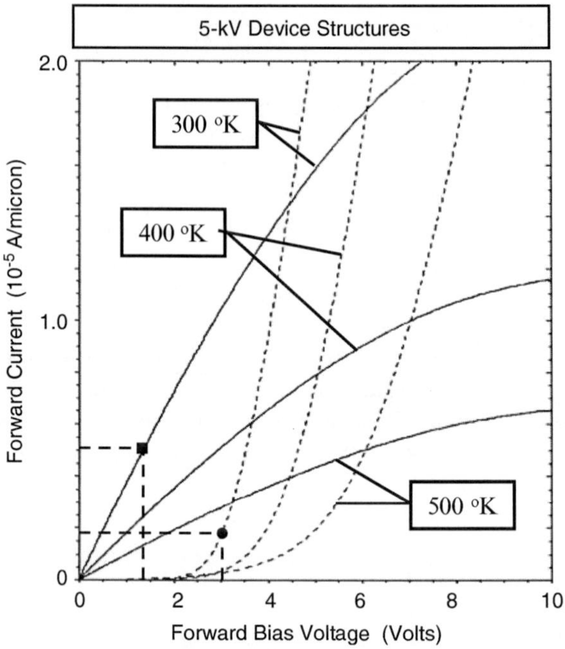

Fig. 6.34 Comparison of 5-kV 4H-SiC shielded inversion-mode MOSFET structure with the 5-kV silicon asymmetric IGBT structure

MOSFET structure degrade much more rapidly with increasing temperature when compared with the 5-kV asymmetric IGBT structure due to a reduction in the electron mobility.

Structure	On-State Current Density (A/cm^2)	On-State Voltage Drop (Volts)	On-State Power Dissipation (W/cm^2)	Energy Loss per Cycle (J/cm^2)	Maximum Operating Frequency (Hz)
SiC MOSFET	100	1.33	66.7	0.051	2,600
Si IGBT	50	3.01	75.2	0.413	300

Fig. 6.35 Comparison of the 5-kV shielded 4H-SiC inversion-mode MOSFET structure with the 5-kV asymmetric IGBT structure

The switching performance of the unipolar 5-kV shielded planar 4H-SiC inversion-mode MOSFET structure can be expected to be superior to that for bipolar 5-kV asymmetric IGBT structure. This is demonstrated in Fig. 6.35 where the turn-off energy loss per cycle is provided for both devices in the case of a power supply voltage of 3,000 V. The 5-kV shielded planar 4H-SiC inversion-mode MOSFET structure has eight-times smaller turn-off energy loss per cycle in spite of handling twice the output power. This allows the 5-kV shielded planar 4H-SiC inversion-mode MOSFET structure to operate at 8.7-times higher frequency if the total power dissipation is 200 W/cm^2 for both devices. Due to the difference in the on-state current density, the chip active area for the 5-kV shielded planar 4H-SiC inversion-mode MOSFET structure is two-times smaller than that of the 5-kV asymmetric IGBT structure.

It has been demonstrated earlier that the specific on-resistance of the 10-kV device is three orders of magnitude smaller than that for the 10-kV silicon MOSFET structure. The channel resistance for the case of a typical channel inversion layer mobility of 20 cm^2/V-s is found to constitute 14% of the total specific on-resistance. The drift resistance is found to constitute 84% of the total specific on-resistance. The total specific on-resistance for the 10-kV shielded planar 4H-SiC inversion-mode MOSFET structure is found to be 39.9 mΩ-cm^2. Consequently, the on-state voltage drop for the 10-kV shielded planar 4H-SiC inversion-mode MOSFET structure is 2.00 V at an on-state current density of 50 A/cm^2.

The on-state *i*–*v* characteristics for the 10-kV shielded planar 4H-SiC inversion-mode MOSFET structure (solid-lines) are compared with those for the 10-kV asymmetric IGBT structure (dashed-lines) in Fig. 6.36. The gate-bias voltage is 10-volts for both devices. The 10-kV IGBT structure has a buffer layer doping concentration of 1×10^{17} cm^{-3} and the high-level lifetime in the N-base region is 20 μs. As discussed in Chap. 5, the on-state current density for the 10-kV asymmetric IGBT structure is 20 A/cm^2. The on-state voltage drop for the 10-kV asymmetric IGBT structure at this current density is 4.46 V. The on-state operating points on the *i*–*v* curves at 300 K are shown in Fig. 6.36 with a square point for the 10-kV

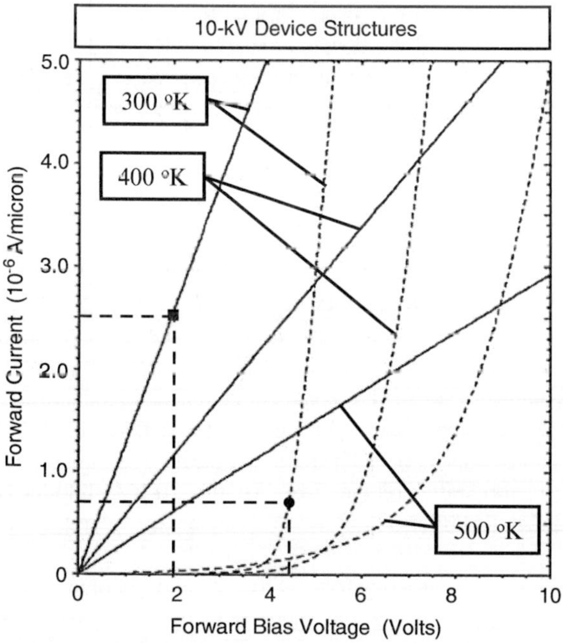

Fig. 6.36 Comparison of 10-kV 4H-SiC shielded inversion-mode MOSFET structure with the 10-kV silicon asymmetric IGBT structure

shielded planar 4H-SiC inversion-mode MOSFET structure and a circular point for the 10-kV asymmetric IGBT structure. From Fig. 6.36, it can be concluded that the on-state *i–v* characteristic for the 10-kV shielded planar 4H-SiC inversion-mode MOSFET structure is much superior to that for the 10-kV asymmetric IGBT structure at 300 K. The on-state characteristics for the 10-kV shielded planar 4H-SiC inversion-mode MOSFET structure degrade much more rapidly with increasing temperature when compared with the 10-kV asymmetric IGBT structure due to a reduction in the electron mobility.

The switching performance of the unipolar 10-kV shielded planar 4H-SiC inversion-mode MOSFET structure can be expected to be superior to that for bipolar 10-kV asymmetric IGBT structure. This is demonstrated in Fig. 6.37 where the turn-off energy loss per cycle is provided for both devices in the case of a power supply voltage of 6,000 V. The 10-kV shielded planar 4H-SiC inversion-mode MOSFET structure has 11-times smaller turn-off energy loss per cycle in spite of handling 2.5-times the output power. This allows the 10-kV shielded planar 4H-SiC inversion-mode MOSFET structure to operate at 11-times higher frequency if the total power dissipation is 200 W/cm^2 for both devices. Due to the difference in the on-state current density, the chip active area for the 10-kV shielded planar 4H-SiC inversion-mode MOSFET structure is 2.5-times smaller than that of the 10-kV asymmetric IGBT structure.

Structure	On-State Current Density (A/cm^2)	On-State Voltage Drop (Volts)	On-State Power Dissipation (W/cm^2)	Energy Loss per Cycle (J/cm^2)	Maximum Operating Frequency (Hz)
SiC MOSFET	50	2.00	48.9	0.059	2600
Si IGBT	20	4.46	44.6	0.675	230

Fig. 6.37 Comparison of 10-kV 4H-SiC shielded inversion-mode MOSFET structure with the 10-kV silicon asymmetric IGBT structure

It has been demonstrated earlier that the specific on-resistance of the 20-kV device is three orders of magnitude smaller than that for the 20-kV silicon MOSFET structure. The channel resistance for the case of a typical channel inversion layer mobility of 20 cm^2/V-s is found to constitute only 3% of the total specific on-resistance. The drift resistance is found to constitute 96.4% of the total specific on-resistance. The total specific on-resistance for the 20-kV shielded planar 4H-SiC inversion-mode MOSFET structure is found to be 189 mΩ-cm^2. Consequently, the on-state voltage drop for the 20-kV shielded planar 4H-SiC inversion-mode MOSFET structure is 4.625 V at an on-state current density of 25 A/cm^2.

The on-state *i–v* characteristics for the 20-kV shielded planar 4H-SiC inversion-mode MOSFET structure are shown in Fig. 6.38 for three operating temperatures. A gate-bias voltage of 10 V was used during the numerical simulations. The on-state characteristics of the 20-kV asymmetric IGBT structure are poor and the operating current density is very low due to the reduced RBSOA boundary. Consequently, these devices are not considered viable. As expected, the on-resistance for the 20-kV shielded planar 4H-SiC inversion-mode MOSFET structure increases rapidly with increasing temperature due to the reduction of the electron mobility.

The excellent switching performance of the unipolar 20-kV shielded planar 4H-SiC inversion-mode MOSFET structure is demonstrated in Fig. 6.39 where the turn-off energy loss per cycle is provided in the case of a power supply voltage of 12,000 V. The low switching power losses allow the 20-kV shielded planar 4H-SiC inversion-mode MOSFET structure to operate at 2,500-Hz if the total power dissipation is 200 W/cm^2.

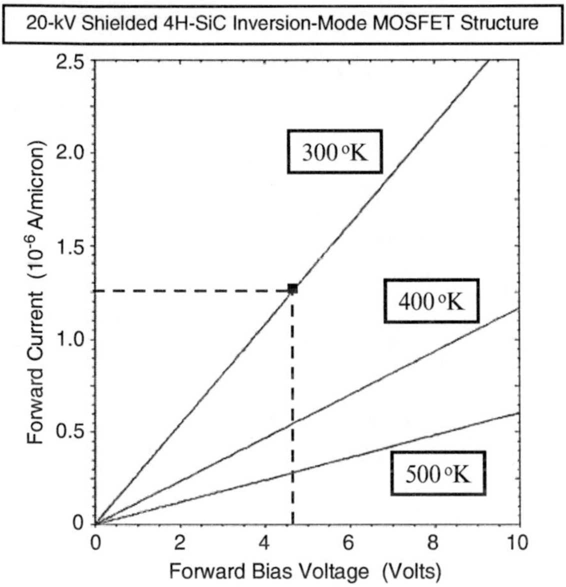

Fig. 6.38 On-state characteristics of the 20-kV 4H-SiC shielded inversion-mode MOSFET structure

Structure	On-State Current Density (A/cm^2)	On-State Voltage Drop (Volts)	On-State Power Dissipation (W/cm^2)	Energy Loss per Cycle (J/cm^2)	Maximum Operating Frequency (Hz)
SiC MOSFET	25	4.63	57.8	0.057	2500

Fig. 6.39 Comparison of 10-kV 4H-SiC shielded inversion-mode MOSFET structure with the 10-kV silicon asymmetric IGBT structure

The successful development of silicon carbide planar inversion-mode power MOSFET structures with very high blocking voltages has been supported mainly for military applications. Devices with 10 kV, 100-A ratings are required for use in future electric ships, more electric aircraft, and all-electric combat vehicles [13].

References

1. B.J. Baliga, "Semiconductors for High Voltage Vertical Channel Field Effect Transistors", Journal of applied Physics, Vol. 53, pp. 1759–1764, 1982.
2. B.J. Baliga, "Power Semiconductor Device Figure of Merit for High Frequency Applications", IEEE Electron Device Letters, Vol. 10, pp. 455–457, 1989.
3. B.J. Baliga, "Silicon Carbide Power Devices", World Scientific Publishers, 2005.
4. B. J. Baliga, "Silicon Carbide Semiconductor Devices having Buried Silicon Carbide Conduction Barrier Layers Therein", U. S. Patent 5,543,637, Issued August 6, 1996.
5. B.J. Baliga, "Advanced Power MOSFET Concepts", Springer-Science, 2010.
6. S. H. Ryu, et al, "10-kV, 123 mΩ-cm^2 4H-SiC Power DMOSFETs", IEEE Electron Device Letters, Vol. 25, pp. 556–558, 2004.
7. S. Sridevan and B.J. Baliga, "Lateral N-channel Inversion Mode 4H-SiC MOSFETs", IEEE Electron Device Letters, Vol. 19, pp. 228–230, 1998.
8. B.J. Baliga, "Fundamentals of Power Semiconductor Devices", Springer-Science, 2008.
9. M. Matin, A. Saha, and J. Cooper, "A Self-Aligned Process for High-Voltage, Short-Channel, Vertical DMOSFETs in 4H-SiC", IEEE Electron Device Transactions, Vol. 51, pp. 1721–1725, 2004.
10. B.J. Baliga and M. Bhatnagar, "Method of Fabricating Silicon Carbide Field Effect Transistor", U.S. Patent 5,322,802, Issued June 21, 1994.
11. Q. Zhang, et al, "Design and Characterization of High-Voltage 4H-SiC p-IGBTs", IEEE Electron Device Transactions, Vol. 55, pp. 1912–1919, 2008.
12. P.M. Shenoy and B.J. Baliga, "The Planar 6H-SiC ACCUFET", IEEE Electron Device Letters, Vol. 18, pp. 589–591, 1997.
13. A. Hefner, et al, "Recent Advances in High-Voltage, High-Frequency, Silicon-Carbide Power Devices", IEEE Industry Applications Society Conference, Vol. 1, pp. 330–337, 2006.

Chapter 7
Silicon Carbide IGBT

In the previous chapter, it has been demonstrated that the silicon carbide planar-shielded inversion-mode power MOSFET structure has excellent on-state resistance for devices with breakdown voltage of up to 10,000 V. However, the specific on-resistance for these devices becomes relatively large when their blocking voltage is scaled to 20,000 V. Consequently, there has been interest in the development of silicon carbide-based high voltage IGBT structures. Due to the high resistivity of P-type substrates in silicon carbide, most of the development work has been focused on p-channel silicon carbide IGBT structures that can utilize heavily doped N-type substrates.

The first p-channel silicon carbide IGBT structures with 10-kV blocking voltage capability were reported in 2005 with the trench-gate architecture [1]. In the case of all MOS-gated power devices formed using silicon carbide, it is necessary to shield the gate oxide from the high electric fields developed in the semiconductor [2]. This can be accomplished for the trench-gate MOSFET or IGBT structures by the incorporation of a junction at the bottom of the trenches [3].

Subsequently, planar p-channel silicon carbide IGBT structures with 9-kV blocking voltage capability were reported in 2007 [4] and 12-kV devices were reported in 2008 [5]. Although this work demonstrated that conductivity modulation of the drift region can be achieved in the silicon carbide IGBT structure similar to that observed in silicon devices, the on-state voltage drop obtained for the silicon carbide IGBT was significantly worse than that shown in Chap. 6 for the 10-kV silicon carbide power MOSFET structure. More recent interest has therefore shifted to even higher blocking voltage ratings from 15 to 20 kV. The on-state characteristics of an asymmetric 4H-SiC p-channel IGBT structure having a 175-μm thick P-base region with doping concentration of 2×10^{14} cm^{-3} were reported for the planar device architecture in 2007 [6]. The devices had on-state voltage drop of 8.5 V at an on-state current density of 25 A/cm^2. The high-level lifetime in the P-base region was measured at 0.46 μs. The measured blocking voltage capability of the devices was only 5 kV although the authors assumed that the parameters for the drift region are suitable for a 20-kV device. The optimization of p-channel asymmetric IGBT structures in 4H-SiC for blocking voltages between 15 and 20 kV

B.J. Baliga, *Advanced High Voltage Power Device Concepts*,
DOI 10.1007/978-1-4614-0269-5_7, © Springer Science+Business Media, LLC 2011

has also been reported [7]. In this analysis, the drift region parameters were obtained like that for the power MOSFET structure. The open-base transistor breakdown analysis appropriate for the IGBT structure was not performed, resulting in underestimation of the thickness of the P-base (drift) region to 175 μm for a 20-kV device structure. A comparison of the high frequency performance of 15-kV asymmetric and symmetric blocking p-channel IGBTs has been reported recently [8]. This study concludes that the lifetime in the buffer layer has to be optimized to obtain the best performance in the asymmetric p-channel 4H-SiC IGBT structure. However, current technology for silicon carbide does not allow selective adjustment of the lifetime in the buffer layer.

In contrast to the literature available on p-channel silicon carbide IGBT structures, very little work has been reported on silicon carbide n-channel IGBT structures. The successful fabrication of 4H-SiC n-channel IGBT structures with 13-kV blocking voltage capability has been reported [9]. The modulation of the conductivity of the drift region was verified in these devices from the on-state characteristics. Despite a low inversion layer mobility of 18 cm^2/V-s observed in the high-voltage structures, an on-state voltage drop of 3.8 V was obtained at an on-state current density of 25 A/cm^2 with a gate bias of 20 V. The devices exhibited a voltage rise-time of 0.25 μs and a current fall-time of 0.1 μs when operated from a 7-kV power supply at a current density of 175 A/cm^2.

The progress in development of high voltage silicon carbide power devices has been reviewed recently [10]. This report concludes that the most appropriate applications for silicon carbide MOSFET structures are in the range of 4–10 kV blocking voltages while that for silicon carbide IGBTs are in the range of 15–30 kV blocking voltages. Based upon this, the discussion of silicon carbide IGBTs is focused on 20-kV rated devices in this chapter. In this chapter, the characteristics of the p-channel and n-channel planar inversion-mode silicon carbide IGBT structures are compared. The voltage drop contributed by the P^+ 4H-SiC substrates of vertical n-channel structure can be made small by reducing the substrate thickness prior to packaging the devices because the drift region thickness is sufficiently large to prevent wafer breakage.

The basic operating principles and characteristics for the silicon IGBT have been described in detail in the textbook [11]. The operating principles of the asymmetric silicon carbide IGBT structure can be expected to be similar to those for the asymmetric silicon IGBT structure. However, the doping concentration in the drift region for the silicon carbide structure is much larger than that for the silicon device structure. Consequently, although the minority carrier concentration in the space-charge region during the switching of the silicon carbide and silicon devices is nearly the same, it is much smaller than the doping concentration in the silicon carbide devices and greater than the doping concentration in the silicon devices. In the case of the silicon asymmetric IGBT structure, it was demonstrated in Chap. 5 that the space-charge region does not reach-through the drift region during the voltage transient, leaving stored charge in the drift region at the end of the voltage transient. In contrast, the space-charge region reaches-through the drift region when the collector voltage is much less than the collector supply voltage in the case of the typical silicon carbide IGBT structure. This alters the shape of the voltage and current transients as discussed in this chapter.

7.1 n-Channel Asymmetric Structure

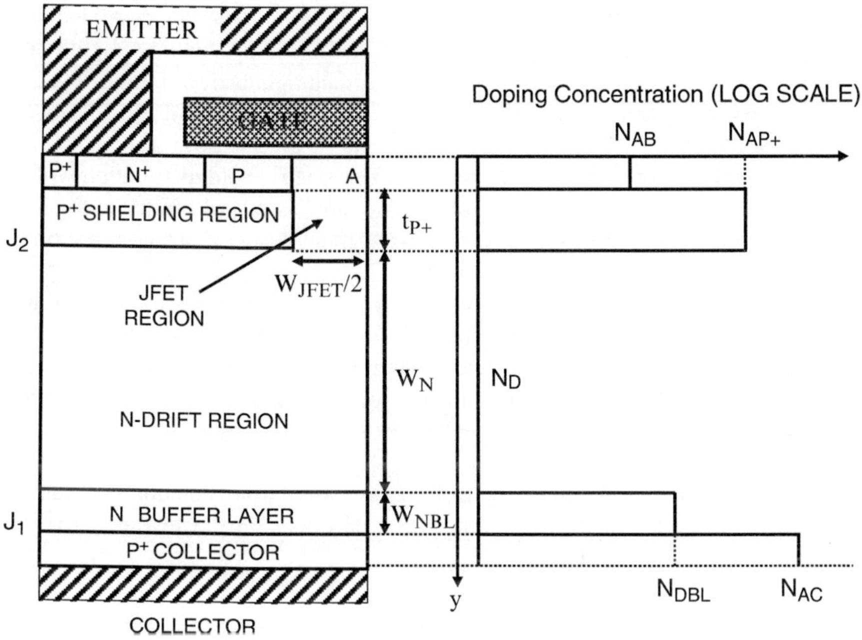

Fig. 7.1 The asymmetric n-channel SiC IGBT structure and its doping profile

The asymmetric n-channel silicon carbide IGBT structure with the planar gate architecture is illustrated in Fig. 7.1 with its doping profile. Since the asymmetric IGBT structure is intended for use in DC circuits, its reverse blocking capability does not have to match the forward blocking capability, allowing the use of an N-buffer layer adjacent to the P^+ collector region. The N-buffer layer has a much larger doping concentration than the lightly doped portion of the N-base region. The electric field in the asymmetric IGBT takes a trapezoidal shape, allowing supporting the forward blocking voltage with a thinner N-base region. This allows achieving a lower on-state voltage drop and superior turn-off characteristics. As in the case of silicon devices discussed in Chap. 5, the doping concentration of the buffer layer and the lifetime in the N-base region must be optimized to perform a trade-off between on-state voltage drop and turn-off switching losses. Unlike the silicon device, the silicon carbide structure has uniform doping concentration for the various layers produced by using either epitaxial growth or by using multiple ion-implantation energies to form a box profile.

7.1.1 Blocking Characteristics

The design of the 20-kV asymmetric 4H-SiC n-channel IGBT structure is discussed in this section. The physics for blocking voltages in the first and third quadrants by the IGBT structure is discussed in detail in the textbook [11]. When a positive bias is applied to the collector terminal of the asymmetric silicon carbide IGBT structure, the junction (J_2) between the P^+ shielding region and the N-base (drift) region becomes reverse-biased while the junction (J_1) between the P^+ collector region and the N-buffer layer becomes forward-biased. The forward blocking voltage is supported across the junction (J_2) between the P^+ shielding region and the N-base (drift) region with a depletion layer extending mostly within the N-base region. The doping concentration and width of the JFET region must be designed to allow suppression of the electric field at point A as discussed in Chap. 6 for the silicon carbide power MOSFET structure. The electric field distribution within the asymmetric IGBT structure is essentially the same as that illustrated in Fig. 4.3 for the asymmetric GTO structure. Consequently, the design procedure described in Chap. 4 can be applied to the asymmetric IGBT structure.

The forward blocking capability of the asymmetric n-channel silicon carbide IGBT structure is determined by the open-base transistor breakdown phenomenon. According to this analysis, the maximum blocking voltage occurs when the common base current gain of the NPN transistor of the n-channel structure becomes equal to unity. For the asymmetric IGBT structure, the emitter injection efficiency is smaller than unity due to the high doping concentration of the N-buffer layer. The emitter injection efficiency for the P^+ collector/N-buffer junction (J_1) can be obtained by using an analysis similar to that described in the textbook for the bipolar power transistor [11]:

$$\gamma_E = \frac{D_{pNBL} L_{nC} N_{AC}}{D_{pNBL} L_{nC} N_{AC} + D_{nC} W_{NBL} N_{DBL}} \tag{7.1}$$

where D_{pNBL} and D_{nC} are the diffusion coefficients for minority carriers in the N-buffer and P^+ collector regions, N_{AC} and L_{nC} are the doping concentration and diffusion length for minority carriers in the P^+ collector region, and N_{DBL} and W_{NBL} are the doping concentration and width of the N-buffer layer. In determining the diffusion coefficients and the diffusion length, it is necessary to account for impact of the high doping concentrations in the P^+ collector region and N-buffer layer on the mobility. In addition, the lifetime within the highly doped P^+ collector region is reduced due to heavy doping effects, which shortens the diffusion length.

The open-base transistor breakdown condition for the asymmetric n-channel silicon carbide IGBT structure is given by:

$$\alpha_{PNP} = (\gamma_E \cdot \alpha_T)_{PNP} M = 1 \tag{7.2}$$

Based upon this expression, it can be concluded that the breakdown voltage for the silicon carbide asymmetric IGBT structure will occur when the multiplication coefficient is only slightly above unity. Using the avalanche breakdown criteria, when the multiplication coefficient becomes equal to infinity, as assumed in some papers [7], will lead to significant error in the design of the drift region for the IGBT structure.

When the collector bias exceeds the reach-through voltage (V_{RT}), the electric field is truncated by the high doping concentration of the N-buffer layer, making the un-depleted width of the NPN transistor base region equal to the width of the N-buffer layer. The base transport factor is then given by:

$$\alpha_T = \frac{1}{\cosh\left(W_{NBL}/L_{pNB}\right)} \tag{7.3}$$

which is independent of the collector bias. Here, $L_{p,NB}$ is the diffusion length for holes in the N-buffer layer. This analysis neglects the depletion region extension within the N-buffer layer. The diffusion length for holes ($L_{p,NB}$) in the N-buffer layer depends upon the diffusion coefficient and the minority carrier lifetime in the N-buffer layer. The diffusion coefficient varies with the doping concentration in the N-buffer layer based upon the concentration dependence of the mobility. In addition, the minority carrier lifetime has been found to be dependent upon the doping concentration [12] in the case of silicon devices. Although this phenomenon has not been verified for silicon carbide, it is commonly used when performing numerical analysis of silicon carbide devices. The effect can be modeled by using the relationship:

$$\frac{\tau_{LL}}{\tau_{p0}} = \frac{1}{1 + (N_D/N_{REF})} \tag{7.4}$$

where N_{REF} is a reference doping concentration whose value will be assumed to be $5 \times 10^{16}\ \text{cm}^{-3}$.

The multiplication factor for a P-N junction is given by:

$$M = \frac{1}{1 - (V_A/BV_{PP})^n} \tag{7.5}$$

with a value of $n = 6$ for the case of a P^+/N junction and the avalanche breakdown voltage of the P-base/N-base junction (BV_{PP}) *without the punch-through phenomenon*. In order to apply this formulation to the punch-through case relevant to the asymmetric silicon carbide IGBT structure, it is necessary to relate the maximum electric field at the junction for the two cases. The electric field at the interface between the lightly doped portion of the N-base region and the N-buffer layer is given by:

$$E_1 = E_m - \frac{qN_D W_N}{\varepsilon_S} \tag{7.6}$$

The voltage supported by the device is given by:

$$V_{C} = \left(\frac{E_{m} + E_{1}}{2}\right) W_{N} = E_{m} W_{N} - \frac{qN_{D}}{2\varepsilon_{S}} W_{N}^{2} \tag{7.7}$$

From this expression, the maximum electric field is given by:

$$E_{m} = \frac{V_{C}}{W_{N}} + \frac{qN_{D} W_{N}}{2\varepsilon_{S}} \tag{7.8}$$

The corresponding equation for the non-punch-through case is:

$$E_{m} = \sqrt{\frac{2qN_{D} V_{NPT}}{\varepsilon_{S}}} \tag{7.9}$$

Consequently, the non-punch-through voltage that determines the multiplication coefficient "M" corresponding to the applied collector bias "V_C" for the punch-through case is given by:

$$V_{NPT} = \frac{\varepsilon_{S} E_{m}^{2}}{2qN_{D}} = \frac{\varepsilon_{S}}{2qN_{D}} \left(\frac{V_{C}}{W_{N}} + \frac{qN_{D} W_{N}}{2\varepsilon_{S}}\right)^{2} \tag{7.10}$$

The multiplication coefficient for the asymmetric silicon carbide IGBT structure can be computed by using this non-punch-through voltage:

$$M = \frac{1}{1 - (V_{NPT}/BV_{PP})^{n}} \tag{7.11}$$

The multiplication coefficient increases with increasing collector bias. The open-base transistor breakdown voltage (and the forward blocking capability of the asymmetric IGBT structure) is determined by the collector voltage at which the multiplication factor becomes equal to the reciprocal of the product of the base transport factor and the emitter injection efficiency.

The silicon carbide n-channel asymmetric IGBT structure must have a forward blocking voltage of 22,000 V for a 20-kV rated device. In the case of avalanche breakdown, there is a unique value of $7.0 \times 10^{14}\ \text{cm}^{-3}$ for the drift region with a width of 186 μm to obtain this blocking voltage. In the case of the asymmetric silicon carbide IGBT structure, it is advantageous to use a much lower doping concentration for the lightly doped portion of the N-base region in order to reduce its width. The strong conductivity modulation of the N-base region during on-state operation favors a smaller thickness for the N-base region independent of its original doping concentration. A doping concentration of $1.5 \times 10^{14}\ \text{cm}^{-3}$ for the N-base region will be assumed for the n-channel asymmetric silicon carbide IGBT structure analyzed in this section.

The doping concentration of the N-buffer layer must be sufficiently large to prevent reach-through of the electric field to the P^+ collector region. Although the electric field at the interface between the N-base region and the N-buffer layer is slightly smaller than that at the blocking junction (J_2), a worse case analysis can be done by assuming that the electric field at this interface is close to the critical electric field for breakdown in the drift region. The minimum charge in the N-buffer layer to prevent reach-through can be then obtained using:

$$N_{DBL}W_{NBL} = \frac{\varepsilon_S E_C}{q} \tag{7.12}$$

Using a critical electric for breakdown in silicon carbide of 2×10^6 V/cm for a doping concentration of 1.5×10^{14} cm^{-3} in the N-base region, the minimum charge in the N-buffer layer to prevent reach-through for a silicon carbide asymmetric IGBT structure is found to be 1.07×10^{13} cm^{-2}. An N-buffer layer with doping concentration of 5×10^{16} cm^{-3} and thickness of 5 μm has a charge of 2.5×10^{13} cm^{-2} that satisfies this requirement.

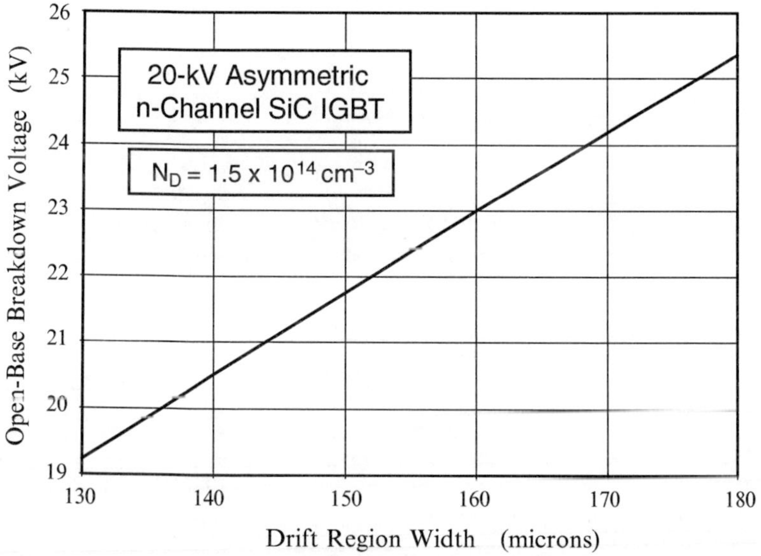

Fig. 7.2 Drift region width optimization for the 20-kV asymmetric n-channel 4H-SiC IGBT structure

The asymmetric n-channel silicon carbide IGBT structure will be assumed to have a P^+ collector region with doping concentration of 1×10^{19} cm^{-3}. It will be assumed that all the acceptors are ionized even at room temperature, although the relatively deep acceptor level in silicon carbide may lead to incomplete dopant ionization. In this case, the emitter injection efficiency computed using Eq. 7.1 is 0.971. When the device is close to breakdown, the entire N-base region is depleted and the base transport factor computed by using Eq. 7.3 in this case is 0.903. In computing these values, a lifetime of 1 μs was assumed for the N-base region resulting in a lifetime of 0.5 μs in the N-buffer layer due to the scaling according to Eq. 7.4. Based upon Eq. 7.3, open-base transistor breakdown will then occur when the multiplication coefficient becomes equal to 1.14 for the above values for the injection efficiency and base transport factor.

The forward blocking capability for the silicon carbide n-channel asymmetric IGBT structure can be computed by using Eq. 7.2 for various widths for the N-base region. The analysis requires determination of the voltage V_{NPT} by using Eq. 7.10 for each width of the N-base region. The resulting values for the forward blocking voltage are plotted in Fig. 7.2. From this graph, the N-base region width required to obtain a forward blocking voltage of 23,000 V is 160 μm.

Simulation Example

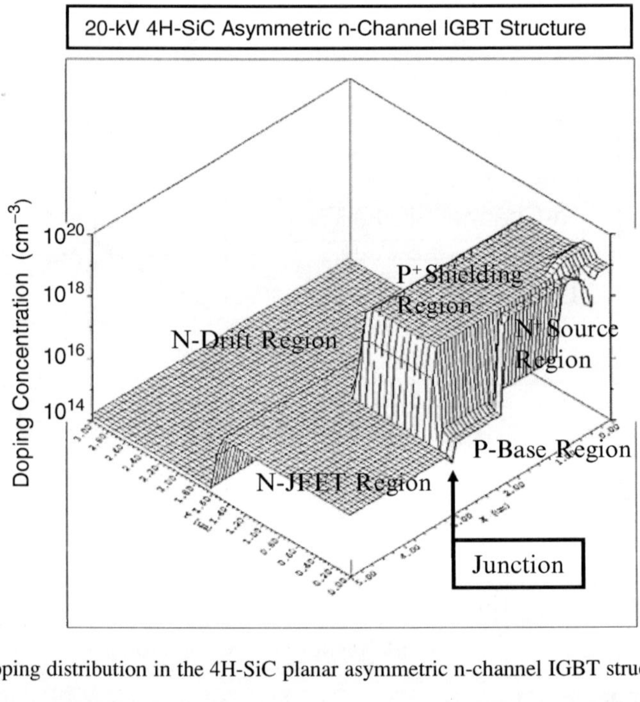

Fig. 7.3 Doping distribution in the 4H-SiC planar asymmetric n-channel IGBT structure

The results of two-dimensional numerical simulations on the 20-kV 4H-SiC asymmetric n-channel IGBT structure are described here to provide a more detailed understanding of the underlying device physics and operation during the blocking mode. For the numerical simulations, the half-cell structure with a width ($W_{Cell}/2$) of 5 μm, as illustrated in Fig. 7.1, was utilized as representative of the structure. The device used for the numerical simulations had a drift region doping concentration of 1.5×10^{14} cm^{-3} and thickness of 160 μm below the P^+ shielding region. The P^+ shielding region extended from a depth of 0.2–1.0 μm with a doping concentration of 1×10^{19} cm^{-3}. The P-base and N^+ source regions were formed within the 0.2 μm of the N-drift region located above the P^+ region. The doping concentration of the P-base region was 2×10^{16} cm^{-3} to achieve the desired threshold voltage. Due to the low doping concentration in the drift region for the 20-kV devices, the uniform doping concentration in the JFET region was enhanced to 1×10^{16} cm^{-3} as is usually required for silicon devices. The enhanced doping concentration was extended to 0.5 μm below the P^+ shielding region. This is similar to the current enhancement layer (CEL) utilized in high voltage silicon carbide IGBT structures [13]. The charge in this layer must be sufficiently small to prevent degradation of the breakdown voltage at the junction between the P^+ shielding region and the drift region.

A three-dimensional view of the doping distribution in the 20-kV 4H-SiC asymmetric n-channel IGBT structure is shown in Fig. 7.3, with the upper surface of the structure located on the right-hand side in order to display the doping concentration in the vicinity of the channel. The highly doped P^+ shielding region with doping concentration of 1×10^{19} cm^{-3} is prominently located just below the surface. The P-base region can be observed to have a much lower doping concentration of 2×10^{16} cm^{-3}. The JFET region can be observed to have a lower doping concentration of 1×10^{16} cm^{-3}. The junction between the P-base region and N-JFET region is indicated in the figure. The doping concentration of the N-drift region is less than that of the JFET region as expected to achieve the desired 20-kV breakdown voltage. It can be seen that the enhanced JFET doping is extended below the P^+ shielding region to a depth of 1.5 μm from the surface.

The lateral doping profile taken along the surface of the 20-kV 4H-SiC asymmetric n-channel IGBT structure is shown in Fig. 7.4. From the profile, it can be observed that the channel extends from 2 to 3 μm creating a channel length of 1 μm in the P-base region. The doping concentration of the JFET region is 1×10^{16} cm^{-3} while that for the P-base region is 2×10^{16} cm^{-3}. The N^+ source region and the P^+ contact region for shorting the source to the base region are visible on the left-hand side. All the regions were defined with uniform doping with abrupt interfaces between them due to the low diffusion rates for dopants in 4H-SiC material.

The vertical doping profile taken through the N^+ source region of the 20-kV 4H-SiC asymmetric n-channel IGBT structure is provided in Fig. 7.5. It can be observed that the N-drift region has a doping concentration of 1.5×10^{14} cm^{-3} and thickness of 160 μm. The N-buffer layer located at the collector junction has a doping concentration of 5×10^{16} cm^{-3} and thickness of 5 μm. The doping concentration of the P^+ collector region is 1×10^{19} cm^{-3}.

The doping profile for the upper 5-μm of the device structure is shown in Fig. 7.6. It can be observed that the N^+ emitter has a doping concentration of 2×10^{19} cm^{-3}. The doping concentration of the P^+ shielding region is 1×10^{19} cm^{-3} and it extends from a depth of 0.2–1.0 μm from the surface. The N-type CEL layer located below the P^+ shielding region has a doping concentration of 1×10^{16} cm^{-3} and extends to a depth of 1.5 μm.

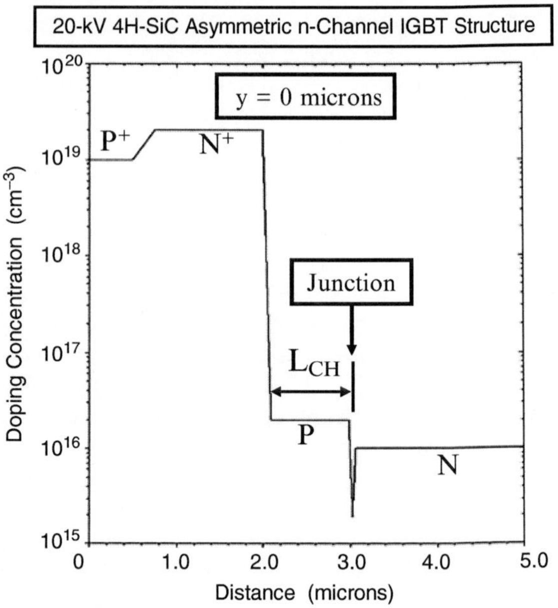

Fig. 7.4 Lateral doping profile for the 4H-SiC asymmetric n-channel IGBT structure

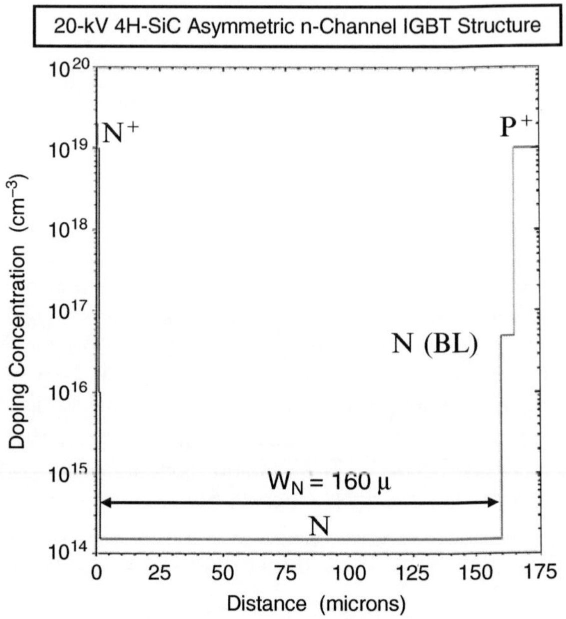

Fig. 7.5 Vertical doping profile in the 4H-SiC asymmetric n-channel IGBT structure

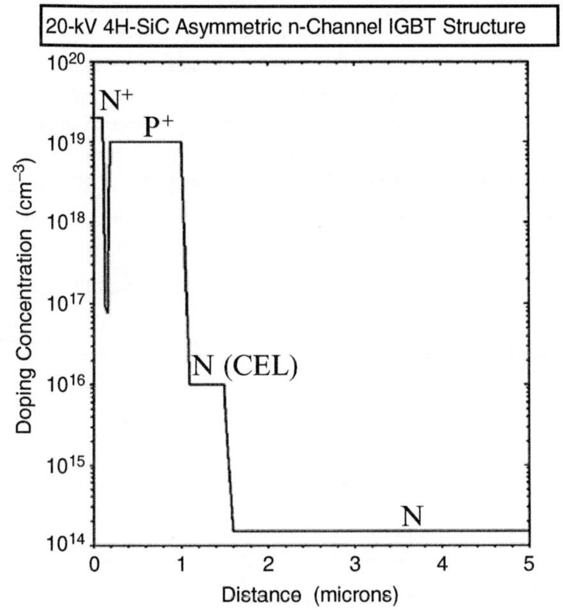

Fig. 7.6 Vertical doping profile in the 4H-SiC asymmetric n-channel IGBT structure

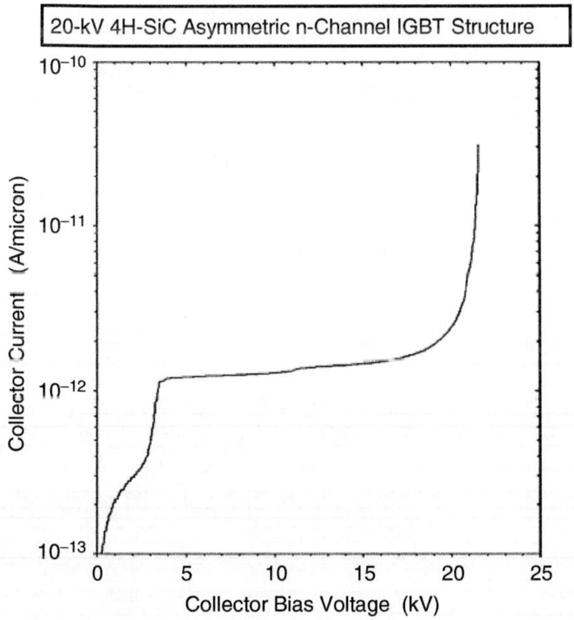

Fig. 7.7 Blocking characteristics of the 4H-SiC asymmetric n-channel IGBT structure

The blocking characteristics for the 20-kV 4H-SiC asymmetric n-channel IGBT structure at room temperature (300 K) cannot be determined by numerical simulations of the cell structure due to the very low intrinsic carrier concentration in silicon carbide. For didactic purposes, the blocking characteristics were therefore obtained at 800 K by increasing the drain voltage while using zero gate bias. The resulting blocking characteristic is shown in Fig. 7.7. It can be seen that leakage current increases with increasing collector bias voltage until it reaches about 3,000 V. This is in excellent agreement with the reach-through voltage of 3,152 V obtained by using Eq. 4.2. Prior to reach-through, the leakage current increases due to the increasing width of the depletion region and the increasing current gain of the NPN transistor in accordance with Eq. 4.5. After reach-through, the leakage current becomes nearly independent of collector voltage.

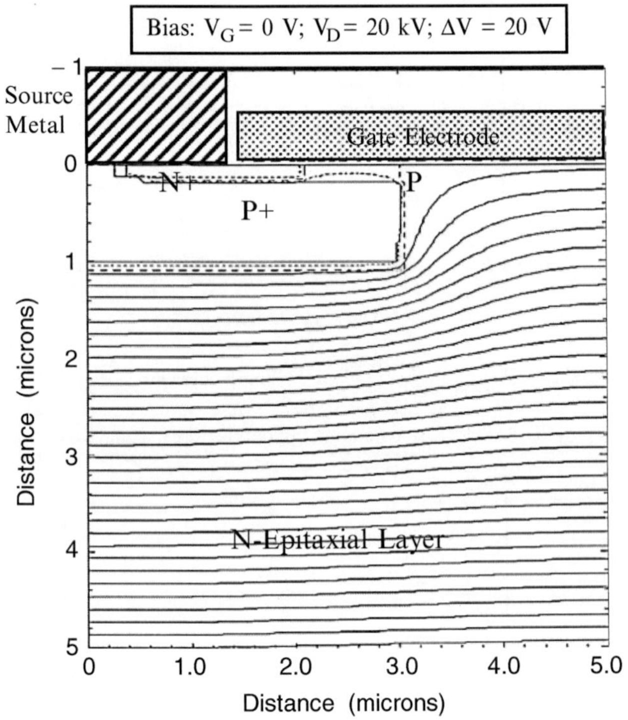

Fig. 7.8 Potential contours in the 4H-SiC asymmetric n-channel IGBT structure

The potential contours within the 20-kV 4H-SiC asymmetric n-channel IGBT structure at a collector bias of 20,000 V are provided in Fig. 7.8 for the upper part of the device structure. It can be observed that the collector voltage is supported below the P^+ shielding region. The potential contours do not extend into the P-base region indicating that it is shielded from the drain potential by the P^+ shielding region. The potential contours are crowding at the edge of the P^+ shielding region indicating an enhanced electric field. This can be clearly observed in Fig. 7.9, which provides a three-dimensional view of the electric field distribution. In this figure, it can also be observed that the electric field in the JFET region, and most importantly at the surface under the gate oxide, has been greatly reduced by the presence of the P^+ shielding region.

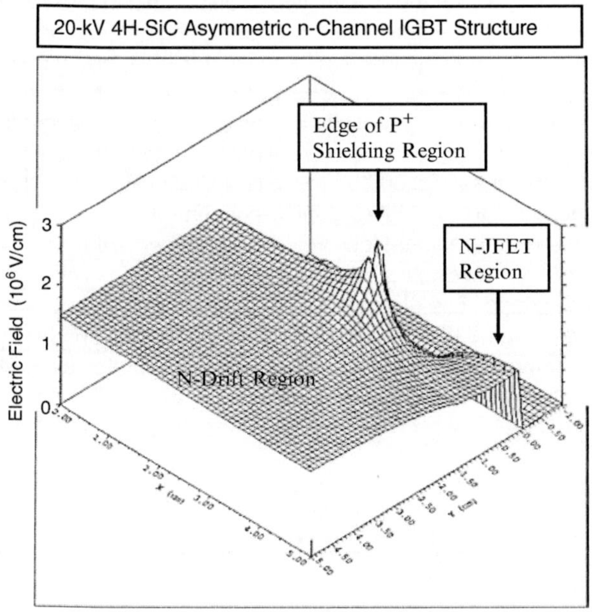

Fig. 7.9 Electric field distribution in the 4H-SiC asymmetric n-channel IGBT structure

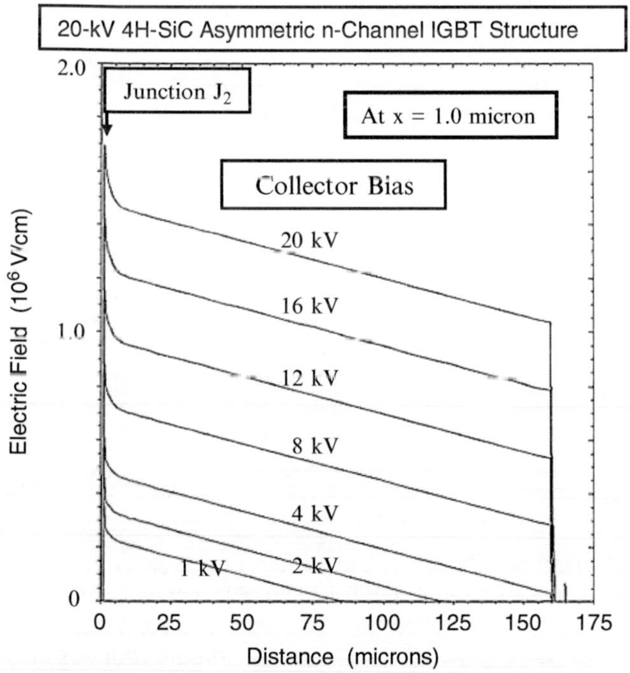

Fig. 7.10 Electric field distribution in the 4H-SiC asymmetric n-channel IGBT structure

It is insightful to examine the electric field profile within the 20-kV 4H-SiC asymmetric n-channel IGBT structure when it is operating in the blocking mode. The electric field profile through the junction between the P^+ shielding region and the N-drift region is provided in Fig. 7.10. The peak of the electric field occurs at the junction as expected and is essentially triangular in shape in accordance with the predictions of Poisson's equation with a uniform doping profile until reach-through occurs at a collector bias of 3,000 V. There is a slight increase in the electric field near the junction due to the higher doping concentration of the CEL layer when compared with the drift region. However, the enhancement in the electric field is small and occurs over a very small distance which results in minimal degradation of the breakdown voltage. The electric field becomes trapezoidal in shape when the collector bias exceeds 3,000 V as expected for an asymmetric blocking structure due to the punch-through of the electric field with the N-buffer layer. The maximum electric field at the junction at a drain bias of 20 kV is 1.7×10^6 V/cm, which is below the critical electric for breakdown for 4H-SiC at the doping concentration of the drift region [11].

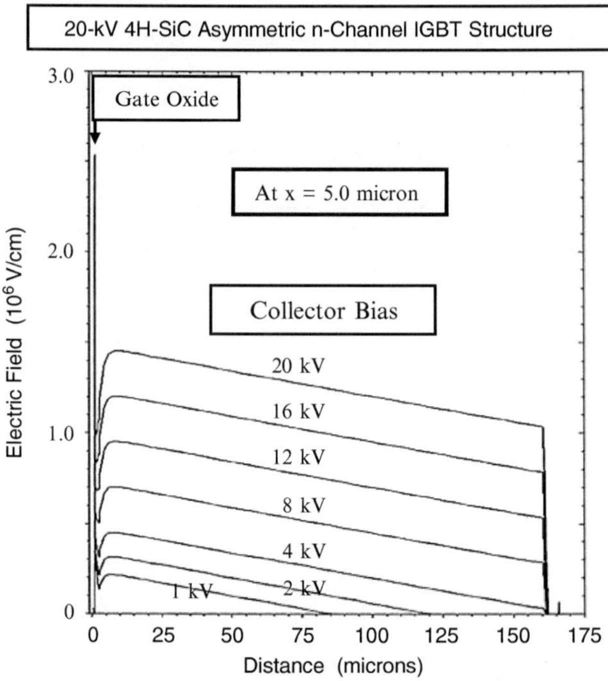

Fig. 7.11 Electric field distribution in the 4H-SiC asymmetric n-channel IGBT structure

The electric field distribution at the middle of the JFET region of the 20-kV 4H-SiC asymmetric n-channel IGBT structure is shown in Fig. 7.11 to allow examination of the electric field in the gate oxide. The suppression of the electric field in the JFET region by the P^+ shielding region greatly reduces the electric field in the semiconductor at its surface to about 1×10^6 V/cm at a collector bias of

20 kV when compared with the maximum electric field of 1.7 × 10^6 V/cm at the junction between the P^+ shielding region and the drift region. This produces a reduced electric field in the gate oxide as well. The electric field in the gate oxide has its highest value of 2.5 × 10^6 V/cm at the middle point of the JFET region. These results clearly demonstrate the importance of using the shielding concept to achieve a practical IGBT device structure in silicon carbide.

7.1.2 On-State Voltage Drop

In Chap. 5, a generally applicable analytical model was developed for the silicon asymmetric IGBT structure which is valid for any injection level in the buffer layer. This analytical model can also be applied to silicon carbide devices. The carrier distribution profiles in the on-state for the asymmetric IGBT structure are shown in Fig. 5.10. The hole and electron concentrations in the N-base region are equal due to charge neutrality and the low doping concentration required for the drift region. The hole concentration in the N-buffer layer is illustrated as less than the doping concentration (N_{DBL}) in the buffer layer but can be greater than the doping level for silicon carbide devices with low buffer layer doping concentrations. In the case of silicon carbide devices, the buffer layer doping is usually uniform as achieved by epitaxial growth.

The hole concentration profile in the N-base region and the N-buffer layer was derived in Chap. 5. Based upon that analysis, an expression for the hole concentration in the buffer layer at junction (J_1) is obtained:

$$p_{NB}^2(0) + \left(\frac{D_{pNB}N_{AP+}L_{nP+} + D_{nP+}N_{DB}L_{pNB}}{D_{nP+}L_{pNB}}\right)p_{NB}(0) - \frac{N_{AP+}L_{nP+}J_C}{qD_{nP+}} = 0 \quad (7.13)$$

The solution of this quadratic equation for the hole concentration in the buffer layer at junction (J_1) is:

$$p_{NB}(0) = \frac{1}{2}\left(\sqrt{b^2 - 4c} - b\right) \quad (7.14)$$

where

$$b = \frac{D_{pNB}N_{AP+}L_{nP+} + D_{nP+}N_{DB}L_{pNB}}{D_{nP+}L_{pNB}} \quad (7.15)$$

and

$$c = -\frac{N_{AP+}L_{nP+}J_C}{qD_{nP+}} \quad (7.16)$$

Since the unified analytical model presented is valid for all injection levels in the N-buffer layer, it can be used to predict the variation of the injected hole concentration with lifetime in the N-base region and the doping concentration in the N-buffer layer for the asymmetric 4H-SiC IGBT structure.

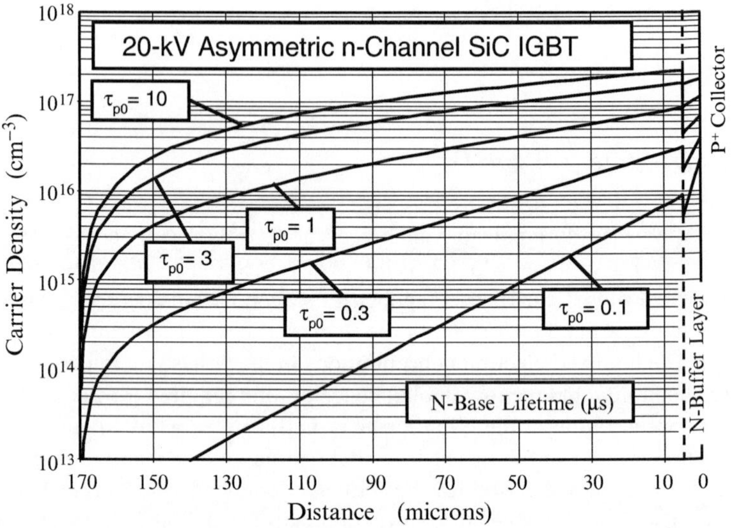

Fig. 7.12 Carrier distribution in the 20-kV n-channel asymmetric SiC IGBT structure: lifetime dependence

The holes diffuse through the buffer layer producing a concentration ($p(W_{NB-})$) inside the buffer layer at the boundary between the N-buffer layer and the N-base region:

$$p(W_{NB-}) = p_{NB}(0)e^{-(W_{NBL}/L_{pNB})} \quad (7.17)$$

where W_{NBL} is the thickness of the buffer layer. The hole concentration ($p(W_{NB-})$) in the N-base region at the boundary between the N-buffer layer and the N-base region can be obtained by equating the hole current density on the two sides of this boundary [11]:

$$p(W_{NB+}) = \frac{L_a \tanh[(W_N + W_{NBL})/L_a]}{2qD_p} J_p(W_{NB-}) \quad (7.18)$$

with

$$J_p(W_{NB-}) = J_p(0)e^{-(W_{NBL}/L_{pNB})} \quad (7.19)$$

The hole concentration profile in the N-base region, as dictated by high-level injection conditions, is given by [11]:

$$p(y) = p(W_{NB+})\frac{\sinh[(W_N + W_{NBL} - y)/L_a]}{\sinh[(W_N + W_{NBL})/L_a]} \quad (7.20)$$

which is valid for $y > W_{NBL}$.

The free carrier distribution obtained by using the above equations is provided in Fig. 7.12 for the case of a 20 kV asymmetric n-channel IGBT structure with an N-base region thickness of 165 μm and a buffer layer thickness of 5 μm. The hole lifetime (τ_{p0}) in the N-base region was varied for these plots from 0.1 to 10 μs. Note that the high-level lifetime (τ_{HL}) in these cases is two times the hole lifetime (τ_{p0}). It can be observed that the hole concentration ($p_{NB}(0)$) decreases at the collector side of the N-buffer layer (at $y = 0$) from 1.8×10^{17} cm^{-3} to 2.25×10^{16} cm^{-3}. In addition, the hole concentration is significantly reduced at the emitter side when the lifetime in the N-base region decreases. The carrier density falls below 1×10^{15} cm^{-3} over a significant portion of the N-base region when the lifetime becomes smaller than 1.0 μs. These results indicate that the on-state voltage drop will increase rapidly when the hole lifetime (τ_{p0}) in the N-base region is reduced below 1 μs. However, the smaller stored charge in the N-base region and buffer layer will reduce the turn-off time and energy loss per cycle.

The on-state voltage drop for the 20-kV asymmetric n-channel IGBT structure can be obtained by using the equations derived in the textbook in Sect. 9.5.5. The on-state voltage drop for the asymmetric IGBT structure can be obtained by using:

$$V_{ON} = V_{P+NBL} + V_B + V_{MOSFET} \quad (7.21)$$

where V_{P+NBL} is the voltage drop across the P$^+$ collector/N-buffer layer junction (J_1), V_B is the voltage drop across the N-base region after accounting for conductivity modulation due to high-level injection conditions, and V_{MOSFET} is the voltage drop across the MOSFET portion. In the asymmetric IGBT structure, the junction (J_1) between the P$^+$ collector region and the N-buffer layer operates at neither high-level nor low-level injection conditions. Consequently, the voltage drop across the junction (J_1) must be obtained using:

$$V_{P+NB} = \frac{kT}{q}\ln\left(\frac{p_{NB}(0)N_{BL}}{n_i^2}\right) \quad (7.22)$$

The voltage drop across the N-base region can be obtained by integrating the electric field inside the N-base region. The voltage drop is obtained by taking the sum of two parts. The first part is given by:

$$V_{B1} = \frac{2L_a J_C \sinh(W_N/L_a)}{qp(W_{NB+})(\mu_n + \mu_p)}\left\{\tanh^{-1}\left[e^{-(W_{ON}/L_a)}\right] - \tanh^{-1}\left[e^{-(W_N/L_a)}\right]\right\} \quad (7.23)$$

The depletion width (W_{ON}) across the P-base/N-base junction (J_2) in the on-state depends on the on-state voltage drop. The voltage drop associated with the second part is given by:

$$V_{B2} = \frac{kT}{q}\left(\frac{\mu_n - \mu_p}{\mu_n + \mu_p}\right)\ln\left[\frac{\tanh(W_{ON}/L_a)\cosh(W_{ON}/L_a)}{\tanh(W_N/L_a)\cosh(W_N/L_a)}\right] \quad (7.24)$$

For the planar gate IGBT structure considered here, the voltage drop across the MOSFET portion includes the channel, accumulation, and JFET regions [11]. The contribution from channel is given by:

$$V_{CH} = \frac{J_C L_{CH} W_{Cell}}{2\mu_{ni} C_{OX}(V_G - V_{TH})} \quad (7.25)$$

The contribution from JFET region is given by:

$$V_{JFET} = \frac{J_C \rho_{JFET}(t_{P+} + W_0) W_{Cell}}{W_{JFET} - 2W_0} \quad (7.26)$$

The contribution from accumulation layer is given by:

$$V_{ACC} = \frac{J_C K_A W_{JFET} W_{Cell}}{4\mu_{nA} C_{OX}(V_G - V_{TH})} \quad (7.27)$$

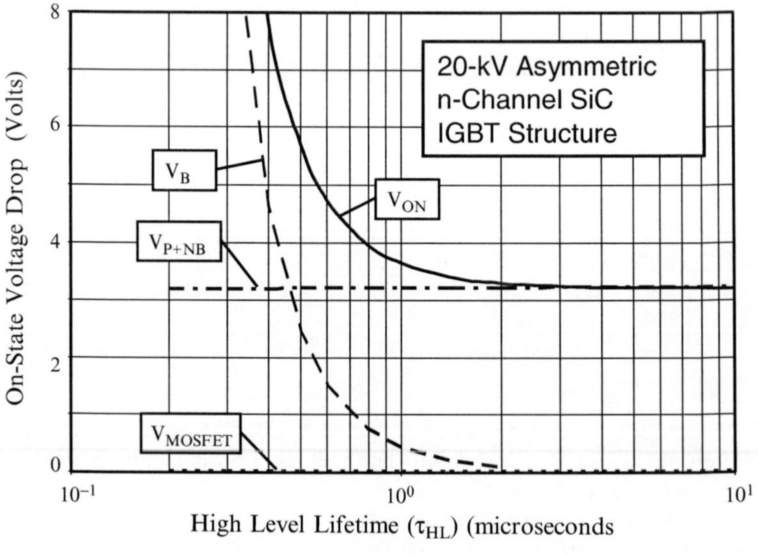

Fig. 7.13 On-state voltage drop for the 20-kV asymmetric n-channel IGBT structure: N-base lifetime dependence

The on-state voltage drop (at an on-state current density of 25 A/cm^2) computed for the 20-kV asymmetric n-channel silicon IGBT structure by using the above equations is provided in Fig. 7.13 as a function of the high-level lifetime in the N-base region. This asymmetric IGBT structure had the optimized N-base region width of 165 μm and N-buffer layer width of 5 μm. The N-buffer layer doping concentration was kept at 5 × 10^{16} cm^{-3} for all the cases. From the figure it can be observed that the on-state voltage drop is close to that of the collector/N-buffer layer junction when the high-level lifetime is greater than 2 μs. This voltage drop is close to 3 V for 4H-SiC devices. The on-state voltage drop increases rapidly when the high-level lifetime is reduced below 0.6 μs due to an increase in the voltage drop across the N-base region. This is consistent with the lack of conductivity modulation of the drift region when the lifetime (τ_{p0}) becomes less than 0.3 μs as shown in Fig. 7.12. The on-state voltage drop is 3.65 V for a high-level lifetime of 1 μs in the N-base region.

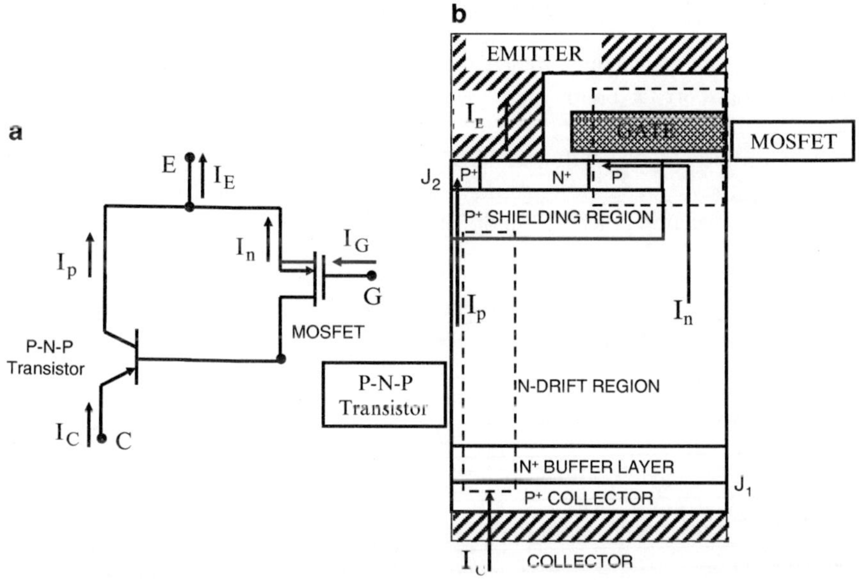

Fig. 7.14 (**a**) Equivalent circuit for the IGBT. (**b**) Current flow within the IGBT

The equivalent circuit for an n-channel IGBT structure [11] based upon the P-N-P Transistor/MOSFET model consists of an n-channel MOSFET providing the base drive current to a P-N-P transistor as shown in Fig. 7.14a. The P-N-P transistor and MOSFET portions are identified by the dashed boxes in cross section shown in Fig. 7.14b. The emitter current for the IGBT structure consists of the hole current flow via the P-N-P transistor and the electron current via the MOSFET portion:

$$I_E = I_p + I_n \tag{7.28}$$

In the IGBT structure, the electron current serves as the base drive current for the P-N-P transistor. Consequently, these currents are interrelated by the common base current gain of the P-N-P transistor:

$$I_p = \alpha_{PNP} I_E = \alpha_{PNP} I_C \tag{7.29}$$

and:

$$I_n = (1 - \alpha_{PNP}) I_E = (1 - \alpha_{PNP}) I_C \tag{7.30}$$

because, under steady-state operating conditions, the gate current for the IGBT structure is zero due to the high impedance of the MOS gate structure.

The current gain of the P-N-P transistor is determined by the product of the emitter injection efficiency and the base transport factor because the multiplication coefficient is unity at the low on-state bias voltages:

$$\alpha_{PNP} = \gamma_{E,ON} \cdot \alpha_{T,NB} \cdot \alpha_{T,NBL} \tag{7.31}$$

where $\gamma_{E,ON}$ is the injection efficiency of transistor emitter in the on-state, $\alpha_{T,NB}$ is the base transport factor for the N-base region, and $\alpha_{T,NBL}$ is the base transport factor for the N-buffer layer.

The injection efficiency for the IGBT structure in the on-state is less than unity due to high-level injection conditions in the N-base region and N-buffer layer region. The injection efficiency in the on-state can be obtained by using:

$$\gamma_{E,ON} = \frac{J_p(J_1)}{J_{C,ON}} \tag{7.32}$$

where $J_p(J_1)$ is the hole current density at junction J_1, which can be computed using:

$$J_p(J_1) = \frac{qD_{pNB} p_{NB}(0)}{L_{pNB}} \tag{7.33}$$

The base transport factor for the N-base region in the on-state can be obtained by using [11]:

$$\alpha_{T,N-Base,0} = \frac{J_p(W_N)}{J_p(W_{NB+})} \tag{7.34}$$

where $J_p(W_N)$ is the hole current density at junction J_2 and $J_p(W_{NB+})$ is the hole current density at interface between the N-base region and the N-buffer layer. These current densities can be obtained by using [11]:

$$J_p(W_{NB+}) = \left[\left(\frac{\mu_p}{\mu_p + \mu_n}\right) + \left(\frac{\mu_n}{\mu_p + \mu_n}\right) K_{AS}\right] J_C \tag{7.35}$$

and

$$J_{\mathrm{p}}(W_{\mathrm{N}}) = \left\{ \frac{\left(\dfrac{\mu_{\mathrm{p}}}{\mu_{\mathrm{p}}+\mu_{\mathrm{n}}}\right) - \left(\dfrac{\mu_{\mathrm{n}}K_{\mathrm{AS}}}{\mu_{\mathrm{p}}+\mu_{\mathrm{n}}}\right)}{\left[\sinh\left(\dfrac{W_{\mathrm{N}}}{L_{\mathrm{a}}}\right)\tanh\left(\dfrac{W_{\mathrm{N}}}{L_{\mathrm{a}}}\right) - \cosh\left(\dfrac{W_{\mathrm{N}}}{L_{\mathrm{a}}}\right)\right]} \right\} J_{\mathrm{C}} \tag{7.36}$$

The base transport factor in the conductivity-modulated lightly doped portion of the N-base region is enhanced by the combination of drift and diffusion due to the high-level injection conditions. The base-transport factor associated with the N-buffer layer can be obtained from the decay of the hole current within the N-buffer layer as given by low-level injection theory:

$$\alpha_{\mathrm{T,N-Buffer}} = \frac{J_{\mathrm{p}}(W_{\mathrm{NB-}})}{J_{\mathrm{p}}(y_{\mathrm{N}})} = e^{-W_{\mathrm{NBL}}/L_{\mathrm{pNB}}} \tag{7.37}$$

Consider the case of the 20-kV asymmetric silicon carbide n-channel IGBT structure with an N-base region width of 165 μm and a low-level lifetime of 1 μs in the N-base region, N-buffer layer doping concentration of 5×10^{16} cm^{-3} and thickness of 5 μm, and P$^+$ collector region (emitter region of the internal PNP transistor) doping concentration of 1×10^{19} cm^{-3}. With these device parameters, the injected concentration of holes [$p_{\mathrm{NB}}(0)$] in the N-buffer layer at the junction (J_1) is found to be 6.9×10^{16} cm^{-3} by using Eq. 7.14. Using this value in Eq. 7.32, the injection efficiency in the on-state is found to be 0.959. With these parameters, the base transport factor for the N-base region is found to be 0.285 by using Eq. 7.34 and the base transport factor for the N-buffer layer is found to be 0.631 by using Eq. 7.37. Combining these values, the common base current gain of the PNP transistor (α_{PNP}) in the on-state is found to be 0.172. Based upon this analysis, it can be concluded that only 17% of the emitter current of the 20-kV asymmetric silicon carbide n-channel IGBT structure is due to the hole current component (I_{p}), and most of the current flow consists of the electron current (I_{n}) at the emitter side.

Simulation Results

The results of two-dimensional numerical simulations for the 20-kV asymmetrical n-channel 4H-SiC IGBT structure are described here. The total width ($W_{\mathrm{Cell}}/2$) of the structure, as shown by the cross section in Fig. 7.1, was 5 μm (area = 5×10^{-8} cm^{-2}). A JFET region width (W_{JFET}) of 4 μm was used with a gate oxide thickness of 500 Å. The device used for the numerical simulations had a drift region doping concentration of 1.5×10^{14} cm^{-3} and thickness of 160 μm below the P$^+$ shielding region. The P$^+$ shielding region extended from a depth of 0.2–1.0 μm with a doping concentration of 1×10^{19} cm^{-3}. The P-base and N$^+$ source regions were formed within the 0.2 μm of the N-drift region located above the P$^+$ region.

The doping concentration of the P-base region was 2×10^{16} cm^{-3} to achieve the desired threshold voltage. Due to the low doping concentration in the drift region for the 20-kV devices, the uniform doping concentration in the JFET region was enhanced to 1×10^{16} cm^{-3} as is usually required for silicon devices. The enhanced doping concentration was extended to 0.5 μm below the P^+ shielding region. The doping profiles for the structure were previously provided in Sect. 7.1.1. The inversion layer mobility for electrons was adjusted to between 15 and 20 cm^2/V-s to match values reported for devices in the literature.

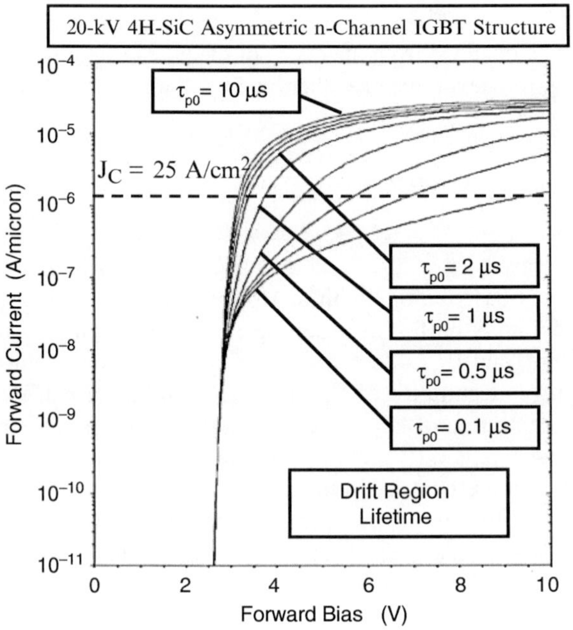

Fig. 7.15 On-state characteristics of the 20-kV asymmetric n-channel IGBT structure: lifetime dependence

The on-state characteristics of the 20-kV asymmetrical n-channel 4H-SiC IGBT structure were obtained by using a gate bias voltage of 10 V for the case of various values for the lifetime in the drift region. This device structure has a buffer layer doping concentration of 5×10^{16} cm^{-3} and thickness of 5 μm. The characteristics obtained from the numerical simulations are shown in Fig. 7.15. The current initially increases exponentially with increasing collector bias. At current densities above 0.2 A/cm^2, the non-state voltage drop begins to increase more rapidly. The on-state voltage drop increases as expected with reduction of the lifetime (τ_{p0}, τ_{n0}) indicated in the figure. The on-state voltage drop at a hole lifetime (τ_{p0}) value of 0.5 μs is found to be 4.54 V at an on-state current density of 25 A/cm^2.

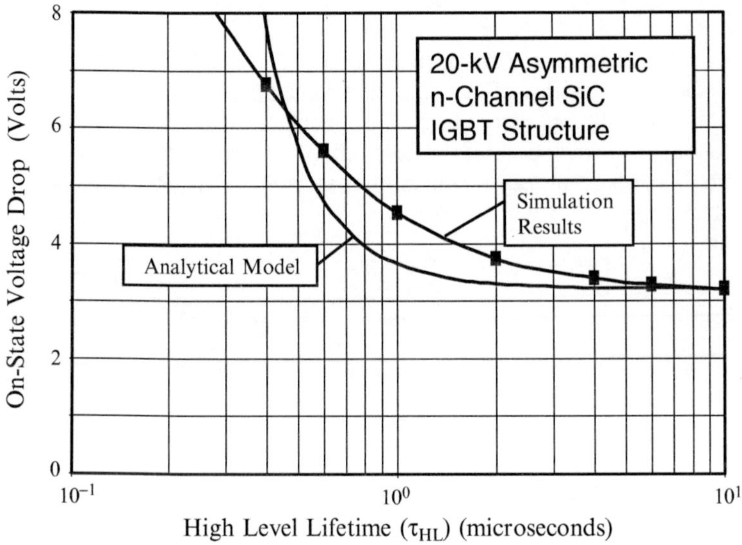

Fig. 7.16 On-state voltage drop for the 20-kV asymmetric n-channel IGBT structure: N-base lifetime dependence

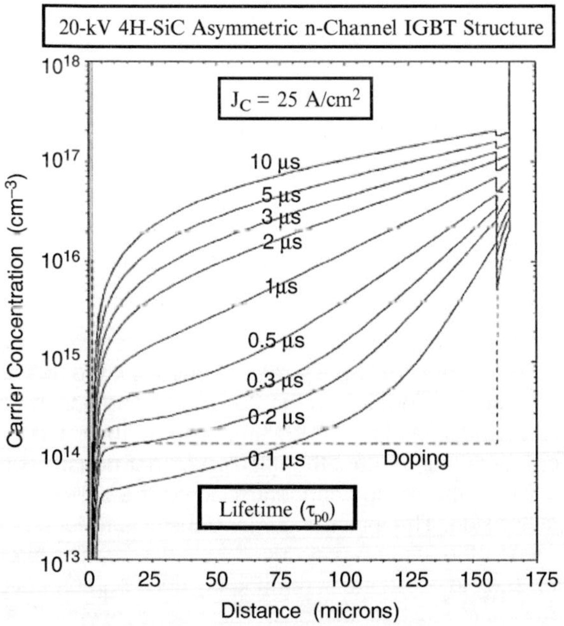

Fig. 7.17 On-state carrier distribution in the 20-kV asymmetric n-channel IGBT structure: lifetime dependence

The variation of the on-state voltage drop as a function of the lifetime in the N-base region predicted by the analytical model is compared with that obtained from the results of the numerical simulation in Fig. 7.16. There is a reasonable agreement between the prediction of the analytical model and the numerical simulations.

The results of the numerical simulations indicate a smaller increase in the on-state voltage drop when the high-level lifetime is reduced below 0.4 μs. This occurs because a larger hole concentration is observed in the N-base region near the emitter side in the results of the numerical simulations that predicted using the analytical model based upon high-level injection. The values obtained from the numerical simulations will be utilized when developing the power loss trade-off curves for the 20-kV asymmetrical n-channel 4H-SiC IGBT structure later in the chapter.

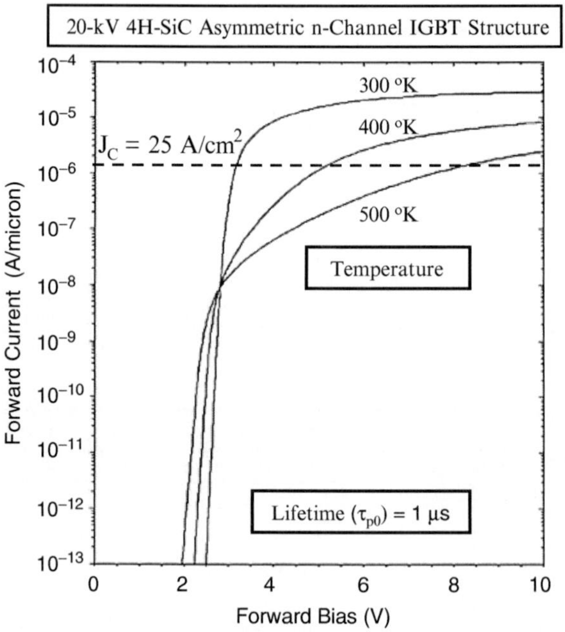

Fig. 7.18 On-state characteristics of the 20-kV asymmetric n-channel IGBT structure: temperature dependence

The on-state voltage drop for the IGBT structure is determined by the distribution of carriers injected into the N-base region producing the desired reduction of its resistance. The hole distribution in the 20-kV asymmetrical n-channel 4H-SiC IGBT structure is provided in Fig. 7.17 for nine cases of the lifetime (τ_{p0}, τ_{n0}) in the drift region. It can be observed that the injected carrier density is three orders of magnitude larger than the doping concentration on the collector side but not as large on the emitter side. The injected carrier density is reduced in the middle of the drift region when the lifetime is reduced below 1 μs. The predictions of the analytical model (see Fig. 7.12) are in remarkably good agreement with the results obtained from the numerical simulations. The hole concentration values at the collector/buffer layer junction and at the various interfaces in the asymmetric IGBT structure are quite well predicted by the analytical model demonstrating the model is applicable not only to silicon devices but silicon carbide devices as well.

The temperature dependence of the on-state characteristics of the 20-kV asymmetrical n-channel 4H-SiC IGBT structure can be observed in Fig. 7.18 for the case of a lifetime of 1 μs in the N-base region. It can be seen that the knee voltage reduces with increasing temperature while the resistance within the device increases [11]. This demonstrates that silicon carbide IGBT structures display the typical behavior for all IGBT structures. The increase in the on-state voltage drop with temperature for the 20-kV asymmetrical n-channel 4H-SiC IGBT structure is quite severe. This is consistent with the observations on 13-kV asymmetrical n-channel 4H-SiC IGBT devices [9].

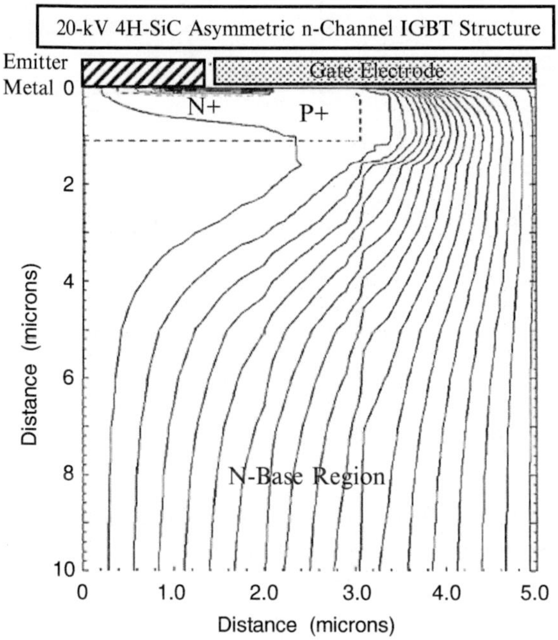

Fig. 7.19 On-state current distribution in the 20-kV asymmetric n-channel IGBT structure

It is also insightful to examine the current distribution within the 20-kV asymmetrical n-channel 4H-SiC IGBT structure during on-state operation. The current flow-lines within the device structure are shown in Fig. 7.19 for the case of a low-level lifetime of 1 μs in the N-base region. It can be observed that most of the current flows via the channel. Only a small fraction of the current flow on the emitter side flows via the P^+ shielding region which is consistent with the conclusions of the analytical model. The current flow via the channel and the P^+ shielding region can also be examined using separate electrodes attached to the N^+ emitter region and the P^+ shielding region, although one electrode is typically used in the actual device structure. These currents are provided in Fig. 7.20 for the case of the above structure. At the on-state current density of 25 A/cm^2, it can be observed that the hole current flowing via the electrode connected to the P^+ shielding region is about 10% of the collector current.

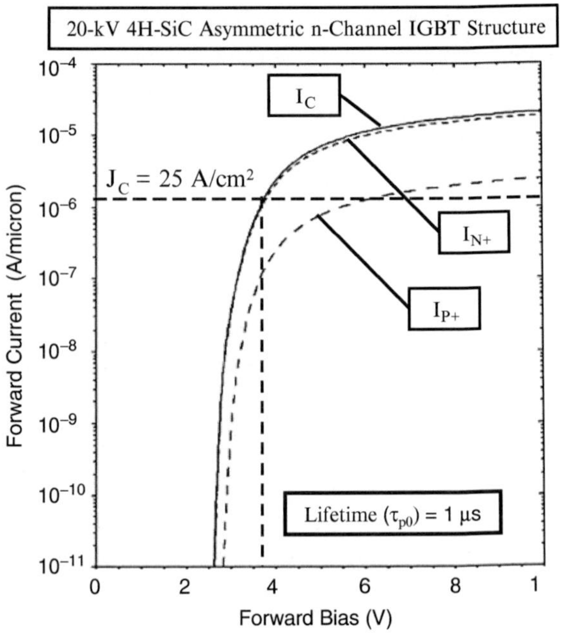

Fig. 7.20 Current distribution in the 20-kV asymmetric n-channel IGBT structure during the on-state

7.1.3 Turn-Off Characteristics

The turn-off behavior for the asymmetric silicon carbide IGBT structure can be expected to be similar to that shown in Chap. 5 for the silicon asymmetric IGBT structure. However, the doping concentration of the drift region in the silicon carbide devices is two orders of magnitude larger than that for the silicon devices. This has a significant impact on the turn-off waveforms as shown in this chapter.

In the case of the silicon device, the hole concentration in the drift region during the voltage rise-time becomes larger than the doping concentration in the drift region. This additional charge in the space-charge region reduces its width to less than the width of the N-base region. Consequently, the space-charge region does not reach-through to the N-buffer layer at the end of the voltage rise-time when the collector voltage becomes equal to the supply voltage. During the current fall-time, the stored charge remaining in the N-base region must be first removed until the space-charge region punches-through to the N-buffer layer. This is followed by the recombination of the stored charge in the buffer layer. For the asymmetric silicon IGBT structure, a single phase is observed for the voltage transient while the current decays in two phases.

In contrast, for the asymmetric silicon carbide IGBT structure, the hole concentration in the space-charge region during the voltage rise-time is much smaller than the doping concentration of the drift region. Consequently, during the voltage

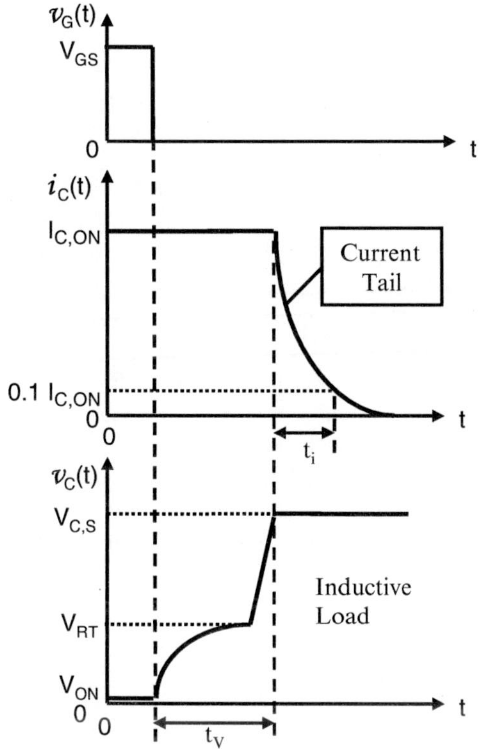

Fig. 7.21 Turn-off waveforms for the asymmetric n-channel SiC IGBT structure

rise-time, the space-charge regions reaches-through to the buffer layer at a collector bias that is well below the collector supply voltage. After the space-charge region reaches-through to the buffer layer, the electric field in the N-base region takes a trapezoidal shape allowing the collector voltage to rise at a much more rapid rate until it reaches the supply voltage. All the stored charge in the N-base region is therefore removed during the voltage rise-time. During the current fall-time, the stored charge in the buffer layer is removed by recombination. This occurs with a single current decay transient.

In order to turn off the IGBT structure, the gate voltage must simply be reduced from the on-state value (nominally 10 V) to zero as illustrated in Fig. 7.21. The magnitude of the gate current can be limited by using a resistance in series with the gate voltage source. The waveform for the gate voltage shown in the figure is for the case of zero gate resistance. Once the gate voltage falls below the threshold voltage, the electron current from the channel ceases. In the case of an inductive load, the collector current for the IGBT structure is then sustained by the hole current flow due to the presence of stored charge in the N-base region. The collector voltage begins to increase in the IGBT structure immediately after the gate voltage reduces below the threshold voltage.

7.1.3.1 Voltage Rise-Time

The analysis of the turn-off waveform for the collector voltage transient for the asymmetric IGBT structure can be performed by using the charge control principle. The analysis of the collector voltage transient was provided in Chap. 5 using the nonlinear hole concentration profile given by Eq. 5.20. The concentration p_{WNB+} in the on-state at the interface between the lightly doped portion of the N-base region and the N-buffer layer was previously derived for the silicon carbide asymmetric IGBT structure in Sect. 7.1.2. To develop the analysis of the collector voltage transient, it will be assumed that the hole concentration profile in the N-base region does not change due to recombination. In this case, the electric field profile in the asymmetric IGBT structure during the collector voltage transient is illustrated in Fig. 5.17. As the space charge region expands toward the collector side, holes are removed from the stored charge region at its boundary. The holes then flow through the space-charge region at their saturated drift velocity due to the high electric field in the space-charge region.

The concentration of holes at the edge of the space-charge region (p_e) increases during the turn-off process as the space-charge width increases:

$$p_e(t) = p(W_{NB+})\frac{\sinh[W_{SC}(t)/L_a]}{\sinh[(W_N + W_{NB})/L_a]} \tag{7.38}$$

According to the charge-control principle, the charge removed by the expansion of the space-charge layer must equal the charge removed due to collector current flow:

$$J_{C,ON} = qp_e(t)\frac{dW_{SC}(t)}{dt} = qp(W_{NB+})\frac{\sinh[W_{SC}(t)/L_a]}{\sinh[(W_N + W_{NB})/L_a]}\frac{dW_{SC}(t)}{dt} \tag{7.39}$$

by using Eq. 7.38. Integrating this equation on both sides and applying the boundary condition of width $W_{SC}(0)$ for the space-charge layer at time zero provides the solution for the evolution of the space-charge region width with time:

$$W_{SC}(t) = L_a\, a\cosh\left\{\frac{J_{C,ON}\sinh[(W_N + W_{NBL})/L_a]}{qL_a p(W_{NB+})}t + \cosh[W_{SC}(0)/L_a]\right\} \tag{7.40}$$

The space-charge layer expands toward the right-hand side as indicated by the horizontal time arrow in Fig. 5.17 with the hole concentration profile in the stored charge region remaining unchanged.

The collector voltage supported by the asymmetric silicon carbide IGBT structure is related to the space-charge layer width by:

$$V_C(t) = \frac{q(N_D + p_{SC})W_{SC}^2(t)}{2\varepsilon_S} \tag{7.41}$$

The hole concentration in the space-charge layer can be related to the collector current density under the assumption that the carriers are moving at the saturated drift velocity in the space-charge layer:

$$p_{SC} = \frac{J_{C,ON}}{q v_{sat,p}} \tag{7.42}$$

The hole concentration in the space-charge region remains constant during the voltage rise-time because the collector current density is constant. Consequently, the slope of the electric field profile in the space-charge region also becomes independent of time. This analytical model for turn-off of the asymmetric IGBT structure under inductive load conditions predicts a nonlinear increase in the collector voltage with time.

The collector voltage increases in accordance with the above model until the space-charge region reaches-through the N-base region. The reach-through voltage during the turn-off of the asymmetric silicon carbide IGBT must be computed with inclusion of the positive charge due to the presence of holes in the space-charge region associated with the collector current flow:

$$V_{RT}\left(J_{C,ON}\right) = \frac{q(N_D + p_{SC})W_N^2}{2\varepsilon_S} \tag{7.43}$$

For the silicon carbide asymmetric n-channel IGBT structure, the hole concentration in the space-charge region (p_{SC}) computed using Eq. 7.42 is 1.8×10^{13} cm^{-3} at an on-state current density of 25 A/cm^2 based upon a saturated velocity of 8.6×10^6 cm/s for holes. For the case of the 20-kV asymmetric silicon carbide IGBT structure with an N-base width of 165 μm and doping concentration of 1.5×10^{14} cm^{-3}, the reach-through voltage is found to be 4,262 V. In contrast, the reach-through voltage under forward blocking operation is only 3,150 V. The time at which reach-through occurs can be derived from Eq. 7.40 by setting the space-charge-region width equal to the width of the N-base region:

$$t_{RT} = \frac{qL_a p(W_{NB+})}{J_{C,ON}} \left\{ \frac{\cosh[W_N/L_a] - \cosh[W_{SC}(0)/L_a]}{\sinh[(W_N + W_{NB})/L_a]} \right\} \tag{7.44}$$

Once the space-charge region reaches-through the N-base region, all the stored charge in the N-base region has been removed by the voltage transient. However, there is still substantial stored charge in the N-buffer layer. The expansion of the space-charge region is now curtailed by the high doping concentration of the N-buffer layer.

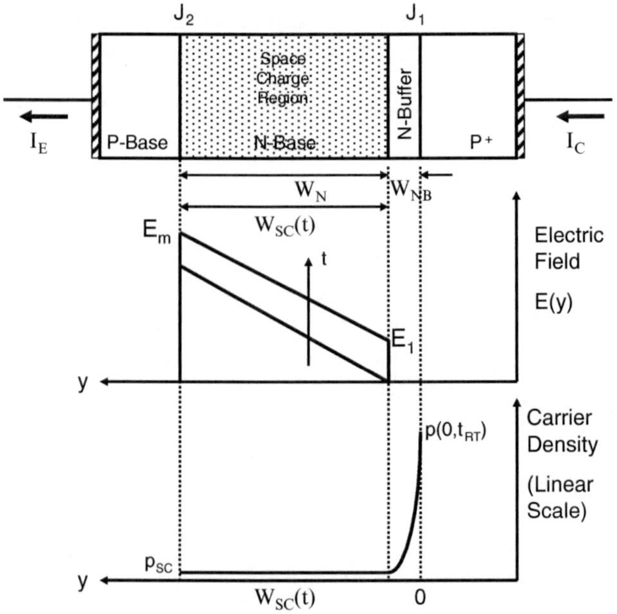

Fig. 7.22 Electric field and free carrier distribution in the asymmetric SiC IGBT structure during second phase of the voltage rise-time

The end of the first phase of the turn-off process occurs when the collector voltages reaches the reach-through voltage (V_{RT}). At this time, the space-charge region has reached the edge of the N-buffer layer. This forces the hole concentration at the edge of the space-charge region (at $y = W_{NB}$ in Fig. 7.22) to the hole concentration (p_{SC}) inside the space-charge region, which is close to zero when compared with the injected hole concentration at the junction (J_1). The hole concentration in the N-buffer layer at junction (J_1) changes abruptly when reach-through occurs in order to maintain the same current density because the collector current density is held fixed during the voltage rise-time.

The hole concentration [$p(0, t_{RT})$] in the N-buffer layer at junction (J_1) during the second phase of the voltage rise-time can be obtained by analysis of current transport in the N-buffer layer. The hole concentration distribution in the N-buffer layer has the same boundary conditions as the base region of a bipolar transistor operating in its active region with finite recombination in the base region [11]:

$$p(y) = p(0, t_{RT}) \left\{ \frac{\sinh\left[(W_{NBL} - y)/L_{pNB}\right]}{\sinh\left(W_{NBL}/L_{pNB}\right)} \right\} \tag{7.45}$$

The hole current in the buffer layer at the junction (J_1) is equal to the collector current density:

$$J_p(0) = qD_{pNB} \left.\frac{dp}{dy}\right|_{y=0} = J_{C,ON} \tag{7.46}$$

Using Eq. 7.45 for the hole carrier distribution:

$$p(0, t_{RT}) = \frac{J_{C,ON} L_{pNB} \tanh(W_{NBL}/L_{pNB})}{q D_{pNB}} \tag{7.47}$$

Substituting into Eq. 7.45:

$$p(y) = \frac{J_{C,ON} L_{pNB}}{q D_{pNB}} \left\{ \frac{\sinh[(W_{NBL} - y)/L_{pNB}]}{\cosh(W_{NBL}/L_{pNB})} \right\} \tag{7.48}$$

This hole distribution is illustrated in Fig. 7.22.

During the second phase of the voltage rise-time, the electric field in the N-base region must increase with a punch-through distribution because of the high doping concentration of the N-buffer layer. As the electric field at the interface between the N-base region and the N-buffer layer grows with increasing collector voltage, a small depletion layer is formed in the N-buffer layer. The formation of the depletion region in the N-buffer layer requires removal of electrons from the donors within the N-buffer layer with an electron current (displacement current) flow toward the collector contact. The electron current available at the interface between the N-base region and the N-buffer layer is determined from the hole current in the N-buffer layer. The hole current at the interface between the N-base region and the N-buffer layer can be obtained from the hole concentration profile given by Eq. 7.48:

$$J_p(W_{NB}) = q D_{pNB} \left.\frac{dp}{dy}\right|_{y=W_{NBL}} = \frac{J_{C,ON}}{\cosh(W_{NBL}/L_{pNB})} \tag{7.49}$$

Consequently, the displacement current is given by:

$$J_D = J_n(W_{NBL}) = J_{C,ON} - J_p(W_{NBL}) = J_{C,ON} \left[1 - \frac{1}{\cosh(W_{NBL}/L_{pNB})} \right] \tag{7.50}$$

The capacitance of the space-charge region during the second phase of the voltage rise-time is independent of the collector voltage because the space-charge-region width is essentially equal to the width of the N-base region because the depletion width in the N-buffer layer is very small due to its high doping concentration. The (specific) capacitance of the space-charge region can be obtained by using:

$$C_{SCR} = \frac{\varepsilon_S}{W_N} \tag{7.51}$$

The rate of rise of the collector voltage based upon charging the space-charge-region capacitance is given by:

$$\frac{dV_C}{dt} = \frac{J_D}{C_{SCR}} \tag{7.52}$$

Using Eq. 7.50:

$$\frac{dV_C}{dt} = \frac{J_{C,ON}}{C_{SCR}}\left[1 - \frac{1}{\cosh(W_{NBL}/L_{pNB})}\right] \tag{7.53}$$

The collector voltage waveform after reach-through is then given by:

$$V_C(t) = V_{RT}(J_{C,ON}) + \frac{J_{C,ON}}{C_{SCR}}\left[1 - \frac{1}{\cosh(W_{NBL}/L_{pNB})}\right]t \tag{7.54}$$

According to this analytical model, the collector voltage should increase linearly with time after the space-charge region reaches-through the N-base region. The end of the voltage rise-time occurs when the collector voltage becomes equal to the collector supply voltage ($V_{C,S}$). Using this criterion in Eq. 7.54, the collector voltage rise-time interval is obtained:

$$t_V = t_{RT} + \frac{\varepsilon_S}{J_{C,ON}W_N}\left[\frac{\cosh(W_{NBL}/L_{pNB})}{\cosh(W_{NBL}/L_{pNB}) - 1}\right]\left[V_{C,S} - V_{RT}(J_{C,ON})\right] \tag{7.55}$$

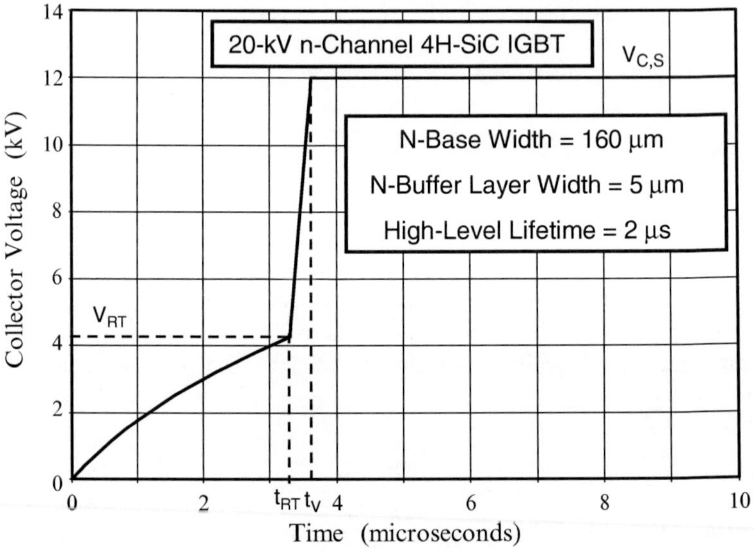

Fig. 7.23 Collector voltage waveform for the asymmetric n-channel SiC IGBT structure during inductive load turn-off

Consider the case of the 20-kV asymmetric silicon carbide n-channel IGBT structure with a N-base region width of 160 μm and a low-level lifetime of 1 μs in the N-base region, N-buffer layer doping concentration of 5×10^{16} cm^{-3} and thickness of 5 μm, and P^+ collector region (emitter region of the internal PNP transistor) doping concentration of 1×10^{19} cm^{-3}. The collector voltage waveform predicted by the above analytical model is provided in Fig. 7.23 for the case of a high-level lifetime of 2 μs and collector supply voltage of 12,000 V. A collector current density of 25 A/cm^2 was used in this example. It can be observed that the collector voltage increases in a nonlinear manner until a reach-through time (t_{RT}) of 3.3 μs. The reach-through collector voltage (V_{RT}) is 4,260 V. After the reach-through of the space-charge region, the collector voltage increases in a linear manner with a high [dV/dt] of 2.43×10^{10} V/s. The collector voltage becomes equal to the collector supply voltage of 12,000 V at time (t_V) of 3.63 μs.

7.1.3.2 Current Fall-Time

At the end of the collector voltage transient, the space-charge region has extended through the entire N-base region, leaving stored charge only in the N-buffer layer. The collector current decays due to the recombination of this stored charge under low-level injection conditions. Unlike in the case of the silicon IGBT described in Chap. 5, the collector current transient occurs in a single phase as described by:

$$J_C(t) = J_{C,ON} e^{-t/\tau_{BL}} \tag{7.56}$$

where t_{BL} is the low-level lifetime in the N-buffer layer.

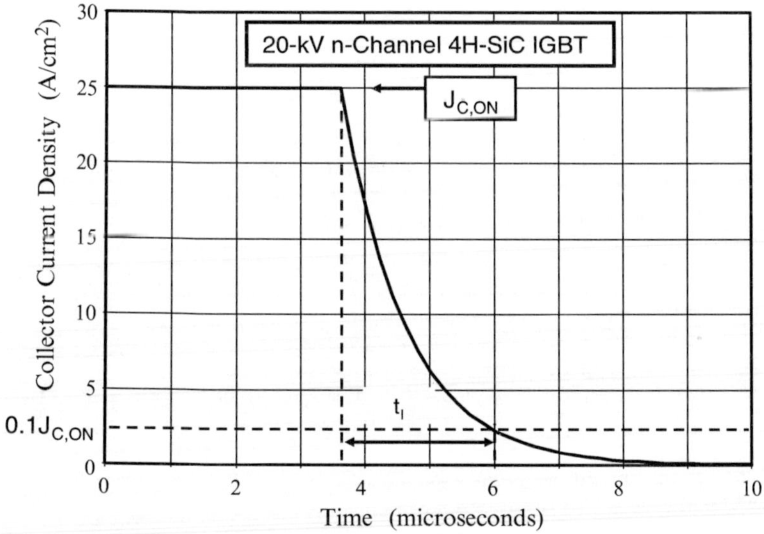

Fig. 7.24 Collector current waveform for the asymmetric n-channel SiC IGBT structure during inductive load turn-off

The collector current waveform for the 20-kV n-channel 4H-SiC asymmetric IGBT structure obtained by using the above model is provided in Fig. 7.24. The current fall-time is defined as the time taken for the current to reduce to 10% of the on-state value. In this case, the current fall-time obtained by using Eq. 7.56 is:

$$t_{\mathrm{I}} = 2.303\ \tau_{\mathrm{BL}} \tag{7.57}$$

For the above example, the current fall-time is found to be 2.3 μs if no scaling of the lifetime with buffer layer doping is taken into account.

Simulation Example

In order to gain insight into the operation of the asymmetric 20-kV n-channel 4H-SiC IGBT structure during its turn-off, the results of two-dimensional numerical simulations for a typical structure are discussed here. The device structure used has the cross section shown in Fig. 7.1 with a cell half-width of 5.0 μm. The doping profile for the silicon carbide IGBT structure used in the numerical simulations was provided in Figs. 7.4 and 7.5. The widths of the uniformly doped N-base region and the diffused N-buffer layer are 160 and 5 μm, respectively. For the typical case discussed here, a high-level lifetime of 2 μs was used in the N-base region.

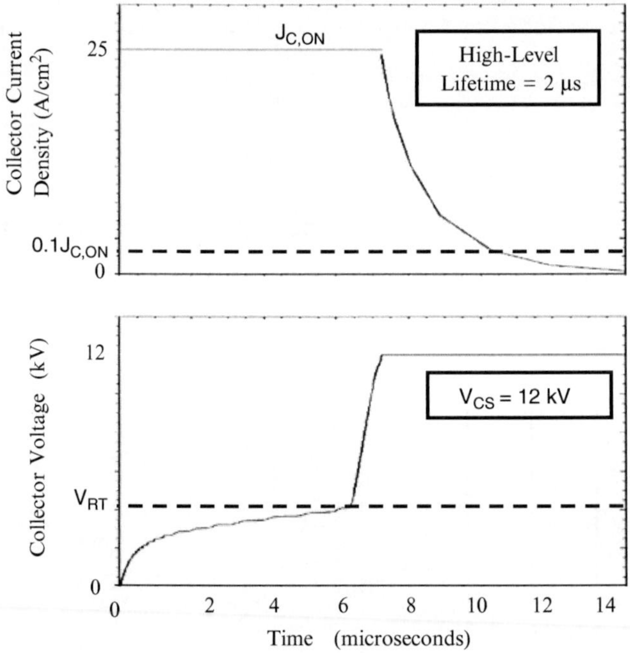

Fig. 7.25 Typical turn-off waveforms for the asymmetric 20-kV n-channel 4H-SiC IGBT structure

The numerical simulations were performed with an abrupt reduction of the gate voltage from 10 to 0 V in 10 ns starting from an on-state current density of 25 A/cm^2. The resulting waveforms obtained from the numerical simulations for the anode voltage and current are shown in Fig. 7.25 for the case of a collector supply voltage of 12,000 V. The collector voltage begins to increase immediately at the end of the gate voltage transient because the P-base/N-base junction (J_2) is already reverse-biased in the on-state. The collector voltage increases non-linearly with the time as predicted by the analytical model until it reaches about 4,000 V. It then increases at a rapid rate as predicted by the analytical model. The rate of rise of the collector voltage during the second part of the transient is found to be 1×10^{10} V/s.

The increase in collector voltage occurs at a much smaller rate than predicted by the analytical model with a reach-through time of 6 μs. This is due to the larger hole concentration in the space-charge region than assumed in the analytical model during the early portion of the transient when the collector voltage is less than 1,000 V as can be observed in Fig. 7.26. The larger hole concentration is due to the smaller hole velocity in the space-charge region when the electric field in the space-charge region is insufficient to produce velocity saturation (as assumed in deriving Eq. 7.42). Even at high collector voltages, the hole concentration in the space-charge region toward the collector side is much larger than that obtained by using Eq. 7.42 because of the low electric fields in this portion. The hole concentration in the space-charge region for the silicon carbide IGBT structures

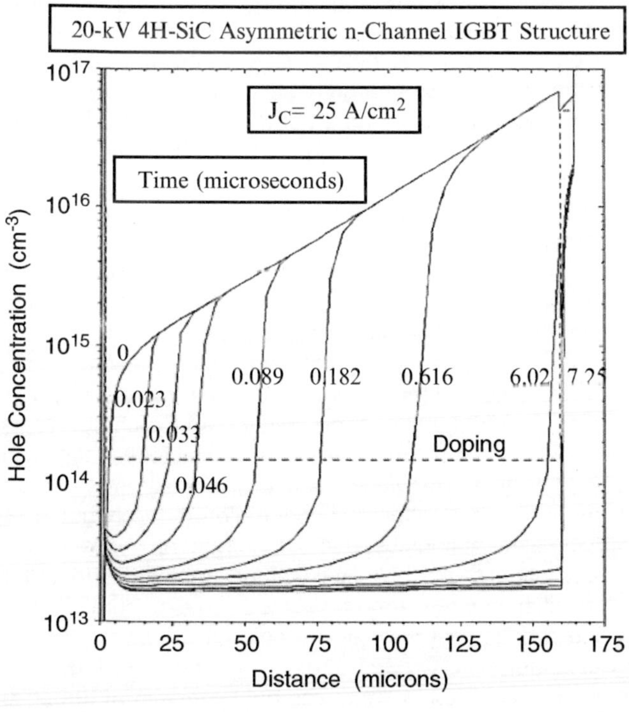

Fig. 7.26 Hole carrier distribution in the 20-kV asymmetric n-channel 4H-SiC IGBT structure for turn-off transient during the voltage rise-time

is much greater than that for silicon devices because the velocity-field curve for silicon carbide exhibits velocity saturation at much larger electric field values. The larger hole concentration in the space-charge region for the silicon carbide IGBT structures alters the shape of the collector voltage versus time as observed in Fig. 7.25.

A one-dimensional view of the minority carrier distribution in the 20-kV asymmetric 4H-SiC IGBT structure is shown in Fig. 7.26 from the initial steady-state operating point ($t = 0$ μs) to the end of the voltage rise-time ($t = 7.25$ μs). These carrier profiles were taken at $x = 1$ μm through the N-base region. The initial carrier profile has the distribution predicted by the analytical model (see Eq. 7.20) for the IGBT structure. It can be observed from Fig. 7.26 that the carrier distribution in the N-base region near the collector does not change during the collector voltage rise phase. A significant space-charge region begins to form immediately during the turn-off and expands toward the right-hand side demonstrating that there is no storage phase for the IGBT structure. At larger collector voltages, the hole concentration in the space-charge region is about 1.8×10^{13} cm^{-3}, which is consistent with the value for p_{SC} obtained using the analytical model (see Eq. 7.42) with the carriers moving at the saturated drift velocity and an on-state current density of 25 A/cm^2. However, at lower collector bias voltages of below 500 V corresponding to time values less than 0.089 μs, the hole concentration in the space-charge region is significantly larger (up to 4×10^{13} cm^{-3}). The larger hole concentration produces a faster rate of rise of the collector voltage (see Eq. 7.41) at lower collector voltages. As the collector voltage increases, the hole concentration in the space-charge region reduces making the collector voltage rise at a slower rate. At large collector voltages, the high hole concentration in the space-charge region toward the collector side, where the electric field is small, shortens the width of the space-charge region. This increases the reach-through voltage and prolongs the reach-through time. The time taken for the collector voltage to increase to the reach-through voltage is found to be 6 μs in the simulations.

Note that the hole concentration at the collector junction (J_1) reduces abruptly when the space-charge region reaches-through. This was taken into account during analytical modeling by development of Eq. 7.47. The hole profile in the N-buffer layer then remains unchanged during the rest of the voltage transient because the collector current density is constant.

The evolution of the electric field profile during the collector voltage rise-time for the 20-kV asymmetric n-channel 4H-SiC IGBT structure is shown in Fig. 7.27. The electric field has a triangular shape until the space-charge region reaches-through to the N-buffer layer at time 6.02 μs. The collector voltage has risen to just above 4,000 V at this time. The electric field profile then takes a trapezoidal form as expected due to the high doping concentration in the N-buffer layer. This is consistent with the assumptions used to develop the analytical model.

After the completion of the collector voltage transient, the collector current decays from the initial on-state current density at an exponential rate as shown in Fig. 7.25. A one-dimensional view of the hole carrier distribution in the 20-kV asymmetric n-channel 4H-SiC IGBT structure is shown in Fig. 7.28 during the current tail time. The collector voltage was held constant at the collector supply voltage of 12,000 V during this transient. The hole concentration in the stored charge region begins to decrease immediately after the end of the voltage

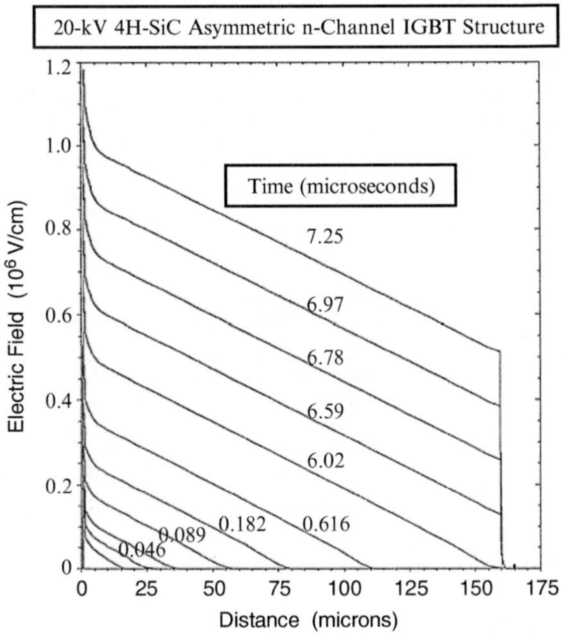

Fig. 7.27 Electric field distribution for the 20-kV asymmetric n-channel 4H-SiC IGBT structure during the voltage rise-time

transient due to the recombination process. The recombination of holes in the N-buffer layer during the collector current transient occurs under low-level injection conditions as assumed in the analytical model. The current fall-time obtained from the numerical simulations (see waveform in Fig. 7.25) is found to be 3 μs which is close to that predicted by the analytical model.

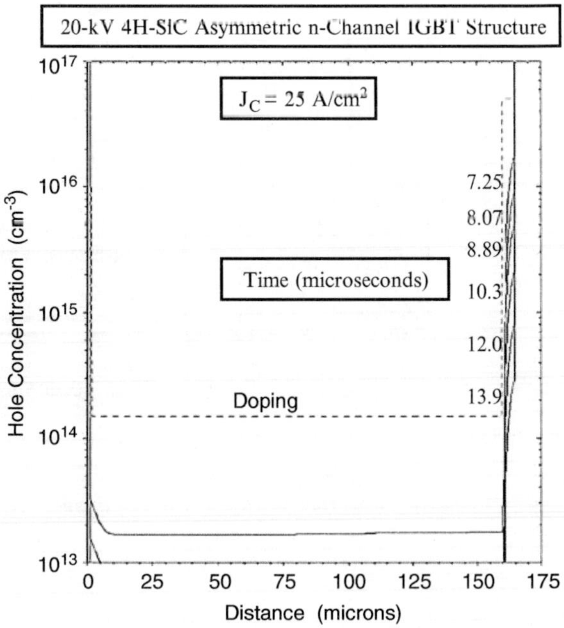

Fig. 7.28 Hole carrier distribution in the 20-kV asymmetric n-channel 4H-SiC IGBT structure for the turn-off transient during the current tail-time

During the current fall-time, unlike in the case of the 5-kV asymmetric silicon IGBT structure, the electric field profile remains essentially the same at the shape corresponding to time 7.25 μs in Fig. 7.27 for the 20-kV asymmetric 4H-SiC IGBT structure. This occurs because the hole concentration in the space-charge region is much smaller than the doping concentration for the 20-kV asymmetric n-channel 4H-SiC IGBT structure.

7.1.4 Lifetime Dependence

From an applications perspective, the optimization of the power losses for the IGBT structure requires performing a trade-off between the on-state voltage drop and the switching losses. One approach to achieve this is by adjusting the lifetime in the drift (N-base) region. A reduction of the lifetime in the drift region also alters the lifetime in the N-buffer layer in the case of silicon devices. However, the relationship between the lifetime in the drift region and the buffer layer has not yet been established for silicon carbide devices. Consequently, it will be assumed that the lifetime in the N-buffer layer is the same as that in the N-base region for silicon carbide structures. The impact of independently optimizing the lifetime in the buffer layer has been analyzed for 15-kV asymmetric 4H-SiC IGBT structures [8].

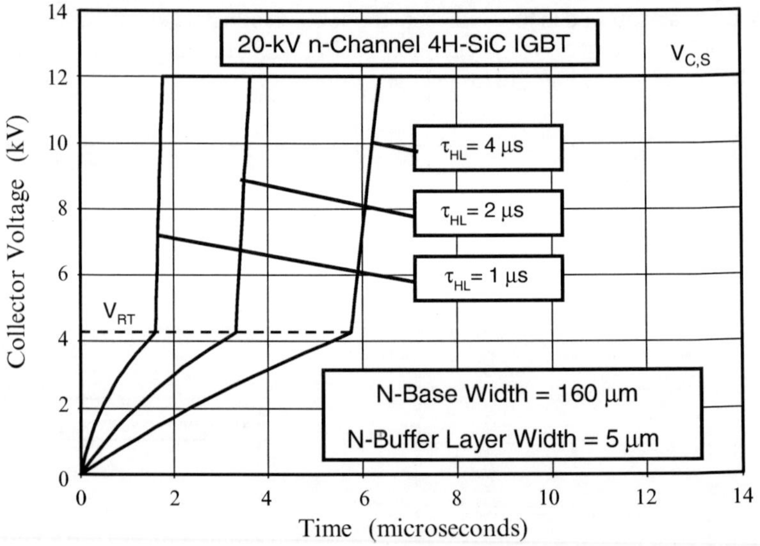

Fig. 7.29 Collector voltage transients during turn-off for the 20-kV asymmetric n-channel 4H-SiC IGBT structure: lifetime dependence

The impact of reducing the lifetime in the drift region on the on-state voltage drop for the 20-kV asymmetric n-channel 4H-SiC IGBT structure was previously discussed in Sect. 7.1.2. As in the case of silicon devices, the on-state voltage drop increases when the lifetime is reduced. The new analytical model developed for turn-off of the asymmetric IGBT structure presented in Chap. 5 can be used to analyze the impact of changes to the lifetime in the drift region on the turn-off characteristics. The collector voltage transients predicted by the new analytical model are shown in Fig. 7.29 for the case of the 20-kV asymmetric 4H-SiC IGBT structure operating with an on-state current density of 25 A/cm^2. The voltage rise-time increases when the lifetime is increased because of the larger concentration for the holes in the N-base region that are being removed during the collector voltage transient. The voltage rise-times obtained by using the analytical model are 1.77, 3.63, and 6.38 μs for high-level lifetime values of 1, 2, and 4 μs, respectively. In all cases, the reach-through voltage has the same value as predicted by Eq. 7.42. However, the rate of increase in the collector voltage [dV/dt] during the second phase of the voltage transient becomes larger when the lifetime is reduced. The collector voltage [dV/dt] increases from 1.24×10^{10} V/s to 4.66×10^{10} V/s when the high-level lifetime is reduced from 4 to 1 μs.

The collector current transients predicted by the new analytical model are shown in Fig. 7.30. It can be observed that the current transient becomes longer when the lifetime in the N-base region increases. The current fall-time increases when the lifetime is increased because of the reduced recombination rate in the N-buffer layer during the current transient. According to the analytical model, the current fall-times obtained by using the analytical model are 1.15, 2.30, and 4.61 μs for high-level lifetime values of 1, 2, and 4 μs, respectively.

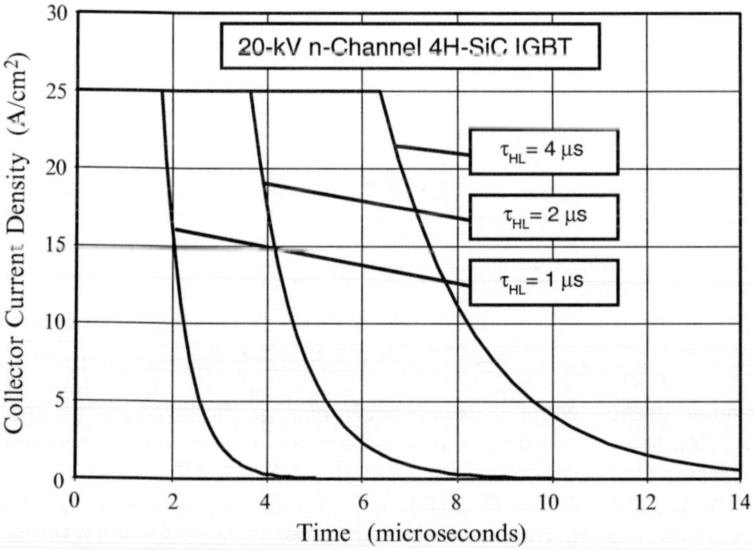

Fig. 7.30 Collector current transients during turn-off for the 20-kV asymmetric n-channel 4H-SiC IGBT structure: lifetime dependence

Simulation Example

In order to gain insight into the impact of the lifetime in the N-base region on the operation of the 20-kV asymmetric 4H-SiC IGBT structure, the results of two-dimensional numerical simulations for a typical structure are discussed here. The device structure used has the cross section shown in Fig. 7.1 with a half-cell width of 5 μm. The widths of the N-base and N-buffer layer regions are 160 and 5 μm, respectively. The high-level lifetime in the N-base region was varied between 0.4 and 4 μs. For turning off the IGBT structures, the numerical simulations were performed with gate voltage rapidly ramped down from 10 to 0 V in 10 ns starting from an on-state current density of 25 A/cm^2. The resulting waveforms obtained from the numerical simulations for the collector voltage and current are shown in Fig. 7.31 for the case of a collector supply voltage of 12,000 V.

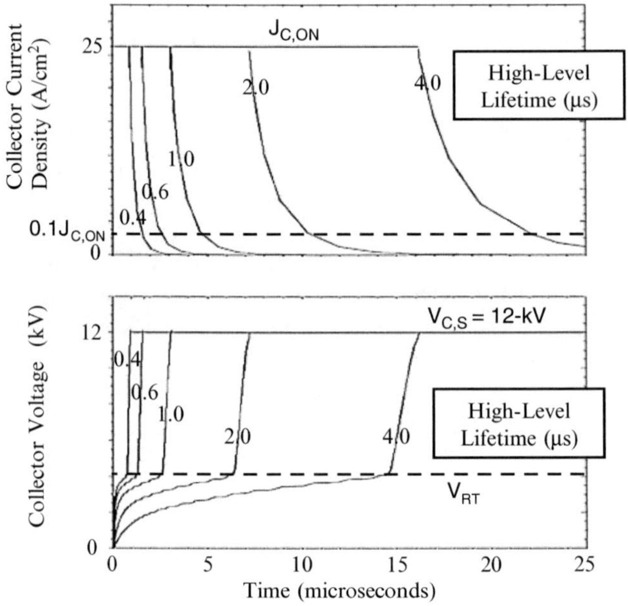

Fig. 7.31 Impact of lifetime on the 20-kV asymmetric n-channel 4H-SiC IGBT turn-off waveforms

The numerical simulations show a decrease in the time taken for the collector voltage to increase to the reach-through voltage when the lifetime in the N-base region is reduced. In the simulation results, the reach-through voltage remains independent of the lifetime in the N-base region as predicted by the analytical model. After reach-through of the space-charge region occurs, the collector voltage increases linearly with time. The [d*V*/d*t*] values for the collector voltage transients increase with reduced lifetime in the drift region as predicted by the analytical model. The [d*V*/d*t*] values are 0.53, 0.84, 1.23, 2.00, and 4.00 10^{10} V/s for high-level lifetime values of 2, 1, 0.5, 0.3, and 0.2 μs, respectively.

The numerical simulations of the 20-kV asymmetric 4H-SiC IGBT structure also show a substantial increase in the collector current fall-time when the lifetime increases. For all the lifetime values, the collector current decays exponentially with time as predicted by the analytical model. The collector current fall-time values are 6.0, 3.3, 1.8, 0.9, and 0.6 μs, for high-level lifetime values of 2, 1, 0.5, 0.3, and 0.2 μs, respectively.

7.1.5 Switching Energy Loss

The power loss incurred during the turn-off switching transient limits the maximum operating frequency for the IGBT structure. Power losses during the turn-on of the IGBT structure are also significant but strongly dependent on the reverse recovery behavior of the fly-back rectifiers in circuits. Consequently, it is common practice to use only the turn-off energy loss per cycle during characterization of IGBT devices. The turn-off losses are associated with the voltage rise-time interval and the current fall-time interval. The energy loss for each event can be computed by integration of the power loss, as given by the product of the instantaneous current and voltage. During the voltage rise-time interval, the anode current is constant while the voltage increases in a nonlinear manner as a function of time until reach-through occurs. In order to simplify the analysis, the energy loss during this interval will be computed using:

$$E_{\mathrm{OFF,V1}} = \frac{1}{2} J_{\mathrm{C,ON}} V_{\mathrm{RT}} t_{\mathrm{RT}} \tag{7.58}$$

During the second phase of the voltage rise-time, the collector voltage increases linearly with time while the collector current is constant. The energy loss during this interval can be computed using:

$$E_{\mathrm{OFF,V2}} = \frac{1}{2} J_{\mathrm{C,ON}} \left(V_{\mathrm{C,S}} - V_{\mathrm{RT}}\right)\left(t_{\mathrm{V}} - t_{\mathrm{RT}}\right) \tag{7.59}$$

For the typical switching waveforms for the 20-kV asymmetric n-channel 4H-SiC IGBT structure shown in Fig. 7.23 with a collector supply voltage of 12,000 V, the energy loss per unit area during the collector voltage rise-time is found to be 0.24 J/cm^2 if the on-state current density is 25 A/cm^2.

During the collector current fall-time interval, the collector voltage is constant while the current decreases exponentially with time. The energy loss during the collector current fall-time interval can be computed using:

$$E_{\mathrm{OFF,I}} = J_{\mathrm{C,ON}} V_{\mathrm{C,S}} \tau_{\mathrm{BL}} \tag{7.60}$$

For the typical switching waveforms for the 20-kV asymmetric n-channel 4H-SiC IGBT structure shown in Fig. 7.24 with a collector supply voltage of 12,000 V, the energy loss per unit area during the collector current fall-time is found to be 0.30 J/cm^2 if the on-state current density is 25 A/cm^2. The total energy loss per unit area ($E_{OFF,V}$ + $E_{OFF,I}$) during the turn-off process for the 20-kV asymmetric n-channel 4H-SiC IGBT structure is found to be 0.54 J/cm^2.

Using the results obtained from the numerical simulations, the on-state voltage drop and the total energy loss per cycle can be computed. These values are plotted in Fig. 7.32 to create a trade-off curve to optimize the performance of the 20-kV asymmetric n-channel 4H-SiC IGBT structure by varying the lifetime in the N-base region. Devices used in lower frequency circuits would be chosen from the left-hand side of the trade-off curve while devices used in higher frequency circuits would be chosen from the right-hand side of the trade-off curve.

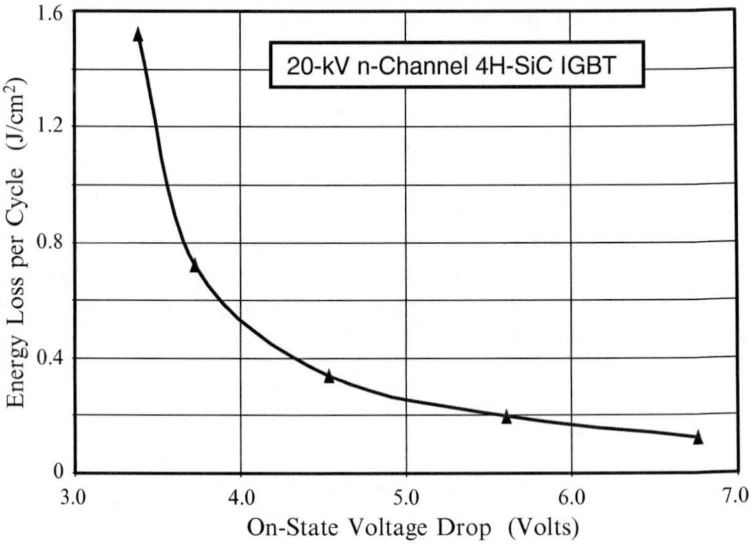

Fig. 7.32 Trade-off curve for the 20-kV asymmetric n-channel 4H-SiC IGBT structure: lifetime in N-base region

7.1.6 Maximum Operating Frequency

The maximum operating frequency for operation of the 20-kV asymmetric n-channel 4H-SiC IGBT structure can be obtained by combining the on-state and switching power losses:

$$P_{D,TOTAL} = \delta P_{D,ON} + E_{OFF} f \tag{7.61}$$

where δ is the duty cycle and f is the switching frequency. In the case of the baseline 20-kV asymmetric n-channel 4H-SiC IGBT device structure with a high-level lifetime of 2 μs in the N-base region, the on-state voltage drop is 3.728 V at an on-state current density of 25 A/cm^2. For the case of a 50% duty cycle, the on-state power dissipation contributes 47 W/cm^2 to the total power loss. For this lifetime value, the energy loss per cycle during the voltage rise-time obtained from the numerical simulations is 0.425 J/cm^2 and the energy loss per cycle during the current fall-time obtained from the numerical simulations is 0.300 J/cm^2. Using a total turn-off energy loss per cycle of 0.725 J/cm^2 in Eq. 7.61 yields a maximum operating frequency of about 210 Hz.

High-Level Lifetime (μs)	On-State Voltage Drop (Volts)	On-State Power Dissipation (W/cm^2)	Energy Loss per Cycle (J/cm^2)	Maximum Operating Frequency (Hz)
4	3.39	42.3	1.523	104
2	3.73	46.6	0.725	212
1	4.54	56.7	0.341	420
0.6	5.61	70.1	0.198	656
0.4	6.76	84.5	0.122	947

Fig. 7.33 Power loss analysis for the 20-kV asymmetric n-channel 4H-SiC IGBT structure

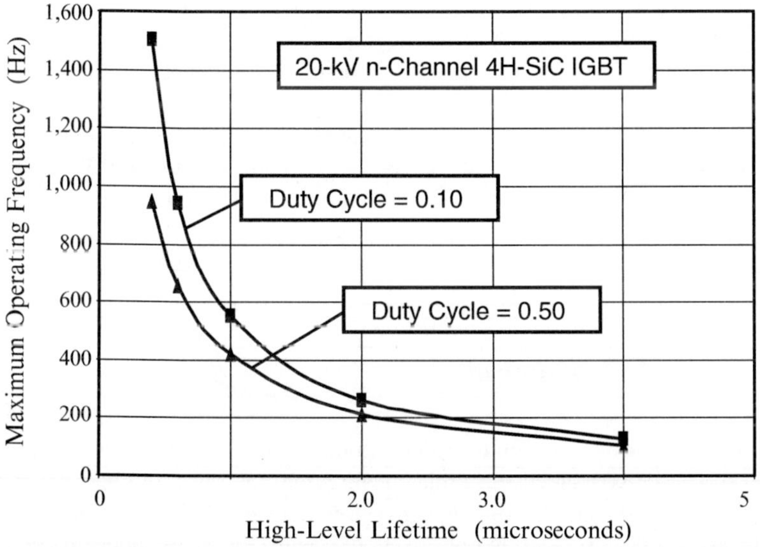

Fig. 7.34 Maximum operating frequency for the 20-kV asymmetric n-channel 4H-SiC IGBT structure

The maximum operating frequency for the 20-kV asymmetric n-channel 4H-SiC IGBT structure can be increased by reducing the lifetime in the N-base region. Using the results obtained from the numerical simulations, the on-state voltage drop and the energy loss per cycle can be computed. These values are provided in

Fig. 7.33 together with the maximum operating frequency as a function of the high level lifetime in the N-base region under the assumption of a 50% duty cycle and a total power dissipation limit of 200 W/cm^2. The maximum operating frequency is plotted in Fig. 7.34 as a function of the high-level lifetime in the N-base region. It can be observed that the maximum operating frequency can be increased up to 950 Hz by reducing the high-level lifetime to 0.4 μs. The IGBT is often operated with pulse-width-modulation to synthesize variable frequency output power for motor control. In these applications, the duty cycle can be much shorter than 50%. In this case, the maximum operating frequency for the 20-kV asymmetric n-channel 4H-SiC IGBT structure can be increased. As an example, the maximum operating frequency for the 20-kV asymmetric n-channel 4H-SiC IGBT structure operated at a 10% duty cycle is included in Fig. 7.34. It can be seen that the maximum operating frequency can now exceed 1,500 Hz.

7.2 Optimized n-Channel Asymmetric Structure

In the previous section, it was found that the 20-kV asymmetric n-channel 4H-SiC IGBT structure exhibits a collector voltage turn-off waveform consisting of two phases. In the first phase, the collector voltage increases gradually up to a reach-through voltage, and then during the second phase the collector voltage increases very rapidly with time until it reaches the collector supply voltage. This produces a high [dV/dt] which is not desirable during circuit operation. The second phase can be prevented from occurring by optimization of the doping concentration and width of the N-base region so that the reach-through of the space-charge region occurs when the collector voltage becomes equal to the collector supply voltage. The design and performance of the optimized 20-kV asymmetric 4H-SiC IGBT structure is discussed in this section.

7.2.1 Structure Optimization

An expression (see Eq. 7.43) for the reach-through voltage was derived in the previous section. It can be concluded from this equation that the reach-through voltage is a function of the width and the doping concentration of the drift region, as well as the on-state current density. During optimization, it is necessary to choose these values after obtaining the hole concentration in the space-charge region using the on-state current density. Although Eq. 7.43 indicates that the reach-through voltage can be increased by solely increasing the doping concentration of the N-base region, this approach results in a reduction of the blocking voltage. Consequently, the width and the doping concentration of the N-base region must be optimized together to simultaneously obtain the desired open-base breakdown voltage of 21 kV and a reach-through voltage equal to a collector supply voltage of 12 kV (as an example).

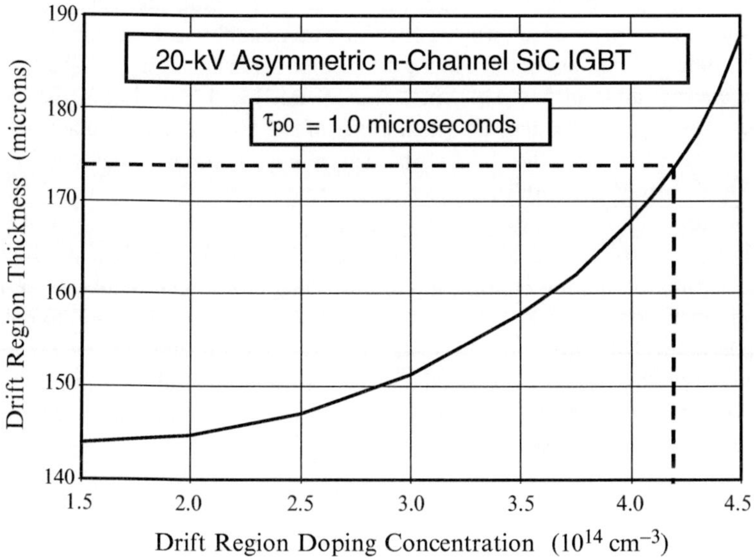

Fig. 7.35 Drift region width optimization for the 20-kV asymmetric 4H-SiC n-channel IGBT structure

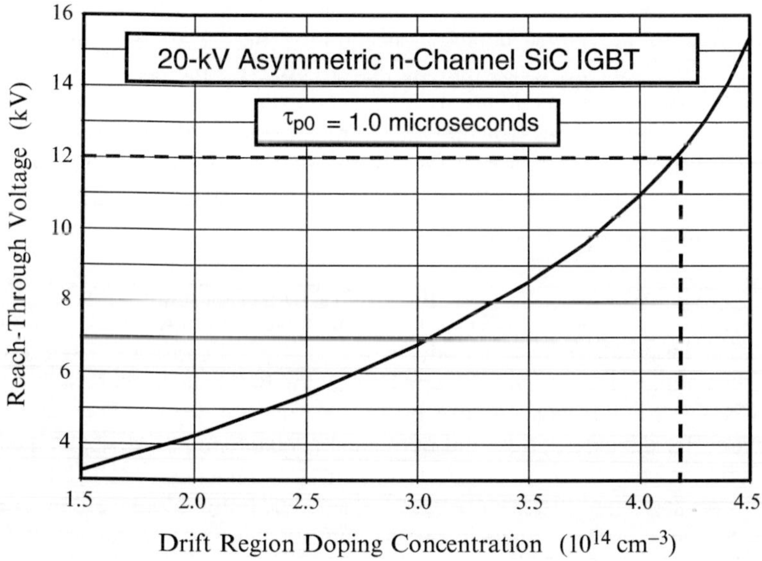

Fig. 7.36 Reach-through voltage for the 20-kV asymmetric 4H-SiC n-channel IGBT structure

The width of the N-base region required to achieve a blocking voltage of 21 kV is shown in Fig. 7.35 based upon using open base transistor breakdown physics. A low-level lifetime of 1 μs was assumed in the drift region for the analysis. A N-buffer layer doping concentration of 5×10^{16} cm^{-3} was assumed with a thickness of 5 μm for this baseline device structure. For each value of the N-base doping concentration, its width was varied until the common-base current gain became equal to unity.

Using the optimum width for the N-base region corresponding to each doping concentration, the reach-through voltage can be computed by using Eq. 7.43. The resulting values for the reach-through voltage are plotted in Fig. 7.36 as a function of the drift region doping concentration. From this plot, it can be observed that a reach-through voltage of 12 kV is obtained when the drift region doping concentration is 4.2×10^{14} cm^{-3}. For this drift region doping concentration, an open-base breakdown voltage of 21 kV is obtained if a drift region width of 175 μm is used according to Fig. 7.35. These values must be chosen for the optimized 20-kV n-channel asymmetric 4H-SiC IGBT structure.

7.2.2 *Blocking Characteristics*

The physics of operation of the optimized 20-kV n-channel asymmetric 4H-SiC IGBT structure is similar to that of the structure discussed in the previous section. However, the electric field profile and the reach-through voltage for the optimized structure are altered due to the larger doping concentration and thickness of the drift region. Due to the larger doping concentration of 4.2×10^{14} cm^{-3} for the optimized 20-kV n-channel asymmetric 4H-SiC IGBT structure, the slope of the electric field profile in the N-base region can be expected to be nearly three times larger than for the previous structure. In addition, the reach-through voltage for the depletion region increases from 4,000 to 12,000 V due to the larger doping concentration and thickness of the N-base region as predicted by using Eq. 4.2.

Simulation Example

The results of two-dimensional numerical simulations on the optimized 20-kV 4H-SiC asymmetric n-channel IGBT structure are described here to provide a more detailed understanding of the underlying device physics and operation during the blocking mode. For the numerical simulations, the half-cell structure with a width ($W_{Cell}/2$) of 5 μm, as illustrated in Fig. 7.1, was utilized as representative of the structure. The device used for the numerical simulations had an optimized drift region doping concentration of 4.2×10^{14} cm^{-3} and thickness of 175 μm below the P^+ shielding region. The rest of the device doping parameters were maintained the same as that of the structure in the previous section. The vertical doping profile taken through the N^+ source region of the 20-kV 4H-SiC asymmetric n-channel IGBT structure is provided in Fig. 7.37. It can be observed that the N-drift region has a doping concentration of 4.2×10^{14} cm^{-3} and thickness of 175 μm. The N-buffer layer located at the collector junction has a doping concentration of 5×10^{16} cm^{-3} and thickness of 5 μm.

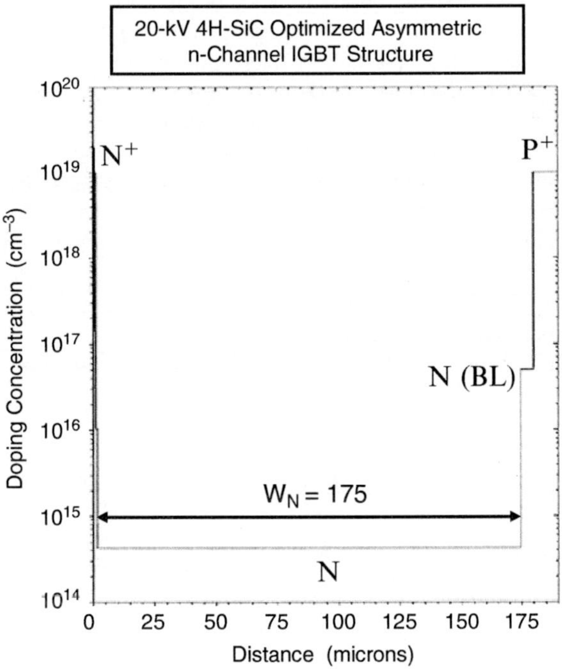

Fig. 7.37 Vertical doping profile in the optimized 4H-SiC asymmetric n-channel IGBT structure

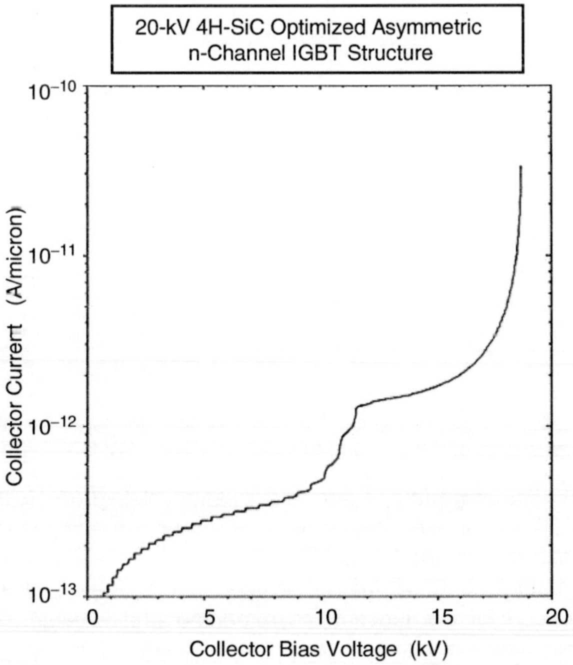

Fig. 7.38 Blocking characteristics of the optimized 4H-SiC asymmetric n-channel IGBT structure

The blocking characteristics of the 20-kV 4H-SiC asymmetric n-channel IGBT structure at room temperature (300 K) cannot be determined by numerical simulations of the cell structure due to the very low intrinsic carrier concentration in silicon carbide. For didactic purposes, the blocking characteristics were therefore obtained at 800 K by increasing the drain voltage while using zero gate bias. The resulting blocking characteristic is shown in Fig. 7.38. It can be seen that leakage current increases with increasing collector bias voltage until it reaches about 12,000 V. This is in excellent agreement with the reach-through voltage of 11,973 V obtained by using Eq. 4.2 with a drift region doping concentration of 4.2×10^{14} cm^{-3} and thickness of 175 μm. Prior to reach-through, the leakage current increases due to the increasing width of the depletion region and the increasing current gain of the PNP transistor in accordance with Eq. 4.5. After reach-through, the leakage current becomes nearly independent of collector voltage. The blocking voltage at 800 K is slightly less than 20 kV.

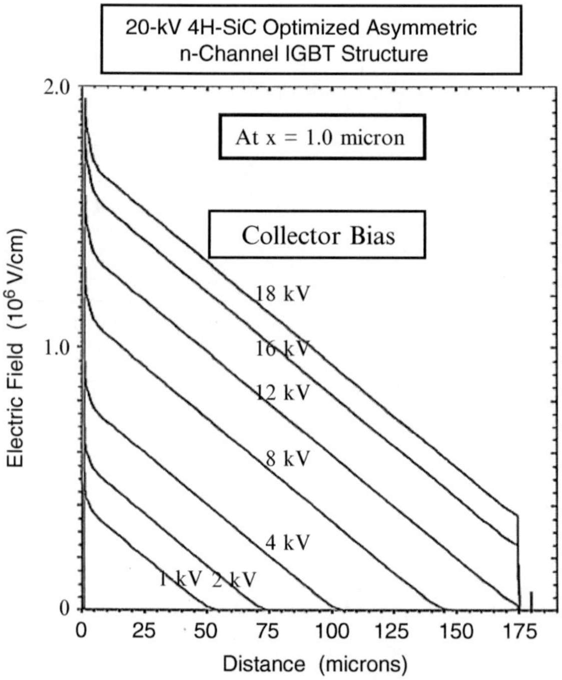

Fig. 7.39 Electric field distribution in the optimized 4H-SiC asymmetric n-channel IGBT structure

It is insightful to examine the electric field profile within the optimized 20-kV 4H-SiC asymmetric n-channel IGBT structure when it is operating in the blocking mode. The electric field profile through the junction between the P^+ shielding region and the N-drift region is provided in Fig. 7.39. The peak of the electric field occurs at the junction as expected and is essentially triangular in shape in accordance with the predictions of Poisson's equation with a uniform doping profile until reach-through occurs at a collector bias of about 12,000 V. There is a slight increase in the electric field near the junction due to the higher doping concentration of the CEL layer when compared with the drift region. However, the enhancement in the electric field is small and occurs over a very small distance, which results in minimal degradation of the breakdown voltage. The electric field then becomes trapezoidal in shape as expected for an asymmetric blocking structure due to the punch-through of the electric field with the N-buffer layer.

The maximum electric field in the drift region below the CEL layer at a drain bias of 18-kV is 1.7×10^6 V/cm which is below the critical electric for breakdown for 4H-SiC at the doping concentration of the drift region [11]. In comparison with the 20-kV 4H-SiC asymmetric n-channel IGBT structure discussed in the previous section, the slope of the electric field profile for the optimized structure is steeper due to the larger doping concentration in the N-base region. In addition, the depletion region reaches-through the N-base region at a much larger collector bias due to the larger doping concentration and thickness of the N-base region.

7.2.3 On-State Voltage Drop

The physics of operation in the on-state for the optimized 20-kV n-channel asymmetric 4H-SiC IGBT structure is identical to that for the structure discussed in the previous section. Based upon the high-level injection model for the IGBT structure, the on-state voltage drop for the optimized structure can be expected to be slightly greater than that for the previous structure for the same lifetime in the N-base region due to its larger width.

Simulation Results

The results of two-dimensional numerical simulations for the optimized 20-kV asymmetrical n-channel 4H-SiC IGBT structure are described here. The total width ($W_{Cell}/2$) of the structure, as shown by the cross section in Fig. 7.1, was 5 μm (area $= 5 \times 10^{-8}$ cm^{-2}). A JFET region width (W_{JFET}) of 4 μm was used with a gate oxide thickness of 500 Å. The device used for the numerical simulations had a drift region doping concentration of 4.2×10^{14} cm^{-3} and thickness of 175 μm below the P^+ shielding region. The P^+ shielding region extended from a depth of 0.2–1.0 μm with a doping concentration of 1×10^{19} cm^{-3}. The P-base and N^+ source regions were formed within the 0.2 μm of the N-drift region located above the P^+ region. The doping concentration of the P-base region was 2×10^{16} cm^{-3} to achieve the desired threshold voltage. Due to the low doping concentration in the drift region, the uniform doping concentration in the JFET region was enhanced to 1×10^{16} cm^{-3} as is usually required for silicon devices. The enhanced doping concentration was extended to 0.5 μm below the P^+ shielding region. The doping profiles for the structure were previously provided in Sect. 7.2.1. The inversion layer mobility for electrons was adjusted to between 15 and 20 cm^2/V-s to match values reported for devices in the literature.

The on-state characteristics of the optimized 20-kV asymmetrical n-channel 4H-SiC IGBT structure were obtained by using a gate bias voltage of 10 V for the case of various values for the lifetime in the drift region. This device structure has a buffer layer doping concentration of 5×10^{16} cm^{-3} and thickness of 5 μm. The characteristics obtained from the numerical simulations are shown in Fig. 7.40. The current initially increases exponentially with increasing collector bias. At current densities above 0.2 A/cm^2, the non-state voltage drop begins to increase more rapidly. The on-state voltage drop increases as expected with reduction of the lifetime (τ_{p0}, τ_{n0}) indicated in the figure. The on-state voltage drop at a hole lifetime (τ_{p0}) value of 10 μs is found to be 3.17 V at an on-state current density of 25 A/cm^2. This value is 0.013 V greater than that of the previous structure in accordance with predictions of the analytical model based upon high-level injection in the N-base region. The on-state voltage drop at a hole lifetime (τ_{p0}) value of 0.5 μs is found to be 4.38 V at an on-state current density of 25 A/cm^2. This value is 0.16 V less than that of the previous structure contrary to the predictions of the analytical model based upon high-level injection in the N-base region. This discrepancy can be explained by examination of the injected hole concentration profile for the two structures.

The variation of the on-state voltage drop as a function of the lifetime in the N-base region obtained from the numerical simulations is provided in Fig. 7.41 for the optimized 20-kV asymmetrical n-channel 4H-SiC IGBT structure. The values for the previous structure are also included in the figure for comparison purposes. It can be observed that the on-state voltage drop for the optimized structure is

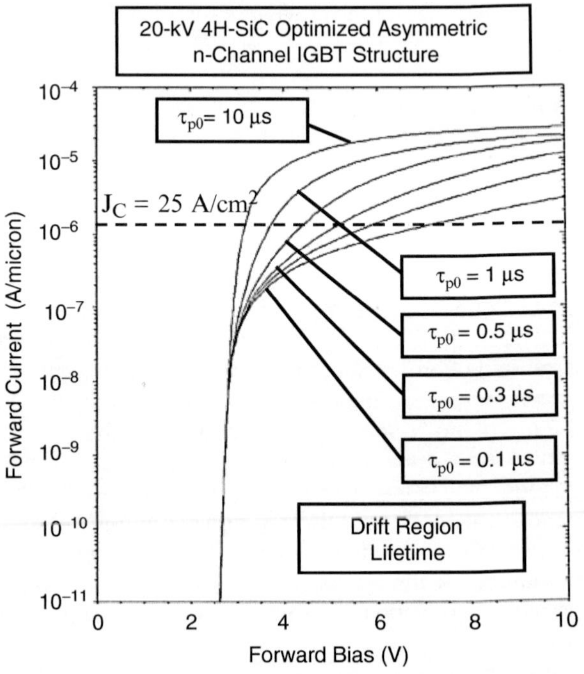

Fig. 7.40 On-state characteristics of the optimized 20-kV asymmetric n-channel IGBT structure: lifetime dependence

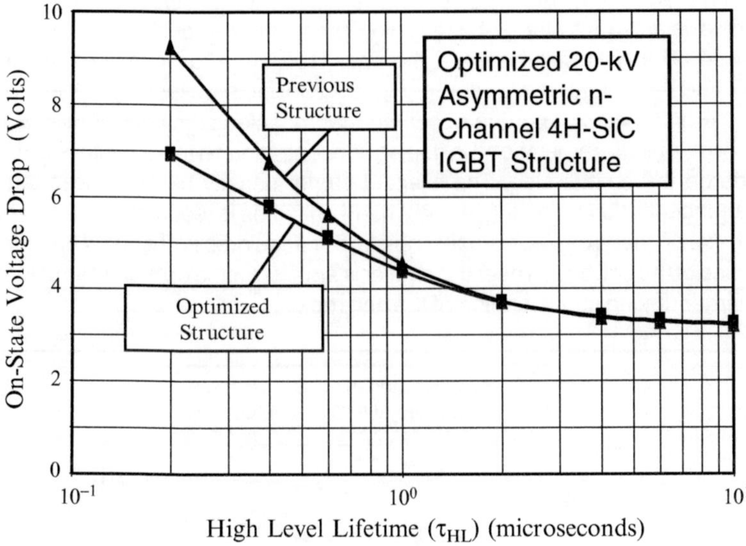

Fig. 7.41 On-state voltage drop for the optimized 20-kV asymmetric n-channel IGBT structure: N-base lifetime dependence

slightly larger when the high-level lifetime in the N-base region is greater than 2 μs. When the high-level lifetime is reduced below 2 μs, the optimized structure has a smaller on-state voltage drop despite the larger thickness of its N-base region.

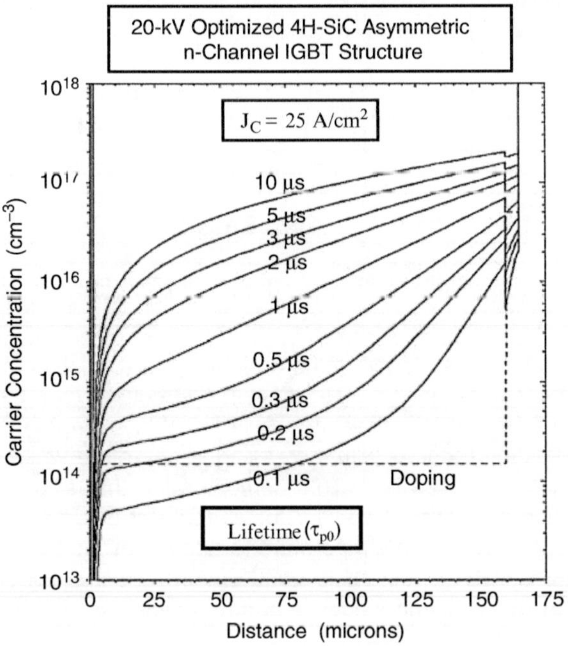

Fig. 7.42 On-state carrier distribution in the optimized 20-kV asymmetric n-channel IGBT structure: lifetime dependence

The on-state voltage drop for the IGBT structure is determined by the distribution of carriers injected into the N-base region producing the desired reduction of its resistance. The hole distribution in the optimized 20-kV asymmetrical n-channel 4H-SiC IGBT structure is provided in Fig. 7.42 for nine cases of the lifetime (τ_{p0}, τ_{n0}) in the drift region. It can be observed that the injected carrier density is three orders of magnitude larger than the doping concentration on the collector side but not as large on the emitter side. The injected carrier density becomes less than the doping concentration of the drift region when the lifetime is reduced below 1 μs on the emitter side. In these cases, a smaller on-state voltage drop is observed because the resistance of the drift region near the emitter is smaller for the optimized structure due to the larger doping concentration of the drift region.

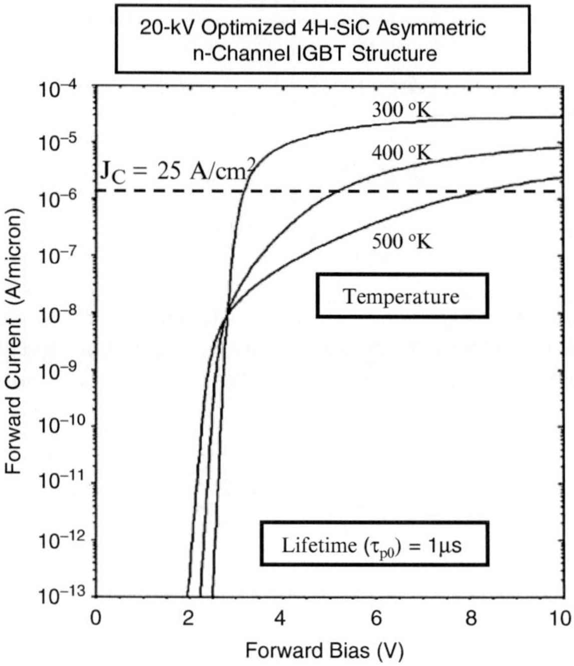

Fig. 7.43 On-state characteristics of the optimized 20-kV asymmetric n-channel IGBT structure: temperature dependence

The temperature dependence of the on-state characteristics of the optimized 20-kV asymmetrical n-channel 4H-SiC IGBT structure can be observed in Fig. 7.43 for the case of a lifetime of 1 μs in the N-base region. It can be seen that the knee voltage reduces with increasing temperature while the resistance within the device increases [11]. This demonstrates that silicon carbide IGBT structures display the typical behavior for all IGBT structures. The increase in the on-state voltage drop with temperature for the optimized 20-kV asymmetrical n-channel 4H-SiC IGBT structure is quite severe. This is consistent with the observations on 13-kV asymmetrical n-channel 4H-SiC IGBT devices [9].

7.2.4 Turn-Off Characteristics

The turn-off behavior for the optimized 20-kV n-channel asymmetric 4H-SiC IGBT structure can be expected to be quite different from that for the 20-kV n-channel asymmetric 4H-SiC IGBT structure discussed in the previous section. In the optimized structure, the collector voltage should increase during a single phase to the collector supply voltage and the collector current should then decay by the recombination of holes in the N-buffer layer. The rise of the collector voltage is described by the physics developed for the previous structure during the first phase (see Eq. 7.41). The current fall occurs with the same physics that governs the recombination of holes in the buffer layer for both structures (see Eq. 7.56).

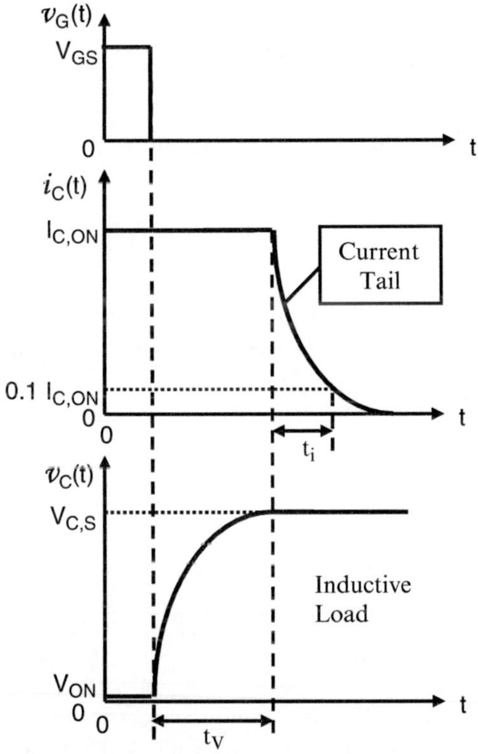

Fig. 7.44 Turn-off waveforms for the optimized asymmetric SiC IGBT structure

In order to turn off the IGBT structure, the gate voltage must simply be reduced from the on-state value (nominally 10 V) to zero as illustrated in Fig. 7.44. The magnitude of the gate current can be limited by using a resistance in series with the gate voltage source. The waveform for the gate voltage shown in the figure is for the case of zero gate resistance. Once the gate voltage falls below the threshold voltage, the electron current from the channel ceases. In the case of an inductive load, the collector current for the IGBT structure is then sustained by the hole current flow due to the presence of stored charge in the N-base region. The collector voltage begins to increase in the IGBT structure immediately after the gate voltage reduces below the threshold voltage.

7.2.4.1 Voltage Rise-Time

The analysis of the turn-off waveform for the collector voltage transient for the optimized asymmetric IGBT structure can be performed by using the same approach as described in the previous section. In the case of the optimized structure, the voltage should increase to the collector supply voltage in a single phase as described by the following equations. The evolution of the space-charge region width with time is given by:

$$W_{SC}(t) = L_a a\cosh\left\{\frac{J_{C,ON}\sinh[(W_N + W_{NBL})/L_a]}{qL_a p(W_{NB+})}t + \cosh[W_{SC}(0)/L_a]\right\} \quad (7.62)$$

The collector voltage supported by the optimized asymmetric silicon carbide IGBT structure is related to the space charge layer width by:

$$V_C(t) = \frac{q(N_D + p_{SC})W_{SC}^2(t)}{2\varepsilon_S} \quad (7.63)$$

The collector voltage increases in accordance with the above model until the space charge region reaches-through the N-base region when the collector voltage becomes equal to the collector supply voltage.

The hole concentration in the space-charge layer can be related to the collector current density under the assumption that the carriers are moving at the saturated drift velocity in the space-charge layer:

$$p_{SC} = \frac{J_{C,ON}}{qv_{sat,p}} \quad (7.64)$$

This analytical model for turn-off of the optimized asymmetric IGBT structure under inductive load conditions predicts a nonlinear increase in the collector voltage with time.

The time at which the collector voltage transient is completed can be derived from Eq. 7.62 by setting the space-charge-region width equal to the width of the N-base region:

$$t_V = \frac{qL_a p(W_{NB}+)}{J_{C,ON}}\left\{\frac{\cosh[W_N/L_a] - \cosh[W_{SC}(0)/L_a]}{\sinh[(W_N + W_{NBL})/L_a]}\right\} \quad (7.65)$$

Once the space-charge region reaches-through the N-base region, all the stored charge in the N-base region is removed by the voltage transient. However, there is still substantial stored charge in the N-buffer layer. The expansion of the space-charge region is now curtailed by the high doping concentration of the N-buffer layer.

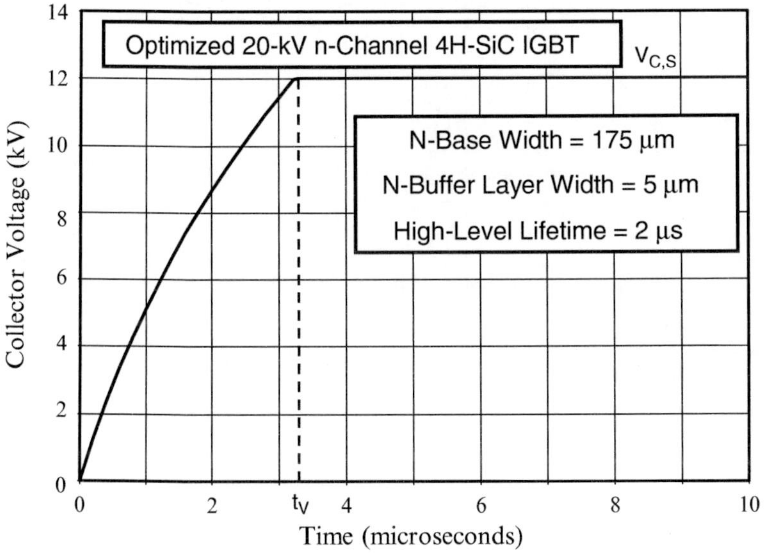

Fig. 7.45 Collector voltage waveform for the optimized asymmetric n-channel SiC IGBT structure during inductive load turn-off

Consider the case of the optimized 20-kV asymmetric silicon carbide n-channel IGBT structure with a N-base region width of 175 μm and a low-level lifetime of 1 μs in the N-base region, N-buffer layer doping concentration of 5×10^{16} cm^{-3} and thickness of 5 μm, and P$^+$ collector region (emitter region of the internal PNP transistor) doping concentration of 1×10^{19} cm^{-3}. The collector voltage waveform predicted by the above analytical model is provided in Fig. 7.45 for this structure. A collector current density of 25 A/cm^2 was used in this example. It can be observed that the collector voltage increases monotonically until a reach-through time (t_{RT}) of 3.3 μs when it becomes equal to the collector supply voltage of 12,000 V.

7.2.4.2 Current Fall-Time

At the end of the collector voltage transient, the space-charge region has extended through the entire N-base region leaving stored charge only in the N-buffer layer in the case of the optimized 20-kV asymmetric silicon carbide n-channel IGBT structure. The collector current therefore decays due to the recombination of this stored charge under low-level injection conditions. As in the case of the 20-kV asymmetric silicon carbide n-channel IGBT structure discussed in the previous chapter, the collector current transient occurs in a single phase as described by:

$$J_C(t) = J_{C,ON} e^{-t/\tau_{BL}} \tag{7.66}$$

where t_{BL} is the low-level lifetime in the N-buffer layer.

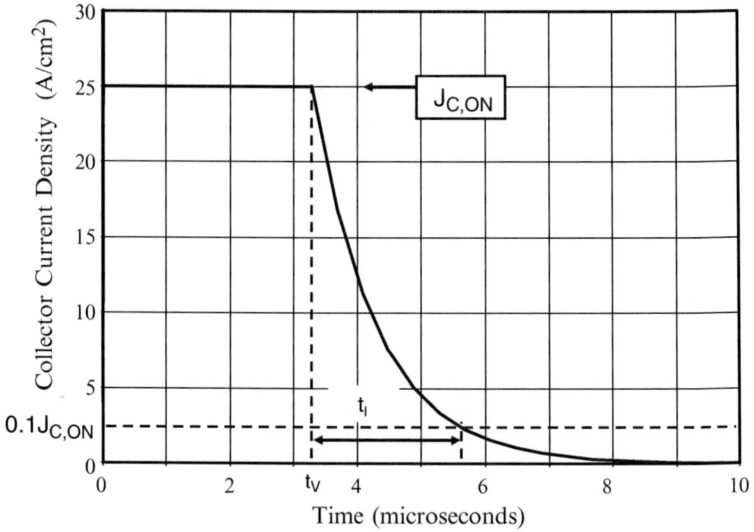

Fig. 7.46 Collector current waveform for the optimized asymmetric n-channel SiC IGBT structure during inductive load turn-off

The collector current waveform for the optimized 20-kV n-channel 4H-SiC asymmetric IGBT structure obtained by using the above model is provided in Fig. 7.46. The current fall-time is defined as the time taken for the current to reduce to 10% of the on-state value. In this case, the current fall time obtained by using Eq. 7.66 is:

$$t_{I,OFF} = 2.303\ \tau_{BL} \tag{7.67}$$

For the above example, the current-fall time is found to be 2.3 μs if no scaling of the lifetime with buffer layer doping is taken into account.

Simulation Example

In order to gain insight into the operation of the optimized asymmetric 20-kV n-channel 4H-SiC IGBT structure during its turn-off, the results of two-dimensional numerical simulations for a typical structure are discussed here. The device structure used has the cross section shown in Fig. 7.1 with a cell half-width of 5.0 μm. The doping profile for the IGBT structure used in the numerical simulations was provided in Fig. 7.37. The widths of the uniformly doped N-base region and the diffused N-buffer layer are 175 and 5 μm, respectively. For the typical case discussed here, a high-level lifetime of 2 μs was used in the N-base region.

The numerical simulations were performed with an abrupt reduction of the gate voltage from 10 to 0 V in 10 ns starting from an on-state current density of 25 A/cm^2. The resulting waveforms obtained from the numerical simulations for the anode voltage and current are shown in Fig. 7.47 for the case of a collector supply voltage of 12,000 V. The collector voltage begins to increase immediately at the end of the gate voltage transient because the P-base/N-base junction (J_2) is already reverse-biased in the on-state. The collector voltage increases non-linearly with the time as predicted by the analytical model until it reaches 12,000 V. This demonstrates that the doping concentration and width of the N-base region are at the optimum value. Unlike the structure discussed in the previous section, no abrupt increase in the collector voltage occurs.

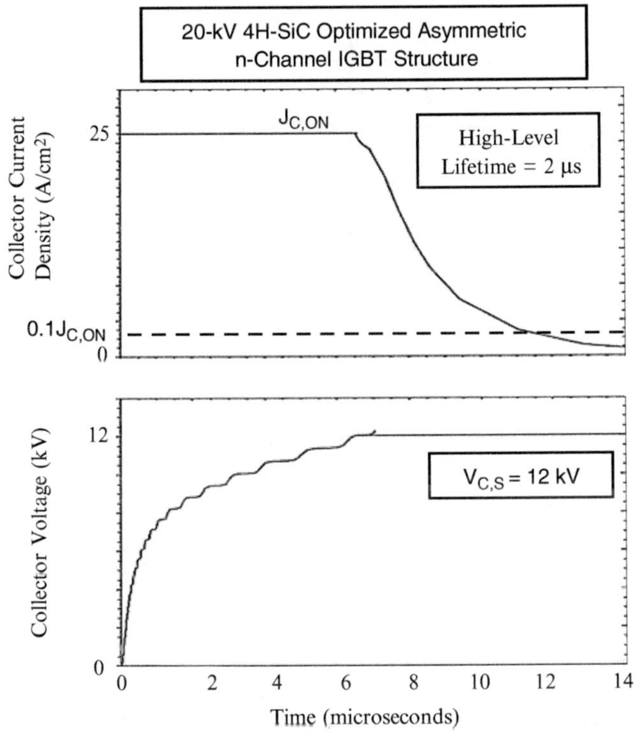

Fig. 7.47 Typical turn-off waveforms for the optimized asymmetric 20-kV n-channel 4H-SiC IGBT structure

The increase in collector voltage occurs at a much smaller rate than predicted by the analytical model with a reach-through time of 6 μs. This is due to the larger hole concentration in the space-charge region than assumed in the analytical model during the early portion of the transient when the collector voltage is less than 1,000 V as can be observed in Fig. 7.48. The larger hole concentration is due to the smaller hole velocity in the space-charge region when the electric field in the space-charge region is insufficient to produce velocity saturation (as assumed in deriving Eq. 7.62). At large collector voltages, the high hole concentration in the space-charge region toward the collector side, where the electric field is small, shortens the width of the space-charge region. This increases the reach-through voltage and prolongs the reach-through time.

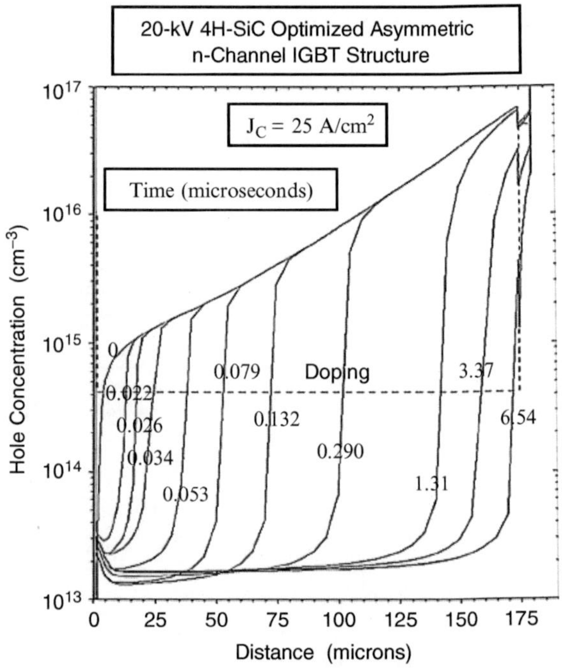

Fig. 7.48 Hole carrier distribution in the optimized 20-kV n-channel asymmetric 4H-SiC IGBT structure for turn-off transient during the voltage rise-time

A one-dimensional view of the minority carrier distribution in the optimized 20-kV asymmetric n-channel 4H-SiC IGBT structure is shown in Fig. 7.48 from the initial steady-state operating point ($t = 0$ μs) to the end of the voltage rise-time ($t = 6.54$ μs). These carrier profiles were taken at $x = 1$ μm through the P-base region. The initial carrier profile has the distribution predicted by the analytical model (see Eq. 7.20) for the IGBT structure. It can be observed from Fig. 7.48 that the carrier distribution in the N-base region near the collector does not change during the collector voltage rise phase. A significant space-charge region begins to form immediately during the turn-off and expands toward the right-hand side demonstrating that there is no storage phase for the IGBT structure. At larger collector voltages, the hole concentration in the space-charge region is about 1.8×10^{13} cm^{-3}, which is consistent with the value for p_{SC} obtained using the analytical model (see Eq. 7.63) with the carriers moving at the saturated drift velocity and an on-state current density of 25 A/cm^2. However, at lower collector bias voltages of below 500 V corresponding to time values less than 0.0853 μs, the hole concentration in the space-charge region is significantly larger (up to 4×10^{13} cm^{-3}). The larger hole concentration produces a faster rate of rise of the collector voltage (see Eq. 7.62) at lower collector voltages. At the collector voltage increases, the hole concentration in the space-charge region reduces making the collector voltage rise at a slower rate. The time taken for the collector voltage to increase to the collector supply voltage is found to be 6.54 μs in the simulations.

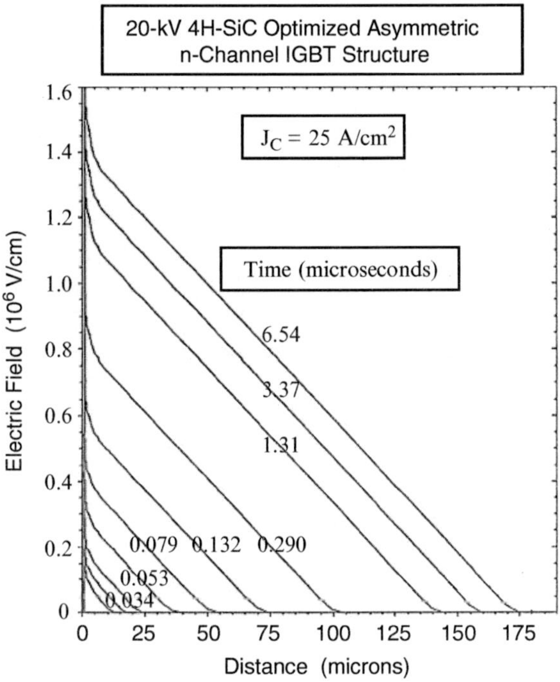

Fig. 7.49 Electric field distribution for the optimized 20-kV asymmetric n-channel 4H-SiC IGBT structure during the voltage rise-time

The evolution of the electric field profile during the collector voltage rise-time for the optimized 20-kV asymmetric 4H-SiC IGBT structure is shown in Fig. 7.49. The electric field has a triangular shape throughout the voltage rise-time. The slope of the electric field profile is greater than that for the structure discussed in the previous chapter due to the larger doping concentration of the N-base region. The space-charge region can be observed to just reach-through the N-base region at time $t = 6.54$ μs when the collector voltage has risen to 12,000 V. This is consistent with the optimized design obtained by using the analytical model.

After the completion of the collector voltage transient, the collector current decays from the initial on-state current density at an exponential rate as shown in Fig. 7.47. A one-dimensional view of the hole carrier distribution in the optimized 20 kV asymmetric n channel 4H-SiC IGBT structure is shown in Fig. 7.50 during the current tail time. The collector voltage was held constant at the collector supply voltage of 12,000-V during this transient. The hole concentration in the stored charge region begins to decrease immediately after the end of the voltage transient due to the recombination process. The recombination of holes in the N-buffer layer during the collector current transient occurs under low-level injection conditions as assumed in the analytical model. The current fall-time obtained from the numerical simulations (see waveform in Fig. 7.47) is found to be 5.0 μs which is twice that predicted by the analytical model. This is due to reduced recombination rate during the initial part of the current decay because the hole injection level is comparable to the doping concentration in the N-buffer layer.

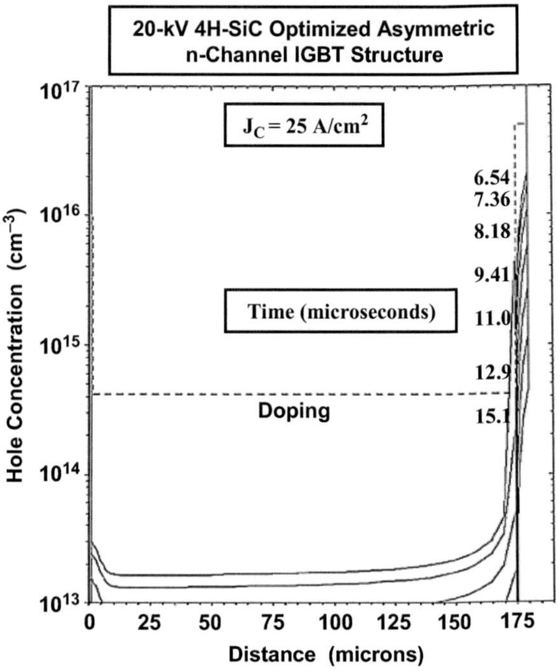

Fig. 7.50 Hole carrier distribution in the optimized 20-kV n-channel asymmetric 4H-SiC IGBT structure for the turn-off transient during the current tail-time

7.2.5 *Lifetime Dependence*

As in the case of the structure discussed in the previous section, it will be assumed that the lifetime in the N-buffer layer is the same as that in the N-base region for the optimized silicon carbide structure. The impact of reducing the lifetime in the drift region on the on-state voltage drop for the optimized 20-kV asymmetric n-channel 4H-SiC IGBT structure was previously discussed in Sect. 7.2.3. The on-state voltage drop increases when the lifetime is reduced as discussed in Sect. 7.2.3. The new analytical model developed for turn-off of the optimized asymmetric IGBT structure presented in Sect. 7.2.4 can be used to analyze the impact of changes to the lifetime in the drift region on the turn-off characteristics.

The collector voltage transients predicted by the analytical model are shown in Fig. 7.51 for the case of the optimized 20-kV asymmetric 4H-SiC IGBT structure operating with an on-state current density of 25 A/cm^2. The voltage rise-time increases when the lifetime is increased because of the larger concentration for the holes in the N-base region that are being removed during the collector voltage transient. The voltage rise-times obtained by using the analytical model are 1.5, 3.3, and 5.8 μs for high-level lifetime values of 1, 2, and 4 μs, respectively. In all cases, the collector voltage increases monotonically to the collector supply voltage of 12,000 V as expected.

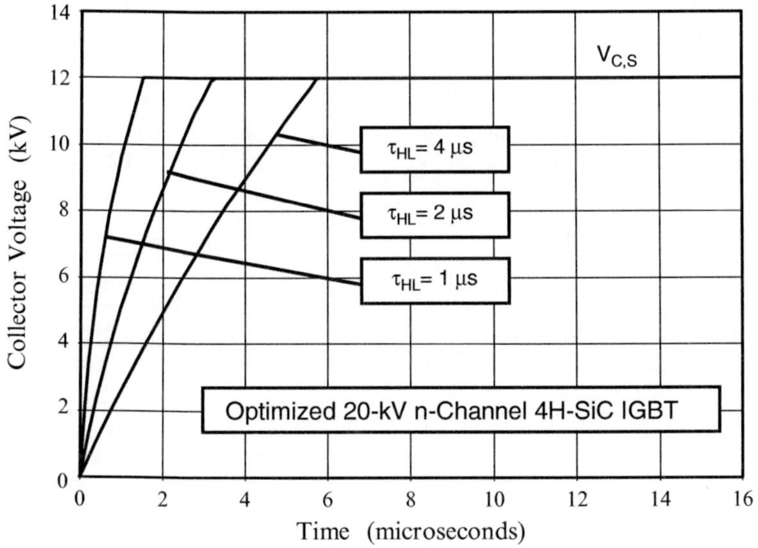

Fig. 7.51 Collector voltage transients during turn-off for the optimized 20-kV asymmetric n-channel 4H-SiC IGBT structure: lifetime dependence

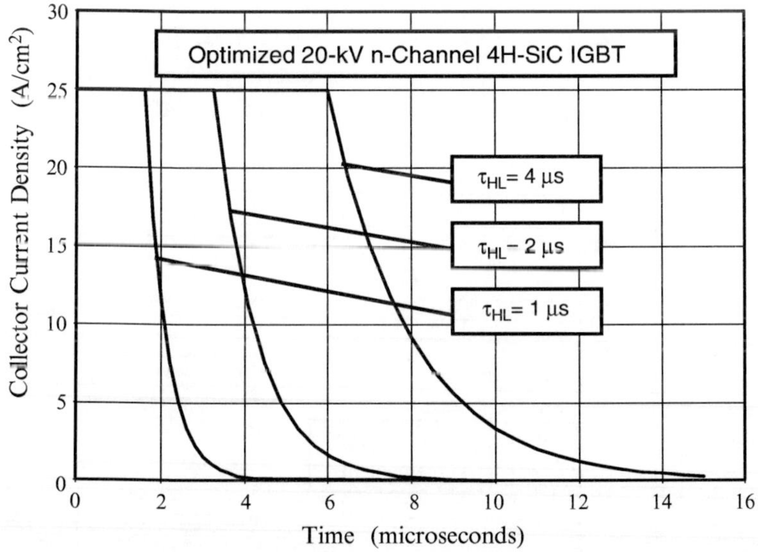

Fig. 7.52 Collector current transients during turn-off for the optimized 20-kV asymmetric n-channel 4H-SiC IGBT structure: lifetime dependence

The collector current transients predicted by the new analytical model are shown in Fig. 7.52. It can be observed that the current transient becomes longer when the lifetime in the N-base region increases. The current fall-time increases when the lifetime is increased because of the reduced recombination rate in the N-buffer layer during the current transient. According to the analytical model, the current fall-times obtained by using the analytical model are 1.2, 2.4, and 4.5 μs for high-level lifetime values of 1, 2, and 4 μs, respectively.

Simulation Example

In order to gain insight into the impact of the lifetime in the N-base region on the operation of the optimized 20-kV asymmetric 4H-SiC IGBT structure, the results of two-dimensional numerical simulations for a typical structure are discussed here. The device structure used has the cross section shown in Fig. 7.1 with a half-cell width of 5 μm. The widths of the N-base and N-buffer layer regions are 175 and 5 μm, respectively. The high-level lifetime in the N-base region was varied between 0.4 and 4 μs. For turning off the IGBT structures, the numerical simulations were performed with gate voltage rapidly ramped down from 10 to 0 V in 10 ns starting from an on-state current density of 25 A/cm^2. The resulting

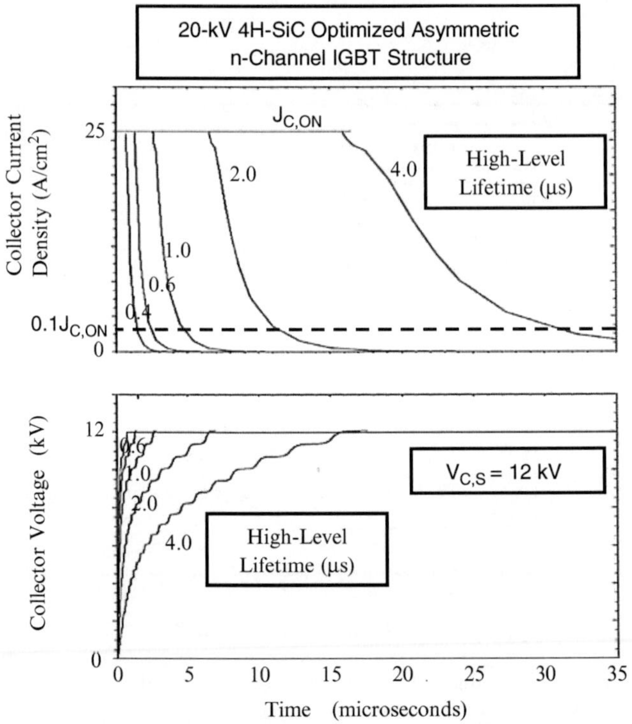

Fig. 7.53 Impact of lifetime on the optimized 20-kV asymmetric n-channel 4H-SiC IGBT turn-off waveforms

waveforms obtained from the numerical simulations for the collector voltage and current are shown in Fig. 7.53 for the case of a collector supply voltage of 12,000 V.

The numerical simulations show a decrease in the time taken for the collector voltage to increase to the collector supply voltage when the lifetime in the N-base region is reduced. In the simulation results, the collector voltage rises to the collector supply voltage of 12,000 V in a single phase as predicted by the analytical model. There is no abrupt change in the collector voltage waveform demonstrating that the structural optimization has eliminated the high [dV/dt] observed for the structure discussed in the previous section.

The numerical simulations of the optimized 20-kV asymmetric n-channel 4H-SiC IGBT structure also show a substantial increase in the collector current fall-time when the lifetime increases. For all the lifetime values, the collector current decays exponentially with time as predicted by the analytical model. The collector current fall-time values obtained from the numerical simulations are 1.0, 1.3, 2.3, 4.7, and 16 μs, for high-level lifetime values of 0.4, 0.6, 1, 2, and 4 μs, respectively. These values are closer to the analytically computed current fall-time if the high-level lifetime in the buffer layer is used.

7.2.6 Switching Energy Loss

The turn-off loss for the optimized asymmetric 4H-SiC IGBT structure during the voltage rise-time interval is different from that for the device structure discussed in the previous section. Since the collector voltage transient for the optimized structure occurs in a single phase until it reaches the collector supply voltage, the energy loss during this interval can be computed using:

$$E_{\mathrm{OFF,V}} = \frac{1}{2} J_{\mathrm{C,ON}} V_{\mathrm{C,S}} t_{\mathrm{V}} \tag{7.68}$$

For the typical switching waveforms for the optimized 20-kV asymmetric n-channel 4H-SiC IGBT structure shown in Fig. 7.47 with a collector supply voltage of 12,000 V, the energy loss per unit area during the collector voltage rise-time is found to be 0.50 J/cm^2 if the on-state current density is 25 A/cm^2.

During the collector current fall-time interval, the collector voltage is constant while the current decreases exponentially with time. The energy loss during the collector current fall-time interval can be computed using:

$$E_{\mathrm{OFF,I}} = J_{\mathrm{C,ON}} V_{\mathrm{S}} \tau_{\mathrm{BL}} \tag{7.69}$$

For the typical switching waveform for the optimized 20-kV asymmetric n-channel 4H-SiC IGBT structure shown in Fig. 7.47 with a collector supply voltage of 12,000 V, the energy loss per unit area during the collector current fall-time is found to be 0.30 J/cm^2 if the on-state current density is 25 A/cm^2. This is same as that for the structure discussed in the previous section. The total energy loss per unit area ($E_{OFF,V} + E_{OFF,I}$) during the turn-off process for the optimized 20-kV asymmetric n-channel 4H-SiC IGBT structure is found to be 0.80 J/cm^2. It can be concluded that, in order to eliminate an abrupt change in the collector voltage for the asymmetric silicon carbide IGBT structure, it is necessary to tolerate an increase in the switching power loss.

Using the results obtained from the numerical simulations, the on-state voltage drop and the total energy loss per cycle can be computed. These values are plotted in Fig. 7.54 to create a trade-off curve to optimize the performance of the optimized 20-kV asymmetric n-channel 4H-SiC IGBT structure by varying the lifetime in the N-base region. Devices used in lower frequency circuits would be chosen from the left-hand side of the trade-off curve while devices used in higher frequency circuits would be chosen from the right-hand side of the trade-off curve. The optimized structure has a superior trade-off curve when compared to the previous structure due to its reduced on-state voltage drop.

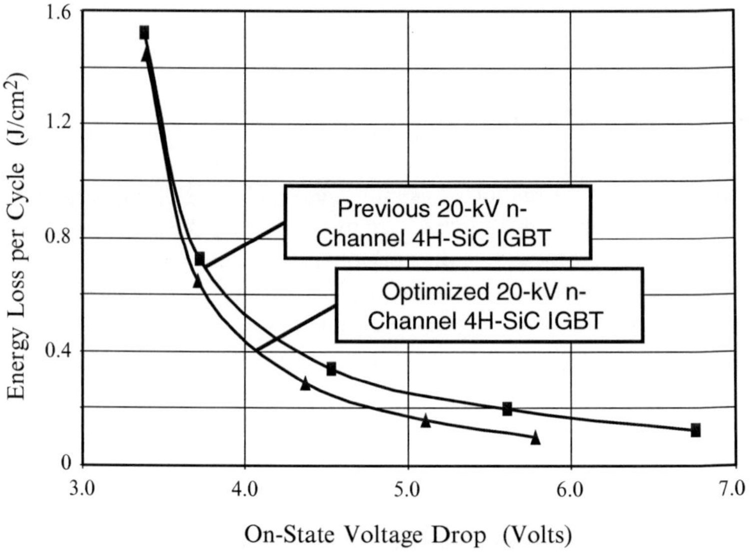

Fig. 7.54 Trade-off curve for the optimized 20-kV asymmetric n-channel 4H-SiC IGBT structure: lifetime dependence. Lifetime in N-base region

7.2.7 *Maximum Operating Frequency*

The maximum operating frequency for operation of the optimized 20-kV asymmetric n-channel 4H-SiC IGBT structure can be obtained by combining the on-state and switching power losses (see Eq. 7.61). In the case of the baseline optimized

20-kV asymmetric n-channel 4H-SiC IGBT device structure with a high-level lifetime of 2 μs in the N-base region, the on-state voltage drop is 3.714 V at an on-state current density of 25 A/cm^2. For the case of a 50% duty cycle, the on-state power dissipation contributes 46 W/cm^2 to the total power loss. For this lifetime value, the energy loss per cycle during the voltage rise-time obtained from the numerical simulations is 0.348 J/cm^2 and the energy loss per cycle during the current fall-time obtained from the numerical simulations is 0.300 J/cm^2. Using a total turn-off energy loss per cycle of 0.648 J/cm^2 in Eq. 7.61 yields a maximum operating frequency of about 240 Hz.

High-Level Lifetime (μs)	On-State Voltage Drop (Volts)	On-State Power Dissipation (W/cm^2)	Energy Loss per Cycle (J/cm^2)	Maximum Operating Frequency (Hz)
4	3.397	42.5	1.45	109
2	3.714	46.4	0.648	237
1	4.376	54.7	0.290	502
0.6	5.106	63.8	0.160	854
0.4	5.781	72.3	0.099	1291

Fig. 7.55 Power loss analysis for the optimized 20-kV asymmetric n-channel 4H-SiC IGBT structure

The maximum operating frequency for the optimized 20-kV asymmetric n-channel 4H-SiC IGBT structure can be increased by reducing the lifetime in the N-base region. Using the results obtained from the numerical simulations, the on-state voltage drop and the energy loss per cycle can be computed. These values are provided in Fig. 7.55 together with the maximum operating frequency as a function of the high level lifetime in the N-base region under the assumption of a 50% duty cycle and a total power dissipation limit of 200 W/cm^2. The maximum operating frequency is plotted in Fig. 7.56 as a function of the high-level lifetime in the N-base region. It can be observed that the maximum operating frequency can be increased up to 1,300 Hz by reducing the high-level lifetime to 0.4 μs. The IGBT is often operated with pulse-width-modulation to synthesize variable frequency output power for motor control. In these applications, the duty cycle can be much shorter than 50%. In this case, the maximum operating frequency for the optimized 20-kV asymmetric n-channel 4H-SiC IGBT structure can be increased. As an example, the maximum operating frequency for the optimized 20-kV asymmetric n-channel 4H-SiC IGBT structure operated at a 10% duty cycle is included in Fig. 7.56. It can be seen that the maximum operating frequency can now approach 2,000 Hz.

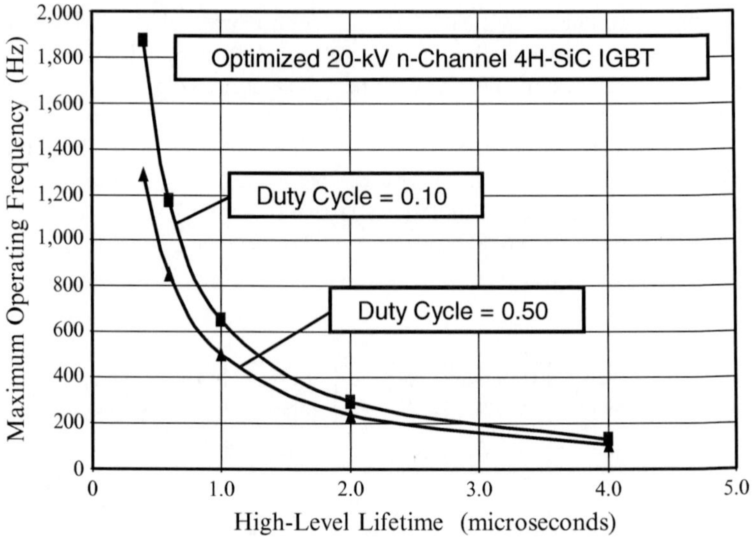

Fig. 7.56 Maximum operating frequency for the optimized 20-kV asymmetric n-channel 4H-SiC IGBT structure

7.3 p-Channel Asymmetric Structure

The asymmetric p-channel silicon carbide IGBT structure with the planar gate architecture is illustrated in Fig. 7.57 with its doping profile. As mentioned at the beginning of this chapter, the asymmetric p-channel silicon carbide IGBT structure has received more attention in the literature than the n-channel structure because of concerns with the high resistance of available P^+ silicon carbide substrates. Since the asymmetric IGBT structure is intended for use in DC circuits, its reverse blocking capability does not have to match the forward blocking capability allowing the use of a P-buffer layer adjacent to the N^+ collector region. The P-buffer layer has a much larger doping concentration than the lightly doped portion of the P-base region. The electric field in the asymmetric IGBT takes a trapezoidal shape allowing supporting the forward blocking voltage with a thinner P-base region. This allows achieving a lower on-state voltage drop and superior turn-off characteristics. As in the case of silicon devices discussed in Chap. 5, the doping concentration of the buffer layer and the lifetime in the P-base (drift-region) region must be optimized to perform a trade-off between on-state voltage drop and turn-off switching losses. Like the asymmetric n-channel silicon carbide IGBT structure, the asymmetric p-channel silicon carbide IGBT structure has uniform doping concentration for the various layers produced by using either epitaxial growth or by using multiple ion-implantation energies to form a box profile.

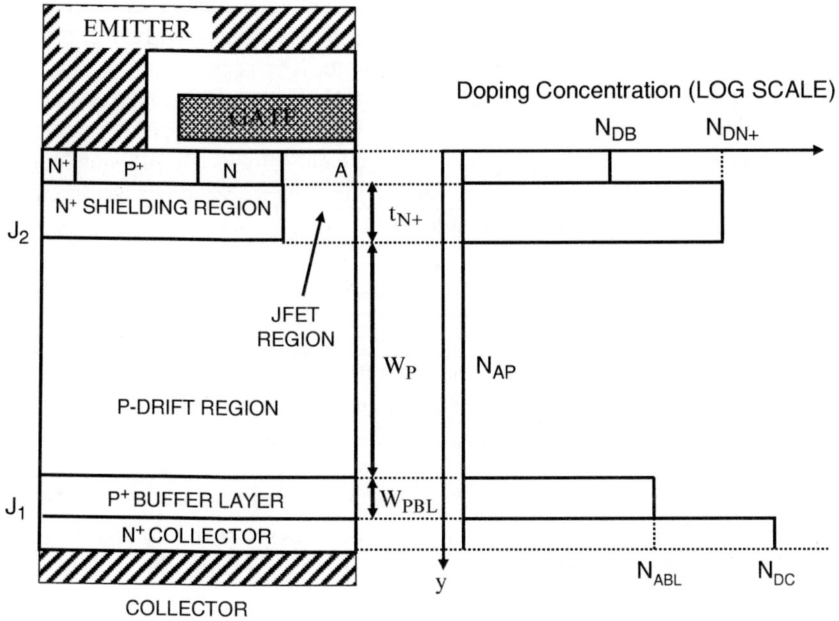

Fig. 7.57 The asymmetric p-channel SiC IGBT structure and its doping profile

7.3.1 Blocking Characteristics

The design of the 20-kV asymmetric 4H-SiC p-channel IGBT structure is discussed in this section. The physics for blocking voltages is similar to the n-channel structure discussed in Sect. 7.1. The electric field distribution within the asymmetric IGBT structure is essentially the same as that illustrated in Fig. 4.3 for the asymmetric GTO structure. Consequently, the design procedure described in Chap. 4 can be applied to the asymmetric IGBT structure. The forward blocking capability of the asymmetric silicon carbide IGBT structure is determined by the open-base transistor breakdown phenomenon. In the case of the asymmetric 4H-SiC p-channel IGBT structure, the maximum blocking voltage occurs when the common base current gain of the NPN transistor becomes equal to unity. For the asymmetric p-channel IGBT structure, the emitter injection efficiency is smaller than unity due to the high doping concentration of the P-buffer layer. The emitter injection efficiency for the N^+ collector/P-buffer junction (J_1) can be obtained by using an analysis similar to that described in the textbook for the bipolar power transistor [11]:

$$\gamma_E = \frac{D_{nPBL} L_{pC} N_{DC}}{D_{nPBL} L_{pC} N_{DC} + D_{pC} W_{PBL} N_{ABL}} \tag{7.70}$$

where D_{nPBL} and D_{pC} are the diffusion coefficients for minority carriers in the P-buffer and N^+ collector regions, N_{DC} and L_{pC} are the doping concentration and diffusion length for minority carriers in the N^+ collector region, and N_{ABL} and W_{PBL} are the doping concentration and width of the P-buffer layer. In determining the diffusion coefficients and the diffusion length, it is necessary to account for impact of the high doping concentrations in the N^+ collector region and P-buffer layer on the mobility. In addition, the lifetime within the highly doped N^+ collector region is reduced due to heavy doping effects, which shortens the diffusion length.

Based upon the above analysis, the open-base transistor breakdown condition for the asymmetric p-channel silicon carbide IGBT structure is given by:

$$\alpha_{NPN} = (\gamma_E \cdot \alpha_T)_{NPN} M = 1 \tag{7.71}$$

Based upon this expression, it can be concluded that the breakdown voltage for the silicon carbide p-channel asymmetric IGBT structure will occur when the multiplication coefficient is slightly above unity. Using the avalanche breakdown criteria, when the multiplication coefficient becomes equal to infinity, as assumed in some papers [7], will lead to significant error in the design of the drift region.

When the collector bias exceeds the reach-through voltage (V_{RT}), the electric field is truncated by the high doping concentration of the N-buffer layer making the un-depleted width of the NPN transistor base region equal to the width of the N-buffer layer. The base transport factor is then given by:

$$\alpha_T = \frac{1}{\cosh(W_{PBL}/L_{nPB})} \tag{7.72}$$

which is independent of the collector bias. Here, $L_{n,PB}$ is the diffusion length for electrons in the P-buffer layer. This analysis neglects the depletion region extension within the P-buffer layer. The diffusion length for electrons (L_{pPB}) in the P-buffer layer depends upon the diffusion coefficient and the minority carrier lifetime in the P-buffer layer. The diffusion coefficient varies with the doping concentration in the P-buffer layer based upon the concentration dependence of the mobility. In addition, the minority carrier lifetime has been found to be dependent upon the doping concentration [12] in the case of silicon devices. Although this phenomenon has not been verified for silicon carbide, it is commonly used when performing numerical analysis of silicon carbide devices. The effect can be modeled by using the relationship:

$$\frac{\tau_{LL}}{\tau_{n0}} = \frac{1}{1 + (N_A/N_{REF})} \tag{7.73}$$

where N_{REF} is a reference doping concentration whose value will be assumed to be $5 \times 10^{16}\ \text{cm}^{-3}$.

The multiplication factor for a P-N junction is given by:

$$M = \frac{1}{1 - (V_C/BV_{PP})^n} \tag{7.74}$$

with a value of $n = 6$ and the avalanche breakdown voltage of the N-base/P-base junction (BV_{PP}) *without the punch-through phenomenon.* In order to apply this formulation to the punch-through case relevant to the asymmetric p-channel silicon carbide IGBT structure, it is necessary to relate the maximum electric field at the junction for the two cases. The electric field at the interface between the lightly doped portion of the P-base region and the P-buffer layer is given by:

$$E_1 = E_m - \frac{qN_A W_P}{\varepsilon_S} \tag{7.75}$$

The applied collector voltage supported by the device is given by:

$$V_C = \left(\frac{E_m + E_1}{2}\right) W_P = E_m W_P - \frac{qN_A}{2\varepsilon_S} W_P^2 \tag{7.76}$$

From this expression, the maximum electric field is given by:

$$E_m = \frac{V_C}{W_P} + \frac{qN_A W_P}{2\varepsilon_S} \tag{7.77}$$

The corresponding equation for the non-punch-through case is:

$$E_m = \sqrt{\frac{2qN_A V_{NPT}}{\varepsilon_S}} \tag{7.78}$$

Consequently, the non-punch-through voltage that determines the multiplication coefficient "M" corresponding to the applied collector bias "V_A" for the punch-through case is given by:

$$V_{NPT} - \frac{\varepsilon_S E_m^2}{2qN_A} = \frac{\varepsilon_S}{2qN_A}\left(\frac{V_C}{W_P} + \frac{qN_A W_P}{2\varepsilon_S}\right)^2 \tag{7.79}$$

The multiplication coefficient for the asymmetric silicon carbide IGBT structure can be computed by using this non-punch-through voltage:

$$M = \frac{1}{1 - (V_{NPT}/BV_{PP})^n} \tag{7.80}$$

The multiplication coefficient increases with increasing collector bias. The open-base transistor breakdown voltage (and the forward blocking capability of the asymmetric IGBT structure) is determined by the collector voltage at which the multiplication factor becomes equal to the reciprocal of the product of the base transport factor and the emitter injection efficiency.

The silicon carbide p-channel asymmetric IGBT structure must have a forward blocking voltage of 22,000 V for a 20-kV rated device. In the case of avalanche breakdown, there is a unique value of 7.0×10^{14} cm^{-3} for the drift region with a width of 186 μm to obtain this blocking voltage. In the case of the asymmetric silicon carbide IGBT structure, it is advantageous to use a much lower doping concentration for the lightly doped portion of the P-base region in order to reduce its width. The strong conductivity modulation of the P-base region during on-state operation favors a smaller thickness for the P-base region independent of its original doping concentration. A doping concentration of 1.5×10^{14} cm^{-3} will be assumed for the P-base region.

The doping concentration of the P-buffer layer must be sufficiently large to prevent reach-through of the electric field to the N^+ collector region. Although the electric field at the interface between the P-base region and the P-buffer layer is slightly smaller than that at the blocking junction (J_2), a worse case analysis can be done by assuming that the electric field at this interface is close to the critical electric field for breakdown in the drift region. The minimum charge in the P-buffer layer to prevent reach-through can be then obtained using:

$$N_{\mathrm{ABL}} W_{\mathrm{PBL}} = \frac{\varepsilon_{\mathrm{S}} E_{\mathrm{C}}}{q} \tag{7.81}$$

Using a critical electric for breakdown in silicon carbide of 2×10^6 V/cm for a doping concentration of 1.5×10^{14} cm^{-3} in the buffer layer, the minimum charge in the P-buffer layer to prevent reach-through for a silicon carbide p-channel asymmetric IGBT structure is found to be 1.07×10^{13} cm^{-2}. A P-buffer layer with doping concentration of 5×10^{16} cm^{-3} and thickness of 5 μm has a charge of 2.5×10^{13} cm^{-2} which satisfies this requirement.

The asymmetric p-channel silicon carbide IGBT structure will be assumed to have an N^+ collector region with doping concentration of 1×10^{19} cm^{-3}. It will be assumed that all the donors are ionized even at room temperature, although the relatively deep donor level in silicon carbide may lead to incomplete dopant ionization. In this case, the emitter injection efficiency computed using Eq. 7.70 is 0.997. When the device is close to breakdown, the entire P-base region is depleted and the base transport factor computed by using Eq. 7.72 in this case is 0.988. In computing these values, a lifetime of 1 μs was assumed for the P-base region resulting in a lifetime of 0.5 μs in the P-buffer layer due to the scaling according to Eq. 7.73. Based upon Eq. 7.71, open-base transistor breakdown will then occur when the multiplication coefficient becomes equal to 1.02 for the above values for the injection efficiency and base transport factor. In comparison with the n-channel asymmetric IGBT structure, the multiplication factor corresponding to open-base transistor breakdown has a much smaller value for the p-channel device. Consequently, it becomes necessary to utilize a thicker drift region for the p-channel device when compared with the n-channel device.

The forward blocking capability for the silicon carbide p-channel asymmetric IGBT structure can be computed by using Eq. 7.71 for various widths for the P-base region. The analysis requires determination of the voltage V_{NPT} by using Eq. 7.79 for each width of the P-base region. The resulting values for the forward blocking voltage are plotted in Fig. 7.58. From this graph, a forward blocking voltage of 22,300 V can be obtained by using a P-base region width of 220 μm. This value is substantially larger than the N-base width of 160 μm for the n-channel device. The larger drift region width required for the p-channel device increases its on-state voltage drop and the stored charge making its trade-off curve worse than for the n-channel device structure.

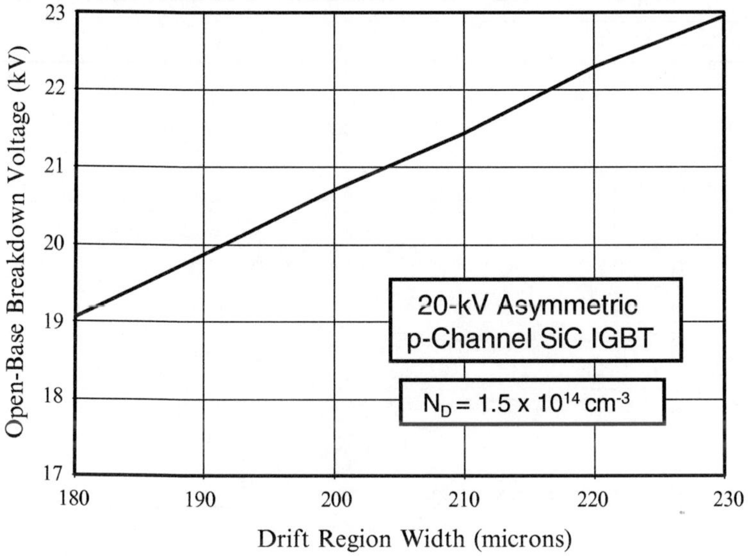

Fig. 7.58 Drift region width optimization for the 20-kV asymmetric p-channel 4H-SiC IGBT structure

Simulation Example

The results of two-dimensional numerical simulations on the 20-kV 4H-SiC asymmetric p-channel IGBT structure are described here to provide a more detailed understanding of the underlying device physics and operation during the blocking mode. For the numerical simulations, the half-cell structure with a width ($W_{Cell}/2$) of 5 μm, as illustrated in Fig. 7.57, was utilized as representative of the structure. The device used for the numerical simulations had a drift region doping concentration of 1.5×10^{14} cm^{-3} and thickness of 220 μm below the N$^+$ shielding region. The N$^+$ shielding region extended from a depth of 0.2–1.0 μm with a doping concentration of 1×10^{19} cm^{-3}. The N-base and P$^+$ source regions were formed within the 0.2 μm of the P-drift region located above the N$^+$ region. The doping concentration of the N-base region was 2×10^{16} cm^{-3} to achieve the desired threshold voltage. Due to the low doping concentration in the drift region for the 20-kV devices, the uniform doping concentration in the JFET region was enhanced to 1×10^{16} cm^{-3} as is usually required for silicon devices. The enhanced doping concentration was extended to 0.5 μm below the N$^+$ shielding region. This is similar to the current enhancement layer (CEL) utilized in high voltage silicon carbide IGBT structures [13]. The charge in this layer must be sufficiently small to prevent degradation of the breakdown voltage at the junction between the N$^+$ shielding region and the drift region.

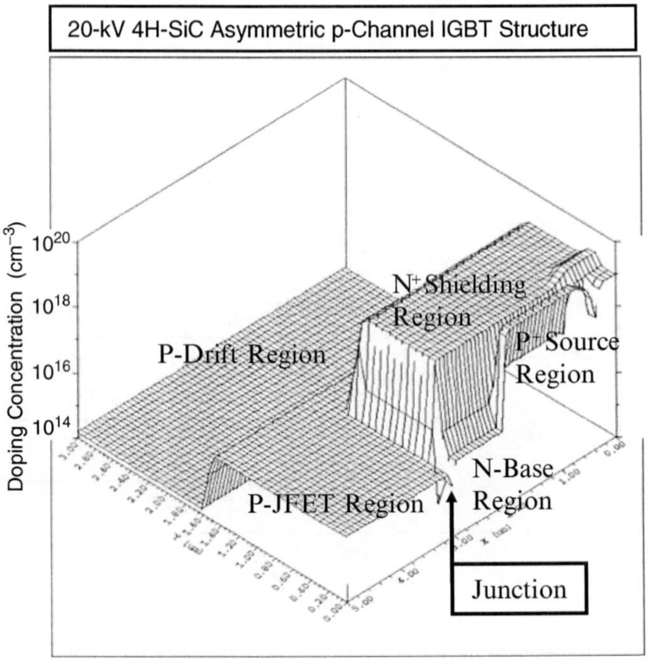

Fig. 7.59 Doping distribution in the 4H-SiC planar asymmetric p-channel IGBT structure

A three-dimensional view of the doping distribution in the 20-kV 4H-SiC asymmetric p-channel IGBT structure is shown in Fig. 7.59 with the upper surface of the structure located on the right-hand side in order to display the doping concentration in the vicinity of the channel. The highly doped N$^+$ shielding region with doping concentration of 1×10^{19} cm^{-3} is prominently located just below the surface. The N-base region can be observed to have a much lower doping concentration of 2×10^{16} cm^{-3}. The JFET region can be observed to have a lower doping concentration of 1×10^{16} cm^{-3}. The junction between the N-base region and P-JFET region is indicated in the figure. The doping concentration of the P-drift region is less than that of the JFET region as expected to achieve the desired 20-kV breakdown voltage. It can be seen that the enhanced JFET doping is extended below the N$^+$ shielding region to a depth of 1.5 μm from the surface.

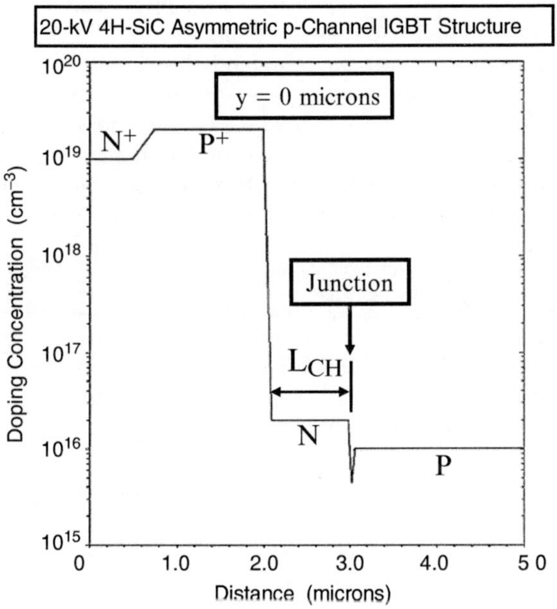

Fig. 7.60 Lateral doping profile for the 4H-SiC asymmetric p-channel IGBT structure

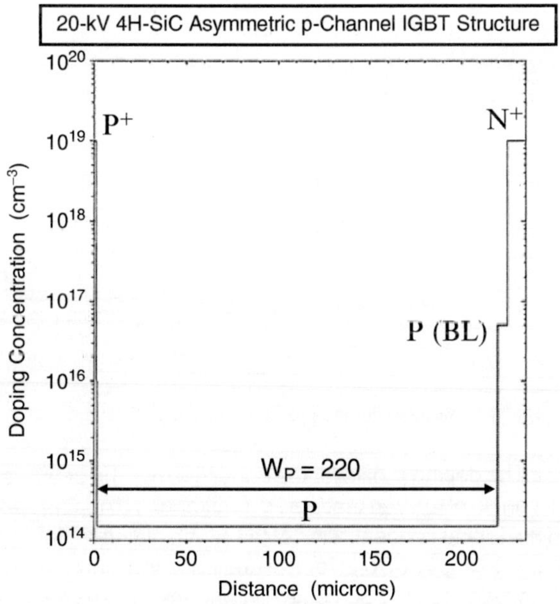

Fig. 7.61 Vertical doping profile in the 4H-SiC asymmetric p-channel IGBT structure

The lateral doping profile taken along the surface of the 20-kV 4H-SiC asymmetric p-channel IGBT structure is shown in Fig. 7.60. From the profile, it can be observed that the channel extends from 2 to 3 μm creating a channel length of 1 μm in the N-base region. The doping concentration of the JFET region is 1×10^{16} cm^{-3} while that for the N-base region is 2×10^{16} cm^{-3}. The P$^+$ source region and the N$^+$ contact region for shorting the source to the base region are visible on the left-hand side. All the regions were defined with uniform doping with abrupt interfaces between them due to the low diffusion rates for dopants in 4H-SiC material.

The vertical doping profile taken through the P$^+$ source region of the 20-kV 4H-SiC asymmetric p-channel IGBT structure is provided in Fig. 7.61. It can be observed that the P-drift region has a doping concentration of 1.5×10^{14} cm^{-3} and thickness of 220 μm. The P-buffer layer located at the collector junction has a doping concentration of 5×10^{16} cm^{-3} and thickness of 5 μm. The doping concentration of the N$^+$ collector region is 1×10^{19} cm^{-3}.

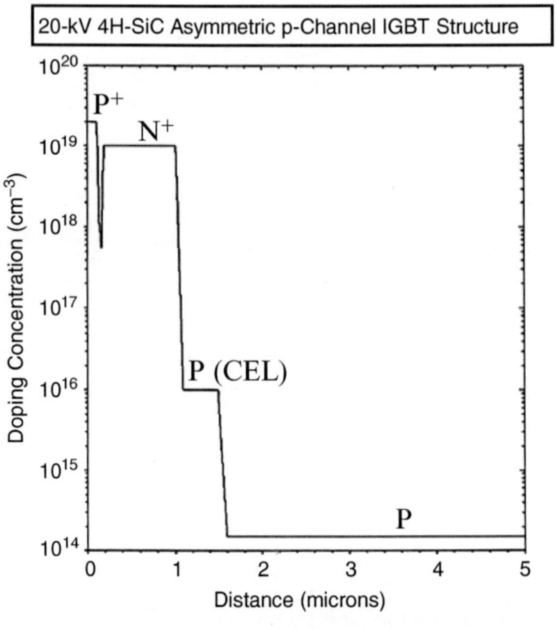

Fig. 7.62 Vertical doping profile in the 4H-SiC asymmetric p-channel IGBT structure

The doping profile for the upper 5-μm of the device structure is shown in Fig. 7.62. It can be observed that the P$^+$ emitter has a doping concentration of 2×10^{19} cm^{-3}. The doping concentration of the N$^+$ shielding region is 1×10^{19} cm^{-3} and it extends from a depth of 0.2 to 1.0 μm from the surface. The N-base region is located between the P$^+$ source region and the N$^+$ shielding region. The P-type CEL layer located below the N$^+$ shielding region has a doping concentration of 1×10^{16} cm^{-3} and extends to a depth of 1.5 μm.

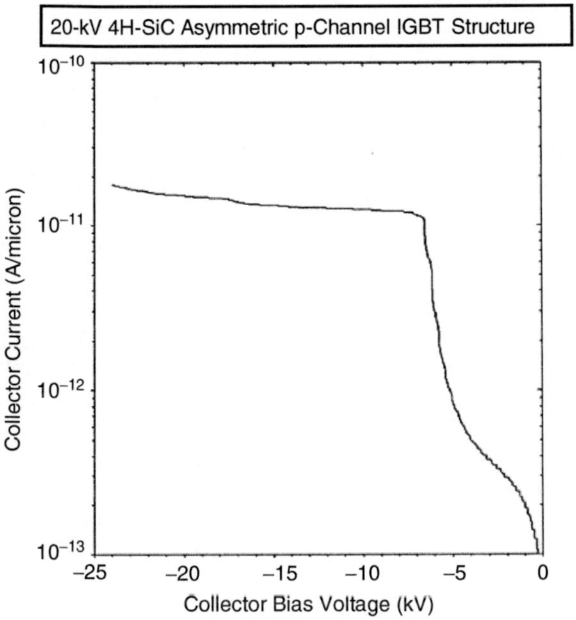

Fig. 7.63 Blocking characteristics of the 4H-SiC asymmetric p-channel IGBT structure

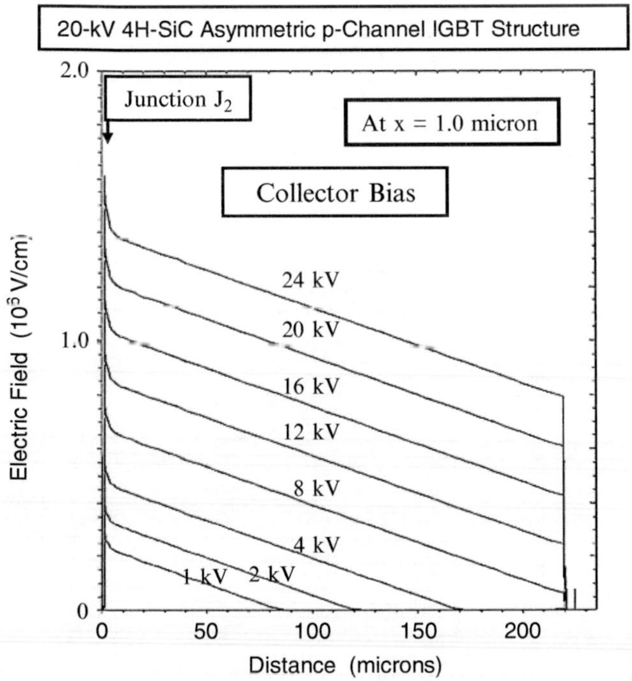

Fig. 7.64 Electric field distribution in the 4H-SiC asymmetric p-channel IGBT structure

The blocking characteristics for the 20-kV 4H-SiC asymmetric p-channel IGBT structure at room temperature (300 K) cannot be determined by numerical simulations of the cell structure due to the very low intrinsic carrier concentration in silicon carbide. For didactic purposes, the blocking characteristics were therefore obtained at 800 K by increasing the collector voltage while using zero gate bias. The resulting blocking characteristic is shown in Fig. 7.63. It can be seen that leakage current increases with increasing collector bias voltage until it reaches about 6,000 V. This is in good agreement with the reach-through voltage of 5,600 V obtained by using Eq. 4.2. Prior to reach-through, the leakage current increases due to the increasing width of the depletion region and the increasing current gain of the PNP transistor in accordance with Eq. 4.5. After reach-through, the leakage current becomes nearly independent of collector voltage.

The potential distribution within the 20-kV 4H-SiC asymmetric p-channel IGBT structure is very similar to that shown previously (see Fig. 7.8) for the n-channel device structure. They are therefore not shown for the p-channel device structure. The electric field distribution in the 20-kV 4H-SiC asymmetric p-channel IGBT structure is shown in Fig. 7.64 through the junction J_2 at various collector bias voltages. It can be observed that the thicker drift region of the p-channel structure has reduced the electric field at junction J_2 which reduces the multiplication factor allowing the device to support the desired blocking voltage. The electric field becomes trapezoidal in shape as expected for an asymmetric blocking structure due to the punch-through of the depletion region with the P-buffer layer when the collector voltage exceeds 6,000 V.

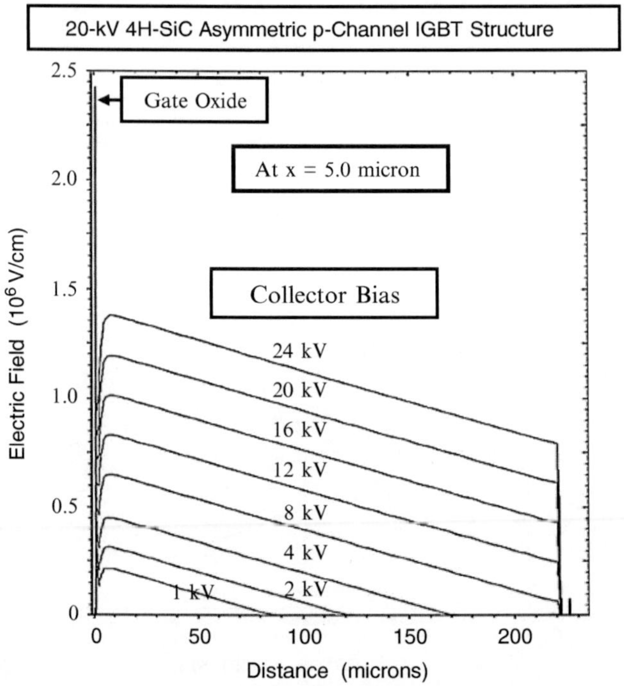

Fig. 7.65 Electric field distribution in the 4H-SiC asymmetric p-channel IGBT structure

The electric field distribution at the middle of the JFET region of the 20-kV 4H-SiC asymmetric p-channel IGBT structure is shown in Fig. 7.65 to allow examination of the electric field in the gate oxide. The suppression of the electric field in the JFET region by the N^+ shielding region greatly reduces the electric field in the semiconductor at its surface to about 1×10^6 V/cm at a collector bias of 20 kV when compared with the maximum electric field of 1.4×10^6 V/cm at the junction between the N^+ shielding region and the drift region. This produces a reduced electric field in the gate oxide as well. The electric field in the gate oxide has its highest value of 2.4×10^6 V/cm at the middle point of the JFET region. These results clearly demonstrate that the p-channel asymmetric silicon carbide IGBT structure operates in a similar manner to the n-channel device structure if the drift region thickness is appropriately designed.

7.3.2 *On-State Voltage Drop*

The minority carrier (electron) distribution profiles for the p-channel asymmetric silicon carbide IGBT structure can be expected to be governed by the same high-level injection physics previously described for the n-channel structure in Sect. 7.1. Consequently, in this section, the characteristics of the 20-kV p-channel asymmetric silicon carbide IGBT structure will be described using only the results of numerical simulations.

Simulation Results

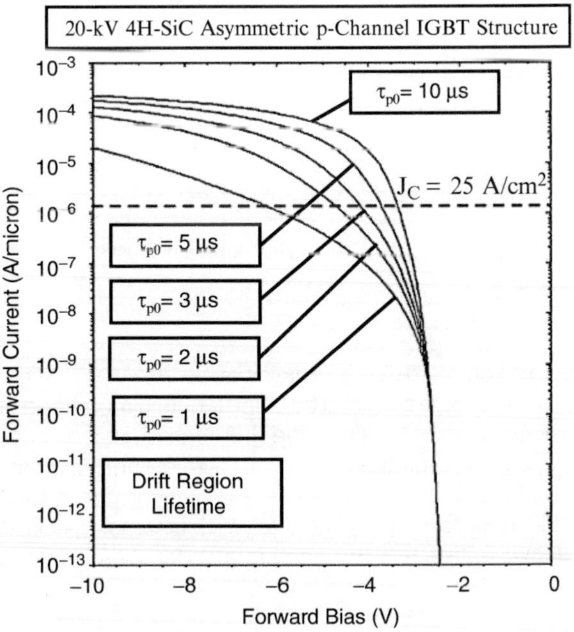

Fig. 7.66 On-state characteristics of the 20-kV asymmetric p-channel IGBT structure: lifetime dependence

The results of two-dimensional numerical simulations for the 20-kV asymmetrical p-channel 4H-SiC IGBT structure are described here. The total width ($W_{CELL}/2$) of the structure, as shown by the cross section in Fig. 7.57, was 5 μm (area = 5 × 10^{-8} cm^{-2}). A JFET region width (W_{JFET}) of 4 μm was used with a gate oxide thickness of 500 Å. The device used for the numerical simulations had a drift region doping concentration of 1.5 × 10^{14} cm^{-3} and thickness of 220 μm below the N^+ shielding region. The N^+ shielding region extended from a depth of 0.2–1.0 μm with a doping concentration of 1 × 10^{19} cm^{-3}. The N-base and P^+ source regions were formed within the 0.2 μm of the P-drift region located above the N^+ region. The doping concentration of the N-base region was 2 × 10^{16} cm^{-3} to achieve the desired threshold voltage. Due to the low doping concentration in the drift region for the 20-kV devices, the uniform doping concentration in the JFET region was enhanced to 1 × 10^{16} cm^{-3} as is usually required for silicon devices. The enhanced doping concentration was extended to 0.5 μm below the N^+ shielding region. The doping profiles for the structure were previously provided in Sect. 7.3.1. The inversion layer mobility for holes was adjusted to between 15 and 20 cm^2/V-s to match values reported for devices in the literature.

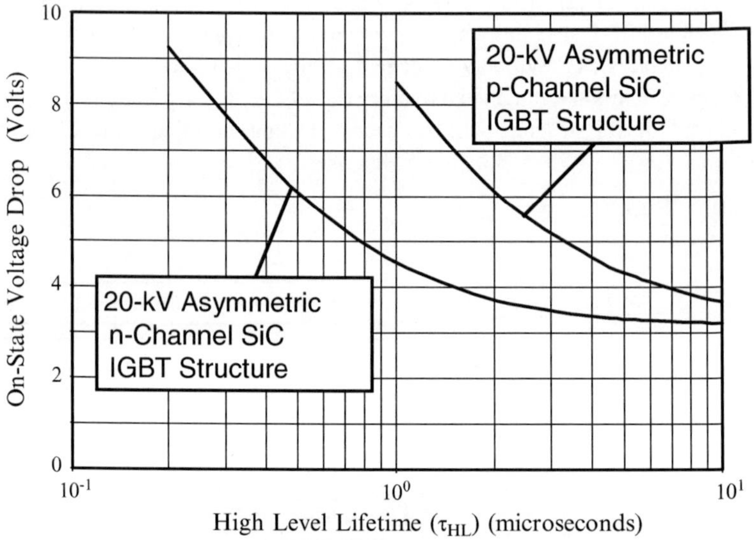

Fig. 7.67 On-state voltage drop for the 20-kV asymmetric p-channel IGBT structure: drift region lifetime dependence

The on-state characteristics of the 20-kV asymmetrical p-channel 4H-SiC IGBT structure were obtained by using a gate bias voltage of 10 V for the case of various values for the lifetime in the drift region. This device structure has a buffer layer doping concentration of 5 × 10^{16} cm^{-3} and thickness of 5 μm. The characteristics obtained from the numerical simulations are shown in Fig. 7.66. The current initially increases exponentially with increasing collector bias. At current densities above 0.2 A/cm^2, the non-state voltage drop begins to increase more rapidly.

The on-state voltage drop increases as expected with reduction of the lifetime (τ_{p0}, τ_{n0}) indicated in the figure. The on-state voltage drop at a hole lifetime (τ_{p0}) value of 2 μs is found to be 4.65 V at an on-state current density of 25 A/cm^2.

The variation of the on-state voltage drop as a function of the lifetime in the drift region obtained by using the numerical simulations is compared with that for the n-channel device structure in Fig. 7.67. It can be clearly seen that the n-channel IGBT structure has a lower on-state voltage drop than that of the p-channel device structure for all lifetime values.

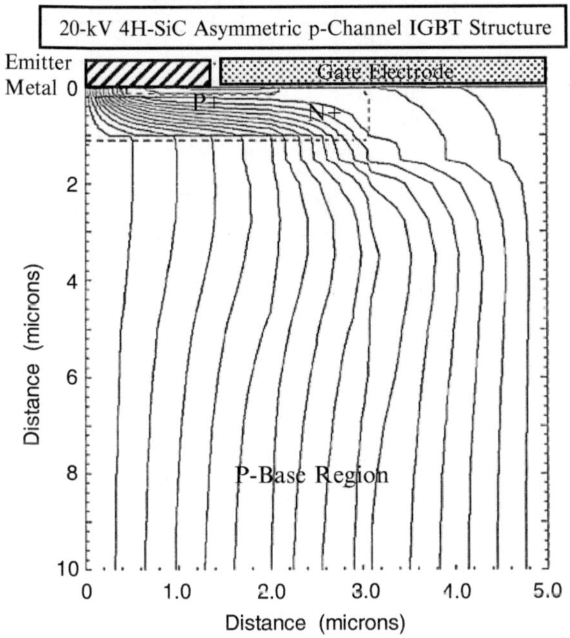

Fig. 7.68 On-state current distribution in the 20-kV asymmetric p-channel IGBT structure

It is insightful to examine the current distribution within the 20-kV asymmetrical p-channel 4H-SiC IGBT structure during on-state operation. The current flow-lines within the device structure are shown in Fig. 7.68 for the case of a low-level lifetime of 2 μs in the P-base region. In contrast with the n-channel structure (see Fig. 7.19), it can be observed that very little current flows through the channel in the 20-kV asymmetrical p-channel 4H-SiC IGBT structure. This is due to the much larger gain of the NPN transistor in the p-channel 4H-SiC IGBT structure due to the larger mobility for electrons. The current flow via the channel and the N$^+$ shielding region can also be examined using separate electrodes attached to the P$^+$ emitter region and the N$^+$ shielding region, although one electrode is typically used in the actual device structure. These currents are provided in Fig. 7.69 for the case of the above structure. At the on-state current density of 25 A/cm^2, it can be observed that the electron current flowing via the electrode connected to the N$^+$ shielding region is about 90% of the collector current.

Fig. 7.69 Current distribution in the 20-kV asymmetric p-channel IGBT structure during the on-state

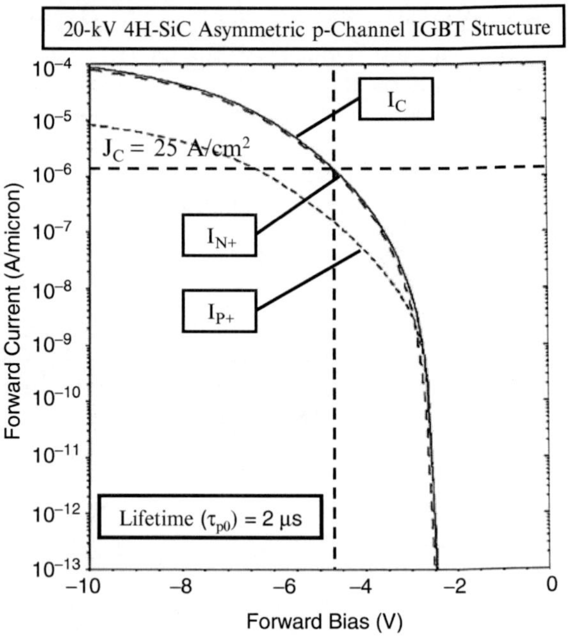

Fig. 7.70 On-state carrier distribution in the 20-kV asymmetric p-channel IGBT structure: lifetime dependence

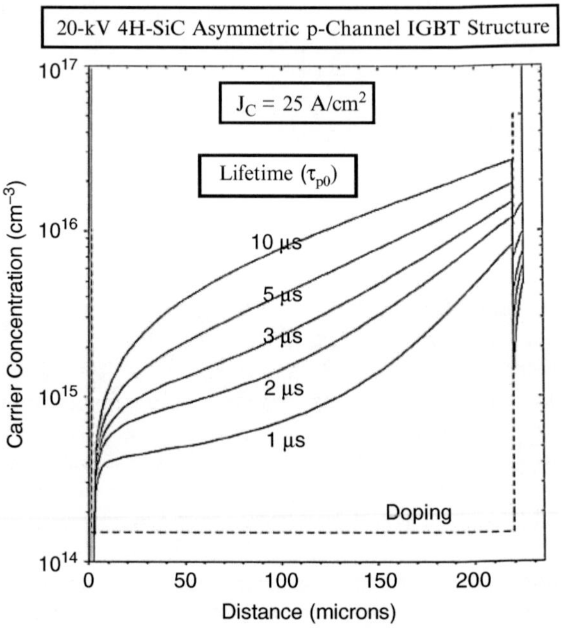

The on-state voltage drop for the IGBT structure is determined by the distribution of carriers injected into the drift region producing the desired reduction of its resistance. The electron distribution in the 20-kV asymmetrical p-channel 4H-SiC IGBT structure is provided in Fig. 7.70 for five cases of the lifetime (τ_{p0}, τ_{n0}) in the drift region. It can be observed that the injected carrier density is two orders of magnitude larger than the doping concentration on the collector side providing the desired conductivity modulation of the drift region. However, the injected carrier density for the p-channel structure is an order of magnitude lower than that observed for the n-channel structure (see Fig. 7.17). This results in a larger on-state voltage drop for the 20-kV asymmetrical p-channel 4H-SiC IGBT structure when compared with the n-channel device structure.

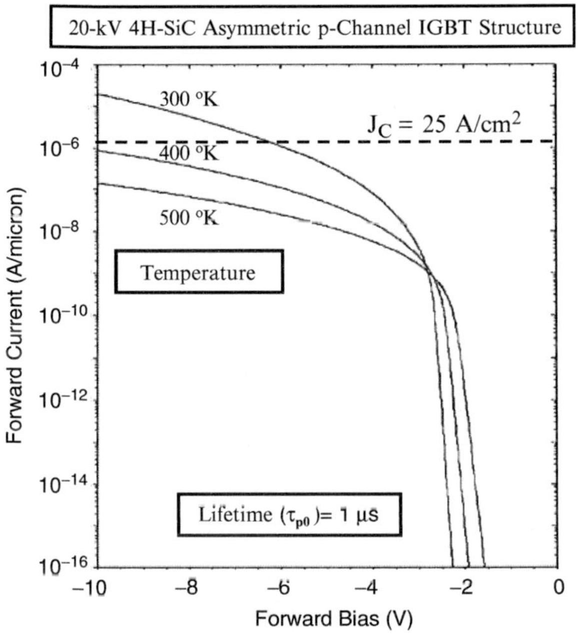

Fig. 7.71 On-state characteristics of the 20-kV asymmetric p-channel IGBT structure: temperature dependence

The temperature dependence of the on-state characteristics of the 20-kV asymmetrical p-channel 4H-SiC IGBT structure can be observed in Fig. 7.71 for the case of a lifetime of 1 μs in the P-base region. It can be seen that the knee voltage reduces with increasing temperature while the resistance within the device increases [11]. This demonstrates that the p-channel silicon carbide IGBT structures display the typical behavior for all IGBT structures. The increase in the on-state voltage drop with temperature for the 20-kV asymmetrical p-channel 4H-SiC IGBT structure is quite severe. However, it has been reported that the on-state characteristics do not change with increasing temperature for 20-kV asymmetrical p-channel 4H-SiC IGBT devices [6] which is contrary to IGBT theory.

7.3.3 Turn-Off Characteristics

The turn-off behavior for the asymmetric p-channel silicon carbide IGBT structure can be expected to be similar to that for the n-channel structure as discussed in Sect. 7.1.3. The same theoretical analysis should therefore be applicable for both devices. Consequently, in this section, the turn-off characteristics of the 20-kV p-channel asymmetric silicon carbide IGBT structure will be described using only the results of numerical simulations.

Simulation Example

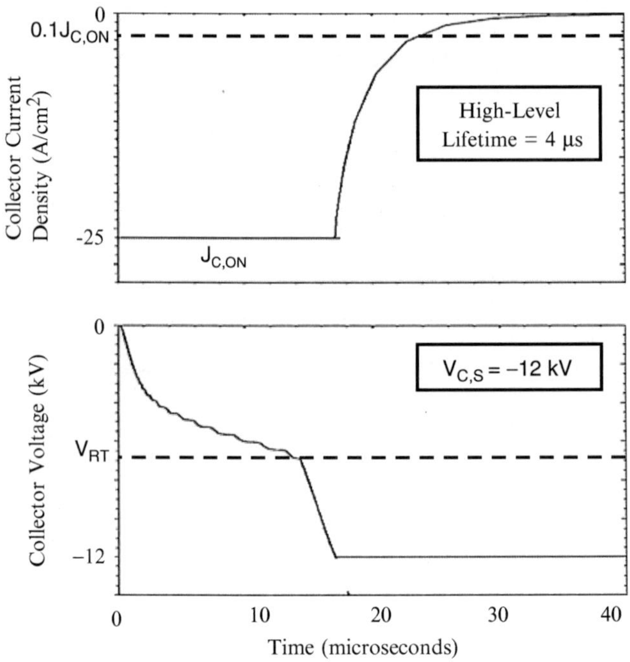

Fig. 7.72 Typical turn-off waveforms for the asymmetric 20-kV p-channel 4H-SiC IGBT structure

In order to gain insight into the operation of the asymmetric 20-kV p-channel 4H-SiC IGBT structure during its turn-off, the results of two-dimensional numerical simulations for a typical structure are discussed here. The device structure used has the cross section shown in Fig. 7.57 with a cell half-width of 5.0 μm. The doping profile for the IGBT structure used in the numerical simulations was provided in Figs. 7.60 and 7.61. The widths of the uniformly doped P-base region and the diffused P-buffer layer are 220 and 5 μm, respectively. For the typical case discussed here, a high-level lifetime of 4 μs was used in the P-base region.

The numerical simulations were performed with by an abrupt reduction of the gate voltage from −10 to 0 V in 10 ns starting from an on-state current density of 25 A/cm^2. The resulting waveforms obtained from the numerical simulations for the anode voltage and current are shown in Fig. 7.72 for the case of a collector supply voltage of −12,000 V. The collector voltage begins to increase immediately at the end of the gate voltage transient because the N-base/P-base junction (J_2) is already reverse biased in the on-state. The collector voltage increases non-linearly with the time as predicted by the analytical model until it reaches about 6,000 V. It then increases at a rapid rate as predicted by the analytical model. The rate of rise of the collector voltage during the second part of the transient is found to be 1.7 × 10^9 V/s.

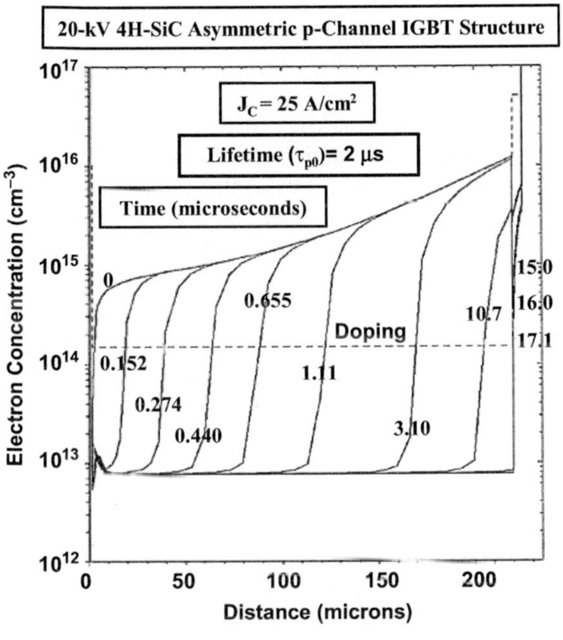

Fig. 7.73 Electron carrier distribution in the 20-kV asymmetric p-channel 4H-SiC IGBT structure for turn-off transient during the voltage rise-time

A one-dimensional view of the minority (electron) carrier distribution in the 20-kV asymmetric p-channel 4H-SiC IGBT structure is shown in Fig. 7.73 from the initial steady-state operating point (t = 0 µs) to the end of the voltage rise-time (t = 17.1 µs). These carrier profiles were taken at x = 1 µm through the N$^+$ shielding region. The initial carrier profile has the distribution predicted by the analytical model (see Eq. 7.20) for the IGBT structure. It can be observed from Fig. 7.73 that the carrier distribution in the P-base region near the collector does not change during the collector voltage-rise phase. A significant space-charge region begins to form immediately during the turn-off and expands toward the right-hand side demonstrating that there is no storage phase for the p-channel IGBT structure. At larger collector voltages, the electron concentration in the space-charge region is about 8×10^{12} cm^{-3}, which is consistent with the value for n_{SC} obtained using the analytical model (see Eq. 7.42) with the electrons moving at the saturated drift velocity of 1.8×10^{7} cm/s and an on-state current density of 25 A/cm^2. However, at lower collector bias voltages of below 500 V corresponding to time values less than 0.44 μs, the electron concentration in the space-charge-region is significantly larger (up to 2×10^{13} cm^{-3}). The larger electron concentration produces a faster rate of rise of the collector voltage (see Eq. 7.41) at lower collector voltages. As the collector voltage increases, the electron concentration in the space-charge region reduces making the collector voltage rise at a slower rate. At large collector voltages, the high electron concentration in the space-charge region toward the collector side, where the electric field is small, shortens the width of the space-charge region. This increases the reach-through voltage and prolongs the reach-through time. The time taken for the collector voltage to increase to the reach-through voltage is found to be 14 μs in the simulations.

Note that the electron concentration at the collector junction (J_1) reduces abruptly when the space-charge region reaches-through. This was taken into account during analytical modeling by development of Eq. 7.47. The electron profile in the P-buffer layer then remains unchanged during the rest of the voltage transient because the collector current density is constant.

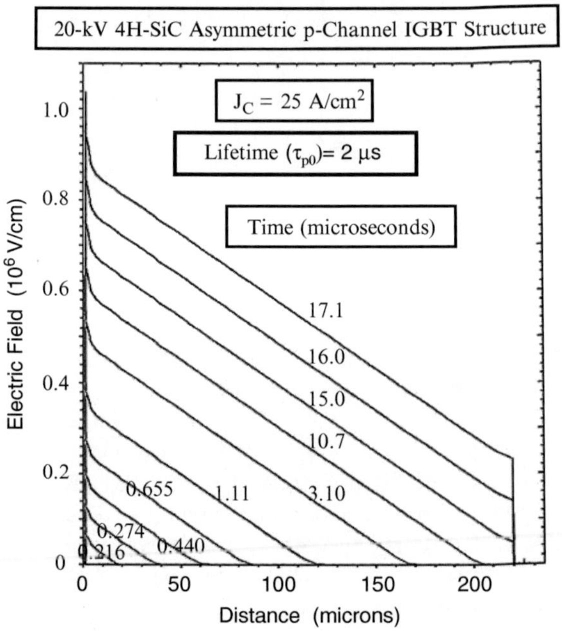

Fig. 7.74 Electric field distribution for the 20-kV asymmetric p-channel 4H-SiC IGBT structure during the voltage rise-time

The evolution of the electric field profile during the collector voltage rise-time for the 20-kV asymmetric p-channel 4H-SiC IGBT structure is shown in Fig. 7.74. The electric field has a triangular shape until the space-charge region reaches-through to the P-buffer layer at time 14 μs. The collector voltage has risen to about 7,000 V at this time. The electric field profile then takes a trapezoidal form as expected due to the high doping concentration in the P-buffer layer. This is consistent with the assumptions used to develop the analytical model.

After the completion of the collector voltage transient, the collector current decays from the initial on-state current density at an exponential rate as shown in Fig. 7.72. A one-dimensional view of the electron carrier distribution in the 20-kV asymmetric p-channel 4H-SiC IGBT structure is shown in Fig. 7.75 during the current tail time. The collector voltage was held constant at the collector supply voltage of −12,000 V during this transient. The electron concentration in the stored charge region begins to decrease immediately after the end of the voltage transient due to the recombination process. The recombination of electrons in the P-buffer layer during the collector current transient occurs under low-level injection conditions as assumed in the analytical model. The current fall-time obtained from the numerical simulations (see waveform in Fig. 7.72) is found to be 6.5 μs.

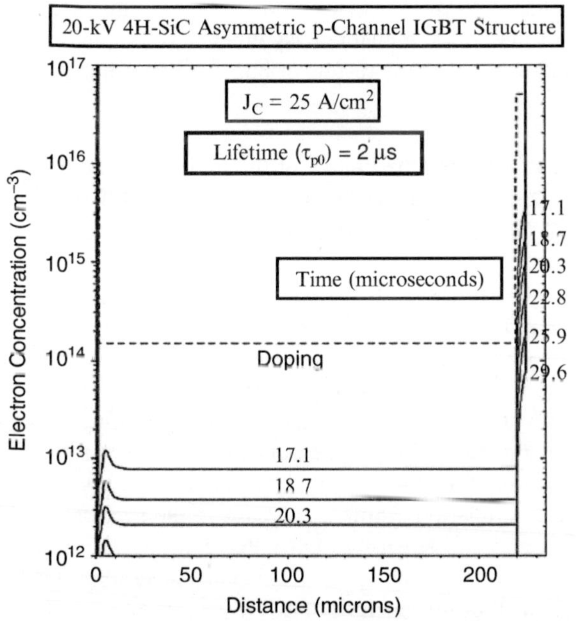

Fig. 7.75 Electron carrier distribution in the 20-kV asymmetric p-channel 4H-SiC IGBT structure for the turn-off transient during the current tail-time

During the current fall-time, unlike in the case of the 5-kV asymmetric silicon IGBT structure, the electric field profile remains essentially the same at the shape corresponding to time 17.1 μs in Fig. 7.74 for the 20-kV asymmetric p-channel 4H-SiC IGBT structure. This occurs because the electron concentration in the space-charge region is much smaller than the doping concentration for the 20-kV asymmetric p-channel 4H-SiC IGBT structure.

7.3.4 Lifetime Dependence

From an applications perspective, the optimization of the power losses for the IGBT structure requires performing a trade-off between the on-state voltage drop and the switching losses. One approach to achieve this is by adjusting the lifetime in the drift (P-base) region. A reduction of the lifetime in the drift region also alters the lifetime in the P-buffer layer in the case of silicon devices. However, the relationship between the lifetime in the drift region and the buffer layer has not yet been established for silicon carbide devices. Consequently, it will be assumed that the lifetime in the P-buffer layer is the same as that in the P-base region for silicon carbide structures. The turn-off behavior for the asymmetric p-channel silicon carbide IGBT structure can be expected to be similar to that for the n-channel structure as discussed in Sect. 7.1.3. The same theoretical analysis should therefore be applicable for both devices. Consequently, in this section, the turn-off characteristics of the 20-kV p-channel asymmetric silicon carbide IGBT structure will be described using only the results of numerical simulations.

Simulation Example

In order to gain insight into the impact of the lifetime in the P-base region on the operation of the 20-kV asymmetric p-channel 4H-SiC IGBT structure, the results of two-dimensional numerical simulations for a typical structure are discussed here. The device structure used has the cross section shown in Fig. 7.57 with a half-cell width of 5 μm. The widths of the P-base and P-buffer layer regions are 220 and 5 μm, respectively. The high-level lifetime in the P-base region was varied between 1 and 10 μs. For turning off the IGBT structures, the numerical simulations were performed with gate voltage rapidly ramped down from −10 to 0 V in 10 ns starting from an on-state current density of 25 A/cm^2. The resulting waveforms obtained from the numerical simulations for the collector voltage and current are shown in Fig. 7.76 for the case of a collector supply voltage of −12,000 V.

The numerical simulations show a decrease in the time taken for the collector voltage to increase to the reach-through voltage when the lifetime in the P-base region is reduced. In the simulation results, the reach-through voltage remains independent of the lifetime in the N-base region as predicted by the analytical model. After reach-through of the space-charge region occurs, the collector voltage increases linearly with time. The [dV/dt] values for the collector voltage transients increase with reduced lifetime in the drift region as predicted by the analytical model. The [dV/dt] values are 0.83, 1.2, 2.5, 3.3, and 5.0 $\times$ 10^9 V/s for high-level lifetime values of 10, 6, 4, 2, and 1 μs, respectively.

The numerical simulations of the 20-kV asymmetric p-channel 4H-SiC IGBT structure also show a substantial increase in the collector current fall-time when the lifetime increases. For all the lifetime values, the collector current decays exponentially with time as predicted by the analytical model. The collector current fall-time values are 15, 10, 7, 3.5, and 1.8 μs, for high-level lifetime values of 10, 6, 4, 2, and 1 μs, respectively.

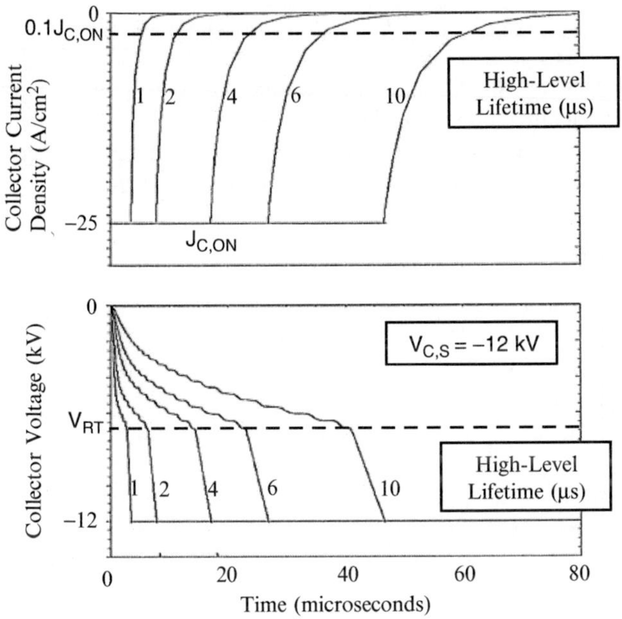

Fig. 7.76 Impact of lifetime on the 20-kV asymmetric p-channel 4H-SiC IGBT turn-off waveforms

7.3.5 Switching Energy Loss

The power loss incurred during the turn-off switching transient limits the maximum operating frequency for the IGBT structure. Power losses during the turn-on of the IGBT structure are also significant but strongly dependent on the reverse recovery behavior of the fly-back rectifiers in circuits. Consequently, it is common practice to use only the turn-off energy loss per cycle during characterization of IGBT devices. The turn-off losses are associated with the voltage rise-time interval and the current fall-time interval. The energy loss for each event can be computed by integration of the power loss, as given by the product of the instantaneous current and voltage. During the voltage rise-time interval, the anode current is constant while the voltage increases in a nonlinear manner as a function of time until reach-through occurs. In order to simplify the analysis, the energy loss during this interval will be computed using:

$$E_{\mathrm{OFF,V1}} = \frac{1}{2} J_{\mathrm{C,ON}} V_{\mathrm{RT}} t_{\mathrm{RT}} \tag{7.82}$$

During the second phase of the voltage rise-time, the collector voltage increases linearly with time while the collector current is constant. The energy loss during this interval can be computed using:

$$E_{\mathrm{OFF,V2}} = \frac{1}{2} J_{\mathrm{C,ON}} \left(V_{\mathrm{C,S}} - V_{\mathrm{RT}}\right)\left(t_{\mathrm{V}} - t_{\mathrm{RT}}\right) \tag{7.83}$$

For the typical switching waveforms for the 20-kV asymmetric p-channel 4H-SiC IGBT structure shown in Fig. 7.72 with a collector supply voltage of −12,000 V, the energy loss per unit area during the collector voltage rise-time is found to be 0.24 J/cm^2 if the on-state current density is 25 A/cm^2.

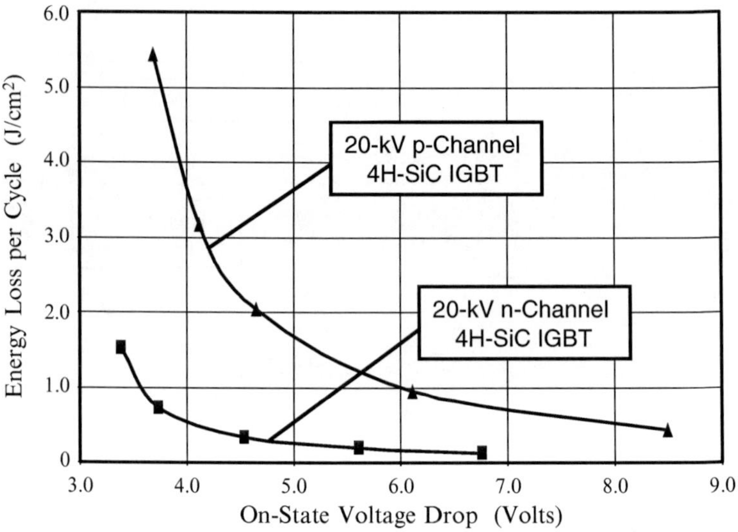

Fig. 7.77 Trade-off curve for the 20-kV asymmetric p-channel 4H-SiC IGBT structure: lifetime in P-base region

During the collector current fall-time interval, the collector voltage is constant while the current decreases exponentially with time. The energy loss during the collector current fall-time interval can be computed using:

$$E_{\mathrm{OFF,I}} = J_{\mathrm{C,ON}} V_{\mathrm{C,S}} \tau_{\mathrm{BL}} \tag{7.84}$$

For the typical switching waveforms for the 20-kV asymmetric p-channel 4H-SiC IGBT structure shown in Fig. 7.76 with a collector supply voltage of −12,000 V, the energy loss per unit area during the collector current fall-time is found to be 0.30 J/cm^2 if the on-state current density is 25 A/cm^2. The total energy loss per unit area ($E_{OFF,V} + E_{OFF,I}$) during the turn-off process for the 20-kV asymmetric p-channel 4H-SiC IGBT structure is found to be 0.54 J/cm^2.

Using the results obtained from the numerical simulations, the on-state voltage drop and the total energy loss per cycle can be computed for the 20-kV p-channel asymmetric IGBT structure. These values are plotted in Fig. 7.77 to create a trade-off curve to optimize the performance of the 20-kV asymmetric p-channel 4H-SiC IGBT structure by varying the lifetime in the N-base region. Devices used in lower frequency circuits would be chosen from the left-hand side of the trade-off curve while devices used in higher frequency circuits would be chosen from the right-hand side of the trade-off curve. For comparison purposes, the trade-off curve for the 20-kV n-channel asymmetric IGBT structure is also shown in Fig. 7.77. It can be concluded that the performance of the 20-kV n-channel asymmetric IGBT structure is far superior to that of the p-channel device.

7.3.6 Maximum Operating Frequency

High-Level Lifetime (μs)	On-State Voltage Drop (Volts)	On-State Power Dissipation (W/cm^2)	Energy Loss per Cycle (J/cm^2)	Maximum Operating Frequency (Hz)
10	3.69	46.1	5.44	28
6	4.12	51.5	3.17	47
4	4.65	58.1	2.04	69
2	6.11	76.3	0.95	131
1	8.49	106	0.43	219

Fig. 7.78 Power loss analysis for the 20-kV asymmetric p-channel 4H-SiC IGBT structure

The maximum operating frequency for operation of the 20-kV asymmetric p-channel 4H-SiC IGBT structure can be obtained by combining the on-state and switching power losses:

$$P_{D,TOTAL} = \delta P_{D,ON} + E_{OFF} f \tag{7.85}$$

where δ is the duty cycle and f is the switching frequency. In the case of the baseline 20-kV asymmetric p-channel 4H-SiC IGBT device structure with a high-level lifetime of 4 μs in the P-base region, the on-state voltage drop is 4.65 V at an on-state current density of 25 A/cm^2. For the case of a 50% duty cycle, the on-state power dissipation contributes 58 W/cm^2 to the total power loss. For this lifetime value, the energy loss per cycle during the voltage rise-time obtained from the numerical simulations is 1.44 J/cm^2 and the energy loss per cycle during the current fall-time obtained from the numerical simulations is 0.60 J/cm^2. Using a total turn-off energy loss per cycle of 2.04 J/cm^2 in Eq. 7.85 yields a maximum operating frequency of only 70 Hz.

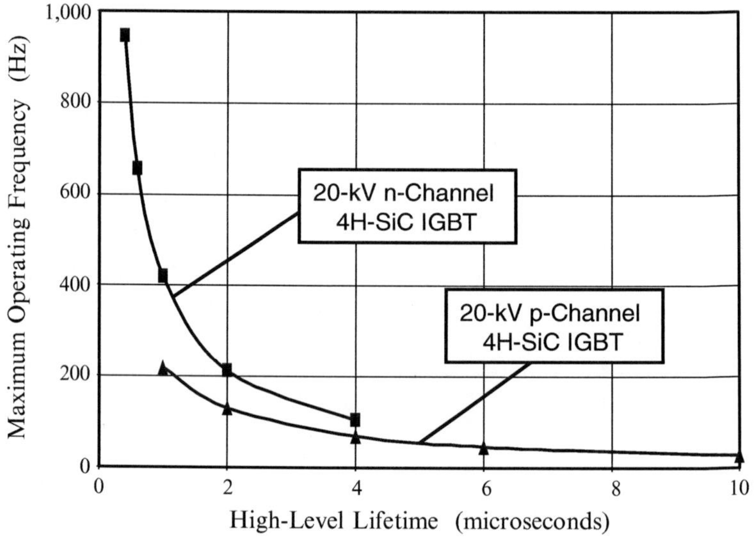

Fig. 7.79 Maximum operating frequency for the 20-kV asymmetric p-channel 4H-SiC IGBT structure

The maximum operating frequency for the 20-kV asymmetric p-channel 4H-SiC IGBT structure can be increased by reducing the lifetime in the P-base region. Using the results obtained from the numerical simulations, the on-state voltage drop and the energy loss per cycle can be computed. These values are provided in Fig. 7.78 together with the maximum operating frequency as a function of the high level lifetime in the N-base region under the assumption of a 50% duty cycle and a total power dissipation limit of 200 W/cm^2. The maximum operating frequency is plotted in Fig. 7.79 as a function of the high-level lifetime in the P-base region. It can be observed that the maximum operating frequency can be increased up to 200 Hz by reducing the high-level lifetime to 1 μs. For comparison purposes, the maximum operating frequency for the 20-kV n-channel asymmetric IGBT structure is also shown in Fig. 7.79. It can be concluded that the maximum operating frequency of the 20-kV n-channel asymmetric IGBT structure is far greater to that of the p-channel device.

7.4 Conclusions

The physics of operation and design principles for the silicon carbide asymmetric IGBT structure have been described in this chapter. The analysis, which includes both n-channel and p-channel structures, provided in this chapter demonstrates that the 20-kV n-channel asymmetric IGBT structure offers excellent characteristics for utility applications. A design methodology is also provided to avoid a high [dV/dt] during the turn-off event for the silicon carbide IGBT structure.

References

1. Q. Zhang, et al, "10kV Trench gate IGBTs on 4H-SiC", IEEE International Symposium on Power Semiconductor Devices and ICs, pp. 159–162, 2005.
2. B.J. Baliga, "Silicon Carbide Switching Device with Rectifying Gate", U.S. Patent 5,396,085, Filed December 28, 1993, Issued March 7, 1995.
3. B.J. Baliga, "Silicon Carbide Power Devices", World Scientific Press, 2006.
4. Q. Zhang, et al, "9 kV 4H-SiC IGBTs with 88 mΩ-cm^2 of $R_{diff,on}$", Material Science Forum Vols. 556–557, pp. 771–774, 2007.
5. Q. Zhang, et al, "12-kV p-Channel IGBTs with Low On-Resistance in 4H-SiC", IEEE Electron Device Letters, Vol. 29, pp. 1027–1029, 2008.
6. Y. Sui, X. Wang, and J.A. Cooper, "High-Voltage Self-Aligned, p-Channel DMOS-IGBTs in 4H-SiC", IEEE Electron Device Letters, Vol. 28, pp. 728–730, 2007.
7. T. Tamaki, et al, "Optimization of On-State and Switching Performances for 15-20 kV 4H-SiC IGBTs", IEEE Transactions on Electron Devices, Vol. 55, pp. 1920–1927.
8. W. Sung, et al, "Design and Investigation of Frequency Capability of 15-kV 4H-SiC IGBT", IEEE International Symposium on Power Semiconductor Devices and ICs, pp. 271–274, 2009.
9. M.K. Das, et al, "A 13-kV 4H-SiC n-channel IGBT with Low $R_{diff,on}$ and Fast Switching", International Conference on Silicon Carbide and Related Materials, October 2007.
10. R.J. Callanan, et al, "Recent Progress in SiC DMOSFETs and JBS Diodes at CREE", IEEE Industrial Electronics Conference, pp. 2885–2890, 2008.
11. B.J. Baliga, "Fundamentals of Power Semiconductor Devices", Springer-Science, New York, 2008.
12. B.J. Baliga and M.S. Adler, "Measurement of Carrier Lifetime Profiles in Diffused Layers of Semiconductors", IEEE Transactions on Electron Devices, Vol. ED-25, pp. 472–477, 1978.
13. Q. Zhang, et al, "Design and Characterization of High-Voltage 4H-SiC p-IGBTs", IEEE Electron Device Transactions, Vol. 55, pp. 1912–1919, 2008.

Chapter 8
Silicon MCT

As discussed in Chap. 5, the silicon insulated gate bipolar transistor (IGBT) has been a highly successful innovation that has been widely accepted by the industry for power control applications with supply voltages ranging from 300 to 6,000 V. As shown in that chapter, the optimization of the IGBT structure from an applications standpoint requires reduction of the lifetime in the drift region to enhance its switching speed. This is accompanied by a significant increase in the on-state voltage drop for the IGBT structure. The large on-state voltage drop in the IGBT structure for smaller lifetime values in the drift region can be traced to poor conductivity modulation of the drift region near the emitter. A superior on-state carrier distribution can be obtained by utilizing thyristor-based on-state current flow as shown in Chap. 2. The gate-turn-off thyristor (GTO) was developed to take advantage of the low on-state voltage drop. However, the gate drive current for the GTO is very large as demonstrated in Chap. 4.

The MOS-controlled thyristor (MCT) structure was proposed [1, 2] to take advantage of thyristor-based on-state current flow under MOS gate control to reduce the gate drive requirements. Historically, the first MCT structures, called the complementary MCT (C-MCT), were fabricated using a p-type drift region to obtain improved turn-off performance with n-channel turn-off MOSFETs integrated into the thyristor structure [1]. Subsequently, MCTs with both n-channel and p-channel MOSFET structures embedded in a thyristor structure with n-type drift region were reported [2]. In the later part of the 1980s, an effort was made to characterize the switching performance [3], scaling [4], and reliability [5] of the concept. A rigorous study to understand the physics of MCT operation and to extend the blocking voltage was conducted in the early 1990s [6–12].

As discussed in previous chapters, the n-channel IGBT structure operates as a wide base PNP transistor driven by an n-channel MOSFET integrated into the structure. The rate of rise of the collector current in the IGBT structure can be regulated by changing the gate resistance [13] as recognized from the inception of the development of the device. This simple circuit solution allows control of the reverse recovery of the antiparallel rectifiers in the typical H-bridge circuits (see Fig. 1.11) preventing very high peak reverse recovery currents that can destroy

B.J. Baliga, *Advanced High Voltage Power Device Concepts*,
DOI 10.1007/978-1-4614-0269-5_8, © Springer Science+Business Media, LLC 2011

both the rectifier and the IGBT switch. In contrast, thyristor-based structures exhibit uncontrolled rapid turn-on due to the internal regenerative action which can lead to extremely high reverse recovery currents in the antiparallel rectifiers leading to the destruction of the rectifier and the switch. In the case of thyristor-based switches, the rate of rise of the current during turn-on must be regulated by using a snubber circuit [14]. The need for snubbers when operating the MCT in a hard-switched circuit has been analyzed for 2.5-kV devices with comparison made to the IGBT performance [15]. This work conclusively demonstrated that the MCT devices are prone to producing very high reverse recovery currents in the antiparallel rectifiers leading to oscillations that produce very high voltages across the components. It was also found that introducing the turn-on snubber (series inductance) produces large oscillations in the current and voltage for the MCT which are unsustainable. It has also been demonstrated that filamentation, where the current constricts to a localized area, is also a significant problem with the MCT structure [16].

Due to the problems encounters with stable operation of the MCT structure in hard-switching circuits that are commonly used for motor control applications, this device fell out of favor in the late 1990s. However, for completeness, the physics of operation of the MCT structure and the characteristics of devices capable of blocking 5 and 10 kV are described in this chapter. It is shown here that the triple-diffused process required for the MCT structure is significantly more complex than the double-diffused process required for the IGBT structure. In addition, the p-channel turn-off MOSFET incorporated into the MCT structure for a device designed with an N-type drift region has high channel resistance which limits its ability to turn-off high on-state current density levels. Consequently, most of the early experimental work was performed using n-channel turn-off MOSFETs incorporated into MCT structures with P-type drift region. These complementary C-MCT structures are not compatible with the commonly used H-bridge architecture with a positive DC bus. In this chapter, only the MCT structure with N-type drift region is analyzed for comparison with the silicon IGBT structures described in Chap. 5 with identical drift regions.

8.1 Basic Structure and Operation

The asymmetric MCT structure with the planar gate architecture [1] is illustrated in Fig. 8.1 with its doping profile. Since the asymmetric MCT structure is intended for use in DC circuits, its reverse blocking capability does not have to match the forward blocking capability allowing the use of an N-buffer layer adjacent to the P^+ anode region. The N-buffer layer has a much larger doping concentration than the lightly doped portion of the N-base region. The electric field in the asymmetric MCT takes a trapezoidal shape allowing supporting the forward blocking voltage with a thinner N-base region. This allows achieving a lower on-state voltage drop and superior turn-off characteristics. The doping concentration of the buffer layer and the lifetime in the N-base region must be optimized to perform a trade-off between on-state voltage drop and turn-off switching losses. The MCT structures are discussed in this chapter with two blocking voltage ratings for comparison with other device structures in the book.

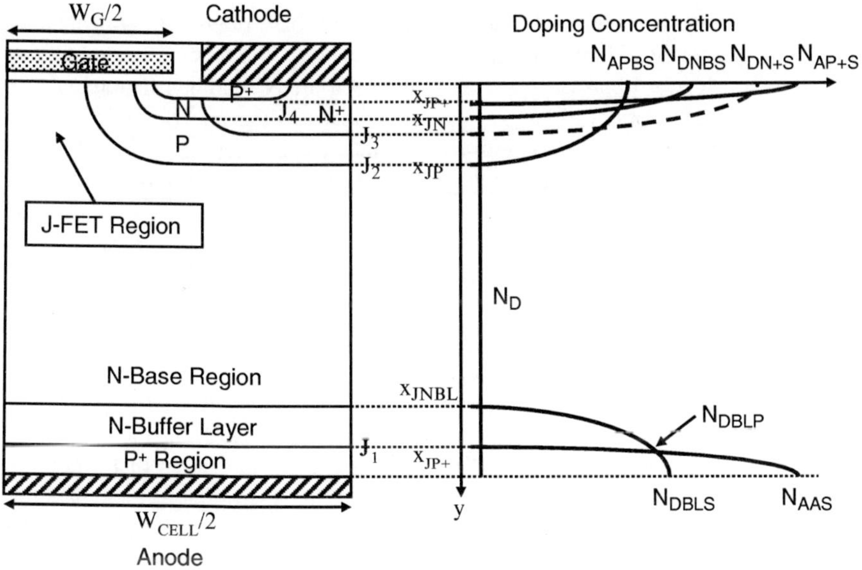

Fig. 8.1 The asymmetric MCT structure and its doping profile

The MCT structure requires two MOSFET regions – one to turn off the thyristor regenerative action and the second to turn on the device structure. The turn-on and turn-off MOSFET regions can be combined as illustrated in Fig. 8.1 by using a triple diffusion process. In the MCT structure shown in the figure, the thyristor structure is achieved between the N^+ cathode region at the top, a narrow P-base region, a wide N-base region with a N-buffer layer to support the high forward blocking voltage, and the P^+ anode region located at the bottom. The n-channel turn-on MOSFET is formed by using an N type diffusion, as its source region, nested inside the P-base region. The doping concentration of this N-type region cannot be very large because it also serves as the body region for the p-channel turn-off MOSFET in the MCT structure. The source of the p-channel MOSFET is formed by diffusion of a P^+ region inside the N-type region.

A schematic illustration of the doping profiles for the various regions is provided in Fig. 8.1 on the right-hand side. The solid lines show the profiles taken vertically through the P^+ source region and the N-type body region. The dashed line shows the doping profile for the N^+ cathode region. It can be observed from this figure that the N-type body region for the turn-off MOSFET is constructed between two other diffusions making the processing difficult for the MCT structure.

When a positive bias is applied to the anode of the MCT structure, junction J_2 between the P-base and N-base regions becomes reverse biased. In principle, this junction is capable of supporting a large voltage with a depletion region formed in the lightly doped N-base region. However, the leakage current generated in the drift region is collected by junction J_2 and flows into the P-base region. If zero gate bias is applied, neither MOSFET is turned on. Consequently, the holes collected by the P-base region forward bias the junction J_3 between the N^+ cathode region and the P-base region. The forward bias on junction J_3 leads to the injection of electrons from the N^+ cathode region into the P-base region. These electrons diffuse through the P-base region and get collected at junction J_2 between the P-base and N-base regions. When the electrons enter the N-base region, they serve as base drive current for the P-N-P transistor leading to strong injection of holes from junction J_1 between the P^+ anode region and N-base region. This sets up the regenerative action that turns on the vertical thyristor in the MCT structure. Consequently, the MCT structure shown in Fig. 8.1 is not capable of supporting a large forward blocking voltage if no (or zero) gate bias is applied.

The high forward blocking voltage can be achieved in the MCT structure shown in Fig. 8.1 by the application of a negative gate bias to turn-on the p-channel MOSFET. When the p-channel MOSFET is turned on, the holes that enter the P-base region due to generation of leakage current in the drift region can bypass the junction J_3 between the N^+ cathode region and the P-base region and get removed by the cathode contact. The MCT structure then behaves like a thyristor structure with cathode shorts as discussed in the textbook [17]. The forward blocking capability of the MCT with a negative gate bias is then determined by open-base P-N-P transistor breakdown.

The MCT structure can be turned on by the application of a positive bias to the MOS-gated structure. In this case, the n-channel MOSFET is turned on providing a path for electrons to flow from the N^+ cathode region into the P-base region. These electrons serve as base drive current for the wide base P-N-P transistor leading to strong injection of holes from junction J_1 between the P^+ anode region and N-base region. This sets up the regenerative action that turns on the vertical thyristor in the MCT structure.

Once the MCT is operating in its on-state, the device can be turned off by switching the gate bias from a positive value to a negative value. The negative gate bias turns on the p-channel MOSFET in the structure providing a path for holes entering the P-base region from the N-base region to be removed to the cathode contact. A simple model for the maximum anode turn-off current density can be formulated using a lumped element approach. In this approach, all the hole current being collected at the junction J_2 is assumed to flow through a lumped shunting resistance for the hole current path. In a simplified model, the hole current path consists of the P-base region and the p-channel MOSFET.

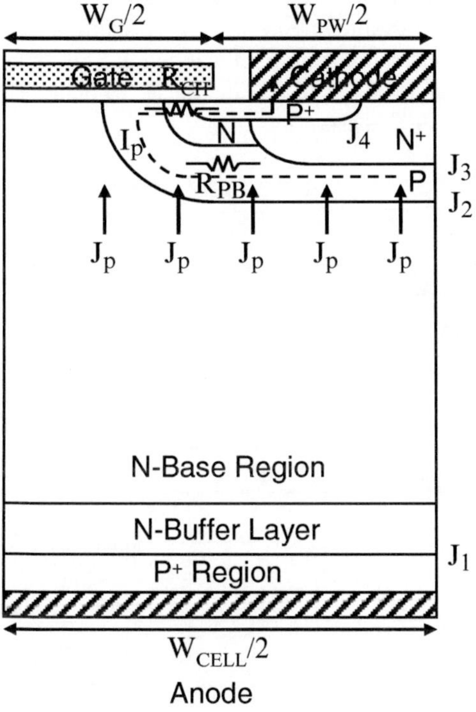

Fig. 8.2 Current flow during turn-off for the MCT structure

Under inductive load operation, the hole current (I_p) collected by the P-base region during turn-off is equal to the initial anode current density ($J_{A,ON}$) multiplied by the cell area:

$$I_p = J_{A,ON}\left(\frac{W_{Cell}Z}{2}\right) \tag{8.1}$$

where Z is the length of the cell in the orthogonal direction to the cross section shown in Fig. 8.2. The voltage drop produced by this current when flowing through the lumped shunting resistance (R_{SH}) must be less than the built-in potential for the junction J_3 between the N^+ cathode region and the P-base region if injection from this junction is to be suppressed to achieve the desired turn-off:

$$V_{bi} = I_p R_{SH} = J_{A,MAX}\left(\frac{W_{Cell}Z}{2}\right)(R_{PB} + R_{CH}) \tag{8.2}$$

The lumped resistance of the P-base region is given by:

$$R_{PB} = \rho_{S,PB}\left(\frac{W_{PW}}{2Z}\right) \tag{8.3}$$

where $\rho_{S,PB}$ is the pinch sheet resistance of the P-base region and W_{PW} is the width of the window in the polysilicon gate region. A single sheet resistance for the P-base region is assumed here for simplicity although a different value may prevail for the P-base region under the N-region and the N^+ cathode region. The resistance of the p-channel MOSFET is given by:

$$R_{CH} = \frac{L_{CH}}{\mu_{pi} C_{OX} (V_G - V_{TH}) Z} \tag{8.4}$$

where L_{CH} is the channel length, μ_{pi} is the mobility for holes in the inversion layer of the p-channel MOSFET, C_{OX} is the gate oxide capacitance, V_G is the gate bias voltage, and V_{TH} is the threshold voltage.

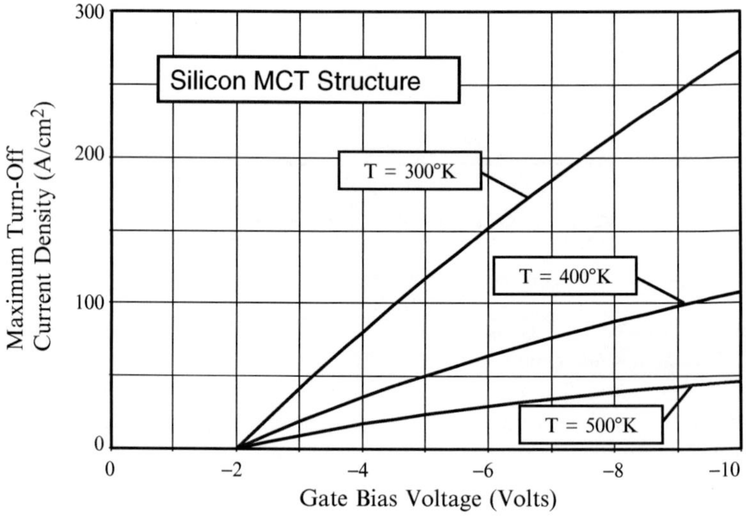

Fig. 8.3 Maximum turn-off current density for the MCT structure

Using the above equations, the maximum turn-off current density is found to be given by:

$$J_{A,MAX} = \frac{4 V_{bi} \mu_{pi} C_{OX} (V_G - V_{TH})}{W_{Cell} \left[\rho_{S,PB} W_{PW} \mu_{pi} C_{OX} (V_G - V_{TH}) + 2 L_{CH} \right]} \tag{8.5}$$

The maximum turn-off current density predicted by the analytical model is plotted in Fig. 8.3 as a function of gate bias voltage under the assumption of a threshold voltage of 2 V. The following values were used in the analytical model: channel mobility for holes of 240 cm^2/V-s, gate oxide thickness of 500 Å, built-in potential of 0.937 V at room temperature using doping concentrations of

1×10^{19} cm^{-3} for the N$^+$ region and 1×10^{17} cm^{-3} for the P-base region, P-base pinch sheet resistance of 650 Ω/sq based upon an average doping concentration of 1×10^{17} cm^{-3} and thickness of 4 μm for the P-base region, cell width of 30 μm and polysilicon window width of 14 μm, and a channel length of 1 μm for the p-channel MOSFET integrated into the MCT structure. It can be observed that the maximum turn-off current density increases with increasing negative gate bias applied to the gate electrode because of a corresponding reduction of the channel resistance. A maximum turn-off current density of 273 A/cm^2 is predicted by the analytical model at room temperature for a gate bias of negative 10 V.

The impact of increasing the temperature on the maximum turn-off capability of the MCT structure is also shown in Fig. 8.3. At 500 K, the maximum turn-off current density is reduced below 50 A/cm^2 for a gate bias of −10 V. This trend can also be observed in Fig. 8.4 where the maximum turn-off current density is plotted as a function of temperature for a fixed gate bias of −10 V. The reduction in the maximum turn-off current density is due to a reduction of the built-in potential and a reduction in the mobility for holes in the P-base region and the inversion layer [17].

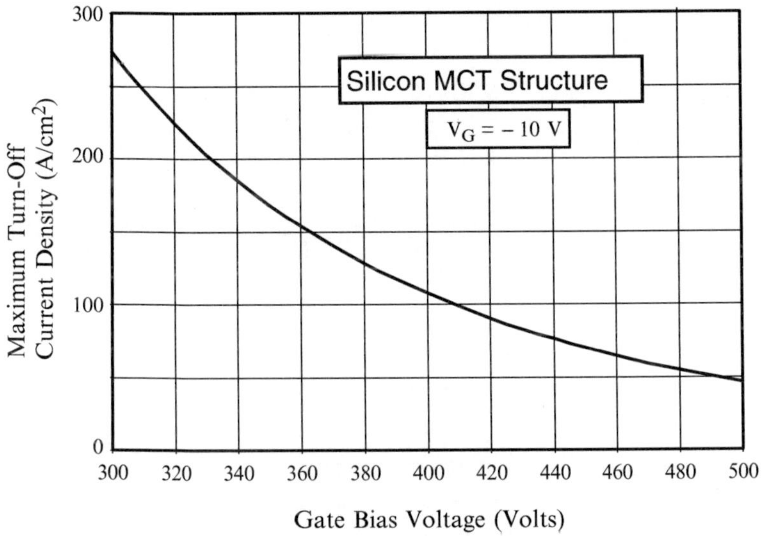

Fig. 8.4 Maximum turn-off current density for the MCT structure

An alternate MCT structure with an N-type high voltage drift region is illustrated in Fig. 8.5. In this structure, an n-channel MOSFET is used to achieve the turn-off of the thyristor within the MCT structure [2]. As in the case of the first MCT structure shown in Fig. 8.1, an n-channel MOSFET is also used to turn on the device. Consequently, the alternate MCT structure requires two separate gate control electrodes, which makes the package more complex because of the four leads.

When a positive bias is applied to gate-1 with zero bias applied to gate-2, electrons can be injected into the N-drift region via the channel of MOSFET-1. This turns on the thyristor formed between the N^+ cathode region and the P^+ anode region by providing the base drive current for the vertical P-N-P transistor. To turn off the structure, the bias to gate-1 is reduced to zero and a positive bias is applied to gate-2 to turn on MOSFET-2. When MOSFET-2 is turned on, the hole current (I_p) collected at junction J_2 can flow to the cathode contact via MOSFET-2 shunting the N^+ cathode/P-base junction (J_4). If the resistance of this path is sufficiently low, injection from the N^+ cathode/P-base junction (J_4) can be suppressed leading to the turn-off of the regenerative thyristor action. The resistance of the path consists of the channel resistance (R_{CH}) of the n-channel MOSFET-2 and the P-base region (R_{PB}). In principle, the alternate MCT structure should have a larger maximum controllable current due to the lower channel resistance of the n-channel MOSFET in its structure when compared with the p-channel MOSFET in the first structure. Despite this advantage, the first MCT structure has been the focus of most of the research investigations because of the convenience of having a single gate electrode for performing both the turn-on and the turn-off functions. In addition, the buried shorting electrode complicates the device processing. For this reason, the alternate MCT structure will not be analyzed in this chapter.

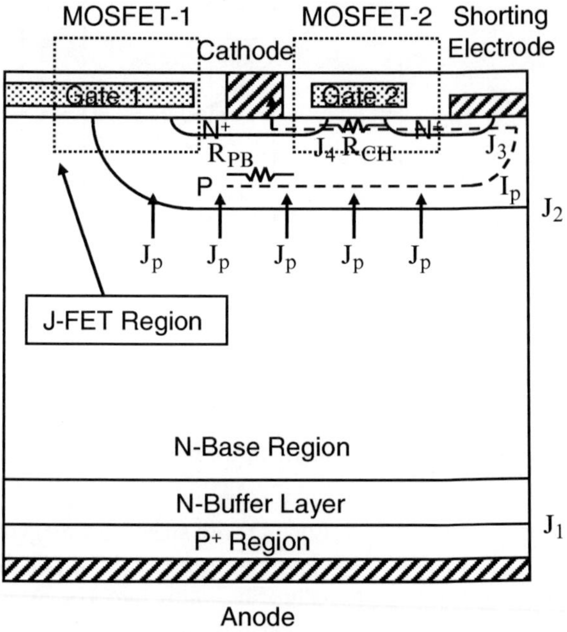

Fig. 8.5 Alternate asymmetric MCT structure

8.2 5,000-V Silicon MCT

The design and characteristics for the 5,000-V asymmetric silicon MCT structure are discussed in this section. The design parameters for the N-base (drift) region required to achieve this blocking voltage are first analyzed. Using the optimum N-base width, the blocking characteristics for the device are then obtained for the case of zero and negative 10 V gate bias. The on-state characteristics for the device are obtained for various lifetime values as well. The gate controlled turn-off behavior of the silicon MCT structure is analyzed including the effect of the lifetime in the drift region.

8.2.1 *Blocking Characteristics*

The physics for blocking voltages in the first and third quadrants by the asymmetric MCT structure are the same as those previously discussed for the silicon IGBT structure if a negative bias is applied to the gate electrode to keep the p-channel MOSFET turned on. The p-channel MOSFET then acts like the cathode short during the forward blocking mode. Without gate bias, there is no path for removal of the leakage current collector by junction J_2 leading to latch-up of the thyristor. This can be problem for operation of the MCT in power circuits during the initial start-up. When power is applied for the first time to the circuit, the gate power supply may not generate the voltage required to turn on the p-channel MOSFET before the anode voltage for the MCT becomes sufficiently large to trigger the thyristor within the MCT structure.

When a positive bias is applied to the anode terminal of the asymmetric MCT structure with a negative bias applied to the gate, the P-base/N-base junction (J_2) becomes reverse biased while the junction (J_1) between the P^+ anode region and the N-base region becomes forward biased. The forward blocking voltage is supported across the P-base/N-base junction (J_2) with a depletion layer extending mostly within the N-base region. The electric field distribution within the asymmetric MCT structure is essentially the same as that illustrated in Fig. 4.3 for the asymmetric GTO structure. Consequently, the design procedure described in Chap. 4 can be applied to the asymmetric MCT structure. From Fig. 4.4, the N-base region width required to obtain a forward blocking voltage of 5,500 V is 470 μm. This width can be slightly reduced when taking into account the voltage supported within the P-base region due to its graded doping profile.

The leakage current in forward blocking mode is produced by space-charge generation within the depletion region. In the case of the asymmetric MCT structure in the forward blocking mode, the space-charge-generation current at the reverse biased P-base/N-base junction (J_2) is amplified by the gain of the internal P-N-P transistor. Initially, the space-generation current increases with increasing anode bias due to expansion of the depletion region. Concurrently, the current gain (α_{PNP}) of the P-N-P transistor is also a function of the anode bias voltage because the base transport factor increases when the anode bias increases. Prior to the complete depletion of the lightly doped portion of the N-base region, the multiplication factor remains close to unity. It is therefore sufficient to account for the increase in the base transport factor with collector bias as given by Eqs. 4.8 and 4.9.

For the case of the silicon asymmetric MCT structure with a width of 450 μm for the lightly doped portion of the N-base region with a doping concentration of 5×10^{12} cm^{-3}, the entire lightly doped portion of the N-base region is completely depleted at a reach-through voltage of 780 V. Once the lightly doped portion of the N-base region becomes completely depleted, the electric field becomes truncated at the interface between the lightly doped portion of the N-base region and the N-buffer layer as illustrated at the bottom of Fig. 4.3. The space charge generation width then becomes independent of the anode bias because the depletion width in the N-buffer layer is small. Under these bias conditions, the base transport factor also becomes independent of the collector bias as given by Eq. 4.10. Consequently, the leakage current becomes independent of the collector bias until the onset of avalanche multiplication. The leakage currents for the silicon asymmetric MCT structure are identical to those provided for the silicon asymmetric IGBT structure in Chap. 5.

Simulation Example

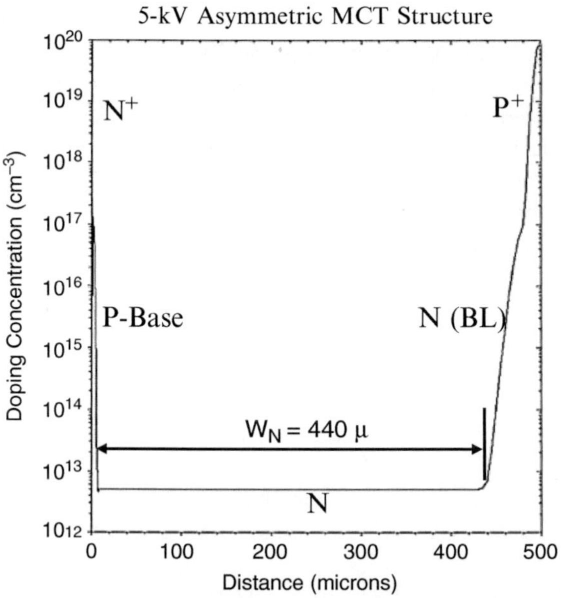

Fig. 8.6 Doping profile for the simulated asymmetric 5-kV MCT structure

The results of two-dimensional numerical simulations are described here to gain insight into the physics of operation for the 5-kV asymmetric MCT structure under voltage blocking conditions. The simulations were performed using a cell with the structure shown in Fig. 8.1 with the gate region located on the right-hand side. This half-cell has a width ($W_{CELL}/2$) of 15 μm (area = 1.5×10^{-7} cm^{-2}).

The asymmetric MCT structure used for the simulations was formed by diffusions performed into a uniformly doped N-type drift region with a doping concentration of 5×10^{12} cm^{-3}. A lifetime (τ_{p0}, τ_{n0}) of 10 μs was used for the baseline device. The N-buffer layer was formed by diffusion from the collector side with a depth of 55 μm. The doping profile in the vertical direction through the N$^+$ cathode region is shown in Fig. 8.6 indicating the net width of the lightly doped portion of the N-base region is 440 μm after accounting for the diffusions. The peak doping concentration of the N-buffer layer is 1.0×10^{17} cm^{-3} and its thickness is 40 μm.

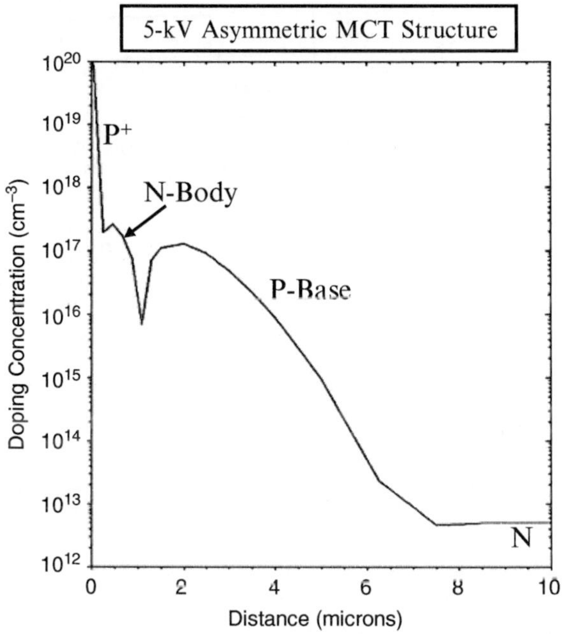

Fig. 8.7 Doping profile for the simulated asymmetric 5-kV MCT structure

The P-base region for the asymmetric MCT structure was formed with a Gaussian doping profile with a surface concentration of 5×10^{17} cm^{-3} and a depth of 6 μm as can be seen in Fig. 8.7 where the vertical doping profile in the upper 10 μm of the structure is provided. The P$^+$ source region was formed with a Gaussian doping profile with a surface concentration of 2×10^{20} cm^{-3} and a depth of 0.2 μm. The N-body region for the p-channel MOSFET is nested between these diffusions and is formed with a Gaussian doping profile with a surface concentration of 8×10^{17} cm^{-3} and a depth of 1 μm. It can be observed that the doping profile for the N-body region of the p-channel turn-off MOSFET will be altered by small changes to the diffusion profiles for the P-base, N-body, and P$^+$ source regions making the fabrication of this MCT structure very challenging.

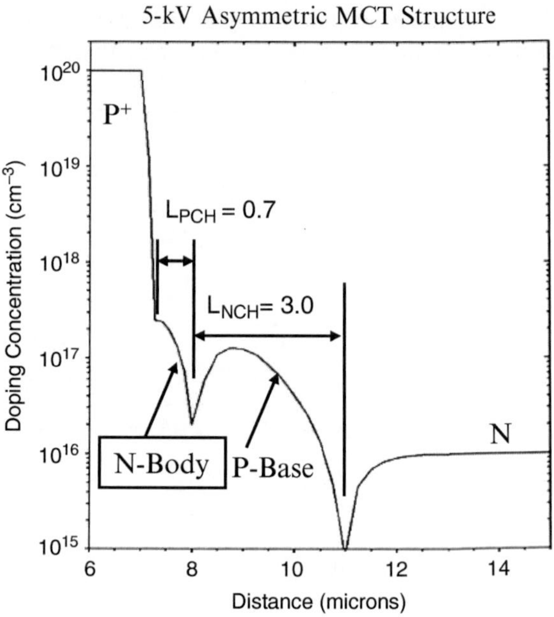

Fig. 8.8 Channel doping profile for the simulated asymmetric 5-kV MCT structure

The doping profile of the channel region for the 5-kV asymmetric MCT structure used for the numerical simulations is provided in Fig. 8.8. This profile was taken along the horizontal line at $y = 0$ μm. It can be seen that the peak doping concentration of the P-base region is 1.5×10^{17} cm^{-3} and the n-channel MOSFET channel length is 3.0 μm. This is sufficient to prevent reach-through limited breakdown in the P-base region. The peak doping concentration of the N-body region is 2.5×10^{17} cm^{-3} and the p-channel MOSFET channel length is 0.7 μm. The relatively high peak doping for the N-body region of the p-channel MOSFET results in a high threshold voltage which degrades the maximum turn-off current capability of the MCT structure.

The forward blocking capability of the silicon asymmetric MCT structure was first obtained using numerical simulations by increasing the anode bias while maintaining the gate electrode at zero volts. The characteristics obtained for a lifetime (τ_{p0}) of 10 μs is provided in Fig. 8.9. It can be observed that the structure cannot support a high voltage. This occurs because there is no direct path for the removal of holes collected at junction J_2 to the cathode electrode when the gate bias is zero. Since the hole current must flow via junction J_3 to the cathode electrode, the upper N-P-N transistor becomes active leading to latch-up of the thyristor.

When a negative bias is applied to the gate electrode, the p-channel MOSFET in the MCT structure is turned on providing a path for the removal of holes collected at junction J_2 to the cathode electrode. The device is then able to support above 5,500 V as desired. The leakage current increases rapidly with increasing anode bias voltage until about 780 V as predicted by the analytical

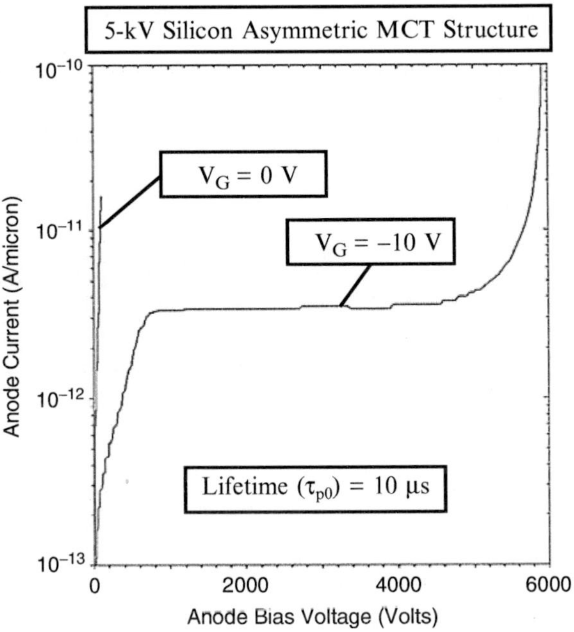

Fig. 8.9 Forward blocking characteristics for the asymmetric MCT structure

model (see Fig. 5.2). This occurs due to the increase in the space-charge generation volume and the increase in the current gain (α_{PNP}) of the open base P-N-P transistor until the anode bias becomes equal to the reach-through voltage obtained using the analytical solution given by Eq. 4.2. The leakage current then becomes independent of the anode voltage until close to the breakdown voltage. This behavior is well described by the analytical model (see Fig. 5.2 for the 5-kV asymmetric IGBT structure). The leakage current density obtained using the analytical model is within a factor of 2 of the values derived from the numerical simulations for all cases. The blocking characteristics for the asymmetric MCT structure are therefore similar to those for the asymmetric IGBT structure. However, the IGBT structure can block voltage with zero gate bias while a negative gate bias is required for the MCT structure. This can be a problem during start-up of power circuits because the gate supply voltage may not be available before the MCT structure is subjected at a high positive anode voltage.

The current flow lines within the asymmetric MCT structure in the blocking mode are provided in Figs. 8.10 and 8.11 for the case of a gate bias of zero and negative 10 V, respectively. From Fig. 8.10, it can be observed that the current collected at the P-base/N-drift region junction J_2 flows via the N^+ cathode region demonstrating the turn-on of the N-P-N transistor. In contrast to this, from Fig. 8.11, it can be observed that the current collected at the P-base/N-drift region junction J_2 flows via the p-channel MOSFET bypassing the N^+ cathode region. This allows the asymmetric MCT structure to support a high forward blocking voltage without turning on the thyristor structure.

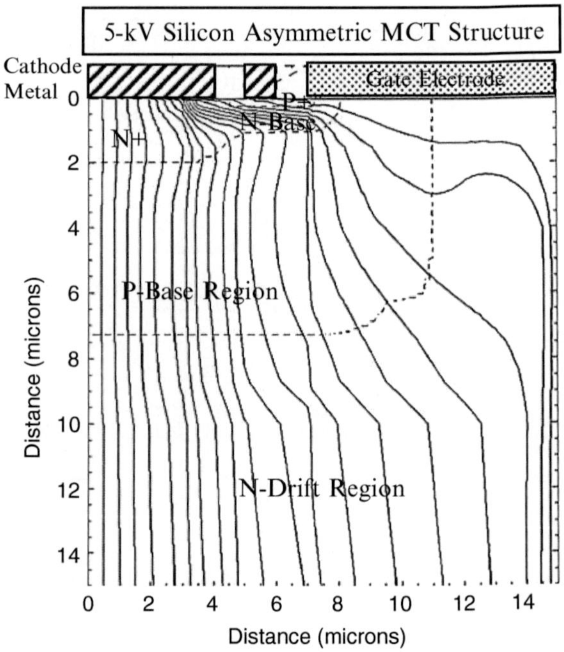

Fig. 8.10 Current flow-lines during the blocking mode for the 5-kV asymmetric MCT structure: $V_G = 0$ V

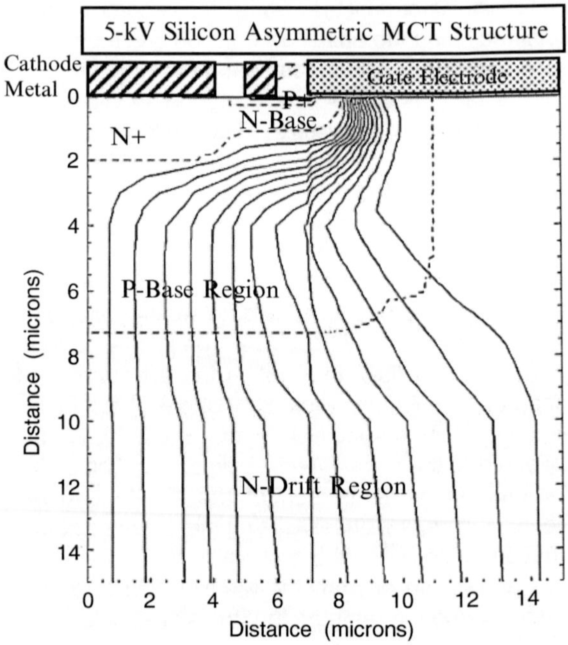

Fig. 8.11 Current flow-lines during the blocking mode for the 5-kV asymmetric MCT structure: $V_G = -10$ V

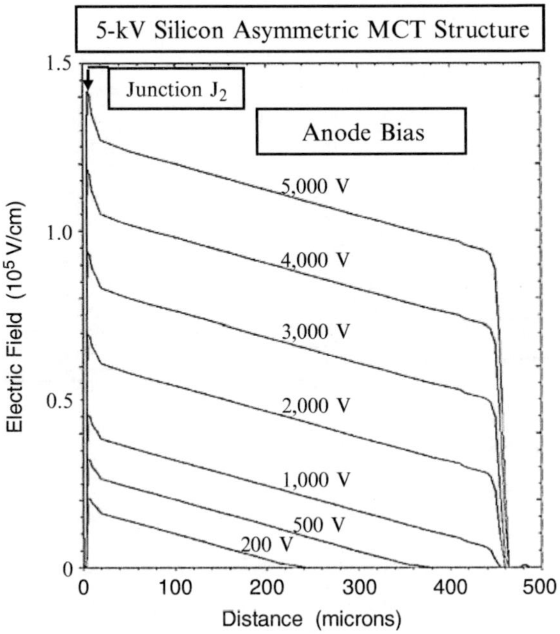

Fig. 8.12 Electric field profiles in the 5-kV asymmetric MCT structure

As in the case of other asymmetric structures, the anode voltage is primarily supported within the lightly doped portion of N-drift region in the asymmetric MCT structure during operation in the forward blocking mode. This is illustrated in Fig. 8.12 where the electric field profiles are shown during operation in the forward blocking mode at several anode bias voltages. It can be observed that the P-Base/N-base junction (J_2) becomes reverse biased during the forward blocking mode with the depletion region extending toward the right-hand side with increasing (positive) collector bias. The electric field has a triangular shape until the entire lightly doped portion of the N-base region becomes completely depleted. This occurs at an anode bias of about 800 V in agreement with the value obtained using the analytical solution (see Eq. 4.2). The electric field profile then takes a trapezoidal shape due to the high doping concentration in the N-buffer layer.

8.2.2 On-State Voltage Drop

The MCT structure operates with latch-up of the thyristor structure within the device. Consequently, the on-state characteristics and the free carrier distribution within the N-drift region can be expected to be similar to those for the thyristor structure (see Chap. 2). However, the incorporation of the p-channel turn-off MOSFET within the thyristor structure degrades the injection efficiency of the cathode junction. This makes the on-state voltage drop for the MCT structure larger than that for the thyristor structure with the same drift region properties.

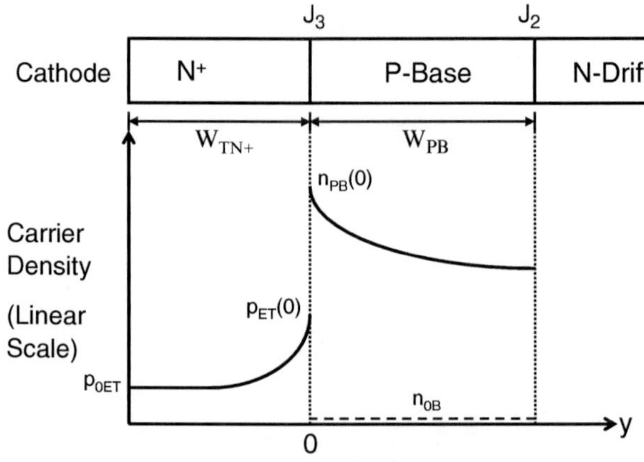

Fig. 8.13 Carrier distribution profiles at the cathode junction for the MCT structure: thyristor region

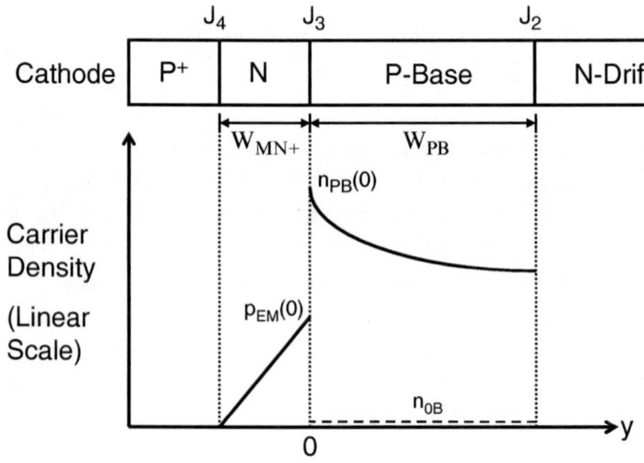

Fig. 8.14 Carrier distribution profiles at the cathode junction for the MCT structure: p-channel MOSFET region

The injected carrier distribution at the cathode junction in the thyristor portion, where the cathode doping concentration is high, is illustrated in Fig. 8.13. It can be observed that the concentration of holes injected into the N^+ cathode region reduces as they diffuse away from the junction until their concentration becomes equal to the equilibrium minority carrier density (p_{0E}) because of the relatively large thickness and high doping concentration of the cathode region in this location. In contrast, the injected carrier distribution at the cathode junction in the portion with the integrated p-channel MOSFET, where the cathode doping concentration is low, is illustrated in Fig. 8.14. It can be observed that the concentration of holes injected into the N^+ cathode region decreases linearly with distance and becomes equal to zero at the junction J_4. Due to the inclusion of the additional P^+ diffusion in this portion of the MCT structure to form the source region of the turn-off MOSFET, the width (W_{MN+}) of the N^+ cathode region is also much smaller than in the thyristor region.

The difference in the injection efficiency in the two portion of the MCT structure can be analyzed by using the carrier profiles shown in the above figures. For the thyristor portion, the electron current at junction J_3 due to injection into the P-base region is given by:

$$J_n(0) = \frac{qD_{nB}}{L_{PB}} n_{PB}(0) = \frac{qD_{nB}}{L_{PB}} n_{0B} e^{qV_{BE}/kT} \tag{8.6}$$

where D_{nB} is the diffusion coefficient for electrons in the P-base region, L_{PB} is the diffusion length for electrons in the P-base region, $n_{PB}(0)$ is the injected concentration of electrons in the P-base region at the junction, n_{0B} is the equilibrium concentration of electrons in the P-base region, and V_{BE} is the voltage across the junction. The hole current at junction J_3 due to injection into the N^+ cathode region is given by:

$$J_p(0) = \frac{qD_{pE}}{L_{PE}} p_{ET}(0) = \frac{qD_{pE}}{L_{PE}} p_{0ET} e^{qV_{BE}/kT} \tag{8.7}$$

where D_{pE} is the diffusion coefficient for holes in the N^+ cathode region, L_{PE} is the diffusion length for holes in the N^+ cathode region, $p_{ET}(0)$ is the injected concentration of holes in the N^+ cathode region at the junction, p_{0ET} is the equilibrium concentration of holes in the N^+ cathode region, and V_{BE} is the voltage across the junction.

The emitter injection efficiency can be obtained by using [17]:

$$\gamma_{ET} = \frac{J_n(0)}{J_n(0) + J_p(0)} \tag{8.8}$$

Using Eqs. 8.6 and 8.7:

$$\gamma_{ET} = \frac{D_{nB} L_{pE} N_{DET}}{D_{nB} L_{pE} N_{DET} + D_{pE} L_{pB} N_{AB}} \tag{8.9}$$

where N_{DET} is the doping concentration in N^+ cathode region at the thyristor portion and N_{AB} is the doping concentration in P-base region. For typical values of an N^+ cathode region doping concentration of 1×10^{20} cm^{-3}, a P-base region doping concentration of 1×10^{17} cm^{-3}, a hole diffusion length of 2 μm in the N^+ cathode region, an electron diffusion length of 4 μm in the P-base region, a diffusion coefficient of 1.3 cm^2/s for holes in the N^+ cathode region, and a diffusion coefficient of 20 cm^2/s for electrons in the P-base region, the emitter injection efficiency in the thyristor portion is found to be very close to unity (0.99985).

For the portion with the p-channel MOSFET, the electron current at junction J_3 due to injection into the P-base region is given by:

$$J_n(0) = \frac{qD_{nB}}{L_{PB}} n_{PB}(0) = \frac{qD_{nB}}{L_{PB}} n_{0B} e^{qV_{BE}/kT} \tag{8.10}$$

which is the same as for the thyristor portion. However, the hole current at junction J_3 due to injection into the N^+ cathode region is given by:

$$J_p(0) = \frac{qD_{pE}}{W_{MN+}} p_{EM}(0) = \frac{qD_{pE}}{W_{MN+}} p_{0EM} e^{qV_{BE}/kT} \tag{8.11}$$

where D_{pE} is the diffusion coefficient for holes in the N^+ cathode region, W_{MN+} is the thickness of the N cathode region, $p_{EM}(0)$ is the injected concentration of holes in the N cathode region at the junction, p_{0EM} is the equilibrium concentration of holes in the N cathode region, and V_{BE} is the voltage across the junction.

The emitter injection efficiency can be obtained by using [17]:

$$\gamma_{EM} = \frac{J_n(0)}{J_n(0) + J_p(0)} \tag{8.12}$$

Using Eqs. 8.10 and 8.11:

$$\gamma_{EM} = \frac{D_{nB} W_{MN+} N_{DEM}}{D_{nB} W_{MN+} N_{DEM} + D_{pE} L_{pB} N_{AB}} \tag{8.13}$$

where N_{DEM} is the doping concentration in N cathode region at the portion with the p-channel MOSFET and N_{AB} is the doping concentration in P-base region. For typical values of an N cathode region doping concentration of 2.5×10^{17} cm^{-3}, a P-base region doping concentration of 1×10^{17} cm^{-3}, a width of 1 μm for the N cathode region, an electron diffusion length of 4 μm in the P-base region, a diffusion coefficient of 1.3 cm^2/s for holes in the N cathode region, and a diffusion coefficient of 20 cm^2/s for electrons in the P-base region, the emitter injection efficiency in the portion with the p-channel MOSFET is found to be 0.89526. This reduced injection efficiency and low doping level for the N cathode region reduces the effective area of the N^+ cathode region in the MCT structure resulting in an increase in the on-state voltage drop when compared with the one-dimensional thyristor structure.

Simulation Results

The results of two-dimensional numerical simulations for the 5-kV asymmetrical silicon MCT structure are described here. The total width ($W_{CELL}/2$) of the structure, as shown by the cross section in Fig. 8.1, was 15 μm (area = 1.5×10^{-7} cm^{-2}). The P-base and N^+ cathode regions were formed by using Gaussian doping profiles defined from the upper surface. In addition, an N-base and P^+ source region is incorporated into the MCT structure by performing diffusions self-aligned to the gate electrode. The N-buffer layer and P^+ collector regions were formed by using Gaussian doping profiles defined from the lower surface. The doping profiles for the baseline device structure were already shown in Figs. 8.6–8.8.

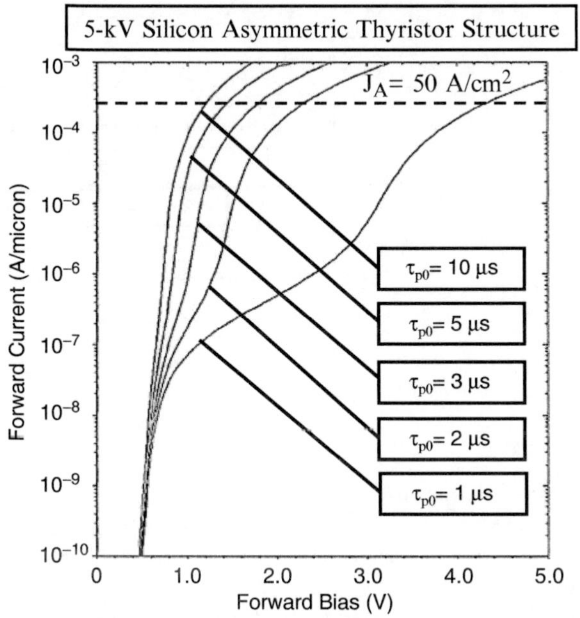

Fig. 8.15 On-state characteristics of the 5-kV asymmetric thyristor structure: lifetime dependence

To understand how closely the MCT structure resembles the thyristor structure in the on-state, the on-state characteristics of the 5-kV silicon asymmetric thyristor structure were obtained for the case of various values for the lifetime in the drift region. This device structure has the same doping profile as the 5-kV asymmetric MCT structure for the drift region, buffer-layer, and anode regions. The P-base region and N^+ cathode regions were formed using the same doping profiles as for the MCT structure. The cross section of the thyristor structure is similar to that shown in Fig. 2.5 but without the cathode short. The width of the thyristor structure used in the simulations was 500 μm for the cell. The on-state characteristics for the thyristor were obtained using numerical simulations with a turn-on gate current density of 0.02 A/cm^2. The resulting on-state characteristics for the 5-kV asymmetric thyristor structure are shown in Fig. 8.15 for the case of various lifetime values in the drift region. The current initially increases exponentially with increasing anode bias as expected. At current densities above 0.001 A/cm^2, the non-state voltage drop begins to increase more rapidly. Consequently, the on-state voltage drop increases as expected with reduction of the lifetime (τ_{p0}, τ_{n0}) indicated in the figure. The on-state voltage drop at a hole lifetime (τ_{p0}) value of 10 μs is found to be 1.202 V at an on-state current density of 50 A/cm^2 and increases to 4.286 V at a reduced hole lifetime (τ_{p0}) value of 1 μs.

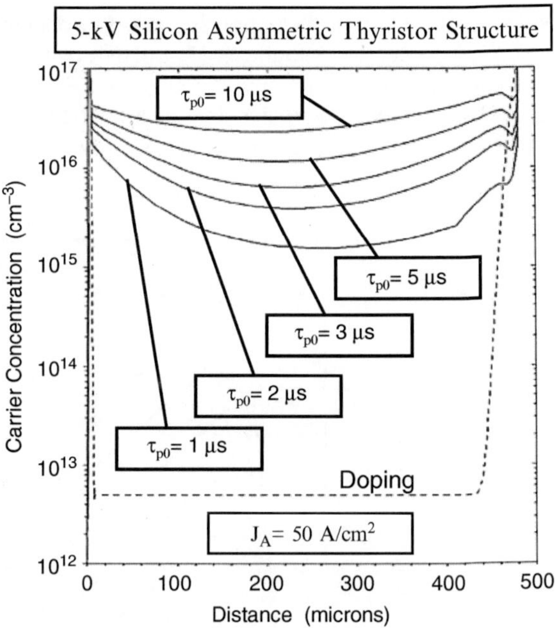

Fig. 8.16 On-state carrier distribution in the 5-kV asymmetric thyristor structure: lifetime dependence

The low on-state voltage drop for the thyristor structure is associated with the distribution of carriers injected into the N-base region producing the desired reduction of its resistance. For comparison with the MCT structure, the hole distribution in the 5-kV asymmetric thyristor structure is provided in Fig. 8.16 for five cases of the lifetime (τ_{p0}, τ_{n0}) in the drift region. It can be observed that the injected carrier density on the anode side is four orders of magnitude larger than the doping concentration on the anode side. The hole concentration is reduced in the middle of the drift region when the lifetime is reduced as expected from the P-i-N rectifier model for the thyristor [17]. It is worth pointing out that the carrier distribution is symmetric for the thyristor structure, i.e., the hole concentration on the cathode side of the drift region is equal to the hole concentration on the anode side of the drift region.

The on-state characteristics of the 5-kV silicon asymmetric MCT structure were obtained by using a positive gate bias voltage of 10 V for the case of various values for the lifetime in the drift region. This device structure has a peak buffer layer doping concentration of 1.0×10^{17} cm^{-3}. The characteristics obtained from the numerical simulations are shown in Fig. 8.17. The current initially increases exponentially with increasing anode bias. At current densities above 0.001 A/cm^2, the non-state voltage drop begins to increase more rapidly. Consequently, the on-state voltage drop increases as expected with reduction of the lifetime (τ_{p0}, τ_{n0}) indicated in the figure. The on-state voltage drop at a hole lifetime (τ_{p0}) value of 10 μs is found to be 1.33 V at an on-state current density of 50 A/cm^2 and increases to 4.92 V at a reduced hole lifetime (τ_{p0}) value of 1 μs.

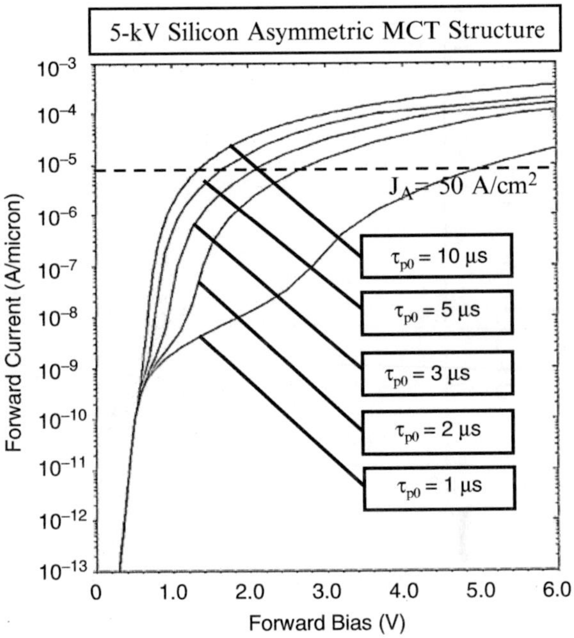

Fig. 8.17 On-state characteristics of the 5-kV asymmetric MCT structure: lifetime dependence

The variation of the on-state voltage drop obtained from the results of the numerical simulation, as a function of the lifetime in the N-base region, is shown in Fig. 8.18 for the case of an anode on-state current density of 50 A/cm^2. For comparison purposes, the on-state voltage drops for the case of the 5-kV asymmetric trench-gate IGBT structure and for the 5-kV asymmetric thyristor structure are also provided in this figure. It can be observed that the MCT structure has a significantly lower on-state voltage drop than the IGBT structure for each lifetime value. This is due to improved carrier distribution in the MCT structure with a high free carrier density near the cathode side of the drift region. However, the on-state voltage drop for the MCT structure is slightly larger than that for the thyristor structure. This is related to the incorporation of the turn-off MOSFET within the MCT structure which degrades the injection efficiency of the cathode junction.

The improved on-state voltage drop for the 5-kV asymmetric MCT structure is determined by the distribution of carriers injected into the N-base region producing the desired reduction of its resistance. The hole distribution in the 5-kV asymmetric MCT structure is provided in Fig. 8.19 for five cases of the lifetime (τ_{p0}, τ_{n0}) in the drift region. It can be observed that the injected carrier density on the anode side is four orders of magnitude larger than the doping concentration (which is similar to that observed for the IGBT structure, see Fig. 5.15). However, the free carrier concentration on the cathode side of the drift region is an order of magnitude larger in the MCT structure because the thyristor regenerative action forward biases the P-base/N-drift junction. In the MCT structure, the injected carrier density is reduced in the middle of the drift region when the lifetime

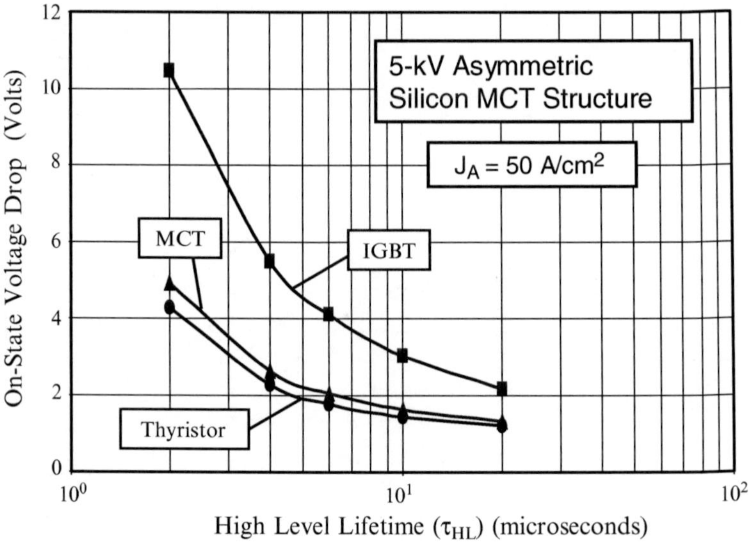

Fig. 8.18 On-state voltage drop for the 5-kV asymmetric MCT structure: N-base lifetime dependence

is reduced resulting in the observed increase in the on-state voltage drop. In comparison with the 5-kV thyristor structure, it can be observed that the hole carrier concentration is reduced by a factor of 2 in the drift region on the cathode side due to reduced injection efficiency of the cathode junction.

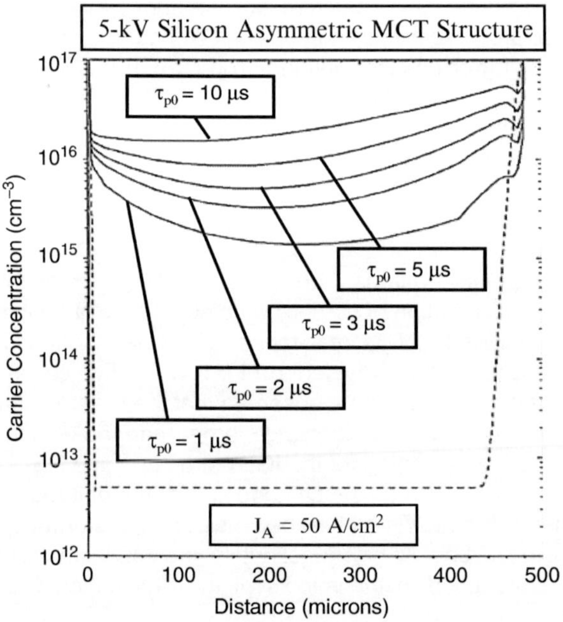

Fig. 8.19 On-state carrier distribution in the 5-kV asymmetric MCT structure: lifetime dependence

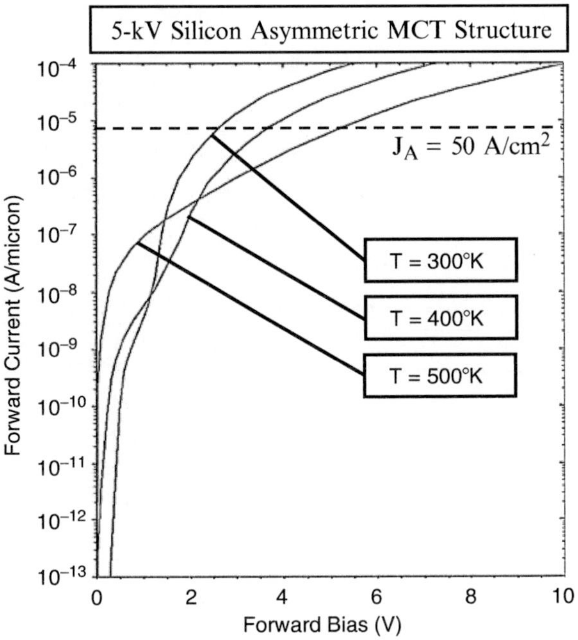

Fig. 8.20 On-state characteristics of the 5-kV asymmetric MCT structure: temperature dependence

The on-state characteristics of the 5-kV silicon asymmetric MCT structure were also obtained as a function of temperature by using a positive gate bias voltage of 10 V for the case of a lifetime of 2 μs in the drift region. This device structure has a peak buffer layer doping concentration of 1×10^{17} cm^{-3}. The characteristics obtained from the numerical simulations are shown in Fig. 8.20. At low anode current densities (below 0.01 A/cm^2), the on-state voltage drop decreases with increasing temperature while at on-state current densities above 2 A/cm^2, it begins to increase with increasing temperature. The on-state voltage drop increases with increasing temperature at an on-state current density of 50 A/cm^2, which is desirable for allowing paralleling of devices and the prevention of hot spots within the device structure.

To understand the operation of the MCT structure in the on-state, it is beneficial to examine the distribution of the current within the structure. The on-state current flow-lines within the 5-kV asymmetric MCT structure are shown in Fig. 8.21 for the case of an on-state current density of 50 A/cm^2. A high-level lifetime of 2 μs was used during this simulation. It can be observed that some of the current flows via the channel formed at the surface of the P-base region due to applied positive gate bias. However, most of the current flows via the heavily doped N$^+$ cathode region as well as the N-base region of the p-channel MOSFET which is acting as a part of the cathode region but with reduced injection efficiency.

Fig. 8.21 On-state current flow-lines for the 5-kV asymmetric MCT structure

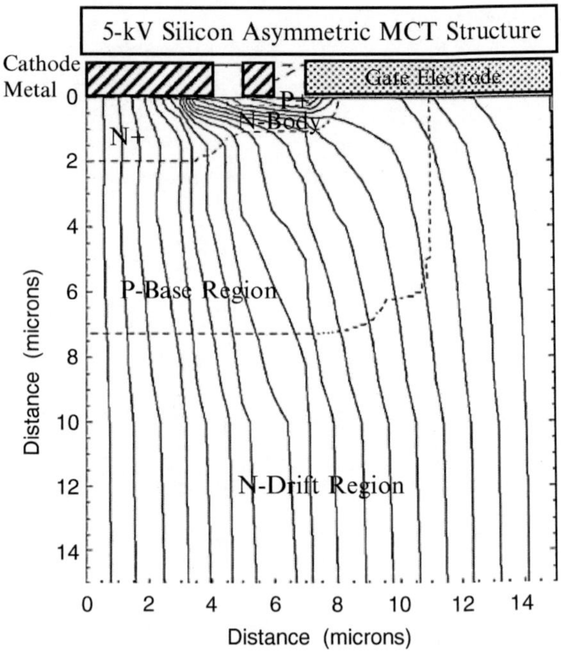

8.2.3 *Turn-Off Characteristics*

Fig. 8.22 Turn-off waveforms for the asymmetric MCT structure

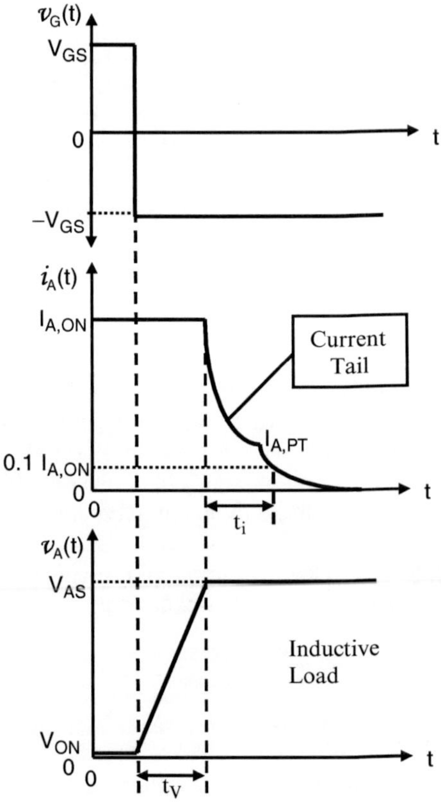

One of the important advantages of the MCT structure, when compared with the GTO structure, is the simplicity of the gate control circuit due to its MOS-gated structure. To turn off the device, the gate voltage must simply be ramped from the on-state value (nominally positive 10 V) to the off-state value (nominally negative 10 V) as illustrated in Fig. 8.22. The magnitude of the gate current can be limited by using a resistance in series with the gate voltage source. The waveform for the gate voltage shown in the figure is for the case of zero gate resistance. Once the gate voltage falls below the threshold voltage, the electron current from the channel ceases. However, the thyristor regenerative action can still continue allowing the MCT structure to remain in its on-state. To turn off the MCT structure, the gate bias must be reversed to turn on the p-channel MOSFET to shunt the hole current entering the P-base region. If the resistance of the shunting path is small, the injection of electrons from the N^+ cathode region then stops and the thyristor regenerative action ceases. In the case of an inductive load, the anode current for the MCT structure is then sustained by the hole current flow due to the presence of stored charge in the N-base region. Unlike the GTO structure, there is no prolonged storage time interval for the MCT structure during its turn-off because the shunting path is very short in length. The anode voltage begins to increase in the MCT structure almost immediately after the gate voltage reaches the negative gate supply voltage.

The anode current decreases once the anode voltage reaches the anode supply voltage as shown in the figure. For the asymmetric MCT structure, the current tail usually occurs in two parts if the anode voltage is insufficient for the space-charge region to extend completely through the N-base region. In this case, there is still some stored charge left in the N-base region near the N-buffer layer after the voltage transient is completed and the anode voltage is equal to the anode supply voltage. During the first part of the anode current decay, the stored charge in the N-base region is removed by recombination, as well the anode current flow. This is a relatively slow decay due to the large high-level lifetime in the N-base region. As the anode current decreases, the hole concentration in the space-charge region decreases allowing the space-charge region to expand even though the anode voltage is constant. Eventually, the space-charge region extends through the entire N-base region when the anode current density becomes equal to the punch-through current density ($J_{A,PT}$). At this point in time, stored charge is present only in the N-buffer layer. The stored charge in the N-buffer layer decreases by recombination at a faster pace due to the smaller lifetime in the N-buffer layer associated with its greater doping concentration than the N-base region. This produces a faster decay of the anode current during the second phase of the current tail as illustrated in the figure.

8.2.3.1 Voltage Rise-Time

The analysis of the turn-off waveform for the anode voltage transient for the asymmetric MCT structure can be performed by using the charge control principle. Since the MCT structure operates like a thyristor in the on-state, its on-state carrier distribution is similar to that for a P-i-N rectifier. The hole concentration inside the N-base region in the on-state is then given by:

$$p(y) = \frac{\tau_{HL} J_{ON}}{2qL_a}\left[\frac{\cosh(y/L_a)}{\sinh(d/L_a)} - \frac{\sinh(y/L_a)}{2\cosh(d/L_a)}\right] \tag{8.14}$$

if $y = 0$ at the middle of the N-base region.

To develop the analysis of the anode voltage transient for the asymmetric MCT structure, it will be assumed that the hole concentration profile in the N-base region does not change due to recombination during the transient. In this case, the electric field profile in the asymmetric MCT structure during the anode voltage transient is illustrated in Fig. 8.23. Although the free carrier distribution has a catenary shape for the MCT structure as given by Eq. 8.14, for simplicity of analysis, it will be assumed that the free carrier concentration in the N-base region is approximately constant. This average hole concentration in the N-base region is given by [17]:

$$p_{AV} = \frac{\tau_{HL} J_{A,ON}}{q(W_N + W_{NB})} \tag{8.15}$$

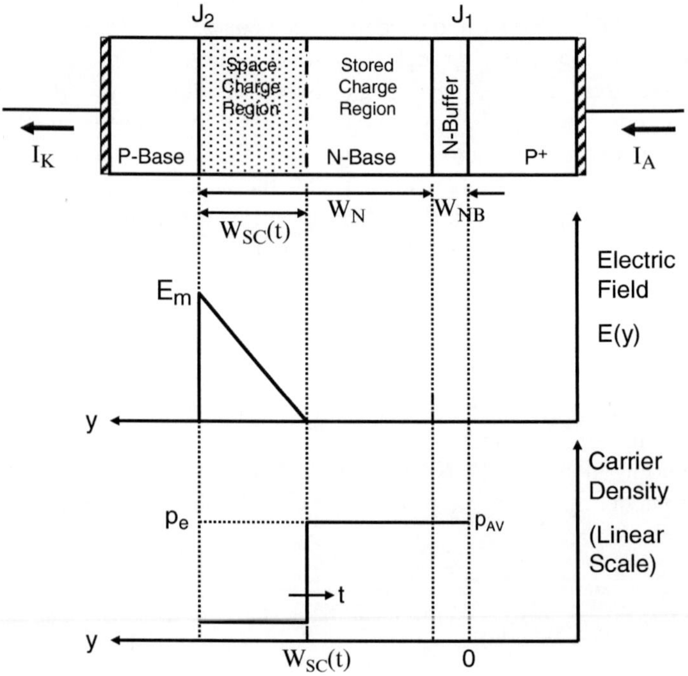

Fig. 8.23 Electric field and free carrier distribution during the voltage rise-time for the asymmetric MCT structure

because the P-base width is very narrow when compared with the width of the N-base region. This equation is derived under the assumption that all the recombination occurs in the N-base region. In the MCT structure, significant recombination also occurs in the end regions making the actual free carrier concentration about half that given by the above equation. As an example, the average free carrier concentration is about 1×10^{16} cm^{-3} for the MCT structure at a high-level lifetime of 4 μs as shown in Fig. 8.19.

As the space-charge region expands toward the anode side, holes are removed from the stored charge region at its boundary. The holes then flow through the space-charge region at their saturated drift velocity due to the high electric field in the space-charge region. Due to the high concentration of holes in the space-charge region associated with the anode current flow, the space-charge layer does not reach-through the N-base region during the voltage transient.

For the analysis of the anode voltage transient, it will be assumed that the hole distribution does not change in the stored charge region of the N-base region during the voltage transient. Consequently, the concentration of holes at the edge of the space-charge region (p_e) during the turn-off process is constant as given by p_{AV}. According to the charge-control principle, the charge removed by the expansion of the space-charge layer must equal the charge removed due to the anode current flow:

$$J_{A,ON} = qp_e(t)\frac{dW_{SC}(t)}{dt} = qp_{AV}\frac{dW_{SC}(t)}{dt} \tag{8.16}$$

Integrating this equation on both sides and applying the boundary condition of width [$W_{SC}(0)$] for the space-charge layer at time zero provides the solution for the evolution of the space-charge region width with time:

$$W_{SC}(t) = \frac{J_{A,ON}}{qp_{AV}}t + W_{SC}(0) \tag{8.17}$$

The initial width of the space-charge region can be obtained by using:

$$W_{SC}(0) = \sqrt{\frac{2\varepsilon_S V_{A,ON}}{q(N_D + p_{SC})}} \tag{8.18}$$

where $V_{A,ON}$ is the on-state voltage drop for the MCT structure. The space-charge layer expands toward the right-hand side as indicated by the horizontal time arrow in Fig. 8.23 with the hole concentration profile in the stored-charge region remaining unchanged.

The anode voltage supported by the asymmetric MCT structure is related to the space-charge layer width by:

$$V_A(t) = \frac{q(N_D + p_{SC})W_{SC}^2(t)}{2\varepsilon_S} \tag{8.19}$$

The hole concentration in the space-charge layer can be related to the anode current density under the assumption that the carriers are moving at the saturated drift velocity in the space-charge layer:

$$p_{SC} = \frac{J_{A,ON}}{qv_{sat,p}} \tag{8.20}$$

The hole concentration in the space-charge region remains constant during the voltage rise-time because the anode current density is constant. Consequently, the slope of the electric field profile in the space-charge region also becomes independent of time.

Applying the solution for the evolution of the space-charge layer from Eq. 8.18 in Eq. 8.19:

$$V_A(t) = \frac{q(N_D + p_{SC})}{2\varepsilon_S}\left[\frac{J_{A,ON}}{qp_{AV}}t + W_{SC}(0)\right]^2 \tag{8.21}$$

The analytical model for turn-off of the asymmetric MCT structure under inductive load conditions predicts an increase in the anode voltage as the square of time. This analytical model does not include the influence of carrier generation due to the impact ionization process at larger anode bias voltages. Impact ionization introduces additional holes and electrons into the space-charge region resulting in a reduction of the rate of rise of the anode voltage prolonging the voltage rise-time.

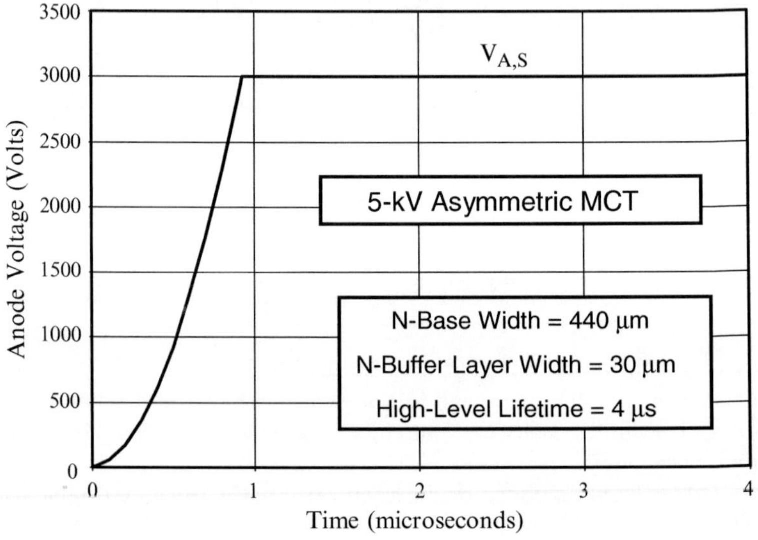

Fig. 8.24 Anode voltage transient during turn-off for the asymmetric MCT structure

The end of the first phase of the turn-off process, where the anode voltage increases while the anode current remains constant, occurs when the anode voltage reaches the anode supply voltage ($V_{A,S}$). This time interval ($t_{V,OFF}$) can be obtained by making the anode voltage equal to the anode supply voltage in Eq. 8.21:

$$t_{V,OFF} = \frac{qp_{AV}}{J_{A,ON}}\left[\sqrt{\frac{2\varepsilon_S V_{A,S}}{q(N_D + p_{SC})}} - W_{SC}(0)\right] \tag{8.22}$$

According to the analytical model, the voltage rise-time is proportional to the square root of the anode bias supply voltage. However, it is only weakly dependent on the on-state current density (through p_{SC}) because the hole concentration p_{AV} is approximately proportional to the on-state current density.

Consider the case of a 5-kV asymmetric MCT structure with N-base and N-buffer layer widths of 440 and 30 μm, respectively. The anode voltage transient obtained using the above analytical model for the case of a high-level lifetime of 4 μs in the N-base region is shown in Fig. 8.24. The voltage increases non-linearly with time with an approximately square-law shape. The time interval for the voltage transient ($t_{V,OFF}$) obtained using the analytical model is 0.92 μs.

The width of the space-charge layer at the end of the voltage transient can be obtained by using the collector supply voltage:

$$W_{SC}\left(t_{V,OFF}\right) = \sqrt{\frac{2\varepsilon_S V_{A,S}}{q(N_D + p_{SC})}} \tag{8.23}$$

It can be concluded that the width of the space charge layer at the end of the first phase depends upon the anode supply voltage and the initial on-state current density (via p_{SC}). In a typical 5-kV asymmetric MCT structure, the space-charge layer width obtained by using the above equation with an on-state current density of 50 A/cm^2 and a collector supply voltage of 3,000 V is 327 μm if the doping concentration of the N-base region is 5×10^{12} cm^{-3}. The space-charge region width at the end of the voltage transient is therefore about 100 μm smaller than the width of the N-base region for a typical anode supply voltage of 3,000 V. Consequently, a substantial amount of stored charge remains in the N-base region after the voltage transient.

8.2.3.2 Current Fall-Time

During the second phase of the turn-off process, the anode current decays while the anode voltage remains fixed at the anode supply voltage. The decay of the anode current occurs in two parts for the asymmetric MCT structure in the same manner as for the IGBT structure (see Chap. 5). At the end of the voltage transient, there is a substantial amount of stored charge in the N-base region. Consequently, during the first part, the anode current flow is governed by the recombination of the excess holes that are trapped within the N-base region under high-level injection conditions. At the same time, holes and electrons are also removed from the stored charge region due to the anode current flow.

As the anode current decreases, the hole concentration in the space-charge region also decreases. Consequently, the space-charge region expands during the first part of the current transient until the space-charge region covers the entire width (W_N) of the N-base region. After this time, the space-charge region width cannot increase any further due to the high doping concentration in the N-buffer layer. Consequently, during the second part of the anode current transient, the anode current flow is governed by the recombination of holes in the N-buffer layer under low-level injection conditions. The second part of the anode current decay occurs at a much faster rate than during the first part due to the smaller lifetime in the N-buffer layer associated with its larger doping concentration.

The anode current decay during the first part occurs until the space-charge region extends completely through the N-base region. The space-charge region punches through to the N-buffer layer at a unique anode current density which is independent of the lifetime in the N-base region. This punch-through anode current density can be derived by equating the space-charge layer width to the width of the N-base region (W_N) and using Eq. 8.20 for the hole concentration in the space-charge region:

$$J_{A,PT} = \frac{2v_{sat,p}\varepsilon_S V_{A,S}}{W_N^2} - qv_{sat,p}N_D \tag{8.24}$$

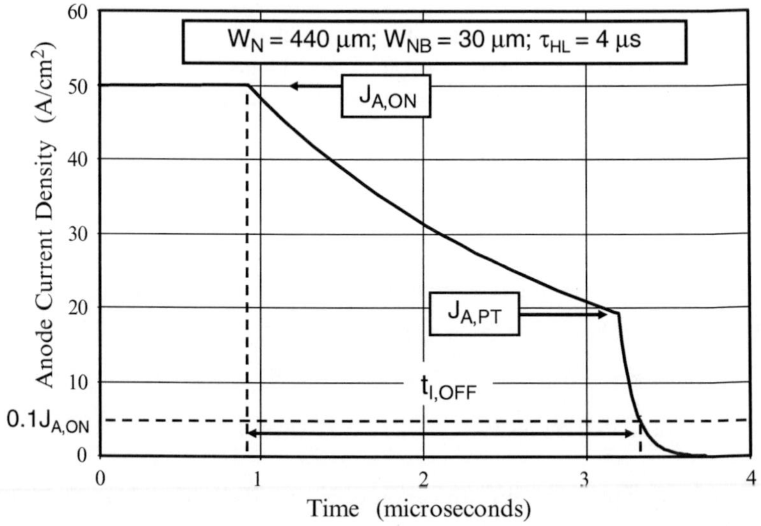

Fig. 8.25 Anode current transient during turn-off for the 5-kV asymmetric MCT structure

Consider the case of a 5-kV asymmetric MCT structure with N-base and N-buffer layer widths of 440 and 30 μm, respectively, and an N-buffer layer doping concentration of 1×10^{17} cm^{-3}. The anode current waveform predicted by the analytical model presented in this section is shown in Fig. 8.25 for a high-level lifetime of 4 μs in the N-base region. It can be observed that the anode current decays in two stages. During the first part, the decay is much faster than predicted by a simple exponential variation with recombination occurring at the high-level lifetime in the N-base region. When the anode current reaches the punch-through current density ($J_{A,PT}$) of 19.3 A/cm^2 predicted by Eq. 8.24, the second part of the anode current transient begins to occur. The current fall time ($t_{I,OFF}$) is usually defined as the time taken for the anode current to reach one-tenth of the on-state value as shown in Fig. 8.25.

Simulation Example

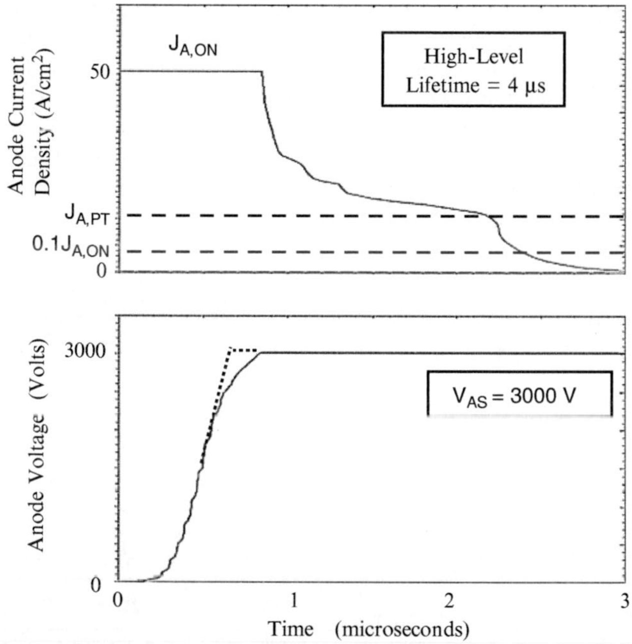

Fig. 8.26 Typical turn-off waveforms for the asymmetric 5-kV MCT structure

To gain insight into the operation of the asymmetric 5-kV MCT structure during its turn-off, the results of two-dimensional numerical simulations for a typical structure are discussed here. The device structure used has the cross section shown in Fig. 8.1 with a cell half-width of 15 μm. The doping profile for the MCT structure used in the numerical simulations was provided in Figs. 8.6–8.8. The widths of the uniformly doped N-base region and the diffused N-buffer layer are 440 and 30 μm, respectively. For the typical case discussed here, a high-level lifetime of 4 μs was used in the N-base region.

The numerical simulations were performed with an abrupt reduction of the gate voltage from positive 10 V to negative 10 V in 20 ns starting from an on-state current density of 50 A/cm^2. The resulting waveforms obtained from the numerical simulations for the anode voltage and current are shown in Fig. 8.26 for the case of an anode supply voltage of 3,000 V. Unlike the GTO structure, there is no storage time associated with the turn-off of the MCT structure due to the small cell structure. The anode voltage increases immediately at the end of the of the gate voltage transient. The anode voltage increases as the square of the time as predicted by the analytical model until it reaches about 2,000 V. It then increases at a slower rate. This is associated with the onset of avalanche multiplication at high anode bias voltages – an effect not included in the analytical model. The dotted line in the figure provides an extrapolation of the anode voltage transient without the effect of impact ionization. The anode voltage rise-time obtained using this extrapolated line is 0.7 μs. The anode voltage rise-time obtained in the numerical simulations for the case of supply voltage of 3,000 V including the effect of impact ionization is larger (0.85 μs).

After the completion of the anode voltage transient, the anode current waveform decays from the initial on-state current density at a rate that decreases with time. The current decays to the punch-through current density (indicated in the figure) in 1.4 μs. A punch-through current density of about 15 A/cm^2 is observed in the numerical simulations. After reaching the punch-through current density, the anode current is observed to decay at a faster rate as described by the analytical model.

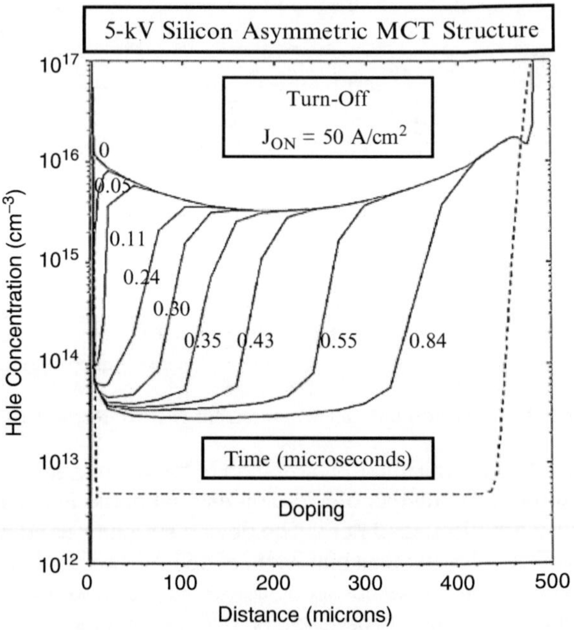

Fig. 8.27 Hole carrier distribution for the 5-kV MCT turn-off transient during the voltage rise-time

A one-dimensional view of the minority carrier distribution in the 5-kV asymmetric MCT structure is shown in Fig. 8.27 from the initial steady-state operating point ($t = 0$ μs) to the end of the voltage rise-time ($t = 0.84$ μs). These carrier profiles were taken at $x = 1$ μm through the P-base region. The initial carrier distribution has the distribution like a P-i-N rectifier. It can be observed from Fig. 8.27 that the carrier distribution in the N-base region near the anode does not change during the anode voltage rise phase. It can be seen that the carrier concentration at the P-base/N-drift junction rapidly reduces to zero within the first 50 ns allowing the junction to support an increase in the anode voltage. A significant space-charge region begins to form immediately during the turn-off and expands toward the right-hand side demonstrating that there is no storage phase for the MCT structure. At larger anode voltages, the hole concentration in the space-charge region is about 3×10^{13} cm^{-3}, which is consistent with the value for p_{SC} obtained using the analytical model with the carriers moving at the saturated drift velocity and an on-state current density of 50 A/cm^2. The width of the space-charge region can be observed to be about 330 μm when the collector voltage reaches 3,000 V, which is close to that predicted by the analytical model.

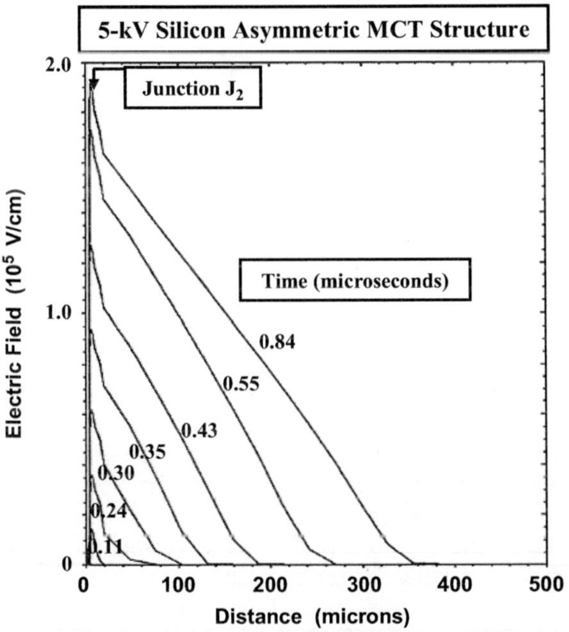

Fig. 8.28 Electric field distribution for the 5-kV asymmetric MCT during the voltage rise-time

The electric field profiles in the 5-kV asymmetric MCT structure obtained from the numerical simulations are shown in Fig. 8.28 for various time instances during the voltage rise-time. It can be observed that the peak electric field occurs at the P-base/N-base junction (J_2) as expected. The peak electric field increases with time due to supporting a larger anode voltage. The electric field is triangular in shape, unlike during the blocking mode (see Fig. 8.12), even at high anode bias voltages due to the large hole charge in the space-charge region. It can also be observed that the slope of the electric field profile decreases when the time exceeds 0.55 μs and the anode voltage reaches 2,000 V, because of the addition of electrons in the space-charge region due to impact ionization. The negative charge of the electrons counteracts the positive charge of the holes producing a larger space-charge region width at the end of the voltage transient than predicted by the analytical model based on just the hole charge.

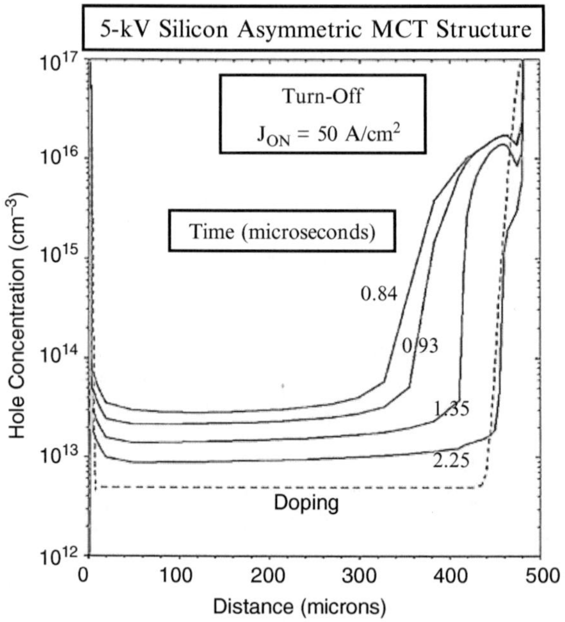

Fig. 8.29 Hole carrier distribution for the 5-kV asymmetric MCT turn-off transient during the current tail-time

A one-dimensional view of the hole carrier distribution in the 5-kV asymmetric MCT structure is shown in Fig. 8.29 during the current tail time. The anode voltage was held constant at the collector supply voltage of 3,000 V during this transient. The hole concentration in the stored charge region begins to decrease immediately after the end of the voltage transient due to the recombination process and the removal of holes and electrons by the anode current flow. At the same time, the space-charge region expands in spite of a constant anode voltage because the hole concentration in the space-charge region reduces. From Fig. 8.29, it can be observed that all the holes in the N-base region have been removed at time $t = 2.2$ μs corresponding to the end of the first phase of the collector current transient (see Fig. 8.26). Subsequently, the holes remaining in the N-buffer layer are at concentrations well below its doping concentration. Consequently, the recombination of holes in the N-buffer layer during the second part of the anode current transient occurs under low-level injection conditions as assumed in the analytical model.

8.2.4 Lifetime Dependence

The optimization of the power losses for the MCT structure requires performing a trade-off between the on-state voltage drop and the switching losses. One approach to achieve this is by adjusting the lifetime in the drift (N-base) region. A reduction of the lifetime in the drift region also alters the lifetime in the N-buffer layer. The impact of reducing the lifetime on the on-state voltage drop was previously shown in Sect. 8.2.2. The on-state voltage drop increases when the lifetime is reduced. The analytical model developed for turn-off of the asymmetric MCT structure presented in the previous section can be used to analyze the impact of changes to the lifetime in the drift region.

Simulation Example

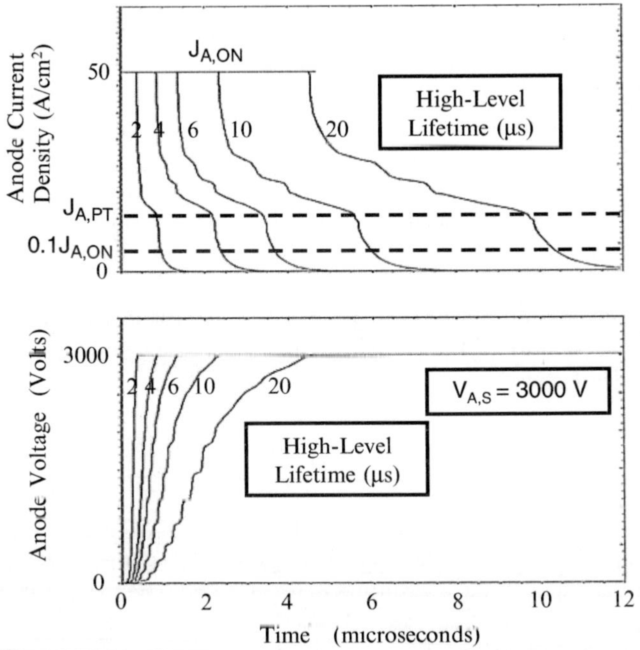

Fig. 8.30 Impact of lifetime on the 5-kV asymmetric MCT turn-off waveforms

To gain insight into the impact of the lifetime in the N-base region on the operation of the 5-kV asymmetric MCT structure, the results of two-dimensional numerical simulations for a typical structure are discussed here. The device structure used has the cross section shown in Fig. 8.1 with a half-cell width of 15 μm. The widths of the N-base and N-buffer layer regions are 440, and 30 μm, respectively. The high-level lifetime in the N-base region was varied between 2 and 20 μs. For turning off the MCT structures, the numerical simulations were performed with gate voltage rapidly ramped down from positive 10 V to negative 10 V in 20 ns starting from an on-state current density of 50 A/cm^2. The resulting waveforms obtained from the numerical simulations for the anode voltage and current are shown in Fig. 8.30 for the case of an anode supply voltage of 3,000 V.

The numerical simulations show a decrease in the voltage rise-time with reduction of the lifetime in the N-base region. The numerical simulations of the 5-kV asymmetrical MCT structure also show a substantial increase in the anode current fall time when the lifetime increases. The numerical simulations show a reduction of the anode current during the first part of the decay to the punch-through anode current ($J_{A,PT}$) which is independent of the lifetime in the N-base region as predicted by the analytical model.

8.2.5 Switching Energy Loss

The power loss incurred during the switching transients limit the maximum operating frequency for the MCT structure. Power losses during the turn-on of the MCT structure are significant but strongly dependent on the reverse recovery behavior of the fly-back rectifiers in circuits. Consequently, it is common practice to use only the turn-off energy loss per cycle during characterization of MCT devices. The turn-off losses are associated with the voltage rise-time interval and the current fall-time interval. The energy loss for each event can be computed by integration of the power loss, as given by the product of the instantaneous current and voltage. During the voltage rise-time interval, the anode current is constant while the voltage increases as the square of time. The energy loss during the voltage rise-time interval can be computed using:

$$E_{\mathrm{OFF,V}} = \frac{1}{3} J_{\mathrm{A,ON}} V_{\mathrm{A,S}} t_{\mathrm{V,OFF}} \tag{8.25}$$

For the typical switching waveforms for the 5-kV asymmetric MCT structure shown in Fig. 8.26 with an anode supply voltage of 3,000 V, the energy loss per unit area during the anode voltage rise-time is found to be 0.042 J/cm^2 if the on-state current density is 50 A/cm^2.

During the anode current fall-time interval, the anode voltage is constant while the current decreases in two phases. To simplify the analysis, the energy loss during the anode current fall-time interval will be computed using:

$$E_{\mathrm{OFF,I}} = \frac{1}{2} J_{\mathrm{A,ON}} V_{\mathrm{AS}} t_{\mathrm{I,OFF}} \tag{8.26}$$

For the typical switching waveforms for the 5-kV asymmetric MCT structure shown in Fig. 8.26 with an anode supply voltage of 3,000 V, the energy loss per unit area during the collector current fall-time is found to be 0.116 J/cm^2 if the on-state current density is 50 A/cm^2. The total energy loss per unit area ($E_{OFF,V}$ + $E_{OFF,I}$) during the turn-off process for the 5-kV asymmetric MCT structure is found to be 0.159 J/cm^2.

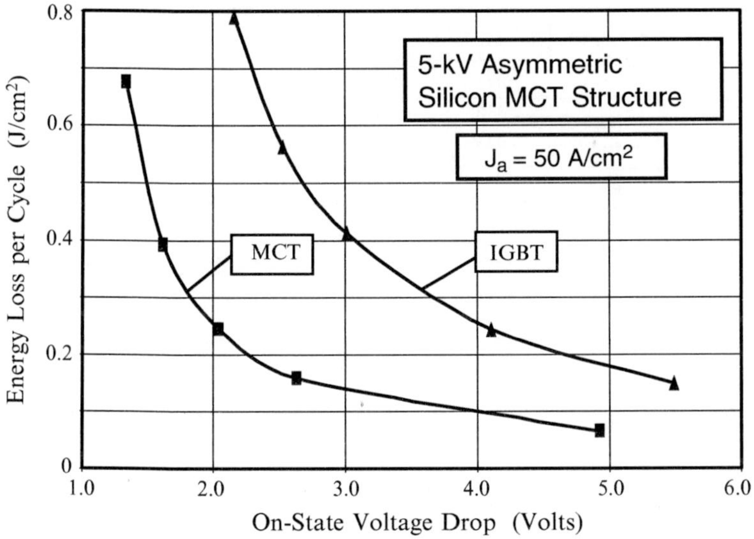

Fig. 8.31 Trade-off curve for the silicon 5-kV asymmetric MCT structure: lifetime in N-base region

Using the results obtained from the numerical simulations, the on-state voltage drop and the total energy loss per cycle can be computed. These values are plotted in Fig. 8.31 to create a trade-off curve to optimize the performance of the silicon 5-kV asymmetric MCT structure by varying the lifetime in the N-base region. Devices used in lower frequency circuits would be chosen from the left-hand side of the trade-off curve while devices used in higher frequency circuits would be chosen from the right-hand side of the trade-off curve. For comparison purposes, the trade-off curve for the 5-kV IGBT structure is also included in the figure. It can be observed from this figure that substantial improvement in the power loss trade-off curve can be obtained by replacing the IGBT with the MCT structure. This was the original basis for the proposal and development of MCT structures. However, in actual practice, it is not possible to simply replace the IGBT with the MCT because of the very limited forward-biased safe operating area (FBSOA) of the MCT structure. The poor FBSOA of the MCT structure requires implementation of snubbers to control the rate of rise of the current when the device is turned on in motor control applications (which increases power losses) [15]. This is essential to prevent extremely large reverse recovery currents in the fly-back diodes used in typical H-bridge circuits.

8.2.6 *Maximum Operating Frequency*

High-Level Lifetime (μs)	On-State Voltage Drop (Volts)	On-State Power Dissipation (W/cm²)	Energy Loss per Cycle (J/cm²)	Maximum Operating Frequency (Hz)
20	1.327	33.2	0.675	247
10	1.619	40.5	0.393	406
6	2.045	51.1	0.245	608
4	2.630	65.8	0.159	846
2	4.923	123	0.065	1183

Fig. 8.32 Power loss analysis for the 5-kV asymmetric MCT structure

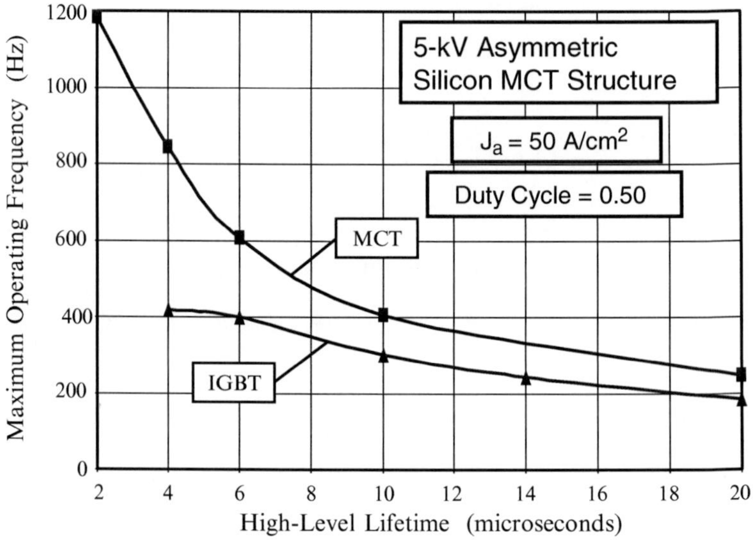

Fig. 8.33 Maximum operating frequency for the 5-kV asymmetric MCT structure

The maximum operating frequency for operation of the 5-kV asymmetric MCT structure can be obtained by combining the on-state and switching power losses:

$$P_{D,TOTAL} = \delta \, P_{D,ON} + E_{OFF} f \qquad (8.27)$$

where δ is the duty cycle and f is the switching frequency. In the case of the baseline asymmetric MCT device structure with a high-level lifetime of 4 μs in the N-base region, the on-state voltage drop is 2.63 V at an on-state current density of 50 A/cm^2. For the case of a 50% duty cycle, the on-state power dissipation contributes 66 W/cm^2 to the total power loss. Using a total turn-off energy loss per cycle of 0.159 J/cm^2 in Eq. 8.27 yields a maximum operating frequency of about 850 Hz if the total power dissipation is 200 W/cm^2.

The maximum operating frequency for the silicon 5-kV asymmetric MCT structure can be increased by reducing the lifetime in the N-base region. Using the results obtained from the numerical simulations, the on-state voltage drop and the energy loss per cycle can be computed. These values are provided in Fig. 8.32 together with the maximum operating frequency as a function of the high level lifetime in the N-base region under the assumption of a 50% duty cycle and a total power dissipation limit of 200 W/cm^2. The maximum operating frequency is plotted in Fig. 8.33 as a function of the high-level lifetime in the N-base region for the case of a duty-cycle of 50%. It can be observed that the maximum operating frequency can be increased up to 1,200 Hz by reducing the high-level lifetime to 2 μs. This is much superior to the maximum operating frequency of 150 Hz for the 5-kV GTO structure and 400-Hz for the 5-kV IGBT structure.

8.3 10,000-V Silicon MCT

The 10-kV silicon asymmetric MCT structure can be expected to function just like the 5-kV device. However, its design and operation is constrained by the larger blocking voltage capability. The lifetime in the N-base region for the 10-kV device must be larger to maintain a reasonable on-state voltage drop. The larger N-base width results in more stored charge within the structure which limits the switching frequency.

In Chap. 4, it was demonstrated that the GTO structure has a limited reverse-biased safe operating area (RBSOA) due to influence of the holes in the space-charge region due to current flow. The analysis of the RBSOA for the MCT structure is identical to that provided in Sect. 4.4. Using the results shown in Fig. 4.57, it can be concluded that to turn off the 10-kV asymmetric IGBT structure with a collector supply voltage of 6 kV, it is necessary to reduce the collector current density to only 20 A/cm^2. However, one of the merits of the MCT structure is the low on-state voltage drop which allows its operation at an on-state current density of 50 A/cm^2. This value will therefore be utilized when determining the on-state voltage drop and switching transients for the 10-kV asymmetric MCT structures. Due to RBSOA limitations, the anode supply voltage for the switching transient must be reduced to 5,000 V.

8.3.1 Blocking Characteristics

The electric field distribution within the asymmetric MCT structure is essentially the same as that illustrated in Fig. 4.3 for the asymmetric GTO structure. Consequently, the design procedure described in Chap. 4 can be applied to the asymmetric MCT structure. From Fig. 4.50, the N-base region width required to obtain a forward blocking voltage of 11,000 V is 1,100 μm. However, the results of the numerical simulation shown in Chap. 4 for the 10-kV GTO structure demonstrate that an N-base width of 800 μm is sufficient.

Simulation Example

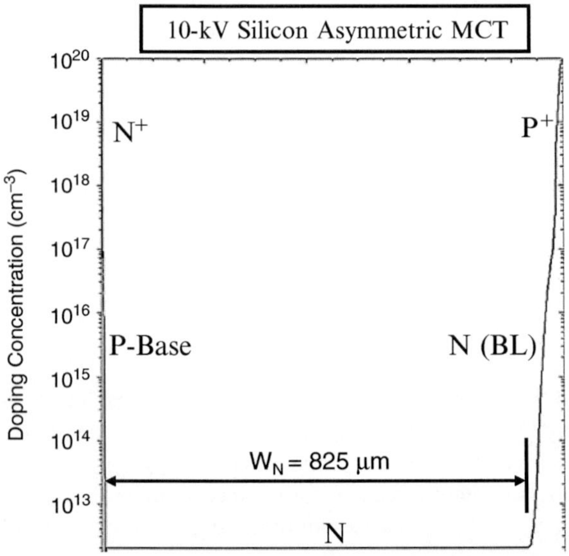

Fig. 8.34 Doping profile for the simulated asymmetric 10-kV MCT structure

To gain insight into the physics of operation for the 10-kV asymmetric MCT structure under voltage blocking conditions, the results of two-dimensional numerical simulations are described here for a device with N-base width of 825 μm. The simulations were performed using a cell with the structure shown in Fig. 8.1. This half-cell has a width of 15 μm (area = 1.5×10^{-7} cm^{-2}). The asymmetric MCT structure used for the simulations was formed by diffusions performed into a uniformly doped N-type drift region with a doping concentration of 2×10^{12} cm^{-3}. All the diffusions in the 10-kV structure had the same parameters as the 5-kV device described in the previous section. The doping profile in the vertical direction through the N$^+$ cathode region is shown in Fig. 8.34 indicating the net width of the lightly doped portion of the N-base region is 825 μm after accounting for the diffusions. The P-base and N$^+$ cathode regions are too shallow to be observed in this figure. Their doping profiles are the same as those for the 5-kV asymmetric MCT structure previously shown in Figs. 8.7 and 8.8.

The forward blocking capability of the 10-kV silicon asymmetric MCT structure was obtained by increasing the anode bias while maintaining the gate electrode at negative 10 V to short the cathode to the P-base region via the p-channel MOSFET. The characteristic obtained for a lifetime (τ_{p0}) of 10 μs is shown in

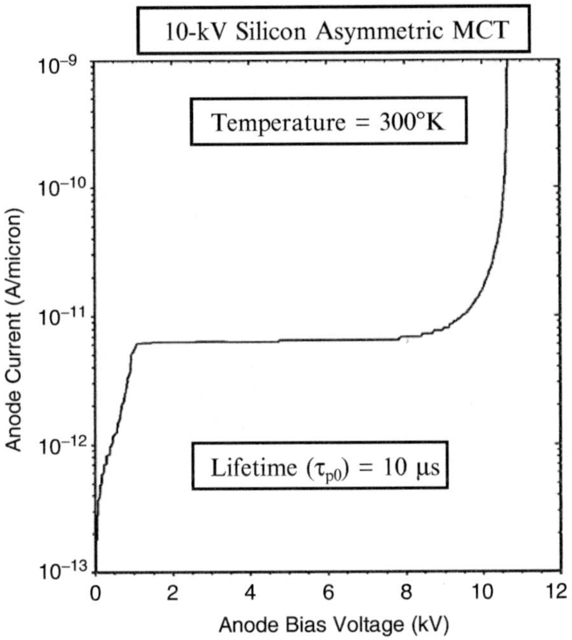

Fig. 8.35 Forward blocking characteristics for the 10-kV asymmetric MCT structure

Fig. 8.35. The leakage current increases rapidly with increasing anode bias voltage until about 1,000 V. This occurs due to the increase in the space-charge generation volume and the increase in the current gain (α_{PNP}) of the open base P-N-P transistor until the anode bias becomes equal to the reach-through voltage of 1,115 V obtained using the analytical solution given by Eq. 4.2. The leakage current then becomes independent of the anode voltage until close to the breakdown voltage. This behavior is well described by the analytical model. The numerical simulations indicate that a breakdown voltage of 10,500 V is possible with an N-base width of only 825 μm.

The voltage is primarily supported within the lightly doped portion of N-base region in the 10-kV asymmetric MCT structure during operation in the forward blocking mode. This is illustrated in Fig. 8.36 where the electric field profiles are shown during operation in the forward blocking mode at several anode voltages. It can be observed that the P-Base/N-base junction (J_2) becomes reverse biased during the forward blocking mode with the depletion region extending toward the right-hand side with increasing (positive) anode bias. The electric field has a triangular shape until the entire lightly doped portion of the N-base region becomes completely depleted. This occurs at an anode bias just above 1,000 V in good agreement with the reach-through voltage of 1,115 V obtained using the analytical solution (see Eq. 4.2). The electric field profile then takes a trapezoidal shape due to the high doping concentration in the N-buffer layer.

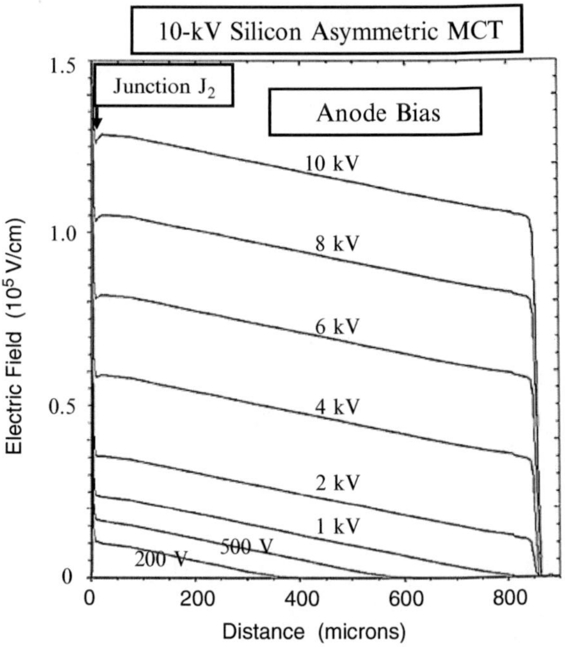

Fig. 8.36 Electric field profiles in the forward blocking mode for the 10-kV asymmetric MCT structure

8.3.2 *On-State Voltage Drop*

The on-state *i–v* characteristics and on-state voltage drop can be computed using the analytical model discussed in Sect. 8.2.2. In general, a larger lifetime is required in the N-base region for the 10-kV device when compared with the 5-kV device due to the larger width for the N-base region.

Simulation Results

The results of two-dimensional numerical simulations for the 10-kV asymmetrical silicon MCT structure are described here. The total half-cell width of the structure, as shown by the cross section in Fig. 8.1, was 15 μm (area $= 1.5 \times 10^{-7}$ cm^{-2}).

The on-state characteristics of the 10-kV silicon asymmetric MCT structure were obtained by using a gate bias voltage of 10 V using various values for the lifetime in the N-base region. The characteristics obtained from the numerical simulations are shown in Fig. 8.37. It can be observed that the on-state voltage drop increases as expected with reduction of the lifetime (τ_{p0}, τ_{n0}) indicated in the figure. The on-state voltage drop for the 10-kV asymmetric MCT structure is substantially smaller than that for the 10-kV asymmetric IGBT structure. For this reason, it is possible to operate the 10-kV asymmetric MCT structure at a larger on-state current density of 50 A/cm^2 from a power loss standpoint. However, due to the limitations of RBSOA, the maximum supply voltage must be reduced to 5,000 V.

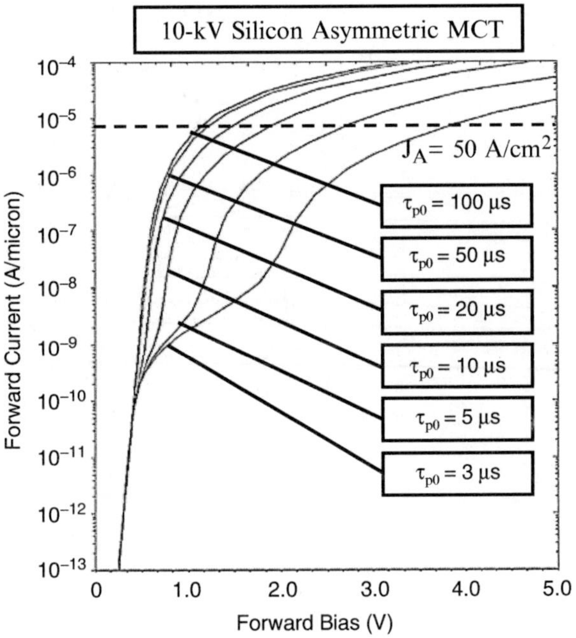

Fig. 8.37 On-state characteristics of the 10-kV asymmetric MCT structure

The good on-state voltage drop for the 10-kV asymmetric MCT structure for larger values of the lifetime in the N-base region is due to the large number of carriers injected into the drift region producing a drastic reduction of its resistance. This is illustrated in Fig. 8.38 where the injected carrier density is shown for seven cases of the lifetime (τ_{p0}, τ_{n0}) in the N-base region of the MCT structure. It can be observed that the injected carrier density is more than three orders of magnitude larger than the doping concentration for the case of a lifetime of 100 μs. The injected carrier density is reduced by a factor of three times near the anode junction when the lifetime is reduced to 3 μs. There is a significant reduction in the injected carrier density in the middle of the drift region when the lifetime is reduced below 10 μs. This is due to the relatively large width for the N-base region when compared with the 5-kV silicon MCT structure. The reduced hole concentration in the drift region produces the observed increase in on-state voltage drop.

The variation of the on-state voltage drop obtained from the results of the numerical simulation, as a function of the lifetime in the N-base region, is shown in Fig. 8.39 for the case of an anode on-state current density of 50 A/cm^2. For comparison purposes, the on-state voltage drop for the case of the 10-kV asymmetric trench-gate IGBT structure at an anode on-state current density of 50 A/cm^2 is also provided in this figure. It can be observed that the MCT structure has a significantly lower on-state voltage drop than the IGBT structure for each lifetime value in spite of the use of the trench-gate structure with high channel density for the IGBT structure. This is due to improved carrier distribution in the MCT structure with a high free carrier density near the cathode side of the drift region.

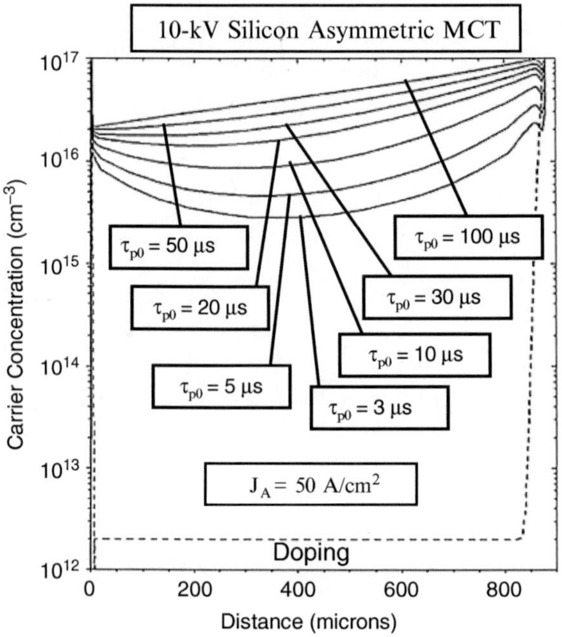

Fig. 8.38 On-state carrier distribution in the 10-kV asymmetric MCT structure

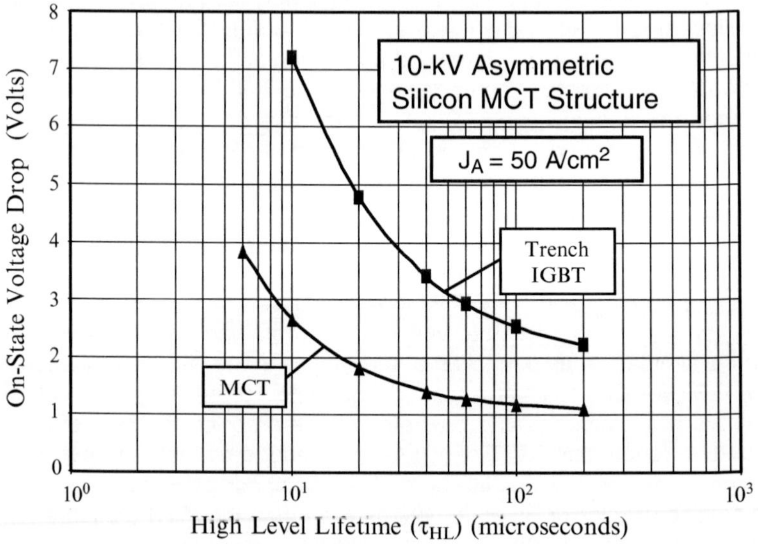

Fig. 8.39 On-state voltage drop for the 10-kV asymmetric MCT structure: N-base lifetime dependence

8.3.3 Turn-Off Characteristics

The physics for turn-off of the 10-kV silicon asymmetric MCT structure can be expected to be the same as that for the 5-kV device structure. Due to limitations with the RBSOA (as discussed in Chap. 4 for the silicon GTO structure), the 10-kV asymmetric MCT structure can be operated at an on-state current density of 50 A/cm^2 only if the anode supply voltage is reduced to 5,000 V. The results of numerical simulations of the 10-kV asymmetric MCT structure under these turn-off conditions are discussed here. For comparison purposes, the results of numerical simulations of the 10-kV asymmetric IGBT structure under the same conditions are also provided.

During the voltage rise-time interval for the 10-kV asymmetric MCT and IGBT structures, the anode/collector current is constant while the voltage increases in a highly nonlinear manner. For simplicity, the energy loss during the voltage rise-time interval can be computed using:

$$E_{\mathrm{OFF,V}} = \frac{1}{2} J_{\mathrm{A,ON}} V_{\mathrm{A,S}} t_{\mathrm{V,OFF}} \tag{8.28}$$

for both devices. Similarly, during the anode/collector current fall-time interval, the anode/collector voltage is constant while the current decreases in two phases. To simplify the analysis, the energy loss during the anode/collector current fall-time interval will be computed using:

$$E_{\mathrm{OFF,I}} = \frac{1}{2} J_{\mathrm{A,ON}} V_{\mathrm{A,S}} t_{\mathrm{I,OFF}} \tag{8.29}$$

for both the devices.

Simulation Results

Numerical simulations of the turn-off for the 10-kV silicon MCT structure with a high-level lifetime of 20 μs were performed by stepping the gate voltage down from positive 10 V to negative 10 V in 20 ns using an on-state current density of 50 A/cm^2. The resulting waveforms obtained from the numerical simulations for the anode voltage and current are shown in Fig. 8.40 for the case of an anode supply voltage of 5,000 V. It can be observed that there is no storage time for the 10-kV asymmetric MCT structure. The anode voltage initially increases non-linearly as described by the analytical model during the early stages of the voltage rise but the rate of rise becomes severely reduced at anode voltages beyond 4,000 V due to the onset of significant impact ionization as the RBSOA boundary is approached. The anode voltage almost saturates at 5,000 V indicating operation of the device close to its RBSOA limit. This is consistent with the predictions of the analytical model for the RBSOA of the MCT structure (see Fig. 4.57). The voltage rise-time is found to be 12 μs from the numerical simulations for the 10-kV asymmetric MCT structure and the corresponding energy loss per cycle is found to be 1.5 J/cm^2.

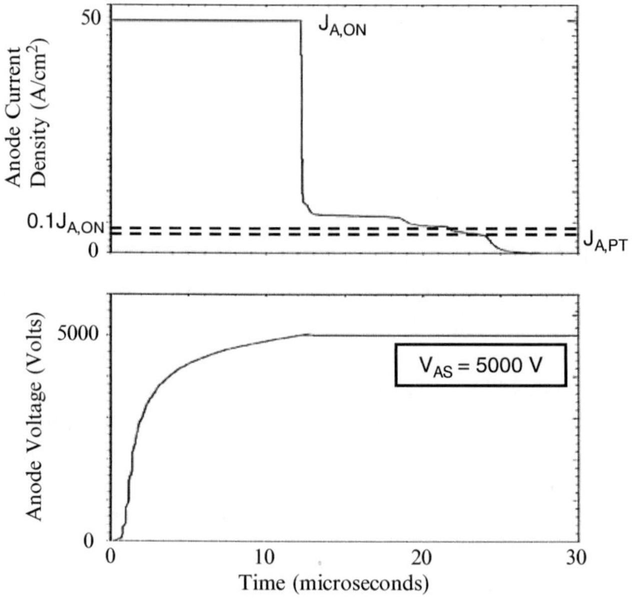

Fig. 8.40 Turn-off waveforms for the 10-kV asymmetric MCT

The anode current turn-off occurs with a rapid initial decrease in current to about 10 A/cm^2 followed by a gradual change as expected from the analytical model. The punch-through anode current density ($J_{A,PT}$) obtained using the analytical model (Eq. 8.24) is 9 A/cm^2 but the simulation results indicate a smaller value of about 4 A/cm^2. Due to the low punch-through anode current density, the anode current turn-off interval, as defined by the anode current reaching 10% of the on-state value, occurs before the space-charge region punches-through to the buffer layer for the 10-kV MCT structure. The current fall-time is found to be 12 μs from the numerical simulations for the 10-kV asymmetric MCT structure and the corresponding energy loss per cycle is found to be 1.25 J/cm^2. The total energy loss per cycle for the 10-kV MCT structure is found to be 2.75 J/cm^2.

For comparison purposes, numerical simulations of the turn-off for the 10-kV silicon IGBT structure with a high-level lifetime of 20 μs were performed by stepping the gate voltage down from positive 10 to 0 V in 10 ns using an on-state current density of 50 A/cm^2. The resulting waveforms obtained from the numerical simulations for the collector voltage and current are shown in Fig. 8.41 for the case of a collector supply voltage of 5,000 V. The collector voltage initially increases linearly as described by the analytical model during the early stages of the voltage rise but the rate of rise becomes severely reduced at collector voltages beyond 4,000 V due to the onset of significant impact ionization as the RBSOA boundary is approached. The collector voltage almost saturates at 5,000 V indicating operation of the device close to its RBSOA limit. This is consistent with the predictions of the analytical model for the RBSOA of the IGBT structure (see Fig. 4.57). The voltage rise-time is found to be 13 μs from

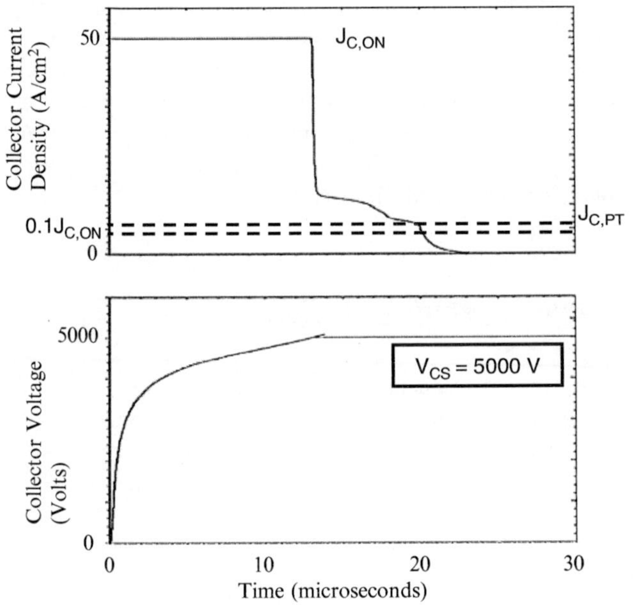

Fig. 8.41 Turn-off waveforms for the 10-kV asymmetric IGBT

the numerical simulations for the 10-kV asymmetric IGBT structure and the corresponding energy loss per cycle is found to be 1.625 J/cm^2.

The collector current turn-off occurs with a rapid initial decrease in current to about 15 A/cm^2 followed by a gradual change as expected from the analytical model. The punch-through collector current density ($J_{C,PT}$) obtained using the analytical model (Eq. 8.24) is 9 A/cm^2 which is close to the simulation results. The switching times for the IGBT structure are similar to those for the MCT structure. The current fall-time is found to be 13 μs from the numerical simulations for the 10-kV asymmetric IGBT structure and the corresponding energy loss per cycle is found to be 0.875 J/cm^2. The total energy loss per cycle for the 10-kV IGBT structure is found to be 2.5 J/cm^2.

8.3.4 Switching Energy Loss

As discussed previously, the maximum operating frequency for the MCT structure is limited by the turn-off losses. The turn-off losses are associated with the voltage rise-time interval and the current fall-time interval. The energy loss for each event can be computed by using the equations previously provided in Sect. 8.3.3. Using this information, the maximum operating frequency for the 10-kV silicon asymmetric MCT structure can be derived using Eq. 8.27. The turn-off energy loss per cycle obtained from the numerical simulations of the silicon 10-kV asymmetric MCT structure can be derived from the waveforms in Fig. 8.40. For case of a high-level lifetime of 20 μs in the N-base region, the energy loss per cycle during the voltage rise-time is 1.5 J/cm^2 while the energy loss per cycle during the current fall time is 1.25 J/cm^2 in the case of an on-state current density of 50 A/cm^2 and an anode supply voltage of 5,000 V. The total energy loss per cycle is 2.75 J/cm^2 for the 10-kV silicon asymmetric MCT structure.

8.3.5 *Maximum Operating Frequency*

The maximum operating frequency for the 10-kV asymmetric MCT structure is limited by the turn-off losses. The turn-off losses are associated with the voltage rise-time interval and the current fall-time interval. The energy loss for each event can be computed by using the equation previously provided in Sect. 8.3.3. Using this information, the maximum operating frequency for the MCT structure can be derived using Eq. 8.27. The data acquired from the numerical simulations of the 10-kV asymmetric MCT and IGBT structures are provided in Fig. 8.42 for the case of an on-state operating current density of 50 A/cm^2.

	High-Level Lifetime (μs)	On-State Voltage Drop (Volts)	On-State Power Dissipation (W/cm^2)	Energy Loss per Cycle (J/cm^2)	Maximum Operating Frequency (Hz)
10-kV Asymmetric IGBT Structure	20	4.766	119	2.50	32
10-kV Asymmetric MCT Structure	20	2.303	57.6	2.75	52

Fig. 8.42 Power loss analysis for the 10-kV asymmetric MCT and IGBT structures with on-state current density of 50 A/cm^2

The maximum operating frequency obtained under the assumption of a 50% duty cycle and a total power dissipation limit of 200 W/cm^2 for the 10-kV asymmetric IGBT and MCT structures are found to be 32 and 52 Hz, respectively. The maximum operating frequency for the silicon 10-kV asymmetric MCT structure is superior to that for the IGBT structure due to its lower on-state voltage drop. However, the MCT structure has not been well received by the power electronics community for typical applications such as motor drives because it lacks good FBSOA. This complicates circuit operation, especially management of reverse recovery of fly-back rectifiers, unless expensive snubber circuits are added to the circuit topology.

8.4 Forward-Biased Safe Operating Area

The MCT structure does not exhibit a substantial region of operation where the anode current can be saturated under gate control. This is demonstrated in this section by using the results of numerical simulations for the 5-kV asymmetric MCT structure.

Simulation Results

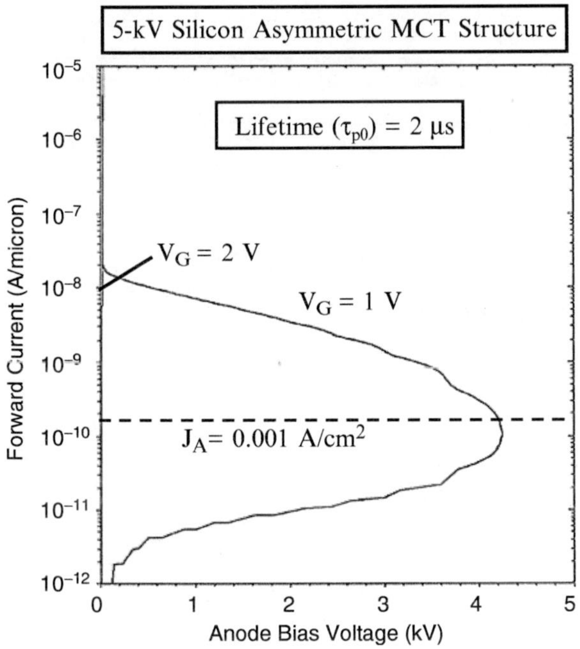

Fig. 8.43 5-kV asymmetric MCT FBSOA boundary

Numerical simulations of the 5-kV silicon asymmetric MCT structure were performed for the case of a high-level lifetime of 4 μs with various values for the gate bias voltage while sweeping the anode voltage. The resulting output characteristics are shown in Fig. 8.43. The device was able to saturate the anode current only at subthreshold gate bias voltages. The trace in the figure, for a gate bias of 1 V, exhibits current saturation up to an anode bias of 4,000 V. At even a slightly greater gate bias voltage, the MCT structure was unable to support large anode bias voltages. From this result, it can be concluded that the MCT structure has essentially no forward-biased safe operating area as mentioned previously in the chapter.

8.5 Reverse-Biased Safe Operating Area

The analytical solution for the reverse-biased safe operating area (RBSOA) for the MCT structure can be obtained by using Eq. 4.97 provided for the GTO structure because the physics of operation is similar. However, the GTO structure suffers from current crowding during the turn-off process. This problem does not occur in the MCT structure. The RBSOA boundary for the 5-kV asymmetric MCT structure obtained by using numerical simulations is provided in this section.

Simulation Results

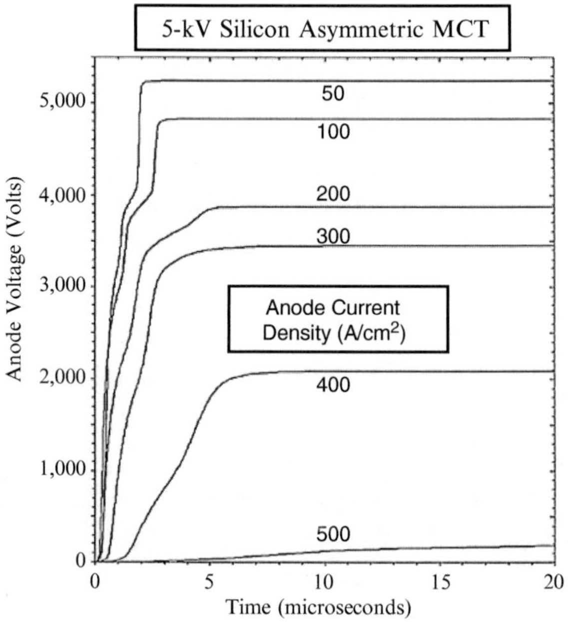

Fig. 8.44 5-kV asymmetric MCT RBSOA turn-off waveforms

The RBSOA boundary for the MCT structure can be obtained by turning off the structure starting with various on-state current densities. The presence of holes in the space-charge region enhances the electric field at the junction between the P-base region and the drift region. The electric field becomes larger for larger initial on-state current densities. Consequently, the collector voltage at which the on-state current density can be sustained by the impact ionization process becomes smaller. During turn-off, the collector voltage becomes limited as a function of time providing the RBSOA limit for each corresponding on-state current density.

Numerical simulations of the 5-kV silicon asymmetric planar-gate MCT structure were performed for the case of a high-level lifetime of 2 μs with various values for the initial on-state current density. The resulting collector voltage waveforms are provided in Fig. 8.44. At anode current densities below 300 A/cm^2, the collector voltage increases and becomes limited by the onset of avalanche breakdown. At larger collector current densities, the maximum sustainable current density for the MCT structure is limited by onset of some injection from the N$^+$ cathode region. Using the collector turn-off waveforms, the RBSOA boundary can be determined as shown in Fig. 8.45. The MCT exhibits a RBSOA boundary that is significantly inferior to that observed for the IGBT structure (see Fig. 5.76).

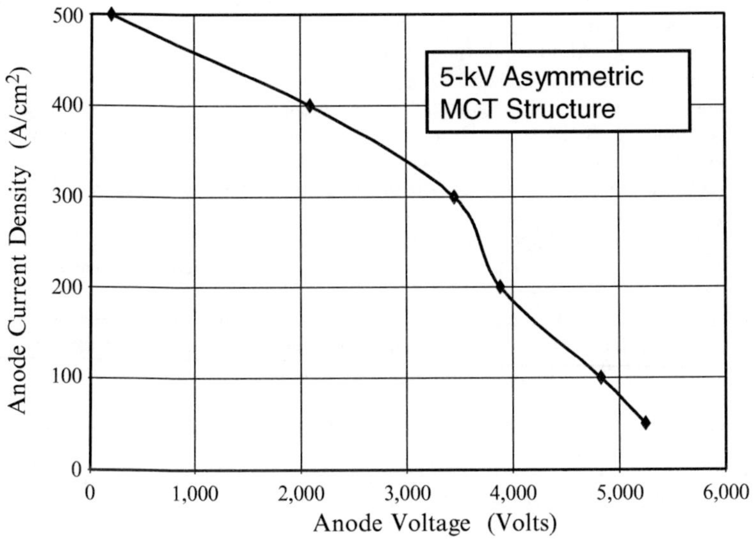

Fig. 8.45 RBSOA boundary for the 5-kV asymmetric MCT structure

8.6 Conclusions

The physics of operation and design principles for the silicon MCT structure have been described in this chapter. When first proposed, this device was touted as being much superior to the IGBT structure due to its lower on-state voltage drop and better power loss trade-off curve. Despite these advantages and significant development effort in the United States and Europe, the MCT structure has not displaced the IGBT in applications because it lacks a forward-biased safe operating area resulting in the addition of expensive and lossy snubbers in applications.

References

1. V.A.K. Temple, "MOS-Controlled Thyristors", IEEE Electron Devices Meeting, Abstract 10.7, pp. 282–285, 1984.
2. M. Stoisiek and H. Strack, "MOS GTO – A Turn-off Thyristor with MOS-Controlled Emitter Shorts", IEEE Electron Devices Meeting, Abstract 6.5, pp. 158–161, 1985.
3. V.A.K. Temple and W. Tantraporn, "Effect of Temperature and Load on MCT Turn-off Capability", IEEE Electron Devices Meeting, Abstract 5.5, pp. 86–121, 1985.
4. M. Stoisiek, et al, "A Large Area MOS-GTO with Wafer-Repair Technique", IEEE Electron Devices Meeting, Abstract 29.3, pp. 666–669, 1987.
5. V.A.K. Temple, S. Arthur, and D.L. Watrous, "MCT (MOS Controlled Thyristor) Reliability Investigation", IEEE Electron Devices Meeting, Abstract 27.4, pp. 618–621, 1988.

6. F. Bauer, et al, "Design Aspects of MOS Controlled Thyristor Elements", IEEE Electron Devices Meeting, Abstract 11.6.1, pp. 297–300, 1989.
7. A. Aemmer, et al, "Multi-Dimensional Simulation of MCT Structures", IEEE International Symposium on Power Semiconductor Devices and ICs, Abstract 2.1, pp. 20–25, 1990.
8. H. Lendenmann, et al, "Switching Behavior and Current Handling Performance of MCT-IGBT Cell Ensembles", IEEE Electron Devices Meeting, Abstract 6.3.1, pp. 149–152, 1991.
9. F. Bauer, et al, "Static and Dynamic Characteristics of High Voltage (3.5 kV) IGT and MCT Devices", IEEE International Symposium on Power Semiconductor Devices and ICs, Abstract 2.1, pp. 22–27, 1992.
10. H. Lendenmann, et al, "Approaching Homogeneous Switching of MCT Devices: Experiment and Simulation", IEEE International Symposium on Power Semiconductor Devices and ICs, Abstract 3.3, pp. 66–70, 1993.
11. H. Dettmer, et al, "A Comparison of the Switching Behavior of IGBT and MCT Power Devices", IEEE International Symposium on Power Semiconductor Devices and ICs, Abstract 3.1, pp. 54–59, 1993.
12. H. Dettmer, et al, "4.5 kV MCT with Buffer Layer and Anode Short Structure", IEEE International Symposium on Power Semiconductor Devices and ICs, Abstract 1.3, pp. 13–17, 1994.
13. B.J. Baliga and M. Smith, "Modulated Conductivity Devices reduce Switching Losses", Electronic Design Magazine, Vol. 28, pp. 153–162, 1983.
14. N. Mohan, T.M. Undeland, and W.P. Robbins, "Power Electronics", John Wiley and Sons, Inc., New York, 1995.
15. F. Bauer, et al, "On the Suitability of BiMOS High Power Devices in Intelligent Snubberless Power Conditioning Circuits", IEEE International Symposium on Power Semiconductor Devices and ICs, Abstract 5.3, pp.201–206, 1994.
16. H. Lendenmann and W. Fichtner, "Turn-off Failure Mechanisms in Large (2.2 kV, 20A) MCT Devices", IEEE International Symposium on Power Semiconductor Devices and ICs, Abstract 5.4, pp.207–212, 1994.
17. B.J. Baliga, "Fundamentals of Power Semiconductor Devices", Springer-Science, New York, 2008.

Chapter 9
Silicon BRT

As discussed in Chap. 8, there was a flurry of activity in the 1990s to explore the development of MOS-gated thyristor structures due to their reduced on-state voltage drop when compared with the IGBT structure. The base-resistance-controlled thyristor (BRT) structure was proposed [1, 2] to take advantage of thyristor-based on-state current flow under MOS gate control to reduce the gate drive requirements. In comparison with the MCT structure discussed in the previous chapter, the BRT structure had the advantage of using a double-diffusion process similar to that used to manufacture IGBT structures. A rigorous study to understand the physics of BRT operation and evaluate the performance of experimental devices with blocking voltages ranging from 600 to 5,000 V was conducted in the 1990s [3–9].

As discussed in the previous chapter, thyristor-based structures exhibit uncontrolled rapid turn-on due to the internal regenerative action that can lead to extremely high reverse recovery currents in the antiparallel rectifiers leading to the destruction of the rectifier and the switch. Like the MCT structure, the BRT structure also exhibits this behavior. The rate of rise of the current during turn-on must be regulated by using a snubber circuit [10]. Due to this problem, The BRT structure has not displaced the IGBT in applications. Dual-gate device structures have been proposed to address this shortcoming in BRT structures [11, 12].

9.1 Basic Structure and Operation

The asymmetric BRT structure with the planar gate architecture [3] is illustrated in Fig. 9.1. The contact shown to the P^+ region (source of the p-channel turn-off MOSFET) at the upper surface of the structure is connected to the cathode contact by forming the cathode metal across the entire surface of the cell. Since the asymmetric BRT structure is intended for use in DC circuits, its reverse blocking capability does not have to match the forward blocking capability allowing the use

DOI 10.1007/978-1-4614-0269-5_9,

of an N-buffer layer adjacent to the P$^+$ anode region. The N-buffer layer has a much larger doping concentration than the lightly doped portion of the N-base region. The electric field in the asymmetric BRT takes a trapezoidal shape allowing supporting the forward blocking voltage with a thinner N-base region. This allows achieving a lower on-state voltage drop and superior turn-off characteristics. The doping concentration of the buffer layer and the lifetime in the N-base region must be optimized to perform a trade-off between on-state voltage drop and turn-off switching losses. BRT structures are discussed in this chapter with two blocking voltage ratings for comparison with other device structures.

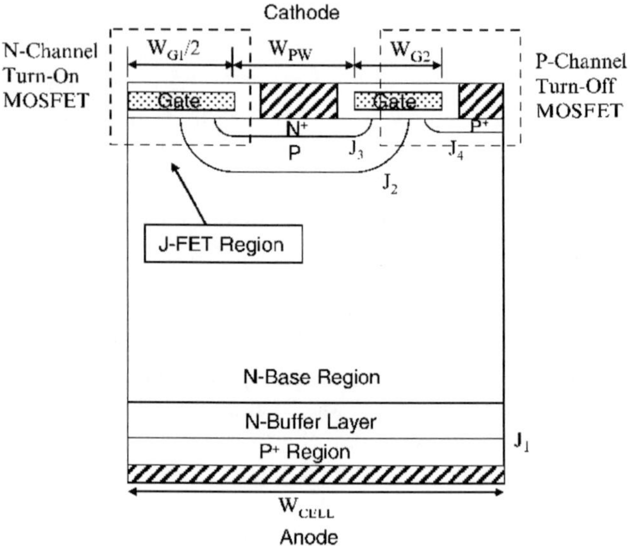

Fig. 9.1 The asymmetric BRT structure

The BRT structure requires two MOSFET regions – one to turn off the thyristor regenerative action and the second to turn on the device structure. The gates of both MOSFETs are interconnected by using a single polysilicon layer to create a three terminal device. The turn-on and turn-off MOSFETs are identified in Fig. 9.1 It can be seen that the BRT structure can be fabricated using a double-diffusion process similar to that used to manufacture IGBTs. In fact, the first BRT structures were made simultaneously with IGBTs on the same wafer for comparison of their performance [3]. In the BRT structure shown in the figure, the thyristor structure is achieved between the N$^+$ cathode region at the top, a narrow P-base region, a wide N-base region to support the high voltage with a N-buffer layer, and the P$^+$ anode region located at the bottom. The doping concentration of the N-drift region under the gate of the n-channel turn-on MOSFET is enhanced using a JFET doping process. The source of the p-channel turn-off MOSFET is formed with a shallow P$^+$ diffusion.

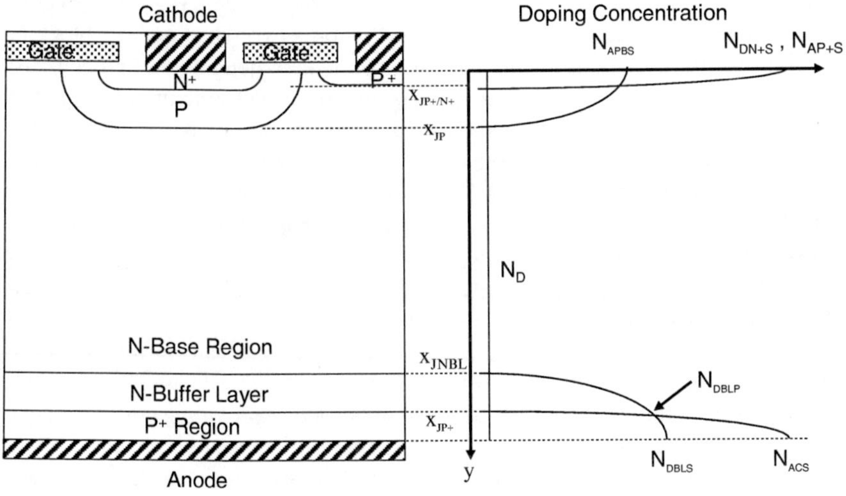

Fig. 9.2 The asymmetric BRT doping profile

A schematic illustration of the doping profiles for the various regions in the asymmetric BRT structure is provided in Fig. 9.2 on the right-hand side. The lines show the profiles taken vertically through the various regions. For convenience, it has been assumed that the doping profile for the P^+ source region of the p-channel turn-off MOSFET is the same as that for the N^+ cathode region. In practice, the doping profiles for both junctions are shallow but not identical.

When a positive bias is applied to the anode of the BRT structure, junction J_2 between the P-base and N-base regions becomes reverse biased. In principle, this junction is capable of supporting a large voltage with a depletion region formed in the lightly doped N-base region. However, the leakage current generated in the drift region is collected by junction J_2 and flows into the P-base region. If zero gate bias is applied, neither MOSFET is turned on. Consequently, the holes collected by the P-base region forward bias the junction J_3 between the N^+ cathode region and the P-base region. The forward bias on junction J_3 leads to the injection of electrons from the N^+ cathode region into the P-base region. These electrons diffuse through the P-base region and get collected at junction J_2 between the P-base and N-base regions. When the electrons enter the N-base region, they serve as base drive current for the P-N-P transistor leading to strong injection of holes from junction J_1 between the P^+ anode region and the N-base region. This sets up the regenerative action that turns on the vertical thyristor in the BRT structure. Consequently, the BRT structure shown in Fig. 9.1 is not capable of supporting a large forward blocking voltage if no gate bias is applied.

The high forward blocking voltage can be achieved in the BRT structure shown in Fig. 9.1 by the application of a negative gate bias to turn on the p-channel MOSFET. When the p-channel MOSFET is turned on, the holes that enter the P-base region due to generation of leakage current in the drift region can bypass the junction J_3 between the N^+ cathode region and the P-base region and get removed by the cathode contact. The BRT structure then behaves like a thyristor structure with cathode shorts as discussed in the textbook [13]. The forward blocking capability of the BRT with a negative gate bias is then determined by open-base transistor breakdown as discussed for the asymmetric GTO structure in Chap. 5.

The BRT structure can be turned on by the application of a positive bias to the MOS gate structure. In this case, the n-channel MOSFET is turned on providing a path for electrons to flow from the N^+ cathode region into the N-base region. These electrons serve as base drive current for the wide base P-N-P transistor leading to strong injection of holes from junction J_1 between the P^+ anode region and the N-base region. This sets up the regenerative action that turns on the vertical thyristor in the BRT structure.

Once the BRT is operating in its on-state, the device can be turned off by switching the gate bias from a positive value to a negative value. The negative gate bias turns on the p-channel MOSFET in the structure providing a path for holes entering the P-base region from the N-base region to be removed to the cathode contact. A simple model for the maximum anode turn-off current density can be formulated using a lumped element approach. In this approach, all of the hole current being collected at the junction J_2 is assumed to flow through a lumped shunting resistance for the hole current path. In a simplified model, the hole current path consists of the P-base region and the p-channel MOSFET.

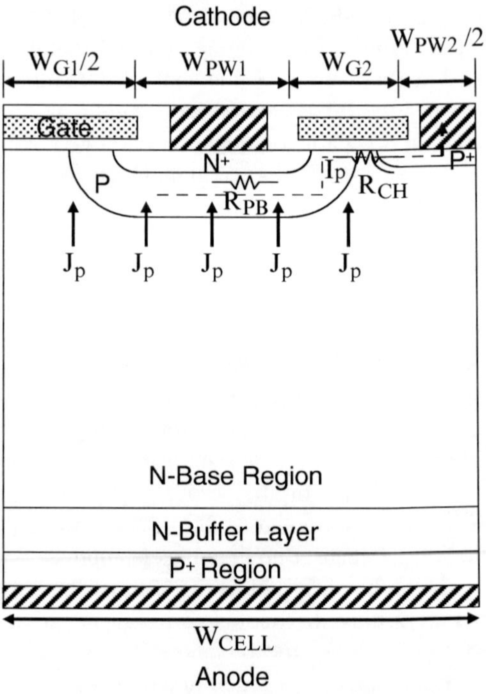

Fig. 9.3 Current flow during turn-off for the BRT structure

In the BRT structure, a part of the anode current flows to the cathode contact via the contact to the P^+ source region of the p-channel turn-off MOSFET structure. The hole current (I_p) collected by the P-base region during turn-off under inductive load operation is equal to the initial anode current density ($J_{A,ON}$) multiplied by the area on the left-hand side of the P-base junction:

$$I_p = J_{A,ON}\left(\frac{W_{G1}}{2} + W_{PW1} + x_{JP}\right)Z \tag{9.1}$$

where Z is the length of the cell in the orthogonal direction to the cross section shown in Fig. 9.3. The voltage drop produced by this current when flowing through the lumped shunting resistance (R_{SH}) must be less than the built-in potential for the junction J_3 between the N^+ cathode region and the P-base region if injection from this junction is to be suppressed to achieve the desired turn-off:

$$V_{bi} = I_p R_{SH} = J_{A,MAX}\left(\frac{W_{G1}}{2} + W_{PW1} + x_{JP}\right)Z(R_{PB} + R_{CH}) \tag{9.2}$$

The lumped resistance of the P-base region is given by:

$$R_{PB} = \rho_{S,PB}\left(\frac{W_{PW1} + 2x_{JP}}{Z}\right) \tag{9.3}$$

where $\rho_{S,PB}$ is the pinch sheet resistance of the P-base region and W_{PW1} is the width of the window in the polysilicon where the cathode is formed. The resistance of the p-channel MOSFET is given by:

$$R_{CH} = \frac{L_{CH}}{\mu_{pi} C_{OX}(V_G - V_{TH})Z} \tag{9.4}$$

where L_{CH} is the channel length, μ_{pi} is the mobility for holes in the inversion layer of the p-channel MOSFET, C_{OX} is the gate oxide capacitance, V_G is the gate bias voltage, and V_{TH} is the threshold voltage.

Using the above equations, the maximum turn-off current density is found to be given by:

$$J_{A,MAX} = \frac{2V_{bi}\mu_{pi}C_{OX}(V_G - V_{TH})}{(W_{G1} + 2W_{PW1} + 2x_{JP})\left[\rho_{S,PB}(W_{PW1} + 2x_{JP})\mu_{pi}C_{OX}(V_G - V_{TH}) + L_{CH}\right]} \tag{9.5}$$

The maximum turn-off current density predicted by the analytical model is plotted in Fig. 9.4 as a function of gate bias voltage under the assumption of a threshold voltage of 2 V. The following values were used in the analytical model: channel mobility for holes of 230 cm^2/V-s, gate oxide thickness of 500 Å, built-in potential of 0.937 V at room temperature using doping concentrations of 1×10^{19} cm^{-3} for the N^+ region and 1.5×10^{17} cm^{-3} for the P-base region, P-base pinch sheet resistance of 600 Ω/sq based upon an average doping concentration of 1.5×10^{17} cm^{-3} and thickness of 3 μm for the P-base region, polysilicon window width of 7 μm, and a channel length of 1.5 μm for the p-channel MOSFET integrated into the BRT structure. It can be observed that the maximum turn-off current density increases with increasing negative gate bias applied to the gate electrode because of a corresponding reduction of the channel resistance. A maximum turn-off current density of 193 A/cm^2 is predicted by the analytical model at room temperature at a gate bias of negative 10 V.

The impact of increasing the temperature on the maximum turn-off capability of the BRT structure is also shown in Fig. 9.4. At 500 K, the maximum turn-off current density is reduced below 50 A/cm^2 for a gate bias of 10 V. The decrease in the maximum turn-off current density is due to a reduction of the built-in potential and a reduction in the mobility for holes in the P-base region and the inversion layer [13]. The predicted maximum turn-off capability for the BRT is lower than that for the MCT structure due to the chosen cell parameters.

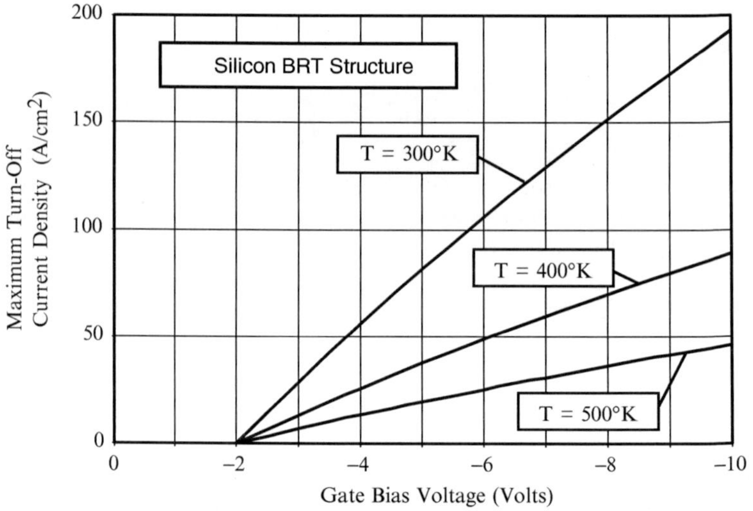

Fig. 9.4 Maximum turn-off current density for the BRT structure

9.2 5,000-V Silicon BRT

The design and characteristics of the 5,000-V asymmetric silicon BRT structure are discussed in this section. The design parameters for the N-base (drift) region required to achieve this blocking voltage are first analyzed. Using the optimum N-base width, the blocking characteristics for the device are then obtained for the case of zero and negative 10-V gate bias. The on-state characteristics for the device are obtained for various lifetime values as well. The gate-controlled turn-off behavior of the silicon BRT structure is analyzed including the effect of the lifetime in the drift region.

9.2.1 Blocking Characteristics

The physics for blocking voltages in the first and third quadrants by the asymmetric BRT structure is the same as those previously discussed for the silicon IGBT structure if a negative bias is applied to the gate electrode in order to keep the p-channel MOSFET turned on. The p-channel MOSFET then acts like the cathode short during the forward blocking mode. Without gate bias, there is no path for removal of the leakage current collected by junction J_2 leading to latch-up of the thyristor. This can be a problem for operation of the BRT in power circuits during the initial start-up. When power is applied for the first time to the circuit, the gate power supply may not generate the voltage required to turn on the p-channel MOSFET before the anode voltage for the BRT becomes sufficiently large to trigger the thyristor within the BRT structure.

When a positive bias is applied to the anode terminal of the asymmetric BRT structure with a negative bias applied to the gate, the P-base/N-base junction (J_2) becomes reverse biased while the junction (J_1) between the P^+ anode region and the N-base region becomes forward biased. The forward blocking voltage is supported across the P-base/N-base junction (J_2) with a depletion layer extending mostly within the N-base region. The electric field distribution within the asymmetric BRT structure is essentially the same as that illustrated in Fig. 4.3 for the asymmetric GTO structure. Consequently, the design procedure described in Chap. 4 can be applied to the asymmetric BRT structure. From Fig. 4.4, the N-base region width required to obtain a forward blocking voltage of 5,500 V is 470 μm. This width can be slightly reduced when taking into account the voltage supported within the P-base region due to its graded doping profile.

The leakage current in forward blocking mode is produced by space-charge generation within the depletion region. In the case of the asymmetric BRT structure in the forward blocking mode, the space-charge generation current at the reverse-biased P-base/N-base junction J_2 is amplified by the gain of the internal P-N-P transistor. Initially, the space-generation current increases with increasing anode bias due to expansion of the depletion region. Concurrently, the current gain (α_{PNP}) of the P-N-P transistor is also a function of the anode bias voltage because the base transport factor increases when the anode bias increases. Prior to the complete depletion of the lightly doped portion of the N-base region, the multiplication factor remains close to unity. It is therefore sufficient to account for the increase in the base transport factor with anode bias as given by Eqs. 4.8 and 4.9.

For the case of the silicon asymmetric BRT structure with a width of 450 μm for the lightly doped portion of the N-base region with a doping concentration of

5×10^{12} cm^{-3}, the entire lightly doped portion of the N-base region is completely depleted at a reach-through voltage of 780 V. Once the lightly doped portion of the N-base region becomes completely depleted, the electric field becomes truncated at the interface between the lightly doped portion of the N-base region and the N-buffer layer as illustrated at the bottom of Fig. 4.3. The space-charge generation width then becomes independent of the anode bias because the depletion width in the N-buffer layer is small. Under these bias conditions, the base transport factor also becomes independent of the collector bias as given by Eq. 4.10. Consequently, the leakage current becomes independent of the collector bias until the onset of avalanche multiplication. The leakage currents for the silicon asymmetric BRT structure are identical to those provided for the silicon asymmetric IGBT structure in Chap. 5.

Simulation Example

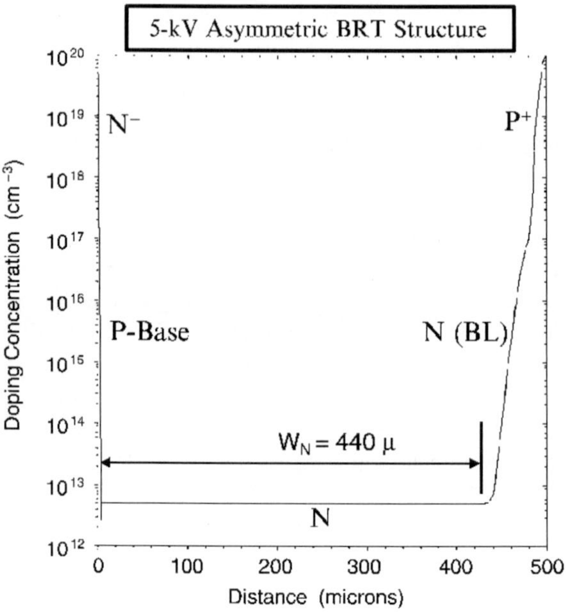

Fig. 9.5 Doping profile for the simulated asymmetric 5-kV BRT structure

The results of two-dimensional numerical simulations are described here in order to gain insight into the physics of operation for the 5-kV asymmetric BRT structure under voltage blocking conditions. The simulations were performed using a cell with the structure shown in Fig. 9.1. This device cell has a width (W_{Cell}) of 20 μm (area = 2.0×10^{-7} cm^{-2}). The asymmetric BRT structure used for the simulations was formed by diffusions performed into a uniformly doped N-type drift region with a doping concentration of 5×10^{12} cm^{-3}. A lifetime (τ_{p0}, τ_{n0}) of 10 μs was used for the baseline device. The N-buffer layer was formed by diffusion from the collector side with a depth of 55 μm. The doping profile in the vertical direction through the N$^+$ cathode region is shown in Fig. 9.5 indicating that the net width of the lightly doped portion of the N-base region is 440 μm after accounting for the diffusions. The peak doping concentration of the N-buffer layer is 1.0×10^{17} cm^{-3} and its thickness is 40 μm.

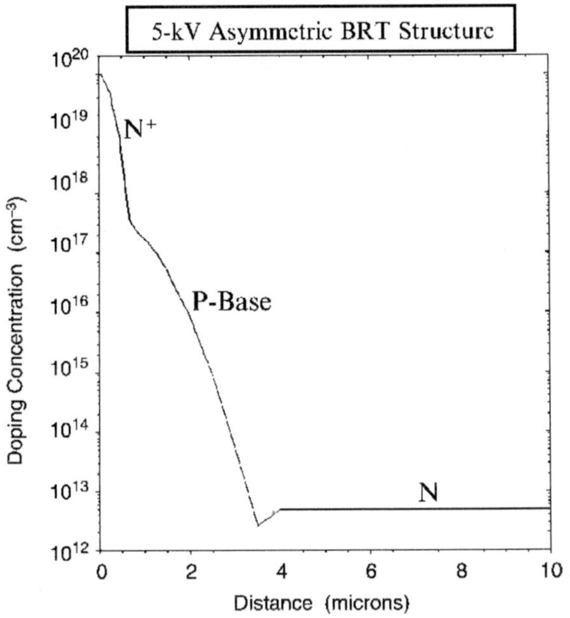

Fig. 9.6 Doping profile for the simulated asymmetric 5-kV BRT structure

The P-base region for the asymmetric BRT structure was formed with a Gaussian doping profile with a surface concentration of 5×10^{17} cm^{-3} and a vertical depth of 3.5 μm as can be seen in Fig. 9.6 where the vertical doping profile in the upper 10 μm of the structure is provided. The N$^+$ cathode region was formed with a Gaussian doping profile with a surface concentration of 1×10^{20} cm^{-3} and a depth of 0.7 μm. The fabrication process for the BRT is similar to that used to manufacture IGBT structures and less complex than that used for the MCT structure.

The doping profile across the surface for the 5-kV asymmetric BRT structure used for the numerical simulations is provided in Fig. 9.7. This profile was obtained along the horizontal line at $y = 0$ μm. It can be seen that the peak doping concentration of the P-base region is 1.5×10^{17} cm^{-3} and the channel length for the n-channel MOSFET is 1.7 μm. This is sufficient to prevent reach-through limited breakdown in the P-base region. The surface doping concentration of the N-body region of the p-channel MOSFET is 1.0×10^{16} cm^{-3} after compensation by the JFET diffusion. The channel length of the p-channel MOSFET is 1.5 μm.

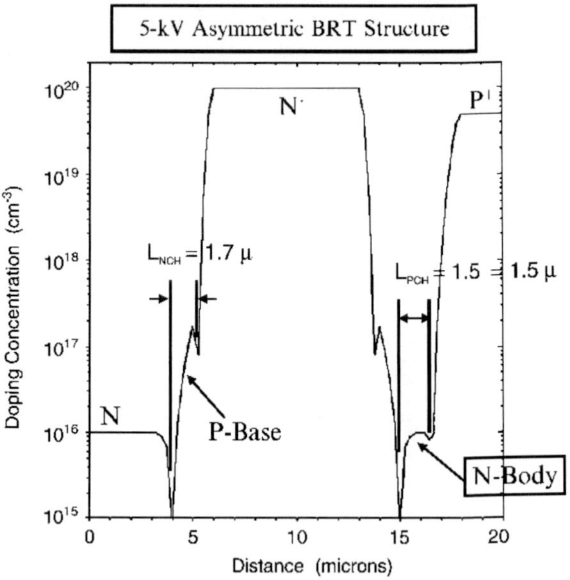

Fig. 9.7 Doping profile across the surface for the simulated asymmetric 5-kV BRT structure

The forward blocking capability of the silicon asymmetric BRT structure was first obtained using numerical simulations by increasing the anode bias while maintaining the gate electrode at zero volts. It was found that the device could not support voltage due to latch-up of the thyristor as observed in the case of the MCT structure with zero gate bias. This occurs because there is no direct path for the removal of holes collected at junction J_2 to the cathode electrode when the gate bias is zero. Since the hole current must flow via junction J_3 to the cathode electrode, the upper N-P-N transistor becomes active leading to latch-up of the thyristor.

When a negative bias is applied to the gate electrode, the p-channel MOSFET in the BRT structure is turned on providing a path for the removal of holes collected at junction J_2 to the cathode electrode. The device is then able to support above 5,500 V as shown in Fig. 9.8. The leakage current increases rapidly with increasing anode bias voltage until about 780 V as predicted by the analytical model (see Fig. 5.2). This occurs due to the increase in the space-charge generation volume and the increase in the current gain (α_{PNP}) of the open-base P-N-P transistor until the anode bias becomes equal to the reach-through voltage obtained using the analytical solution given by Eq. 4.2. The leakage current then becomes independent of the anode voltage until close to the breakdown voltage. This behavior is well described by the analytical model (see Fig. 5.2). The leakage current density obtained using the analytical model is within a factor of 2 of the values derived from the numerical simulations for all cases. The blocking characteristics for the asymmetric BRT structure are therefore similar to those for the asymmetric IGBT structure. However, the IGBT structure can block voltage with zero gate bias while a negative gate bias is required for the BRT structure. This can be a problem during start-up of power circuits because the gate supply voltage may not be available before the BRT structure is subjected to a high positive anode voltage.

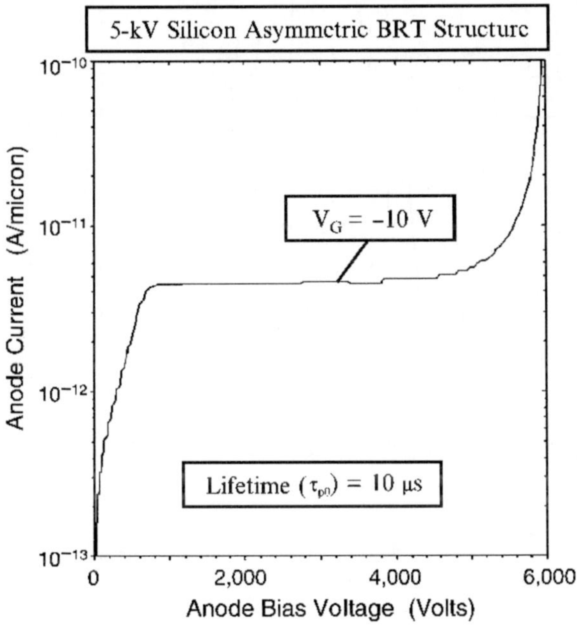

Fig. 9.8 Forward blocking characteristics for the asymmetric BRT structure

The current flow lines within the asymmetric BRT structure in the blocking mode are provided in Fig. 9.9 for the case of a gate bias of negative 10 V. From Fig. 9.9, it can be observed that the current collected at the P-base/N-drift region junction J_2 flows via the p-channel MOSFET bypassing the N^+ cathode region. This allows the asymmetric BRT structure to support a high forward blocking voltage without turning on the thyristor structure.

As in the case of other asymmetric structures, the anode voltage is primarily supported within the lightly doped portion of the N-drift region in the asymmetric BRT structure during operation in the forward blocking mode. The electric field profile within the asymmetric BRT structure is very similar to that shown previously for the asymmetric MCT structure and is not included here in the interest of conserving space.

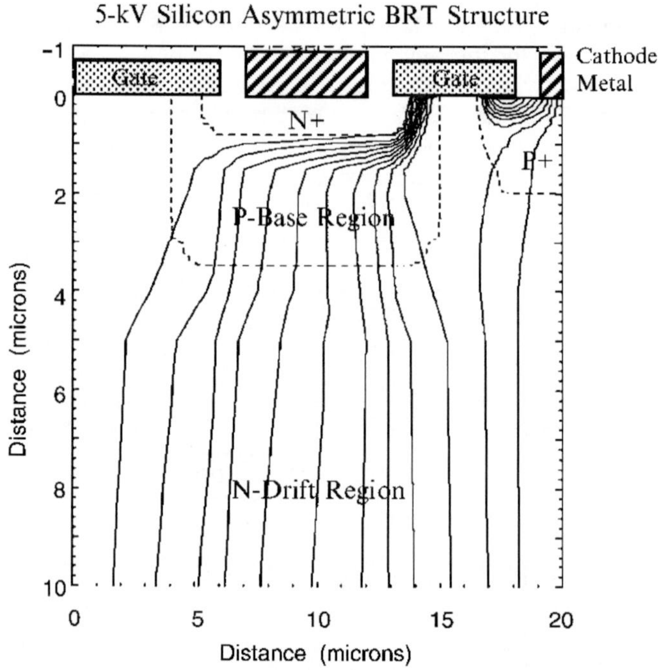

Fig. 9.9 Current flow lines during the blocking mode for the 5-kV asymmetric BRT structure: $V_G = -10$ V

9.2.2 *On-State Voltage Drop*

The BRT structure operates with latch-up of the thyristor structure within the device. Consequently, the on-state characteristics and the free carrier distribution within the N-drift region can be expected to be similar to those for the thyristor structure (see Chap. 2). However, the incorporation of the p-channel turn-off MOSFET within the thyristor structure degrades the injection efficiency of the cathode junction because some of the hole current in the P-base region is diverted to the P^+ source region of the p-channel turn-off MOSFET. This makes the on-state voltage drop for the BRT structure larger than that for the thyristor structure with the same drift region properties.

Simulation Results

The results of two-dimensional numerical simulations for the 5-kV asymmetrical silicon BRT structure are described here. The total width (W_{Cell}) of the structure, as shown by the cross section in Fig. 9.1, was 20 μm (area = 2.0×10^{-7} cm^{-2}). The doping profiles for the baseline device structure were already shown in Figs. 9.5–9.7.

In order to understand how closely the BRT structure resembles the thyristor structure in the on-state, the on-state characteristics of the 5-kV silicon asymmetric thyristor structure were obtained for the case of various values for the lifetime in the drift region. This thyristor structure was discussed in the previous chapter and its on-state characteristics were shown in Fig. 8.15 for the case of various lifetime values in the drift region. The current initially increases exponentially with increasing anode bias as expected. At current densities above 0.001 A/cm^2, the non-state voltage drop begins to increase more rapidly. Consequently, the on-state voltage drop increases as expected with reduction of the lifetime (τ_{p0}, τ_{n0}) indicated in the figure. The on-state voltage drop at a hole lifetime (τ_{p0}) value of 10 μs is found to be 1.202 V at an on-state current density of 50 A/cm^2 and increases to 4.286 V at a reduced hole lifetime (τ_{p0}) value of 1 μs. The hole distribution in the 5-kV asymmetric thyristor structure was provided in Fig. 8.16 for five cases of the lifetime (τ_{p0}, τ_{n0}) in the drift region. It can be observed that the injected carrier density is four orders of magnitude larger than the doping concentration on the anode side. The hole concentration is reduced in the middle of the drift region when the lifetime is reduced as expected from the P-i-N rectifier model for the thyristor [13]. It is worth pointing out that the carrier distribution is symmetric for the thyristor structure, i.e., the hole concentration on the cathode side of the drift region is equal to the hole concentration on the anode side of the drift region.

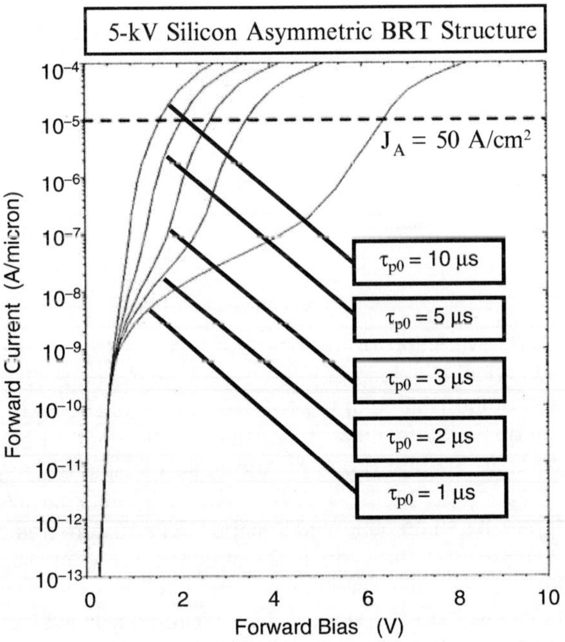

Fig. 9.10 On-state characteristics of the 5-kV asymmetric BRT structure: lifetime dependence

The on-state characteristics of the 5-kV silicon asymmetric BRT structure were obtained by using a positive gate bias voltage of 10 V for the case of various values for the lifetime in the drift region. This device structure has a peak buffer layer doping concentration of 1×10^{17} cm^{-3}. The characteristics obtained from the numerical simulations are shown in Fig. 9.10. The current initially increases exponentially with increasing anode bias. At current densities above 0.001 A/cm^2, the non-state voltage drop begins to increase more rapidly. Consequently, the on-state voltage drop increases as expected with reduction of the lifetime (τ_{p0}, τ_{n0}) indicated in the figure. The on-state voltage drop at a hole lifetime (τ_{p0}) value of 10 μs is found to be 1.636 V at an on-state current density of 50 A/cm^2 and increases to 6.486 V at a reduced hole lifetime (τ_{p0}) value of 1 μs.

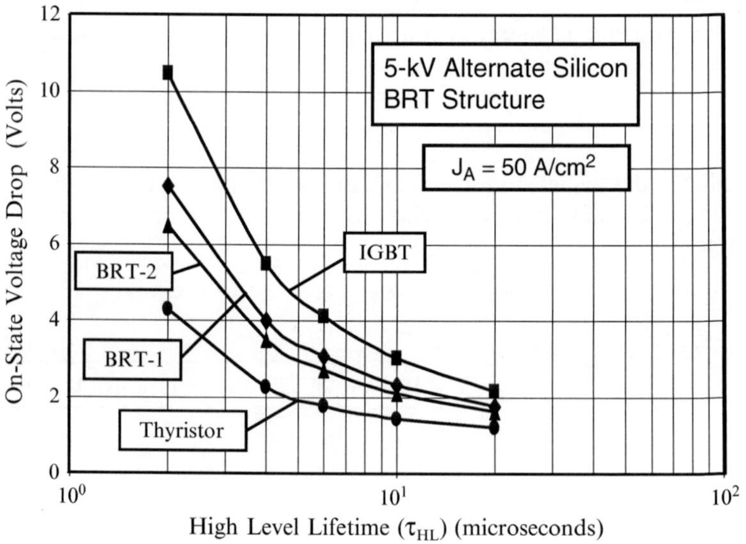

Fig. 9.11 On-state voltage drop for the 5-kV asymmetric BRT structure: N-base lifetime dependence

The variation of the on-state voltage drop obtained from the results of the numerical simulation for the 5-kV asymmetric BRT structure (BRT-1), as a function of the lifetime in the N-base region, is shown in Fig. 9.11 for the case of an anode on-state current density of 50 A/cm^2. For comparison purposes, the on-state voltage drops for the case of the 5-kV asymmetric trench-gate IGBT structure and for the 5-kV asymmetric thyristor structure are also provided in this figure. It can be observed that the BRT structure has a significantly lower on-state voltage drop than the IGBT structure for each lifetime value. This is due to improved carrier distribution in the BRT structure with a high free carrier density near the cathode side of the drift region. However, the on-state voltage drop for the BRT structure is significantly larger than that for the thyristor structure.

This is related to the incorporation of the turn-off MOSFET within the BRT structure, which degrades the injection efficiency of the cathode junction. The on-state voltage drop for the BRT structure can be reduced by increasing the width of the N^+ cathode region. This is demonstrated in Fig. 9.11 for the case of structure BRT-2, which has an N^+ cathode region with a width of 16 μm. Even in this case, the on-state voltage drop for the BRT structure is significantly larger than that observed for the thyristor structure with the same lifetime in the drift region.

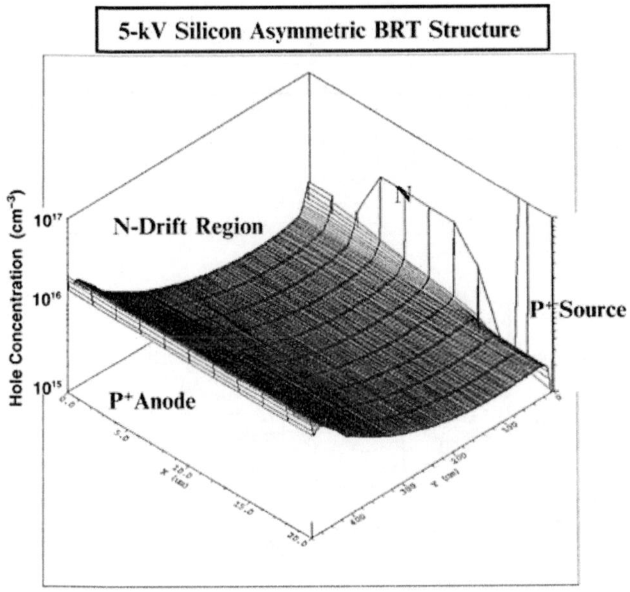

Fig. 9.12 On-state carrier distribution in the 5-kV asymmetric BRT structure

A three-dimensional view of the injected hole concentration within the 5-kV asymmetric BRT structure is provided in Fig. 9.12 at an on-state current density of 50 A/cm^2. It can be observed that the hole distribution is very uniform at the anode side of the drift region. However, the hole concentration has the catenary distribution similar to the thyristor structure only on the left-hand side of the structure. At the P^+ source region of the p-channel MOSFET on the right-hand side of the structure, the hole concentration is forced to zero by the reverse-biased junction. This has the adverse impact of reducing the hole concentration under the cathode region for about half of the cell width. The reduced hole concentration on the cathode side is responsible for the larger on-state voltage drop observed for the BRT structure.

The on-state voltage drop for the 5-kV asymmetric BRT structure is determined by the distribution of carriers injected into the N-base region producing the desired reduction of its resistance. The hole distribution in the 5-kV asymmetric BRT structure is provided in Fig. 9.13 for five cases of the lifetime (τ_{p0}, τ_{n0})

in the drift region. It can be observed that the injected carrier density is four orders of magnitude larger than the doping concentration on the anode side, which is similar to that observed for the IGBT structure (see Fig. 5.15). In comparison with the 5-kV thyristor structure (see Fig. 8.16), it can be observed that the hole carrier concentration is reduced by an order of magnitude in the drift region on the cathode side for the BRT structure. This produces a larger on-state voltage drop for the BRT structure when compared with the thyristor structure. However, the free carrier concentration on the cathode side of the drift region in the BRT structure is larger than that observed in the IGBT structure.

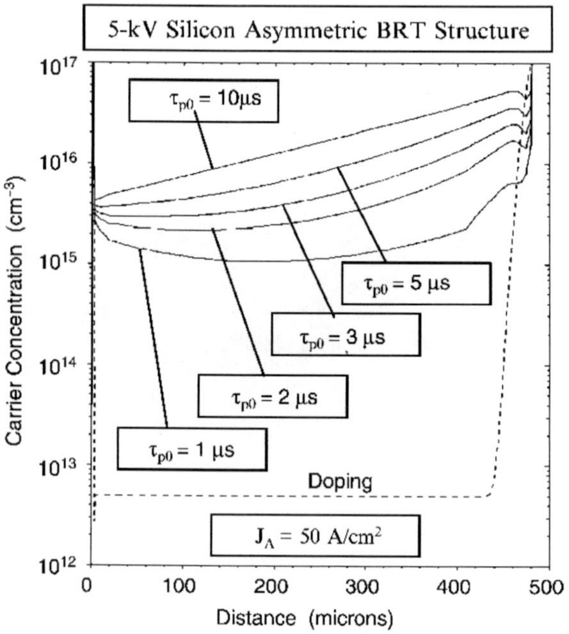

Fig. 9.13 On-state carrier distribution in the 5-kV asymmetric BRT structure: lifetime dependence

The on-state characteristics of the 5-kV silicon asymmetric BRT structure were also obtained as a function of temperature by using a positive gate bias voltage of 10 V for the case of a lifetime of 2 μs in the drift region. The characteristics obtained from the numerical simulations are shown in Fig. 9.14. At low anode current densities (below 0.05 A/cm^2), the on-state voltage drop decreases with increasing temperature while at on-state current densities above 5 A/cm^2, it begins to increase with increasing temperature. The on-state voltage drop exhibits a positive temperature coefficient at an on-state current density of 50 A/cm^2, which is desirable for allowing paralleling of devices and the prevention of hot spots within the device structure.

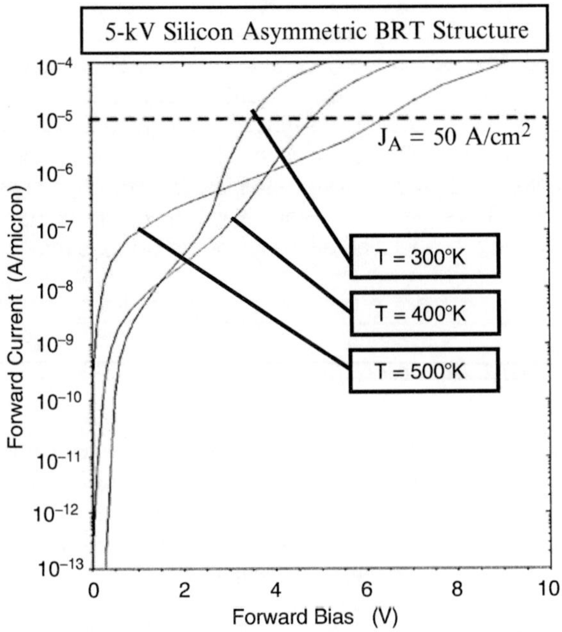

Fig. 9.14 On-state characteristics of the 5-kV asymmetric BRT structure: temperature dependence

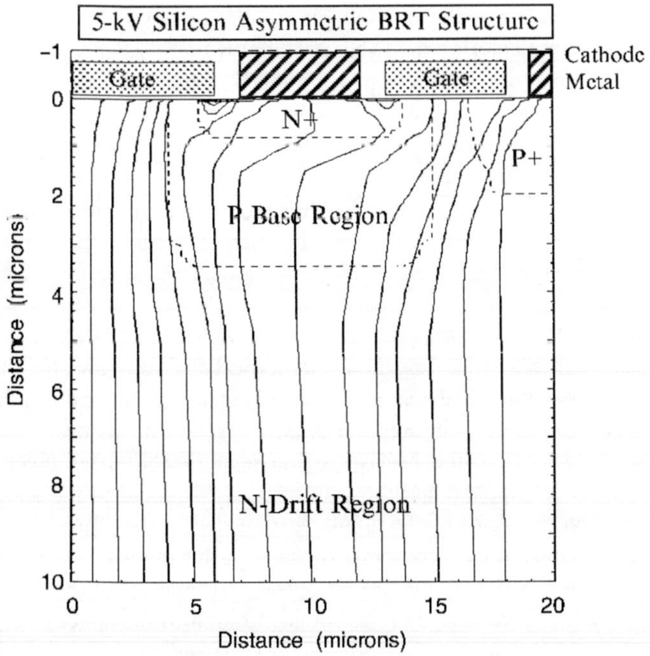

Fig. 9.15 On-state current flow lines for the 5-kV asymmetric BRT structure

In order to understand the operation of the BRT structure in the on-state, it is beneficial to examine the distribution of the current within the structure. The on-state current flow lines within the 5-kV asymmetric BRT structure are shown in Fig. 9.15 for the case of an on-state current density of 50 A/cm^2. A high-level lifetime of 2 μs was used during this simulation. It can be observed that some of the current flows via the JFET region and the channel formed at the surface of the P-base region due to applied positive gate bias. In addition, a significant part of the anode current flows via the P$^+$ source region of the p-channel MOSFET resulting in degrading the injection efficiency of the cathode region.

9.2.3 Turn-Off Characteristics

One of the important advantages of the BRT structure, when compared with the GTO structure, is the simplicity of the gate control circuit due to its MOS-gated structure. In order to turn off the device, the gate voltage must simply be ramped from the on-state value (nominally positive 10 V) to the off-state value (nominally negative 10 V) as illustrated in Fig. 8.22 for the MCT structure. The magnitude of the gate current can be limited by using a resistance in series with the gate voltage source. The waveform for the gate voltage shown in the figure is for the case of zero gate resistance. Once the gate voltage falls below the threshold voltage, the electron current from the channel ceases. However, the thyristor regenerative action can still continue allowing the BRT structure to remain in its on-state. In order to turn off the BRT structure, the gate bias must be reversed to turn on the p-channel MOSFET to shunt the hole current entering the P-base region. If the resistance of the shunting path is small, the injection of electrons from the N$^+$ cathode region then stops and the thyristor regenerative action ceases. In the case of an inductive load, the anode current for the BRT structure is then sustained by the hole current flow due to the presence of stored charge in the N-base region. Unlike the GTO structure, there is no prolonged storage time interval for the BRT structure during its turn-off because the shunting path is very short in length. The anode voltage begins to increase in the BRT structure almost immediately after the gate voltage reaches the negative gate supply voltage.

The anode current decreases once the anode voltage reaches the anode supply voltage as shown in Fig. 8.22. For the asymmetric BRT structure, the current tail usually occurs in two parts if the anode voltage is insufficient for the space-charge region to extend completely through the N-base region. In this case, there is still some stored charge left in the N-base region near the N-buffer layer after the voltage transient is completed and the anode voltage is equal to the anode supply voltage. During the first part of the anode current decay, the stored charge in the N-base region is removed by recombination, as well the anode current flow. This is a relatively slow decay due to the large high-level lifetime in the N-base region. As the anode current decreases, the hole concentration in the space-charge region decreases allowing the space-charge region to expand even though the anode voltage is constant. Eventually, the space-charge region extends through the entire N-base region when the anode current density becomes equal to the punch-through current density ($J_{A,PT}$). At this point in time, stored charge is present only in the N-buffer layer. The stored charge in the N-buffer layer decreases by recombination at a faster pace due to the smaller lifetime in the N-buffer layer associated with its greater doping concentration than the N-base region. This produces a faster decay of the anode current during the second phase of the current tail as illustrated in the figure.

Based upon the above description of the turn-off process in the BRT structure, it can be concluded that the turn-off waveforms for the BRT structure are similar to those already discussed in Chap. 8 for the MCT structure. The analytical model developed in Chap. 8 for the MCT structure can therefore be applied to the BRT structure.

Simulation Example

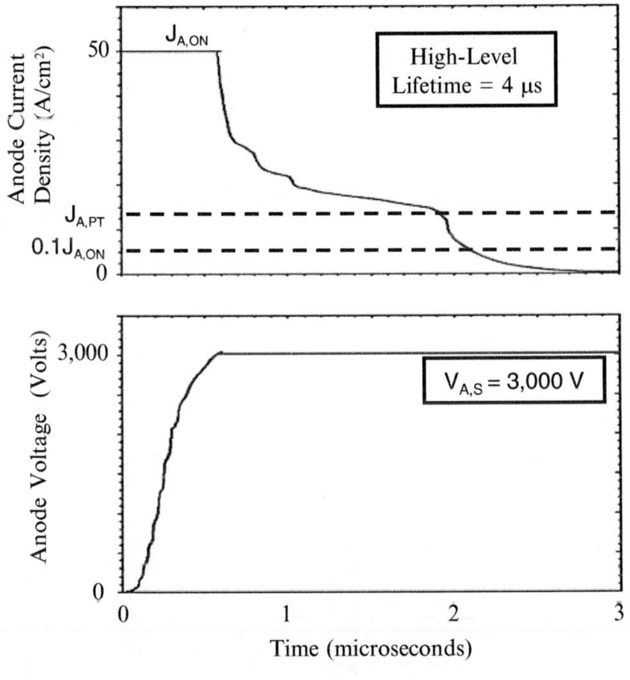

Fig. 9.16 Typical turn-off waveforms for the asymmetric 5-kV BRT structure

In order to gain insight into the operation of the asymmetric 5-kV BRT structure during its turn-off, the results of two-dimensional numerical simulations for a typical structure are discussed here. The device structure used has the cross section shown in Fig. 9.1 with a cell half-width of 20 μm. The doping profiles for the BRT structure used in the numerical simulations were provided in Figs. 9.5–9.7. The widths of the uniformly doped N-base region and the diffused N-buffer layer are 440 and 30 μm, respectively. For the typical case discussed here, a high-level lifetime of 4 μs was used in the N-base region.

The numerical simulations were performed with an abrupt reduction of the gate voltage from positive 10 V to negative 10 V in 20 ns starting from an on-state current density of 50 A/cm^2. The resulting waveforms obtained from the numerical simulations for the anode voltage and current are shown in Fig. 9.16 for the case of an anode supply voltage of 3,000 V. Unlike the GTO structure, there is no storage time associated with the turn-off of the BRT structure due to the small cell structure. The anode voltage increases immediately at the end of the gate voltage transient. The anode voltage increases as the square of the time as predicted by the analytical model until it reaches about 2,000 V. It then increases at a slower rate. This is associated with the onset of avalanche multiplication at high anode bias voltages – an effect not included in the analytical model. The waveforms for the BRT structure are very similar to those shown in Chap. 8 for the MCT structure. The anode voltage rise time for the BRT structure obtained in the numerical simulations for the case of supply voltage of 3,000 V (0.59 μs) is smaller than that observed for the MCT structure due to the reduced injected hole concentration on the cathode side.

After the completion of the anode voltage transient, the anode current waveform decays from the initial on-state current density at a rate that decreases with time. The current decays to the punch-through current density (indicated in the figure) in 1.3 μs. The punch-through current density of about 15 A/cm^2 is observed in the numerical simulations. After reaching the punch-through current density, the collector current is observed to decay at a faster rate as described by the analytical model.

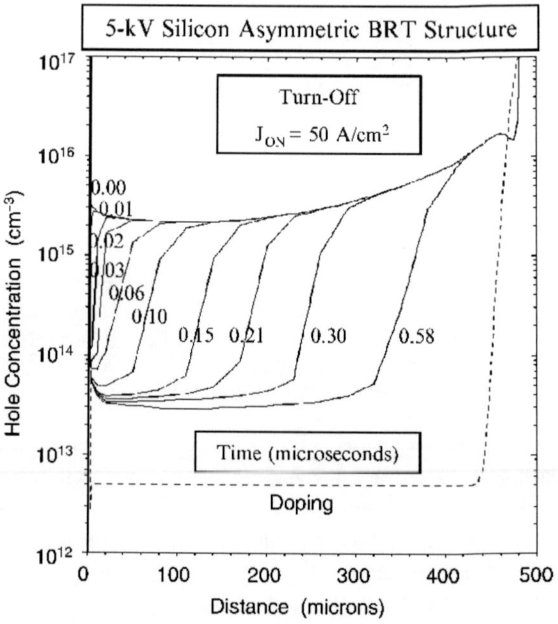

Fig. 9.17 Hole carrier distribution for the 5-kV BRT turn-off transient during the voltage rise time

A one-dimensional view of the minority carrier distribution in the 5-kV asymmetric BRT structure is shown in Fig. 9.17 from the initial steady-state operating point ($t = 0$ μs) to the end of the voltage rise time ($t = 0.58$ μs). These carrier profiles were taken at $x = 10$ μm through the P-base and N^+ cathode regions. The initial carrier distribution has the distribution like a P-i-N rectifier but the hole concentration on the cathode side is much lower than on the anode side. It can be observed from Fig. 9.17 that the carrier distribution in the N-base region near the anode does not change during the anode voltage rise phase. The carrier concentration at the P-base/N-drift junction rapidly reduces to zero within the first 25 ns allowing the junction to support an increase in the anode voltage. This time is shorter than for the MCT structure because the hole concentration on the cathode side is smaller for the BRT structure. A significant space-charge region begins to form immediately during the turn-off and expands toward the right-hand side demonstrating that there is no prolonged storage phase for the BRT structure. At larger anode voltages, the hole concentration in the space-charge region is about 3×10^{13} cm^{-3}, which is consistent with the value for p_{SC} obtained using the analytical model with the carriers moving at the saturated drift velocity and an on-state current density of 50 A/cm^2. The width of the space-charge region can be observed to be about 330 μm when the collector voltage reaches 3,000 V, which is close to that predicted by the analytical model.

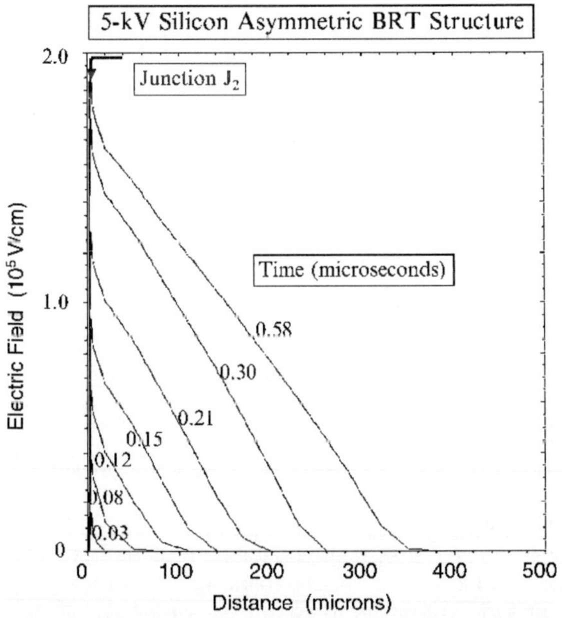

Fig. 9.18 Electric field distribution for the 5-kV asymmetric BRT during the voltage rise time

The electric field profiles in the 5-kV asymmetric BRT structure obtained from the numerical simulations are shown in Fig. 9.18 for various time instances during the voltage rise time. It can be observed that the peak electric field occurs at the

P-base/N-base junction (J_2) as expected. The peak electric field increases with time due to supporting a larger anode voltage. The electric field is triangular in shape, unlike during the blocking mode, even at high anode bias voltages due to the large hole charge in the space-charge region. It can also be observed that the slope of the electric field profile decreases when the time exceeds 0.30 μs and the anode voltage reaches 2,000 V, because of the addition of electrons in the space-charge region due to impact ionization. The negative charge of the electrons counteracts the positive charge of the holes producing a slightly larger space-charge region width at the end of the voltage transient than predicted by the analytical model based on just the hole charge.

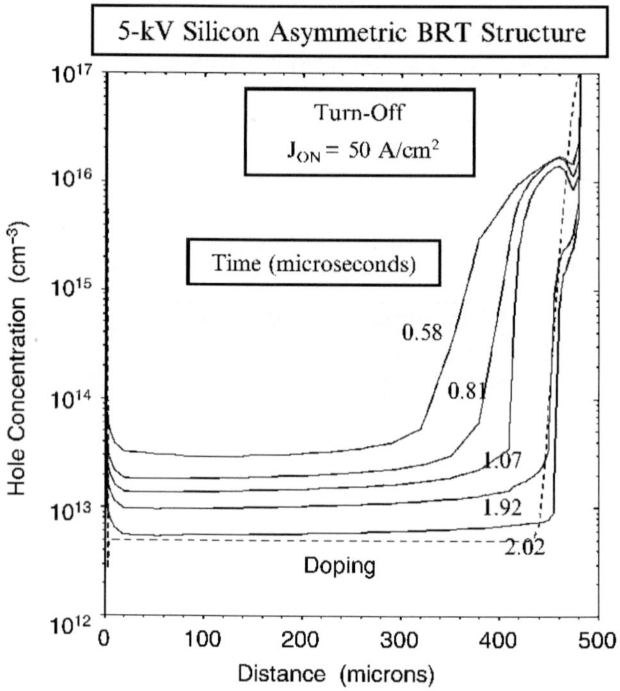

Fig. 9.19 Hole carrier distribution for the 5-kV asymmetric BRT turn-off transient during the current tail time

A one-dimensional view of the hole carrier distribution in the 5-kV asymmetric BRT structure is shown in Fig. 9.19 during the current tail time. The anode voltage was held constant at the anode supply voltage of 3,000 V during this transient. The hole concentration in the stored charge region begins to decrease immediately after the end of the voltage transient due to the recombination process and the removal of holes and electrons by the anode current flow. At the same time, the space-charge region expands in spite of a constant anode voltage because the hole concentration in the space-charge region reduces. From Fig. 9.19, it can be observed that all the holes in the N-base region have been removed at time $t = 1.92$ μs corresponding to the end of the first phase of the collector current transient (see Fig. 9.16). Subsequently, the holes remaining in the N-buffer layer are at concentrations well below its doping concentration. Consequently, the recombination of holes in the N-buffer layer during the second part of the anode current transient occurs under low-level injection conditions as assumed in the analytical model.

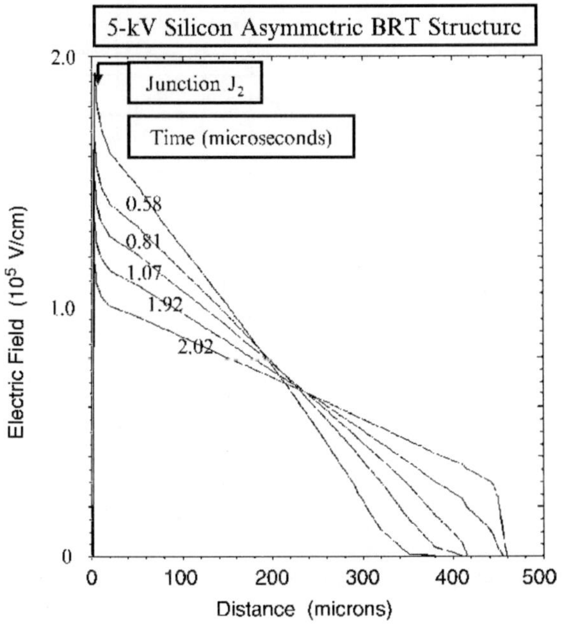

Fig. 9.20 Electric field distribution for the 5-kV asymmetric BRT during the current fall time

The electric field profiles in the 5-kV asymmetric BRT structure obtained from the numerical simulations are shown in Fig. 9.20 for various time instances during the current fall time. It can be observed that the peak electric field occurs at the P-base/N-base junction (J_2) as expected. The peak electric field decreases with time due to the spreading of the space charge layer as the anode current density becomes smaller leading to a reduced hole concentration in it. The electric field is triangular in shape until $t = 1.92$ μs. At this time, the space-charge region punches-through to the N-buffer layer. After this time, the electric field takes a trapezoidal shape.

It is instructive to examine the current flow pattern within the BRT structure during the turn-off process. The current flow lines within the 5-kV asymmetric BRT structure are provided in Fig. 9.21 at the end of the voltage rise time when the anode voltage is 3,000 V while the anode current density is at 50 A/cm^2. It can be observed that all of the hole current collected by the P-base region is shunted to the P$^+$ source of the p-channel turn-off MOSFET structure via the surface channel. At the same time, the P$^+$ source of the p-channel turn-off MOSFET structure is also directly collecting some of the hole current because the P$^+$ source/N-drift region junction is reverse biased during the turn-off.

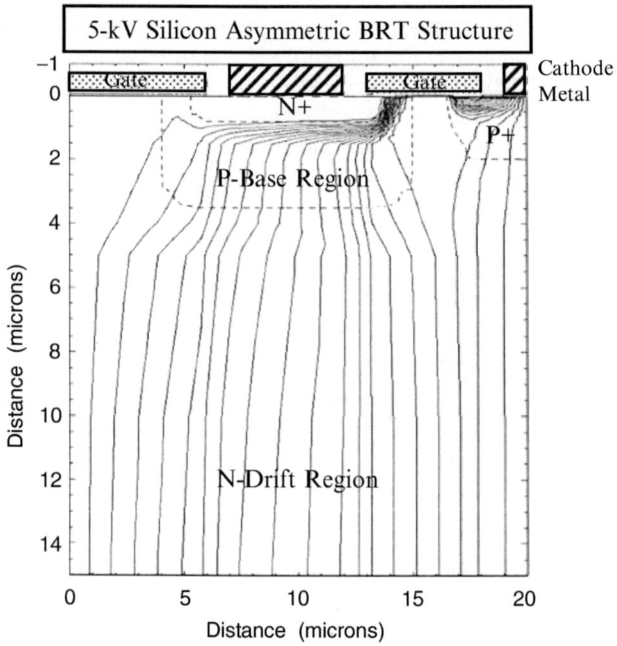

Fig. 9.21 Current distribution for the 5-kV asymmetric BRT during turn-off

9.2.4 Lifetime Dependence

The optimization of the power losses for the BRT structure requires performing a trade-off between the on-state voltage drop and the switching losses. One approach to achieve this is by adjusting the lifetime in the drift (N-base) region. A reduction of the lifetime in the drift region also alters the lifetime in the N-buffer layer. The impact of reducing the lifetime on the on-state voltage drop was previously shown in Sect. 9.2.2. The on-state voltage drop increases when the lifetime is reduced. The analytical model developed for turn-off of the asymmetric MCT structure presented in the previous chapter can be used to analyze the impact of changes to the lifetime in the drift region for the asymmetric BRT structure.

Simulation Example

In order to gain insight into the impact of the lifetime in the N-base region on the operation of the 5-kV asymmetric BRT structure, the results of two-dimensional numerical simulations for a typical structure are discussed here. The device structure used has the cross section shown in Fig. 9.1 with a half-cell width of 20 μm. The widths of the N-base and N-buffer layer regions are 440 and 30 μm, respectively. The high-level lifetime in the N-base region was varied between 2 and 20 μs. For turning off the BRT structures, the numerical simulations were performed with gate voltage rapidly ramped down from positive 10 V to negative

10 V in 20 ns starting from an on-state current density of 50 A/cm^2. The resulting waveforms obtained from the numerical simulations for the anode voltage and current are shown in Fig. 9.22 for the case of an anode supply voltage of 3,000 V.

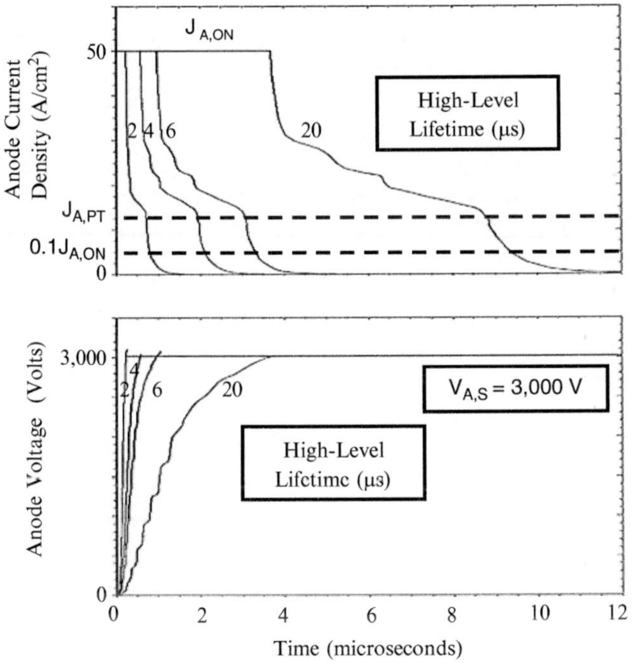

Fig. 9.22 Impact of lifetime on the 5-kV asymmetric BRT turn-off waveforms

The numerical simulations show a decrease in the voltage rise time with reduction of the lifetime in the N-base region. The numerical simulations of the 5-kV asymmetrical BRT structure also show a substantial decrease in the anode current fall time when the lifetime is reduced. The numerical simulations show a reduction of the anode current during the first part of the decay to the punch-through anode current ($J_{C,PT}$), which is independent of the lifetime in the N-base region as predicted by the analytical model.

9.2.5 Switching Energy Loss

The power loss incurred during the switching transients limit the maximum operating frequency for the BRT structure. Power losses during the turn-on of the BRT structure are significant but strongly dependent on the reverse recovery behavior of the fly-back rectifiers in circuits. Consequently, it is common practice to use only the turn-off energy loss per cycle during characterization of BRT devices. The turn-off losses are associated with the voltage rise-time interval and the current fall-time interval. The energy loss for each event can be computed by integration of the power loss, as given by the product of the instantaneous current and voltage. During the voltage rise-time interval, the anode current is constant while the voltage increases as the square of time. Since the turn-off switching waveforms for the BRT structure have the same shape as those for the MCT structure, the energy loss during the voltage rise-time interval and the current fall-time interval can be computed using the equations previously provided in Chap. 8. For the typical switching waveforms for the 5-kV asymmetric BRT structure shown in Fig. 9.16 with an anode supply voltage of 3,000 V, the energy loss per unit area during the anode voltage rise time is found to be 0.029 J/cm^2 if the on-state current density is 50 A/cm^2; and the energy loss per unit area during the collector current fall time is found to be 0.115 J/cm^2 if the on-state current density is 50 A/cm^2. The total energy loss per unit area ($E_{OFF,V}$ + $E_{OFF,I}$) during the turn-off process for the 5-kV asymmetric BRT structure is found to be 0.144 J/cm^2.

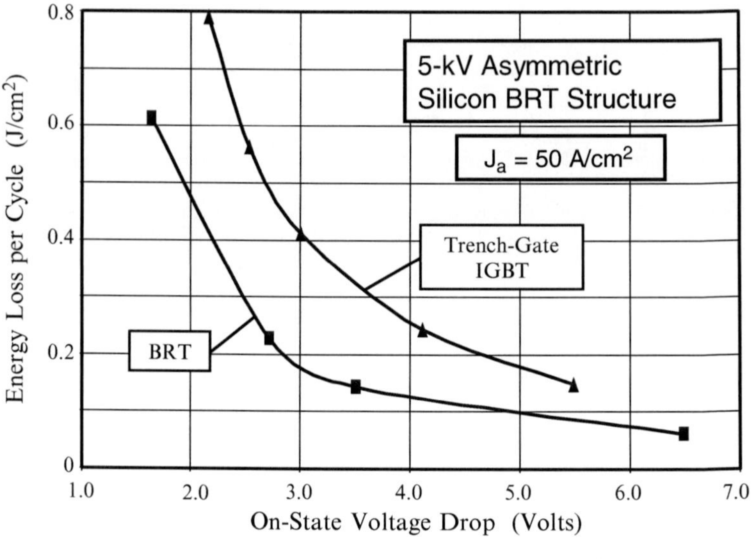

Fig. 9.23 Trade-off curve for the silicon 5-kV asymmetric BRT structure: lifetime in N-base region

Using the results obtained from the numerical simulations, the on-state voltage drop and the total energy loss per cycle can be computed. These values are plotted in Fig. 9.23 to create a trade-off curve to optimize the performance of the silicon 5-kV asymmetric BRT structure by varying the lifetime in the N-base region. Devices used in lower frequency circuits would be chosen from the left-hand side of the trade-off curve while devices used in higher frequency circuits would be chosen from the right-hand side of the trade-off curve. For comparison purposes, the trade-off curve for the 5-kV trench-gate IGBT structure is also included in the figure. It can be observed from this figure that substantial improvement in the power loss trade-off curve can be obtained by replacing the IGBT with the BRT structure. However, in actual practice, it is not possible to simply replace the IGBT with the BRT because of the very limited forward-biased safe operating area (FBSOA) of the BRT structure. The poor FBSOA of the BRT structure requires implementation of snubbers to control the rate of rise of the current when the device is turned on in motor control applications. This is essential in order to prevent extremely large reverse recovery currents in the fly-back diodes used in typical H-bridge circuits.

9.2.6 Maximum Operating Frequency

High-Level Lifetime (μs)	On-State Voltage Drop (Volts)	On-State Power Dissipation (W/cm²)	Energy Loss per Cycle (J/cm²)	Maximum Operating Frequency (Hz)
20	1.636	40.9	0.613	259
6	2.717	67.92	0.227	581
4	3.510	87.75	0.144	781
2	6.486	162.2	0.062	613

Fig. 9.24 Power loss analysis for the 5-kV asymmetric BRT structure

The maximum operating frequency for operation of the 5-kV asymmetric BRT structure can be obtained by combining the on-state and switching power losses:

$$P_{D,TOTAL} = \delta P_{D,ON} + E_{OFF} f \tag{9.6}$$

where δ is the duty cycle and f is the switching frequency. In the case of the baseline asymmetric BRT device structure with a high-level lifetime of 4 μs in the N-base region, the on-state voltage drop is 3.51 V at an on-state current density of 50 A/cm^2. For the case of a 50% duty cycle, the on-state power dissipation contributes 88 W/cm^2 to the total power loss. Using a total turn-off energy loss per cycle of 0.144 J/cm^2 in Eq. 9.6 yields a maximum operating frequency of about 800 Hz.

Using the results obtained from the numerical simulations, the on-state voltage drop and the energy loss per cycle can be computed. These values are provided in Fig. 9.24 together with the maximum operating frequency as a function of the high level lifetime in the N-base region under the assumption of a 50% duty cycle and a total power dissipation limit of 200 W/cm^2. The maximum operating frequency is plotted in Fig. 9.25 as a function of the high-level lifetime in the N-base region for the case of a duty cycle of 50%. It can be observed that the maximum operating frequency can be increased up to 800 Hz by reducing the high-level lifetime to 2 μs. This is much superior to the maximum operating frequency of 150 Hz for the 5-kV GTO structure and 400 Hz for the 5-kV trench-gate IGBT structure.

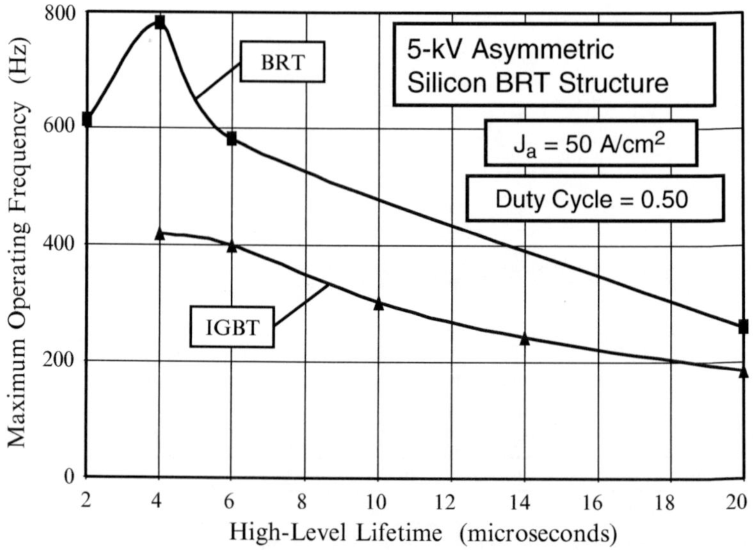

Fig. 9.25 Maximum operating frequency for the 5-kV asymmetric BRT structure

9.3 Alternate Structure and Operation

In the previous section, it was shown that the BRT structure illustrated in Fig. 9.1 does not support voltage with zero gate bias. This is a serious problem for fail-safe operation in power circuits. An alternate BRT structure is discussed in this section that does not suffer from this shortcoming.

The alternate asymmetric BRT structure is illustrated in Fig. 9.26. In this structure, a depletion-mode p-channel MOSFET is used for turning off the thyristor regenerative action. The depletion-mode p-channel MOSFET can be created by adding a shallow lightly doped P-body region between the P-base region of the thyristor and the P^+ source region of the p-channel MOSFET structure. In the absence of a gate bias, the lightly doped P-body region provides a shunt path of holes collected by the P-base region preventing sufficient forward basing of the cathode/P-base junction to allow thyristor turn-on by the leakage current flowing during the forward blocking mode. When a positive bias is applied to the gate, a turn-on channel is formed on the n-channel MOSFET while the P-body region becomes depleted increasing its resistance. This turns-on the thyristor allowing the BRT to operate in its on-state with a low on-state voltage drop. In order to turn off the device, the gate bias is switched from positive to negative. This shuts off the n-channel MOSFET and reduces the resistance of the p-channel MOSFET by the formation of an accumulation layer on the surface of the P-body region.

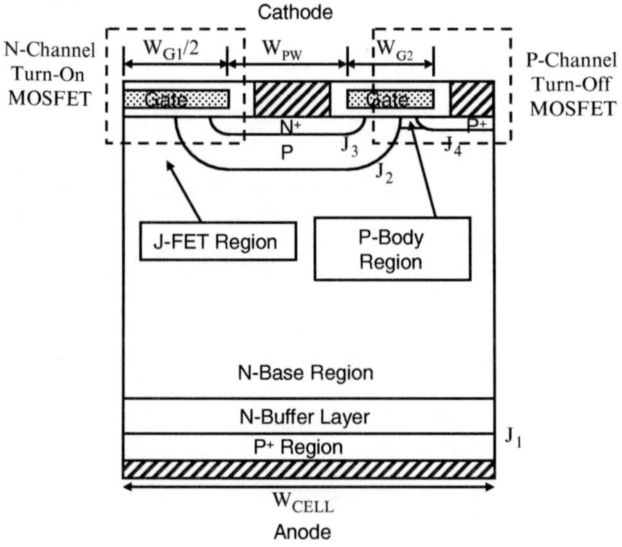

Fig. 9.26 An alternate asymmetric BRT structure

The maximum turn-off current density for the alternate BRT structure can be derived using the same methodology as described in the previous section. The only difference is that the channel resistance of the turn-off MOSFET is determined by the accumulation layer mobility for holes that is larger than the inversion layer mobility. The resistance of the accumulation-mode p-channel MOSFET is given by:

$$R_{CH} = \frac{L_{CH}}{\mu_{pa} C_{OX}(V_G - V_{THA})Z} \tag{9.7}$$

where L_{CH} is the channel length, μ_{pa} is the mobility for holes in the accumulation layer of the p-channel MOSFET, C_{OX} is the gate oxide capacitance, V_G is the gate bias voltage, and V_{THA} is the accumulation threshold voltage (can be assumed to be close to zero). In addition, the resistance of the p-channel MOSFET is further reduced by the P-body region:

$$R_{PBody} = \rho_{S,PBody} \frac{L_{CH}}{Z} \tag{9.8}$$

where $\rho_{S,PBody}$ is the sheet resistance of the P-body region. The shunting resistance for the alternate BRT structure is given by:

$$R_{SH} = R_{PB} + \frac{R_{CH} \cdot R_{PBody}}{R_{CH} + R_{PBody}} \tag{9.9}$$

Using this shunting resistance, the maximum turn-off current density is found to be given by:

$$J_{A,MAX} = \frac{2V_{bi}}{R_{SH}(W_{G1} + 2W_{PW1} + 2x_{JP})} \tag{9.10}$$

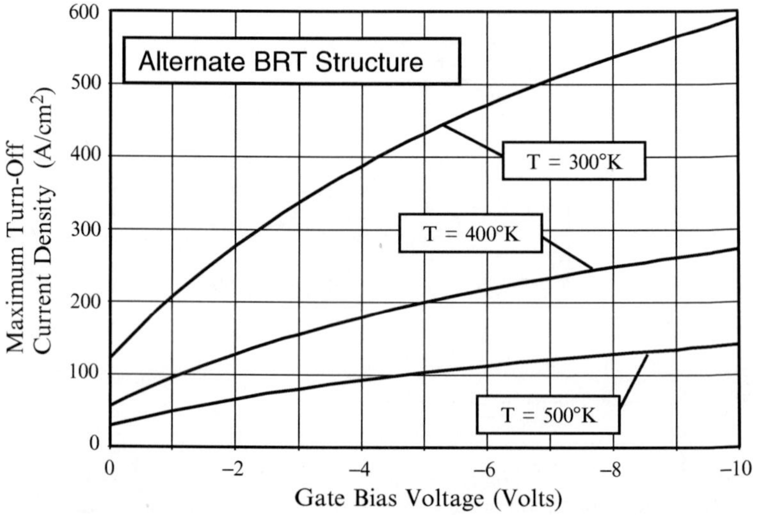

Fig. 9.27 Maximum turn-off current density for the alternate BRT structure

The maximum turn-off current density predicted by the analytical model is plotted in Fig. 9.27 as a function of gate bias voltage under the assumption of a threshold voltage of 0 V. The following values were used in the analytical model: channel mobility for holes of 400 cm^2/V-s, gate oxide thickness of 500 Å, built-in potential of 0.937 V at room temperature using doping concentrations of 1×10^{19} cm^{-3} for the N^+ region and 1.5×10^{17} cm^{-3} for the P-base region, P-base pinch sheet resistance of 600 Ω/sq based upon an average doping concentration of 1.5×10^{17} cm^{-3} and thickness of 3 μm for the P-base region, polysilicon window width of 7 μm, and a channel length of 1.5 μm for the p-channel MOSFET integrated into the BRT structure. The P-body region was assumed to have a doping concentration of 2×10^{16} cm^{-3} and thickness of 1 μm. A striking difference between the alternate BRT structure and the BRT structure discussed in the previous section is the substantial turn-off current density at even zero gate bias. This is due to the introduction of the current path provided by the P-body region in the alternate BRT structure. It can be observed that the maximum turn-off current density increases with increasing negative gate bias applied to the gate electrode because of a corresponding reduction of the channel resistance. A maximum turn-off current density of 600 A/cm^2 is predicted by the analytical model at room temperature at a gate bias of negative 10 V. This value is much larger than for the BRT structure discussed in the previous section due to the larger mobility for holes in the accumulation layer.

The impact of increasing the temperature on the maximum turn-off capability of the BRT structure is also shown in Fig. 9.27. At 500 K, the maximum turn-off current density is reduced to 140 A/cm^2 for a gate bias of 10 V. The reduction in the maximum turn-off current density is due to a reduction of the built-in potential and a reduction in the mobility for holes in the P-base region and the accumulation layer [13]. The predicted maximum turn-off capability for the alternate BRT is much larger than that for the MCT structure due to the enhanced conductivity of the p-channel turn-off MOSFET.

9.3.1 Blocking Characteristics

The physics for blocking voltages in the first and third quadrants by the alternate BRT structure is similar to that previously discussed for the BRT structure in the previous chapter. However, the alternate BRT structure can support a large forward blocking voltage with zero gate bias. This eliminates any problem for operation of the alternate BRT structure in power circuits during the initial start-up. The operation of the alternate BRT structure is demonstrated using the results of numerical simulations for a 5-kV asymmetric structure.

Simulation Example

The results of two-dimensional numerical simulations are described here in order to gain insight into the physics of operation for the 5-kV asymmetric alternate BRT structure under voltage blocking conditions. The simulations were performed using a cell with the structure shown in Fig. 9.26. This device cell has a width

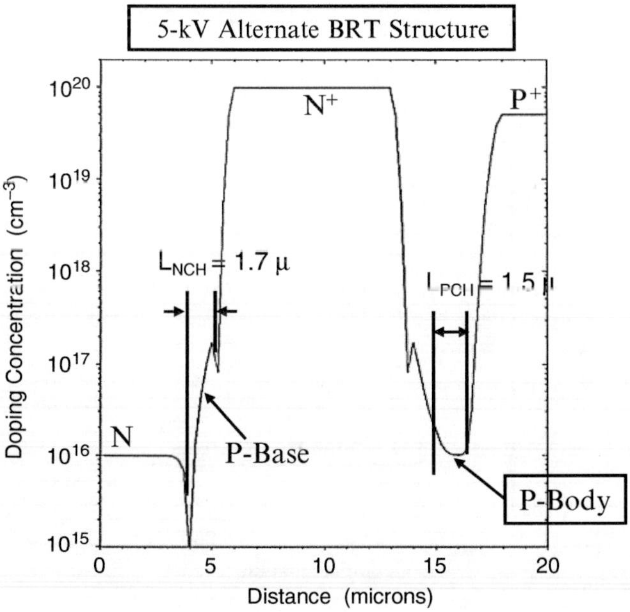

Fig. 9.28 Doping profile across the surface for the simulated asymmetric 5-kV alternate BRT structure

(W_{Cell}) of 20 μm (area = 2.0 × 10^{-7} cm^{-2}). The asymmetric alternate BRT structure used for the simulations was formed by using the same diffusions as the BRT structure in the previous section but a p-type diffusion was added to form the P-body region with a depth of 0.5 μm.

The doping profile across the surface for the 5-kV asymmetric alternate BRT structure used for the numerical simulations is provided in Fig. 9.28. This profile was obtained along the horizontal line at $y = 0$ μm. Comparing with Fig. 9.7, it can be seen that the profile is same on the left-hand side but on the right-hand side, the N-base region has been replaced with the P-body region.

The forward blocking capability of the silicon asymmetric alternate BRT structure was obtained using numerical simulations by increasing the anode bias while maintaining the gate electrode at zero volts. It was found that the device could support up to 6,000 V as shown in Fig. 9.29. It is therefore not necessary to apply a gate bias to the alternate BRT structure to block a high voltage. This allows fail-safe operation for the alternate BRT structure in power circuits during start-up.

The current flow lines within the alternate BRT structure in the blocking mode are provided in Fig. 9.30 for the case of a gate bias of zero volts. From Fig. 9.30, it can be observed that the current collected at the P-base/N-drift region junction J_2 flows via the P-body region bypassing the N^+ cathode region. This allows the alternate BRT structure to support a high forward blocking voltage without turning on the thyristor structure even when the p-channel MOSFET is not turned on. Although a strong shunting path is desirable for the leakage current, this can degrade the on-state characteristics as discussed in the next section.

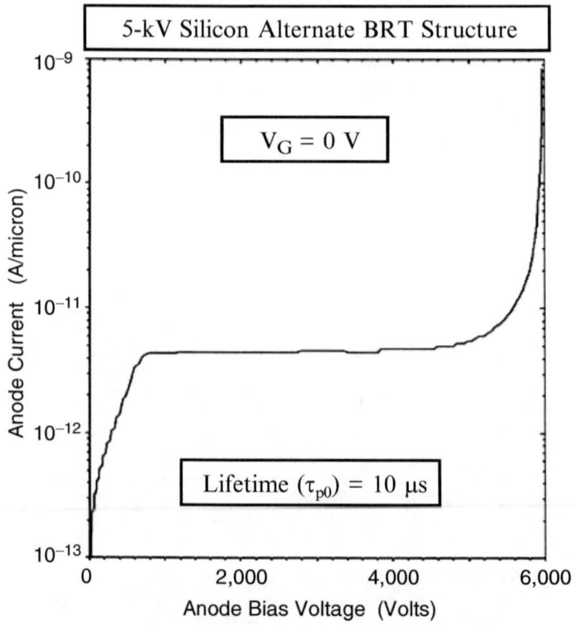

Fig. 9.29 Forward blocking characteristics for the alternate BRT structure

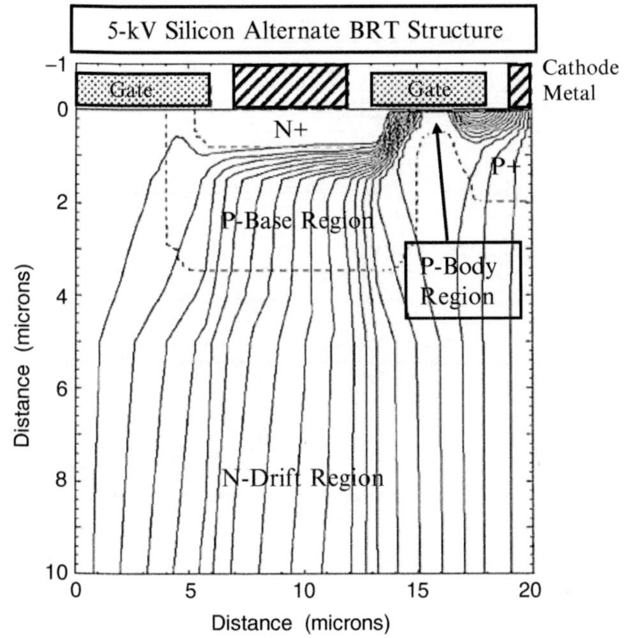

Fig. 9.30 Current flow lines during the blocking mode for the 5-kV alternate BRT structure: $V_G = 0$ V

9.3.2 *On-State Voltage Drop*

The alternate BRT structure operates with latch-up of the thyristor structure within the device. Consequently, the on-state characteristics and the free carrier distribution within the N-drift region can be expected to be similar to those for the thyristor structure (see Chap. 2). However, the incorporation of the p-channel turn-off MOSFET within the thyristor structure degrades the injection efficiency of the cathode junction because some of the hole current is diverted to the P^+ source region of the p-channel turn-off MOSFET. This can become a serious problem when the doping concentration of the P-body region is made too large as discussed in this section with the aid of numerical simulations.

Simulation Results

The results of two-dimensional numerical simulations for the 5-kV asymmetrical alternate BRT structure are described here. The total width (W_{Cell}) of the structure, as shown by the cross section in Fig. 9.26, was 20 μm (area $= 2.0 \times 10^{-7}$ cm^{-2}). The doping profiles were discussed in the previous section. The on-state characteristics of the 5-kV silicon asymmetric alternate BRT structure at a positive gate bias voltage of 10 V are shown in Fig. 9.31 for two cases of P-body doping concentrations. It can be observed that the thyristor has difficulty in latching-up when the surface concentration of the P-body region is increased from 2 to 5×10^{16} cm^{-3}.

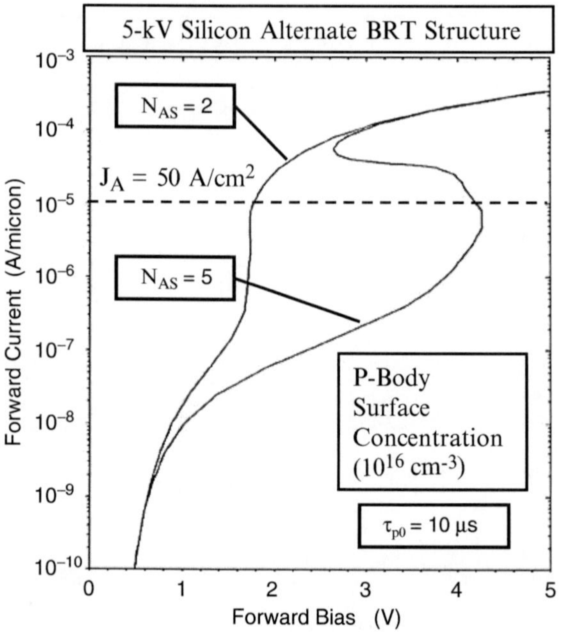

Fig. 9.31 On-state characteristics of the 5-kV alternate BRT structures

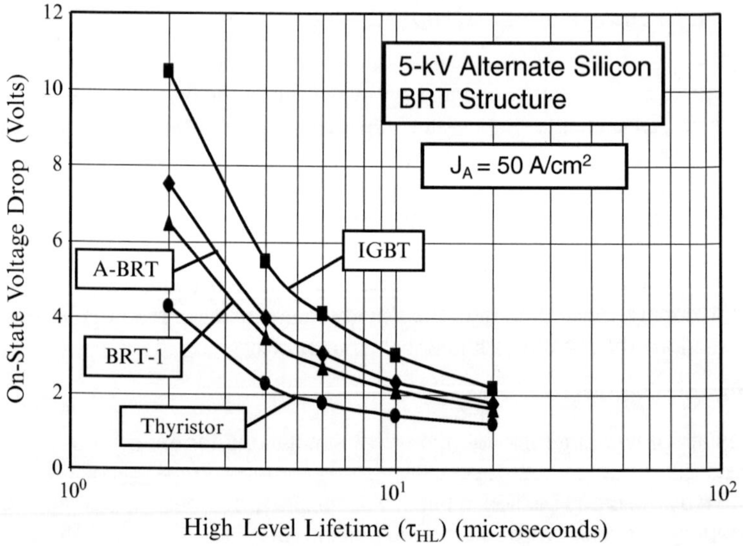

Fig. 9.32 On-state voltage drop for the 5-kV asymmetric BRT structure: N-base lifetime dependence

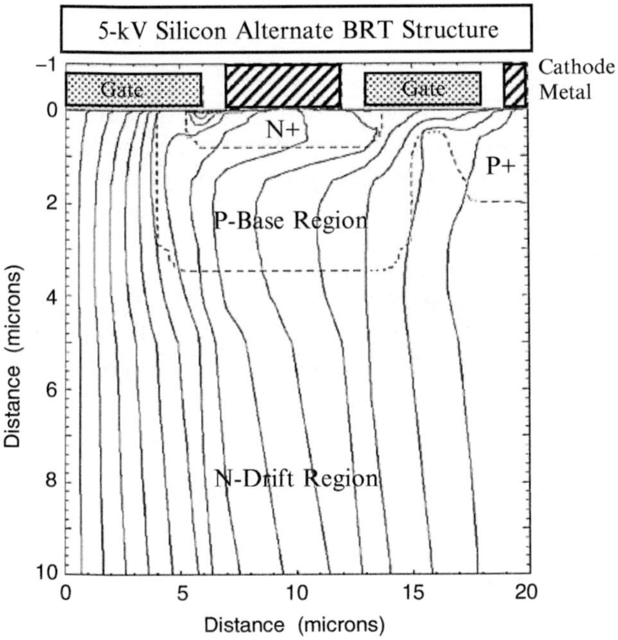

Fig. 9.33 On-state current flow lines for the 5-kV alternate BRT structure

The variation of the on-state voltage drop obtained from the results of the numerical simulation for the 5-kV asymmetric alternate BRT structure (A-BRT), as a function of the lifetime in the N-base region, is shown in Fig. 9.32 for the case of an anode on-state current density of 50 A/cm^2. For comparison purposes, the on-state voltage drops for the case of the 5-kV asymmetric trench-gate IGBT structure and for the 5-kV asymmetric thyristor structure are also provided in this figure. It can be observed that the alternate BRT structure has a slightly larger on-state voltage drop than the BRT-1 structure for each lifetime value.

In order to understand the operation of the alternate BRT structure in the on-state, it is beneficial to examine the distribution of the current within the structure. The on-state current flow lines within the 5-kV asymmetric alternate BRT structure are shown in Fig. 9.33 for the case of an on-state current density of 50 A/cm^2. A high-level lifetime of 2 μs was used during this simulation. It can be observed that some of the current entering the P-base region flows via the P-body region of the p-channel turn-off MOSFET degrading the injection efficiency of the cathode region.

9.3.3 Turn-Off Characteristics

The turn-off process in the alternate BRT structure is very similar to that described for the previous BRT structure. Consequently, the results of numerical simulation for the alternate BRT structure are not provided here in the interest of conserving space.

9.4 10,000-V Silicon BRT

The 10-kV silicon asymmetric BRT structure can be expected to function just like the 5-kV device. However, its design and operation is constrained by the larger blocking voltage capability. The lifetime in the N-base region for the 10-kV device must be larger to maintain a reasonable on-state voltage drop. The larger N-base width results in more stored charge within the structure, which limits the switching frequency.

In Chap. 4, it was demonstrated that the GTO structure has a limited reverse-biased safe operating area (RBSOA) due to the influence of the holes in the space-charge region due to current flow. The analysis of the RBSOA for the BRT structure is identical to that provided in Sect. 4.4. Using the results shown in Fig. 4.57, it can be concluded that in order to turn off the 10-kV asymmetric BRT structure with a collector supply voltage of 6 kV, it is necessary to reduce the collector current density to only 20 A/cm^2. However, one of the merits of the BRT structure is the low on-state voltage drop, which allows its operation at an on-state current density of 50 A/cm^2. This value will therefore be utilized when determining the on-state voltage drop and switching transients for the 10-kV asymmetric BRT structures. Due to RBSOA limitations, the anode supply voltage for the switching transient must be reduced to 5,000 V.

9.4.1 Blocking Characteristics

The electric field distribution within the asymmetric BRT structure is essentially the same as that illustrated in Fig. 4.3 for the asymmetric GTO structure. Consequently, the design procedure described in Chap. 4 can be applied to the asymmetric BRT structure. From Fig. 4.50, the N-base region width required to obtain a forward blocking voltage of 11,000 V is 1,100 μm. However, the results of the numerical simulation shown in Chap. 4 for the 10-kV GTO structure demonstrate that an N-base width of 800 μm is sufficient.

Simulation Example

In order to gain insight into the physics of operation for the 10-kV asymmetric BRT structure under voltage blocking conditions, the results of two-dimensional numerical simulations are described here for a device with N-base width of 825 μm. The simulations were performed using a cell with the structure shown in Fig. 9.1. This half-cell has a width of 20 μm (area = 2.0×10^{-7} cm^{-2}). The asymmetric BRT structure used for the simulations was formed by diffusions performed into a uniformly doped N-type drift region with a doping concentration of 2×10^{12} cm^{-3}. All the diffusions in the 10-kV structure had the same parameters as the 5-kV device described in the previous section. The doping profile in the vertical direction through the N$^+$ cathode region is shown in Fig. 9.34

Fig. 9.34 Doping profile for the simulated asymmetric 10-kV BRT structure

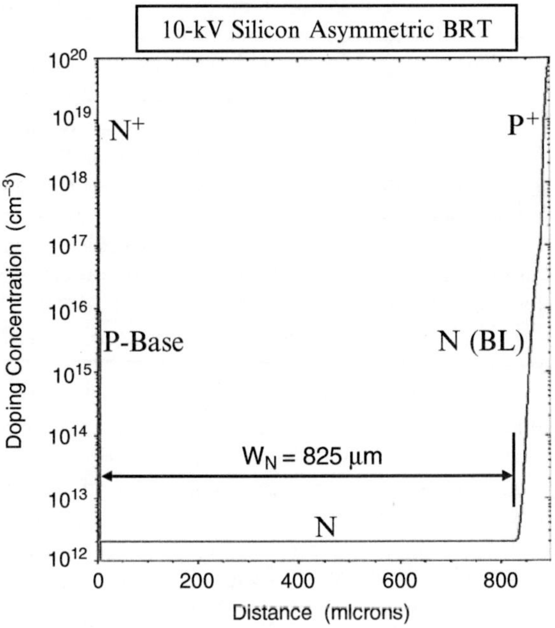

indicating that the net width of the lightly doped portion of the N-base region is 825 μm after accounting for the diffusions. The P-base and N^+ cathode regions are too shallow to be observed in this figure. Their doping profiles are the same as those for the 5-kV asymmetric BRT structure previously shown in Figs. 9.6 and 9.7.

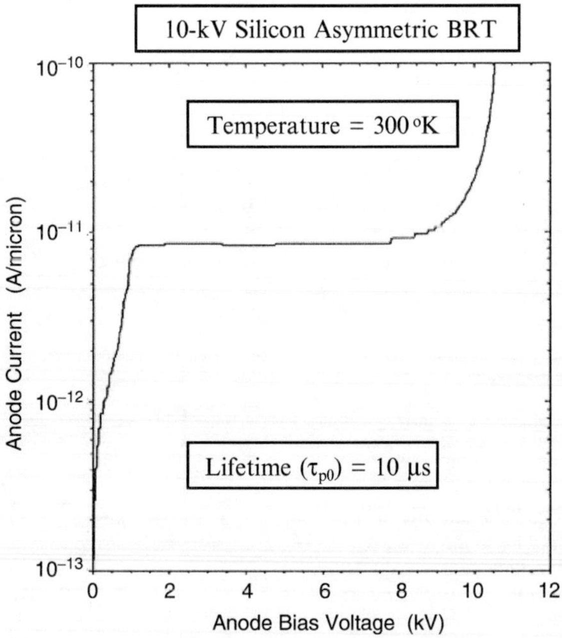

Fig. 9.35 Forward blocking characteristics of the 10-kV asymmetric BRT structure

The forward blocking capability of the 10-kV silicon asymmetric BRT structure was obtained by increasing the anode bias while maintaining the gate electrode at negative 10 V in order to short the cathode to the P-base region via the p-channel MOSFET. The characteristic obtained for a lifetime (τ_{p0}) of 10 μs is shown in Fig. 9.35. The leakage current increases rapidly with increasing anode bias voltage until about 1,000 V. This occurs due to the increase in the space-charge generation volume and the increase in the current gain (α_{PNP}) of the open-base P-N-P transistor until the anode bias becomes equal to the reach-through voltage of 1,115 V obtained using the analytical solution given by Eq. 4.2. The leakage current then becomes independent of the anode voltage until close to the breakdown voltage. This behavior is well described by the analytical model. The numerical simulations indicate that a breakdown voltage of 10,500 V is possible with an N-base width of only 825 μm.

The voltage is primarily supported within the lightly doped portion of the N-base region in the 10-kV asymmetric BRT structure during operation in the forward blocking mode. This is illustrated in Fig. 9.36 where the electric field profiles are shown during operation in the forward blocking mode at several collector voltages. It can be observed that the P-base/N-base junction (J_2) becomes reverse biased during the forward blocking mode with the depletion region extending toward the right-hand side with increasing (positive) collector bias. The electric field has a triangular shape until the entire lightly doped portion of the N-base region becomes completely depleted. This occurs at a collector bias just above 1,000 V in good agreement with the reach-through voltage of 1,115 V obtained using the analytical solution (see Eq. 4.2). The electric field profile then takes a trapezoidal shape due to the high doping concentration in the N-buffer layer.

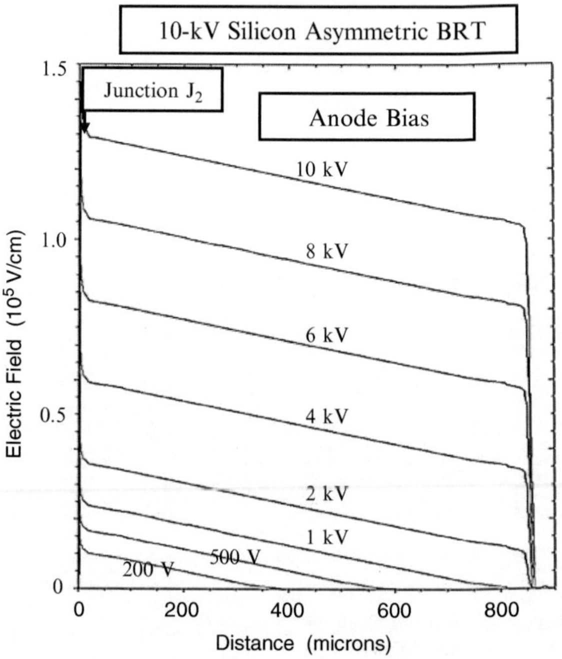

Fig. 9.36 Electric field profiles in the forward blocking mode for the 10-kV asymmetric BRT structure

9.4.2 On-State Voltage Drop

The on-state *i-v* characteristics and on-state voltage drop can be computed using the analytical model discussed in Sect. 9.2.2. In general, a larger lifetime is required in the N-base region for the 10-kV device when compared with the 5-kV device due to the larger width for the N-base region.

Simulation Results

The results of two-dimensional numerical simulations for the 10-kV asymmetrical silicon BRT structure are described here. The total half-cell width of the structure, as shown by the cross section in Fig. 9.1, was 20 μm (area $= 2.0 \times 10^{-7}$ cm^{-2}).

The on-state characteristics of the 10-kV silicon asymmetric BRT structure were obtained by using a gate bias voltage of 10-V using various values for the lifetime in the N-base region. The characteristics obtained from the numerical simulations are shown in Fig. 9.37. It can be observed that the on-state voltage drop increases as expected with reduction of the lifetime (τ_{p0}, τ_{n0}) indicated in the figure. The on-state voltage drop for the 10-kV asymmetric BRT structure is substantially smaller than that for the 10-kV asymmetric IGBT structure. For this reason, it is possible to operate the 10-kV asymmetric BRT structure at a larger on-state current density of 50 A/cm^2 from a power loss standpoint. However, due to the limitations of RBSOA, the maximum supply voltage must be reduced to 5,000 V.

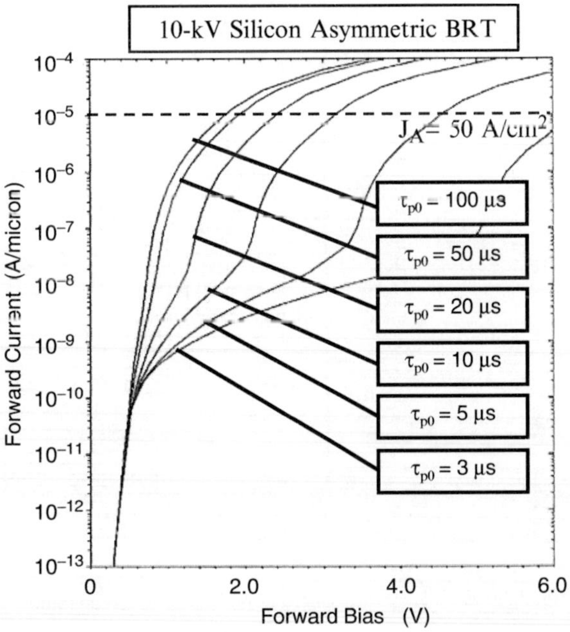

Fig. 9.37 On-state characteristics of the 10-kV asymmetric BRT structure

The good on-state voltage drop for the 10-kV asymmetric BRT structure for larger values of the lifetime in the N-base region is due to the large number of carriers injected into the drift region producing a drastic reduction of its resistance. This is illustrated in Fig. 9.38 where the injected carrier density is shown for six cases of the lifetime (τ_{p0}, τ_{n0}) in the N-base region of the BRT structure. It can be observed that the injected carrier density in the drift region on the anode side is more than four orders of magnitude larger than the doping concentration for the case of a lifetime of 100 μs. The injected carrier density is reduced by a factor of 5 times near the anode junction when the lifetime is reduced to 3 μs. There is a significant reduction in the injected carrier density in the middle of the drift region when the lifetime is reduced below 10 μs. This is due to the relatively large width for the N-base region when compared with the 5-kV silicon BRT structure. The reduced hole concentration in the drift region produces the observed increase in on-state voltage drop.

The variation of the on-state voltage drop obtained from the results of the numerical simulation, as a function of the lifetime in the N-base region, is shown in Fig. 9.39 for the case of an anode on-state current density of 50 A/cm^2. For comparison purposes, the on-state voltage drop for the case of the 10-kV asymmetric trench-gate IGBT structure is also provided in this figure at a collector on-state current density of 50 A/cm^2. It can be observed that the BRT structure has a lower on-state voltage drop than the IGBT structure for each lifetime value in

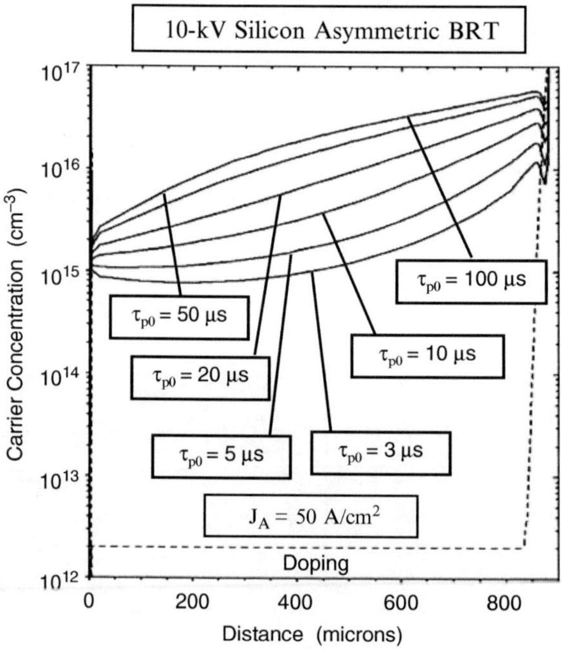

Fig. 9.38 On-state carrier distribution in the 10-kV asymmetric BRT structure

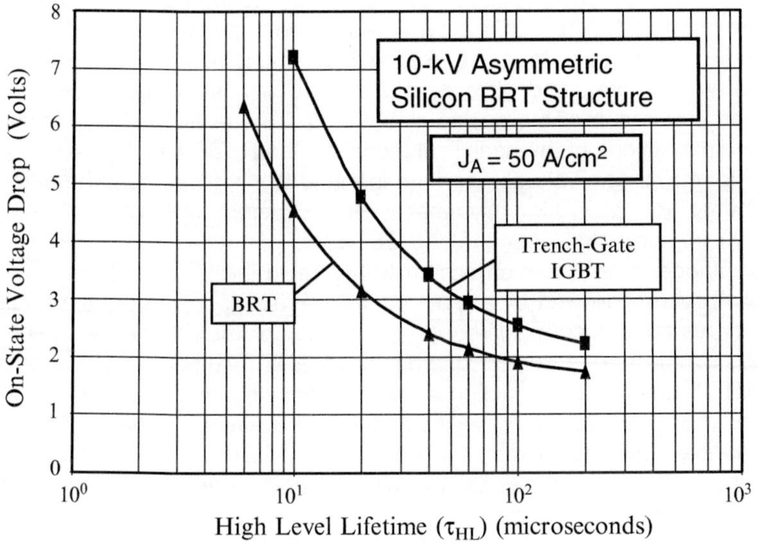

Fig. 9.39 On-state voltage drop for the 10-kV asymmetric BRT structure: N-base lifetime dependence

spite of the use of the trench-gate structure with high channel density for the IGBT structure. This is due to improved carrier distribution in the BRT structure with a high free carrier density near the cathode side of the drift region.

9.4.3 Turn-Off Characteristics

The physics for turn-off of the 10-kV silicon asymmetric BRT structure can be expected to be the same as that for the 5-kV device structure. Due to limitations with the RBSOA (as discussed in Chap. 4 for the silicon GTO structure), the 10-kV asymmetric BRT structure can be operated at an on-state current density of 50 A/cm^2 only if the anode supply voltage is reduced to 5,000 V. The results of numerical simulations of the 10-kV asymmetric BRT structure under these turn-off conditions are discussed here.

Simulation Results

Numerical simulations of the turn-off for the 10-kV silicon BRT structure with a high-level lifetime of 20 μs were performed by stepping the gate voltage down from positive 10 V to negative 10 V in 20 ns using an on-state current density of 50 A/cm^2. The resulting waveforms obtained from the numerical simulations for the anode voltage and current are shown in Fig. 9.40 for the case of an anode supply voltage of 5,000 V. It can be observed that there is no storage time for the 10-kV asymmetric BRT structure. The anode voltage initially increases non-linearly as described by the analytical model during the early stages of the voltage rise but the rate of rise becomes severely reduced at anode voltages beyond 4,000 V due to the onset of significant impact ionization as the RBSOA boundary is approached. The anode voltage almost saturates at 5,000 V indicating operation of the device close to its RBSOA limit. This is consistent with the predictions of the analytical model for the RBSOA of the BRT structure (see Fig. 4.57). The voltage rise time is found to be 10 μs from the numerical simulations for the 10-kV asymmetric BRT structure and the corresponding energy loss per cycle is found to be 1.25 J/cm^2.

The anode current turn-off occurs with a rapid initial decrease in current to about 10 A/cm^2 followed by a gradual change as expected from the analytical model. This rapid decrease is due to a reduction of impact ionization generated carriers as the anode current decreases (due to recombination of the stored charge) because of the reduction in the electric field in the space-charge region (as shown in Fig. 9.20 for the 5-kV structure). This effect is stronger in the 10-kV structure because it is operating very close to the RBSOA boundary. The punch-through anode current density ($J_{A,PT}$) obtained using the analytical model (Eq. 8.24) is 9 A/cm^2 and the simulation results indicate a slightly smaller value of about 8 A/cm^2. The current fall time is found to be 8 μs from the numerical simulations for the 10-kV asymmetric BRT structure and the corresponding energy loss per cycle is found to be 1.00 J/cm^2. The total energy loss per cycle for the 10-kV BRT structure is found to be 2.25 J/cm^2.

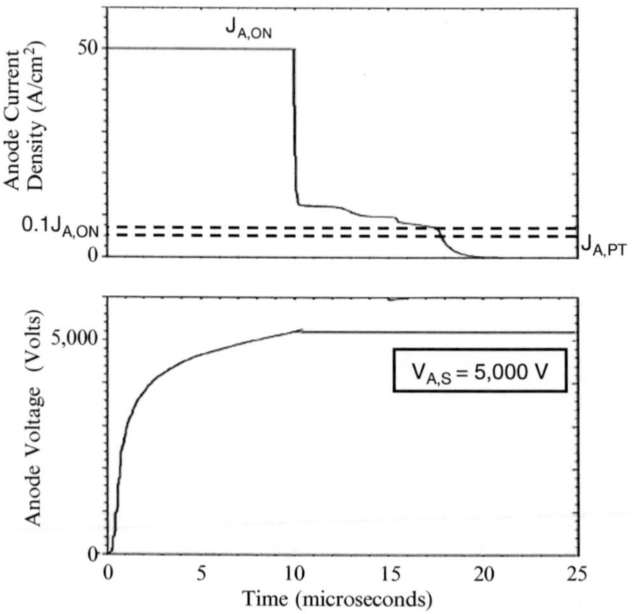

Fig. 9.40 Turn-off waveforms for the 10-kV asymmetric BRT structure

9.4.4 Switching Energy Loss

As discussed previously, the maximum operating frequency for the BRT structure is limited by the turn-off losses. The turn-off losses are associated with the voltage rise-time interval and the current fall-time interval. The energy loss for each event can be computed by using the equations previously provided in Sect. 9.2.5. Using this information, the maximum operating frequency for the 10-kV silicon asymmetric BRT structure can be derived using Eq. 9.6. The turn-off energy loss per cycle obtained from the numerical simulations of the silicon 10-kV asymmetric BRT structure can be derived from the waveforms in Fig. 9.40. For the case of a high-level lifetime of 20 μs in the N-base region, the energy loss per cycle during the voltage rise time is 2.25 J/cm^2 while the energy loss per cycle during the current fall time is 1.00 J/cm^2 in the case of an on-state current density of 50 A/cm^2 and an anode supply voltage of 5,000 V. The total energy loss per cycle is 2.25 J/cm^2 for the 10-kV silicon asymmetric BRT structure.

9.4.5 Maximum Operating Frequency

The maximum operating frequency for the 10-kV asymmetric BRT structure is limited by the turn-off losses. The turn-off losses are associated with the voltage rise-time interval and the current fall-time interval. The energy loss for each event can be computed by using the equation previously provided in Sect. 9.2.6. Using this information, the maximum operating frequency for the BRT structure can be derived using Eq. 9.6. The data acquired from the numerical simulations of the 10-kV asymmetric BRT and IGBT structures are provided in Fig. 9.41 for the case of an on-state operating current density of 50 A/cm^2.

	High-Level Lifetime (μs)	On-State Voltage Drop (Volts)	On-State Power Dissipation (W/cm^2)	Energy Loss per Cycle (J/cm^2)	Maximum Operating Frequency (Hz)
10-kV Asymmetric IGBT Structure	20	4.766	119	2.50	32
10-kV Asymmetric BRT Structure	20	3.158	78.9	2.25	54

Fig. 9.41 Power loss analysis for the 10-kV asymmetric BRT and IGBT structures with on-state current density of 50 A/cm^2

The maximum operating frequency obtained under the assumption of a 50% duty cycle and a total power dissipation limit of 200 W/cm^2 for the 10-kV asymmetric IGBT and BRT structures are found to be 32 and 54 Hz, respectively.

The maximum operating frequency for the silicon 10-kV asymmetric BRT structure is superior to that for the IGBT structure due to its lower on-state voltage drop. However, the BRT structure has not been well received by the power electronics community for typical applications such as motor drives because it lacks good FBSOA. This complicates circuit operation, especially management of reverse recovery of fly-back rectifiers, unless expensive snubber circuits are added to the circuit topology.

9.5 Forward-Biased Safe Operating Area

The BRT structure does not exhibit a substantial region of operation where the anode current can be saturated under gate control. This is demonstrated in this section by using the results of numerical simulations for the 5-kV asymmetric BRT structure. The BRT structure discussed in Sect. 9.2 does not exhibit any FBSOA because of the absence of a hole bypass path when positive gate bias is applied. However, the alternate BRT structure discussed in Sect. 9.3 can be expected to exhibit a limited FBSOA due to the presence of a hole bypass path through the P-body region even when small positive gate bias is applied. The results of numerical simulations of the FBSOA of this structure are discussed here.

Simulation Results

Numerical simulations of the 5-kV silicon asymmetric alternate BRT structure were performed for the case of a high-level lifetime of 4 μs with various values

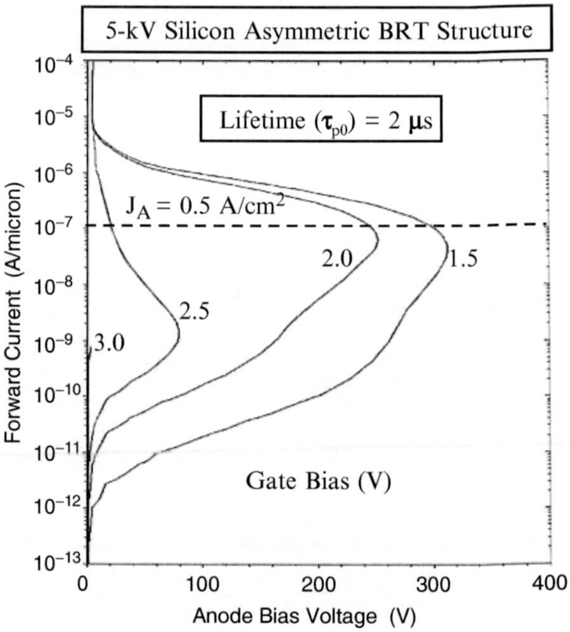

Fig. 9.42 5-kV alternate BRT FBSOA boundary

for the gate bias voltage while sweeping the anode voltage. The resulting output characteristics are shown in Fig. 9.42. The device was able to saturate the anode current at low gate bias voltages. However, latch-up of the thyristor occurs whenever the anode current density exceeds 0.5 A/cm^2. This demonstrates that even the alternate BRT structure has a poor FBSOA.

9.6 Reverse-Biased Safe Operating Area

The analytical solution for the RBSOA for the BRT structure can be obtained by using Eq. 4.97 provided for the GTO structure because the physics of operation is similar. However, the GTO structure suffers from current crowding during the turn-off process. This problem does not occur in the BRT structure. The RBSOA boundaries for the 5-kV asymmetric BRT structures obtained by using numerical simulations are provided in this section.

Simulation Results

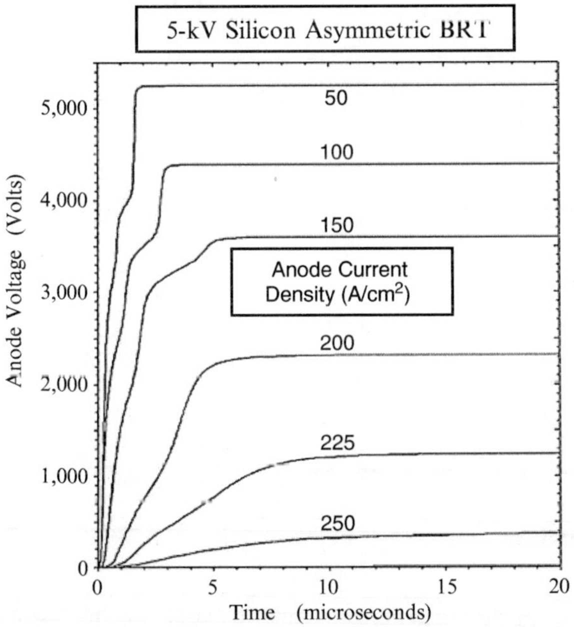

Fig. 9.43 5-kV asymmetric BRT RBSOA turn-off waveforms

The RBSOA boundary for the BRT structure can be obtained by turning off the structure starting with various on-state current densities. The presence of holes in the space-charge region enhances the electric field at the junction between the P-base region and the drift region. The electric field becomes larger for larger initial on-state current densities. Consequently, the anode voltage at which the on-state current density can be sustained by the impact ionization process becomes smaller. During turn-off, the anode voltage becomes limited as a function of time providing the RBSOA limit for each corresponding on-state current density.

Numerical simulations of the 5-kV silicon asymmetric BRT structure were performed for the case of a high-level lifetime of 2 μs with various values for the initial on-state current density. The resulting anode voltage waveforms are provided in Fig. 9.43. At anode current densities below 200 A/cm^2, the anode voltage increases and becomes limited by the onset of avalanche breakdown. At larger anode current densities, the turn-off is limited by injection from the N$^+$ cathode region. Using the anode voltage waveforms, the RBSOA boundary can be determined as shown in Fig. 9.44. The BRT exhibits a much inferior RBSOA boundary to that of the IGBT structure (see Fig. 5.76), which has made it a poor substitute for the IGBT in spite of the lower on-state power loss.

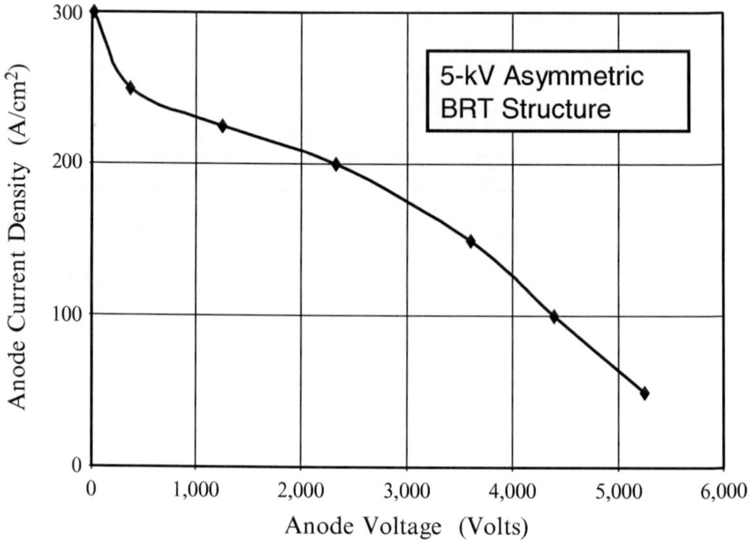

Fig. 9.44 RBSOA boundary for the5-kV asymmetric BRT structure

9.7 Conclusions

The physics of operation and design principles for the silicon BRT structure have been described in this chapter. This structure was proposed as an alternative to the MCT structure due to its simpler fabrication process that is close to the manufacturing process for IGBT devices. Unlike the MCT, the BRT structure can be configured to provide good forward blocking capability without any gate bias. Despite these advantages, the BRT structure has not displaced the IGBT in applications because it lacks a FBSOA resulting in the need for snubbers and it has much inferior RBSOA.

References

1. B.J. Baliga, "Gated Base Controlled Thyristor", U.S. Patent Number 5,099,300, Filed June 14, 1990, Issued March 24, 1992.
2. B.J. Baliga, "Base Resistance Controlled Thyristor with Single-Polarity Turn-on and Turn-off Control", U.S. Patent Number 5,198,687, Filed July 23, 1992, Issued March 30, 1993.
3. M. Nandakumar, et al, "A New MOS-Gated Power Thyristor Structure with Turn-Off Achieved by Controlling the Base Resistance", IEEE Electron Device Letters, Vol. 12, pp. 227-229, 1991.
4. M. Nandakumar, et al, "The Base Resistance Controlled Thyristor (BRT): A New MOS Gated Power Thyristor", IEEE International Symposium on Power Semiconductor Devices and ICs, Abstract 6.4, pp. 138-141, 1991.
5. M. Nandakumar, et al, "Fast Switching Power MOS-Gated (EST and BRT) Thyristors", IEEE International Symposium on Power Semiconductor Devices and ICs, Abstract 10.1, pp. 256-260, 1992.
6. B.J. Baliga, "Trench-Gate Base Resistance-Controlled Thyristors (UMOS-BRTs)", IEEE Electron Device Letters, Vol. 13, pp. 597-599, 1992.
7. M. Nandakumar and B.J. Baliga, "Modeling the Turn-Off Characteristics of the Base Resistance Controlled Thyristor (BRT)", Solid State Electronics, Vol. 38, pp. 703-713, 1995.
8. B.J. Baliga and R. Kurlagunda, "Floating Base Thyristor", Electronics Letters, Vol. 31, No. 18, pp. 1613-1615, 31st August 1995.
9. S. Sridhar and B.J. Baliga, "Improved BRT Structures Fabricated using SIMOX Technology", IEEE Electron Device Letters, Vol. 17, pp. 512-514, 1996.
10. N. Mohan, T.M. Undeland, and W.P. Robbins, "Power Electronics", John Wiley and Sons, Inc., New York, 1995.
11. R. Kurlagunda and B.J. Baliga, "The Dual-Gate BRT", IEEE International Symposium on Power Semiconductor Devices and ICs, Abstract 2.3, pp. 29-33, 1995.
12. T. Yamazaki and B.J. Baliga, "New Dual-Gate BRT Structures with Enhanced FBSOA", IEEE International Symposium on Power Semiconductor Devices and ICs, pp. 293-296, 1997.
13. B.J. Baliga, "Fundamentals of Power Semiconductor Devices", Springer-Science, New York, 2008.

Chapter 10
Silicon EST

As discussed in Chap. 8, there was a flurry of activity in the 1990s to explore the development of MOS-gated thyristor structures due to their reduced on-state voltage drop when compared with the IGBT structure. The base-resistance-controlled thyristor (BRT) structure was proposed [1, 2] to take advantage of thyristor-based on-state current flow under MOS gate control to reduce the gate drive requirements. In comparison with the MCT structure discussed in chapter 8, the BRT structure had the advantage of using a double-diffusion process similar to that used to manufacture IGBT structures. However, as discussed in Chap. 9, the BRT structure has not been found to be favorable for applications because it lacks a forward-biased safe operating area (FBSOA) as in the case of the MCT structure. Thyristor-based structures exhibit uncontrolled rapid turn-on due to the internal regenerative action that can lead to extremely high reverse recovery currents in the antiparallel rectifiers leading to the destruction of the rectifier and the switch. The MCT and BRT structures exhibit this behavior. The emitter switched thyristor (EST) concept was proposed [3] to create a device structure in which the on-state current flow occurs via a latched-up four-layer thyristor region while the device still exhibits current saturation under gate control like the IGBT structure. With the gate-controlled output characteristics of the EST structure, it is possible to control the rate of rise of the anode current without the need for snubber circuits. The idea was first described in the literature in 1990 [4, 5]. A rigorous study to understand the physics of EST operation and evaluate the performance of experimental devices with blocking voltages ranging from 600 to 5,000 V was conducted in the 1990s [6–19].

Since the characteristics of the EST structure resemble those of the IGBT structure, it is possible to replace the IGBT with the EST structure in motor control circuits. The lower on-state voltage drop for the EST structure allows reduction of the overall power losses leading to an improvement in the efficiency. Two basic EST structures, the single channel (SC) and dual channel (DC), are discussed in this chapter.

B.J. Baliga, *Advanced High Voltage Power Device Concepts*,
DOI 10.1007/978-1-4614-0269-5_10, © Springer Science+Business Media, LLC 2011

10.1 Basic Structure and Operation

The single-channel asymmetric EST (SC-EST) structure with the planar gate architecture [4] is illustrated in Fig. 10.1. The SC-EST structure contains a floating N^+ cathode region to which no metal contact is provided. The P-body region of the turn-off MOSFET is formed by merging two P-base diffusions using the D-MOS process. When the main thyristor formed between the floating N^+ cathode region, the P-base region, the N-drift region, and the P^+ anode region latches up, the current flowing through the floating N^+ cathode region must go through the channel of MOSFET-2 before it enters the cathode terminal. Consequently, the thyristor current flow can be controlled by the bias applied to the gate of the n-channel MOSFET-2. The on-state voltage drop for the EST structure is above that for a thyristor structure because of the additional voltage drop in the channel of MOSFET-2. In the EST structure, current saturation is possible if the gate bias for the MOSFET is reduced close to the threshold voltage so that the MOSFET operates in its saturation mode. However, the current saturation capability may become limited by breakdown of the short channel MOSFET-2 at high anode current densities.

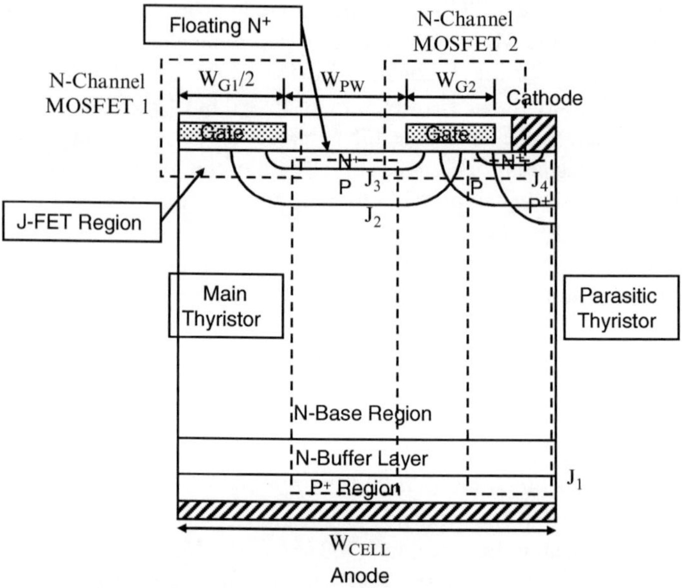

Fig. 10.1 The asymmetric SC-EST structure

Since the asymmetric EST structure is intended for use in DC circuits, its reverse blocking capability does not have to match the forward blocking capability allowing the use of an N-buffer layer adjacent to the P^+ anode region. The N-buffer layer has a much larger doping concentration than the lightly doped portion of the N-base region. The electric field in the asymmetric EST takes a trapezoidal shape allowing supporting the forward blocking voltage with a thinner N-base region.

This allows achieving a lower on-state voltage drop and superior turn-off characteristics. The doping concentration of the buffer layer and the lifetime in the N-base region must be optimized to perform a trade-off between on-state voltage drop and turn-off switching losses. EST structures are discussed in this chapter with two blocking voltage ratings for comparison with other device structures.

The EST structure shown in Fig. 10.1 also contains a second MOSFET structure to allow turning on the device. For a three-terminal device, the gates of both MOSFET-1 and MOSFET-2 are connected together when they are formed using a single interconnected polysilicon layer. When a positive bias is applied to the gate, the channels of both MOSFETs are turned on. Electrons can then flow from the N^+ region on the right-hand side, through the channel of MOSFET-2 and the N^+ floating region, and via the channel of MOSFET-1 into the N-drift region. This provides the base drive current for the vertical P-N-P transistor. The injected holes from the P^+ anode region (emitter of the vertical P-N-P transistor) diffuse through the N-drift region (base of the vertical P-N-P transistor) and are collected at junction J_2 between the P-base region and the N-drift region. The holes can be removed by flowing through the P-base region into the P^+ region on the right-hand side, which is connected to the cathode metal. This behavior is similar to that of an IGBT structure. The EST structure therefore exhibits an IGBT mode at lower anode current levels. However, the holes collected at junction J_2 also serve as the base drive current for the vertical N-P-N transistor formed between the floating N^+ region, the P-base region, and the N-drift region. When the hole current increases, the junction J_3 becomes forward biased due to the resistance of the P-base region. The thyristor latches up when this forward bias exceeds the built-in potential.

The EST structure contains a parasitic thyristor formed between the N^+ region on the right-hand side, the P-base region under it, the N-drift region, and the P^+ anode. If this thyristor latches up, the anode current can no longer be regulated by the gate bias on the MOSFETs. Since this parasitic thyristor is similar to that within the IGBT structure, its latch-up can be suppressed by using the P^+ diffusion shown in Fig. 10.1. The latch-up of the parasitic thyristor determines the maximum turn-off current capability of the EST structure.

The dual-channel asymmetric EST (DC-EST) structure with the planar gate architecture [7] is illustrated in Fig. 10.2. The DC-EST structure contains two channels between the floating N^+ cathode region and the cathode contact metal. When the main thyristor formed between the floating N^+ cathode region, the P-base region, the N-drift region, and the P^+ anode region latches up, the current flowing through the floating N^+ cathode region must go through the channel of the MOSFET before it enters the cathode terminal. Consequently, the thyristor current flow can be controlled by the bias applied to the gate of the n-channel MOSFET. The on-state voltage drop for the EST structure is above that for a thyristor structure because of the additional voltage drop in the channels of MOSFET. In the EST structure, current saturation is possible if the gate bias for the MOSFET is reduced close to the threshold voltage so that the MOSFET operates in its saturation mode. The current saturation capability of the DC-EST extends to high voltages and is similar to that for the IGBT structure. Its excellent FBSOA with a lower on-state voltage drop makes it a possible replacement for high-voltage IGBT structures.

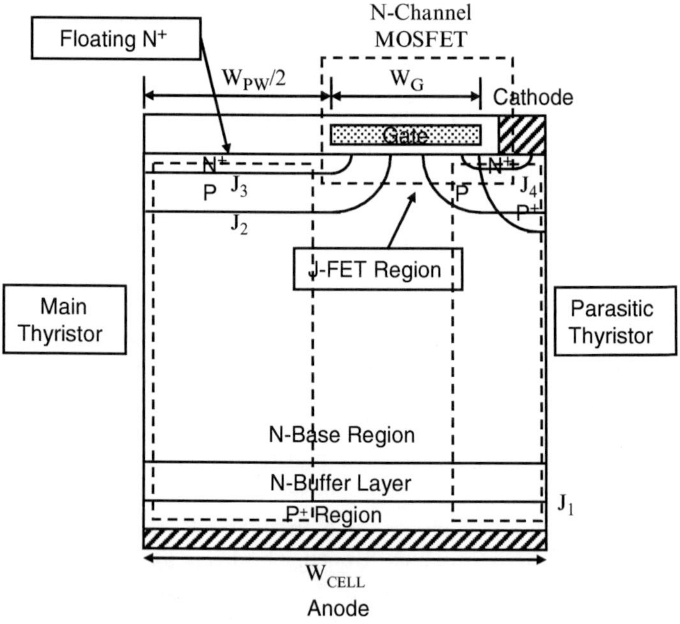

Fig. 10.2 The asymmetric DC-EST structure

The P-base region under the floating N$^+$ region in the DC-EST structure is usually connected to the cathode metal through a high resistance path (shorting resistance). This can be done in practice by shorting the N$^+$ floating region and the P-base region at a point orthogonal to the cross section shown in Fig. 10.2. The resistance of this path must be carefully designed to allow the thyristor to latch up at a low anode current density.

When a positive bias is applied to the anode of the EST structure, junction J_2 between the P-base and N-base regions becomes reverse biased. This junction is capable of supporting a large voltage with a depletion region formed in the lightly doped N-base region. In the SC-EST structure, the leakage current generated in the drift region is collected by junction J_2 and flows into the P-base region. It is then removed via the P$^+$ region on the right-hand side. In the DC-EST structure, the leakage current generated in the drift region is collected by junction J_2 and flows into the P-base region. It is then removed via the shorting resistance allowing the structure to support a high voltage in the forward blocking mode.

Once the EST is operating in its on-state, the device can be turned off by switching the gate bias from a positive value to zero. This stops the electron transport that serves to drive the vertical P-N-P transistor. At the same time, the floating N$^+$ region is isolated from the cathode contact extinguishing the thyristor regenerative action. A simple model for the maximum anode turn-off current density can be formulated using a lumped element approach. In this approach, all of the hole current being collected at the junction J_2 is assumed to flow through a lumped shunting resistance for the hole current path. In a simplified model, the maximum turn-off current is limited by latch-up of the parasitic thyristor.

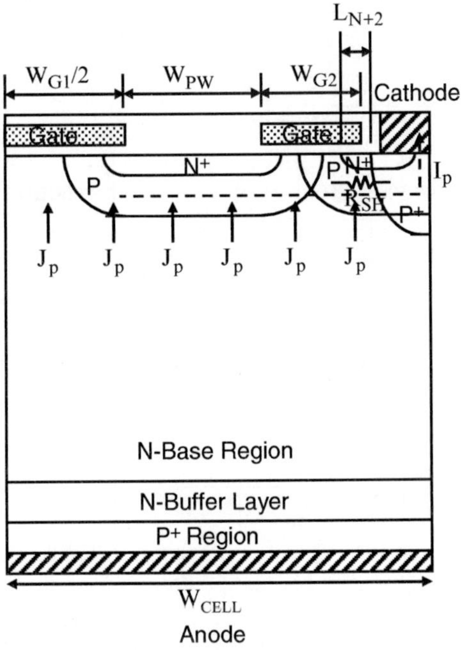

Fig. 10.3 Current flow during turn-off for the SC-EST structure

The current flow within the SC-EST structure during turn-off is illustrated in Fig. 10.3. A part of the anode current flows to the cathode contact via the contact to the P^+ region. The hole current responsible for latch-up of the parasitic thyristor (I_p) is that collected by the P-base region on the left-hand side of the gate region with width W_{G2}. During turn-off under inductive load operation, hole current responsible for latch-up of the parasitic thyristor (I_p) is equal to the initial anode current density ($J_{A,ON}$) multiplied by the area on the left-hand side of the gate region with width W_{G2}:

$$I_p = J_{A,ON}\left(\frac{W_{G1}}{2} + W_{PW} + W_{G2}\right)Z \tag{10.1}$$

where Z is the length of the cell in the orthogonal direction to the cross section shown in Fig. 10.3. The voltage drop produced by this current when flowing through the lumped shunting resistance (R_{SH}) must be less than the built-in potential for the junction J_4 between the N^+ cathode region and the P-base region if injection from this junction is to be suppressed to achieve the desired turn-off:

$$V_{bi} = I_p R_{SH} = J_{A,MAX} R_{SH}\left(\frac{W_{G1}}{2} + W_{PW} + W_{G2}\right)Z \tag{10.2}$$

The lumped resistance of the P-base region is given by:

$$R_{SH} = \rho_{S,PB}\frac{L_{N+2}}{Z} \tag{10.3}$$

where $\rho_{S,PB}$ is the pinch sheet resistance of the P-base region and L_{N+2} is the length of the N^+ region beyond the P^+ diffusion.

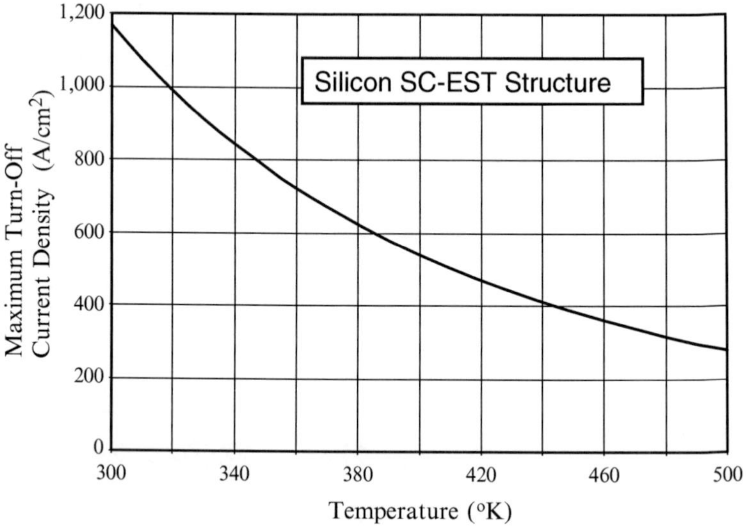

Fig. 10.4 Maximum turn-off current density for the SC-EST structure

Using the above equations, the maximum turn-off current density is found to be given by:

$$J_{A,MAX} = \frac{2V_{bi}}{\rho_{S,PB} L_{N+2} (W_{G1} + 2W_{PW1} + 2W_{G2})} \quad (10.4)$$

The maximum turn-off current density predicted by the analytical model is plotted in Fig. 10.4 as a function of temperature. The following values were used in the analytical model: built-in potential of 0.937 V at room temperature using doping concentrations of 1×10^{19} cm^{-3} for the N^+ region and 1.5×10^{17} cm^{-3} for the P-base region, P-base pinch sheet resistance of 1,400 Ω/sq based on an average doping concentration of 1.0×10^{17} cm^{-3} and thickness of 2 μm for the P-base region, polysilicon window width of 50 μm, and a gate width of 3.0 μm for the n-channel MOSFET integrated into the EST structure. It can be observed that the maximum turn-off current density at room temperature is much larger than that for the MCT and BRT structures. The maximum turn-off current density decreases with increasing temperature due to a reduction of the built-in potential and a reduction in the mobility for holes in the P-base region [20]. The predicted maximum turn-off capability for the SC-EST structure is much larger than that for the MCT and BRT structures at high temperatures as well.

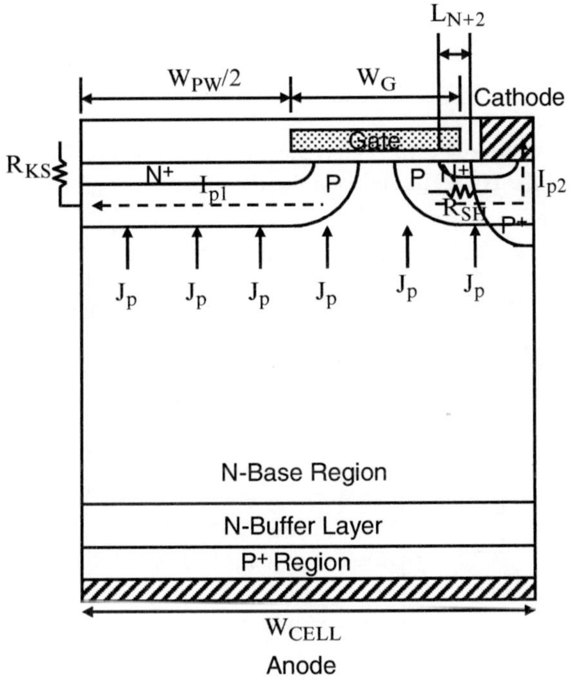

Fig. 10.5 Current flow during turn-off for the DC-EST structure

The current flow within the DC-EST structure during turn-off is illustrated in Fig. 10.5. A part of the anode current (I_{p1}) flows to the cathode short (resistance R_{KS}) via the contact to the P-base region. The hole current responsible for latch-up of the parasitic thyristor (I_{p2}) is that collected by the P-base region on the right-hand side. During turn-off under inductive load operation, hole current density responsible for latch-up of the parasitic thyristor (J_p) is equal to the initial anode current density ($J_{A,ON}$) multiplied by the area on the right-hand side of the gate region with width W_G:

$$I_{p2} = J_{A,ON}\left(\frac{W_G}{2}\right)Z \tag{10.5}$$

where Z is the length of the cell in the orthogonal direction to the cross section shown in Fig. 10.5. The voltage drop produced by this current when flowing through the lumped shunting resistance (R_{SH}) must be less than the built-in potential for the junction J_4 between the N^+ cathode region and the P-base region if injection from this junction is to be suppressed to achieve the desired turn-off:

$$V_{bi} = I_{p2}R_{SH} = J_{A,MAX}R_{SH}\left(\frac{W_G}{2}\right)Z \tag{10.6}$$

The lumped resistance of the P-base region is given by:

$$R_{SH} = \rho_{S,PB} \frac{L_{N+2}}{Z} \tag{10.7}$$

where $\rho_{S,PB}$ is the pinch sheet resistance of the P-base region and L_{N+2} is the length of the N^+ region beyond the P^+ diffusion.

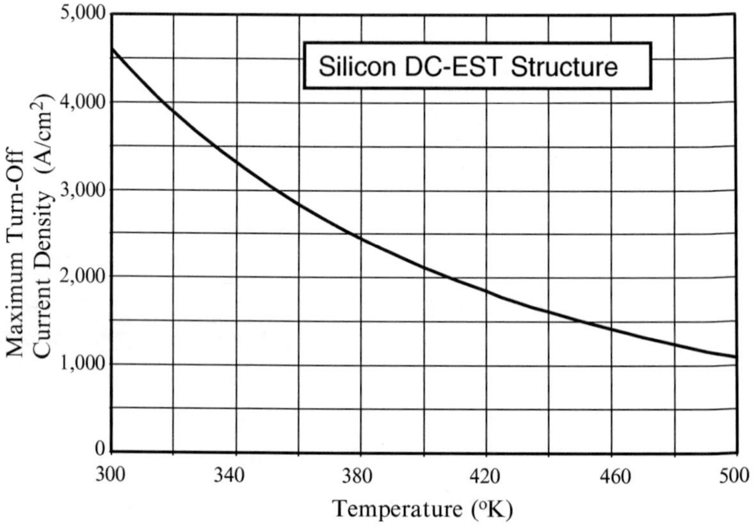

Fig. 10.6 Maximum turn-off current density for the DC-EST structure

Using the above equations, the maximum turn-off current density is found to be given by:

$$J_{A,MAX} = \frac{2V_{bi}}{\rho_{S,PB} L_{N+2} W_G} \tag{10.8}$$

The maximum turn-off current density predicted by the analytical model is plotted in Fig. 10.6 as a function of temperature. The following values were used in the analytical model: built-in potential of 0.937 V at room-temperature using doping concentrations of 1×10^{19} cm^{-3} for the N^+ region and 1.5×10^{17} cm^{-3} for the P-base region, P-base pinch sheet resistance of 1,400 Ω/sq based on an average doping concentration of 1.0×10^{17} cm^{-3} and thickness of 2 μm for the P-base region, polysilicon window width of 40 μm, and a gate width of 5.0 μm for the n-channel MOSFET integrated into the EST structure. It can be observed that the maximum turn-off current density at room temperature for the DC-EST structure is much larger than that for the SC-EST structure. This is due to the smaller width for the floating N^+ region in the DC-EST structure and the availability of an alternate path for removal of current via the shunt resistance. The maximum turn-off current density decreases with increasing temperature due to a reduction of the built-in potential and a reduction in the mobility for holes in the P-base region [20]. The predicted maximum turn-off capability for the DC-EST is much larger than that for the MCT and BRT structures.

10.2 5,000-V Silicon SC-EST

The design and characteristics for the 5,000-V asymmetric SC-EST structure are discussed in this section. The design parameters for the N-base (drift) region required to achieve this blocking voltage are first analyzed. Using the optimum N-base width, the blocking characteristics for the device are then obtained. The on-state characteristics for the device are obtained for various lifetime values as well. The gate-controlled turn-off behavior of the silicon SC-EST structure is analyzed including the effect of the lifetime in the drift region.

10.2.1 Blocking Characteristics

The physics for blocking voltages in the first and third quadrants by the asymmetric SC-EST structure is the same as that previously discussed for the silicon IGBT structure. When a positive bias is applied to the anode terminal of the asymmetric SC-EST structure with zero bias applied to the gate, the P-base/N-base junction (J_2) becomes reverse biased while the junction (J_1) between the P^+ anode region and the N-base region becomes forward biased. The forward blocking voltage is supported across the P-base/N-base junction (J_2) with a depletion layer extending mostly within the N-base region. The electric field distribution within the asymmetric SC-EST structure is essentially the same as that illustrated in Fig. 4.3 for the asymmetric GTO structure. Consequently, the design procedure described in Chap. 4 can be applied to the asymmetric SC-EST structure. From Fig. 4.4, the N-base region width required to obtain a forward blocking voltage of 5,500 V is 470 μm. This width can be slightly reduced when taking into account the voltage supported within the P-base region due to its graded doping profile.

The leakage current in forward blocking mode is produced by space-charge generation within the depletion region. In the case of the asymmetric SC-EST structure in the forward blocking mode, the space-charge generation current at the reverse-biased P-base/N-base junction J_2 is amplified by the gain of the internal P-N-P transistor. Initially, the space-generation current increases with increasing anode bias due to expansion of the depletion region. Concurrently, the current gain (α_{PNP}) of the P-N-P transistor is also a function of the anode bias voltage because the base transport factor increases when the anode bias increases. Prior to the complete depletion of the lightly doped portion of the N-base region, the multiplication factor remains close to unity. It is therefore sufficient to account for the increase in the base transport factor with anode bias as given by Eqs. 4.8 and 4.9.

For the case of the silicon asymmetric SC-EST structure with a width of 450 μm for the lightly doped portion of the N-base region with a doping concentration of 5×10^{12} cm^{-3}, the entire lightly doped portion of the N-base region is completely depleted at a reach-through voltage of 780 V. Once the lightly doped portion of the N-base region becomes completely depleted, the electric field becomes truncated at the interface between the lightly doped portion of the N-base region and the N-buffer layer as illustrated at the bottom of Fig. 4.3. The space-charge generation width then becomes independent of the anode bias because the depletion width in the N-buffer layer is small. Under these bias conditions, the base transport factor also becomes independent of the anode bias as given by Eq. 4.10. Consequently, the leakage current becomes independent of the anode bias until the onset of avalanche multiplication. The leakage currents for the silicon asymmetric SC-EST structure are identical to those provided for the silicon asymmetric IGBT structure in Chap. 5.

Simulation Example

The results of two-dimensional numerical simulations are described here in order to gain insight into the physics of operation for the 5-kV asymmetric SC-EST structure under voltage blocking conditions. The simulations were performed using a cell with the structure shown in Fig. 10.1 with a width (W_{Cell}) of 65 μm (area = 6.5×10^{-7} cm^{-2}). A large floating N^{+} cathode region width of 50 μm was required to create a transition from the IGBT mode to the EST mode at a sufficiently low on-state current density as discussed in the next section.

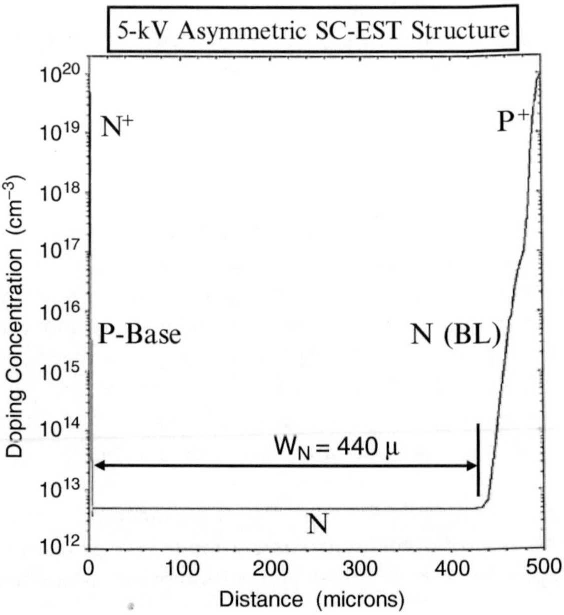

Fig. 10.7 Doping profile for the simulated asymmetric 5-kV SC-EST structure

The asymmetric SC-EST structure used for the simulations was formed by diffusions performed into a uniformly doped N-type drift region with a doping concentration of 5×10^{12} cm^{-3}. The N-buffer layer was formed by diffusion from the anode side with a depth of 55 μm. The doping profile in the vertical direction through the N$^+$ cathode region is shown in Fig. 10.7 indicating that the net width of the lightly doped portion of the N-base region is 440 μm after accounting for the diffusions. The peak doping concentration of the N-buffer layer is 1.0×10^{17} cm^{-3} and its thickness is 40 μm.

The P-base region for the asymmetric SC-EST structure was formed with a Gaussian doping profile with a surface concentration of 3×10^{17} cm^{-3} and a vertical depth of 3.5 μm as can be seen in Fig. 10.8 where the vertical doping profile in the upper 10 μm of the structure is provided. The N$^+$ cathode region was formed with a Gaussian doping profile with a surface concentration of 1×10^{20} cm^{-3} and a depth of 0.7 μm. The P$^+$ region on the right-hand side for suppressing latch-up of the parasitic thyristor was formed with a Gaussian doping profile with a surface concentration of 5×10^{19} cm^{-3} and a depth of 5 μm. The fabrication process for the EST is similar to that used to manufacture IGBT structures and less complex than that used for the MCT structure.

The doping profile across the surface for the 5-kV asymmetric SC-EST structure used for the numerical simulations is provided in Fig. 10.9. This profile was obtained along the horizontal line at $y = 0$ μm. It can be seen that the peak doping concentration of the P-body region of the n-channel MOSFET-1 is 1.0×10^{17} cm^{-3} and its channel length is 1.0 μm. The surface doping concentration of the P-body region (formed by merging two P-base diffusions) of the n-channel MOSFET-2 is 1.0×10^{17} cm^{-3} and its channel length is 1.5 μm. These values control the on-state voltage drop and current saturation behavior of the SC-EST.

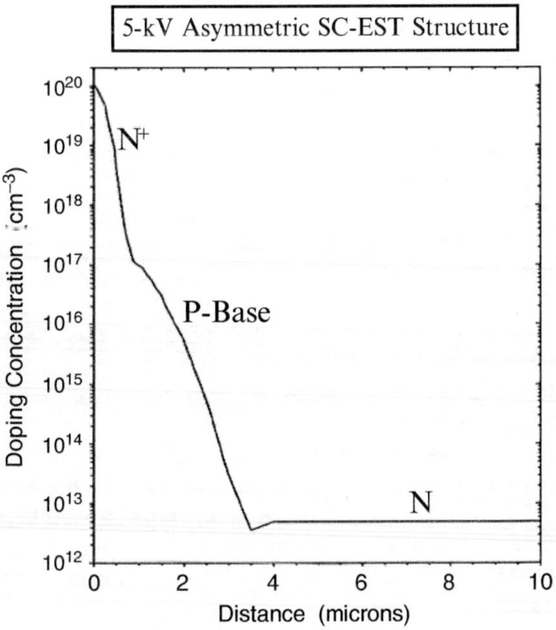

Fig. 10.8 Doping profile for the simulated asymmetric 5-kV SC-EST structure

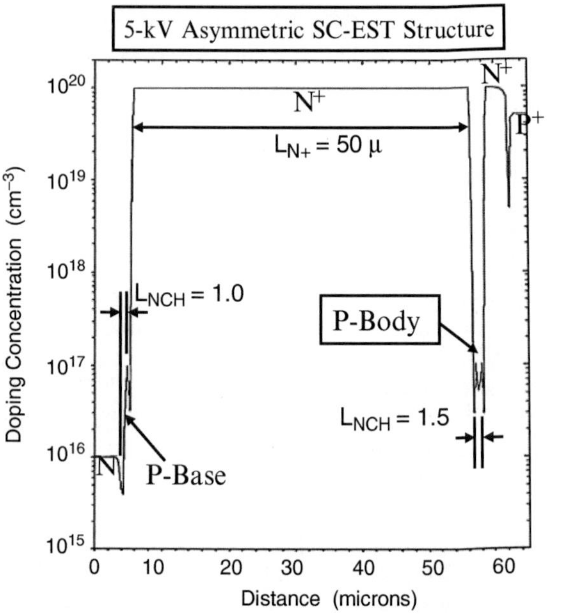

Fig. 10.9 Doping profile across the surface for the simulated 5-kV SC-EST structure

The forward blocking capability of the silicon asymmetric SC-EST structure was obtained using numerical simulations by increasing the anode bias while maintaining the gate electrode at zero volts. It was found that the device could support more than 6,000 V as shown in Fig. 10.10 at a temperature of 400°K. The leakage current increases rapidly with increasing anode bias voltage until about 780 V as predicted by the analytical model (see Fig. 5.2). This occurs due to the increase in the space-charge generation volume and the increase in the current gain (α_{PNP}) of the open base P-N-P transistor until the anode bias becomes equal to the reach-through voltage obtained using the analytical solution given by Eq. 4.2. The leakage current then becomes independent of the anode voltage until close to the breakdown voltage. This behavior is well described by the analytical model (see Fig. 5.2). The leakage current density obtained using the analytical model is within a factor of 2 of the values derived from the numerical simulations for all cases. The blocking characteristics for the asymmetric SC-EST structure are therefore similar to those for the asymmetric IGBT structure.

The current flow lines within the asymmetric SC-EST structure in the blocking mode are provided in Fig. 10.11 for the case of a gate bias of zero volts. From Fig. 10.11, it can be observed that the current collected at the P-base/N-drift region junction J_2 flows via the merged P-base regions and the P^+ region to the cathode contact. This allows the asymmetric SC-EST structure to support a high forward blocking voltage at zero gate bias without turning on the thyristor structure.

As in the case of other asymmetric structures, the anode voltage is primarily supported within the lightly doped portion of the N-drift region in the asymmetric SC-EST structure during operation in the forward blocking mode. The electric field profile within the asymmetric SC-EST structure is very similar to that shown previously for the asymmetric IGBT structure and is not included here in the interest of space.

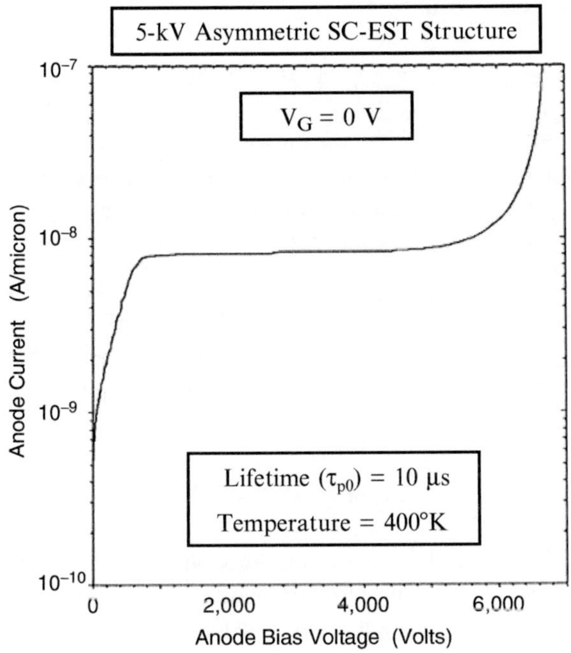

Fig. 10.10 Forward blocking characteristics for the asymmetric SC-EST structure

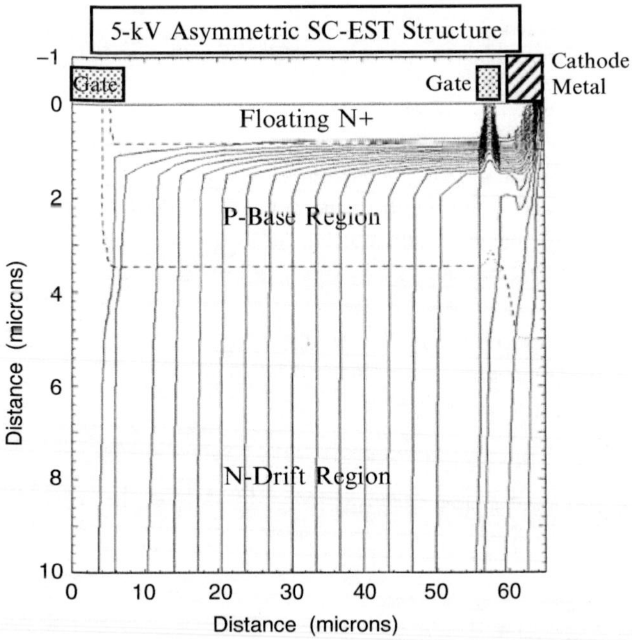

Fig. 10.11 Current flow lines during the blocking mode for the 5-kV asymmetric SC-EST structure: $V_G = 0$ V, $V_A = 5{,}000$ V

10.2.2 On-State Voltage Drop

The SC-EST structure operates with latch-up of the thyristor structure within the device. Consequently, the free carrier distribution within the N-drift region can be expected to be similar to those for the thyristor structure (see Chap. 2). However, the incorporation of the n-channel MOSFET-2 in series with the thyristor structure increases the on-state voltage drop. The on-state voltage drop for the SC-EST structure can be analytically calculated using:

$$V_{\mathrm{ON}} = V_{\mathrm{THY}} + V_{\mathrm{MOSFET}} \tag{10.9}$$

The voltage drop across the thyristor (V_{THY}) can be computed using:

$$V_{\mathrm{THY}} = \frac{2kT}{q}\ln\left[\frac{J_{\mathrm{A,ON}}d}{2qD_{\mathrm{a}}n_{\mathrm{i}}F(d/L_{\mathrm{a}})}\right] \tag{10.10}$$

where the "d" is half the thickness of the i-region, D_a is the ambipolar diffusion coefficient, n_i is the intrinsic concentration, and L_a is the ambipolar diffusion length in the drift region. The function $F(d/L_a)$ varies with the lifetime in the drift region [20].

The voltage drop across the MOSFET-2 can be computed using [20]:

$$V_{\mathrm{MOSFET}} = I_{\mathrm{CH}}R_{\mathrm{CH}} = I_{\mathrm{CH}}\left[\frac{L_{\mathrm{CH}}}{Z\mu_{\mathrm{ni}}C_{\mathrm{OX}}(V_{\mathrm{G}} - V_{\mathrm{TH}})}\right] \tag{10.11}$$

where I_{CH} is the channel current and R_{CH} is the channel resistance. The channel resistance is determined by L_{CH} the channel length, μ_{ni} the electron mobility in the inversion layer, C_{OX} the gate oxide capacitance, V_G the gate bias voltage, V_{TH} the threshold voltage, and Z the width of the device orthogonal to the cross section. The channel current consists of the current flowing through the main thyristor via the floating N^+ cathode region:

$$I_{\mathrm{CH}} = (W_{\mathrm{PW}}Z)J_{\mathrm{A,ON}} \tag{10.12}$$

Substituting into Eq. 10.11:

$$V_{\mathrm{MOSFET}} = J_{\mathrm{A,ON}}\left[\frac{W_{\mathrm{PW}}L_{\mathrm{CH}}}{\mu_{\mathrm{ni}}C_{\mathrm{OX}}(V_{\mathrm{G}} - V_{\mathrm{TH}})}\right] \tag{10.13}$$

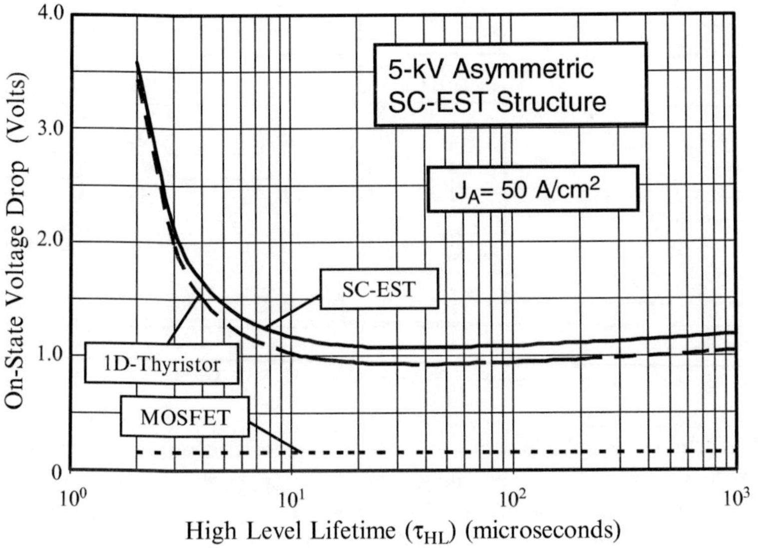

Fig. 10.12 On-state voltage drop for the SC-EST structure

The on-state voltage drop for the 5,000 V asymmetric SC-EST computed by using the above equations is provided in Fig. 10.12 as a function of the high-level lifetime in the drift region. The drift region of the structure consists of a lightly doped portion with a thickness of 440 μm and a buffer layer with a thickness of 30 μm. This device structure had a polysilicon window (W_{PW}) of 50 μm, a channel length of 1.5 μm, and a gate oxide thickness of 500 Å. As in the case of previous structures, the inversion layer mobility was assumed to be 450 cm^2/V-s. At an on-state current density of 50 A/cm^2, the MOSFET contributes only 0.153 V to the total on-state voltage drop independent of the high-level lifetime. When the high-level lifetime is reduced below 10 μs, the voltage drop across the thyristor portion begins to increase rapidly resulting in an increase in the on-state voltage drop for the SC-EST structure. This analytical model assumed that the entire length of the floating N^+ region operates like a one-dimensional thyristor.

The SC-EST has several operating modes in the on-state. At low on-state current densities, the device operates like an IGBT prior to the latch up of the main thyristor. Once the main thyristor latches up, the device operates in the EST mode. At large on-state current densities, the parasitic thyristor latches up resulting in loss of gate control. The transition points between these regimes of operation can be analytical modeled by analysis of the hole current flow through the structure.

When the SC-EST structure is operating in the IGBT mode, electrons are supplied via the gate on the left-hand side as base drive current for the P-N-P transistor. The resulting hole current flow is illustrated in Fig. 10.13. The hole current is collected by the P-base/N-drift junction and removed via the cathode contact on the right-hand side. The hole current (I_p) flows via the resistance of the P-base region (R_{PB}) under the floating N^+ cathode region and the resistance of the merged P-base regions (R_{MPB})

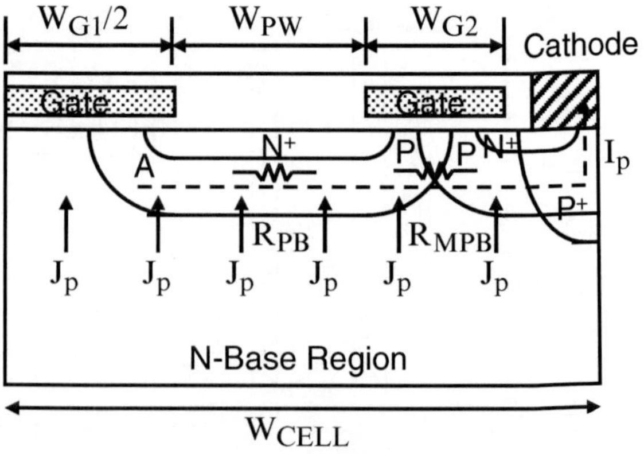

Fig. 10.13 Current flow in the SC-EST structure during the IGBT mode

before reaching the cathode contact via the P^+ region. This produces a voltage drop that creates a forward bias across the junction between the floating N^+ cathode region and the P-base region. The highest forward bias occurs at point A located furthest away from the cathode contact. In a lumped element analysis, the forward bias voltage at point A is given by:

$$V_A = I_p(R_{PB} + R_{MPB}) \tag{10.14}$$

The lumped hole current can be computed by using:

$$I_P = \alpha_{PNP} J_{A,ON}\left(\frac{W_{G1}}{2} + W_{PW}\right)Z \tag{10.15}$$

where Z is the width of the device orthogonal to the cross section shown in Fig. 10.13 and α_{PNP} is the common base current gain of the PNP transistor.

The lumped resistance of the P-base region is given by:

$$R_{PB} = \rho_{S,PB}\frac{W_{PW}}{Z} \tag{10.16}$$

where $\rho_{S,PB}$ is the sheet resistance of the P-base region. Similarly, the lumped resistance of the merged P-base regions is given by:

$$R_{MPB} = \rho_{S,MPB}\frac{W_{G2}}{Z} \tag{10.17}$$

where $\rho_{S,MPB}$ is the sheet resistance of the merged P-base region.

The SC-EST structure transitions from the IGBT mode to the EST mode when the voltage at point A becomes larger than the junction built-in potential (V_{bi}) because the injection of electrons begins to occur from the floating N^+ region at point A. Using the above equations with this criterion, the IGBT/EST mode transition anode current density is obtained:

$$J_{\mathrm{A}}(\mathrm{IGBT/EST}) = \frac{2V_{\mathrm{bi}}}{\alpha_{\mathrm{PNP}}(W_{\mathrm{G1}} + 2W_{\mathrm{PW}})(\rho_{\mathrm{S,PB}}W_{\mathrm{PW}} + \rho_{\mathrm{S,MPB}}W_{\mathrm{G2}})} \quad (10.18)$$

In the case of an SC-EST structure with turn-on gate width (W_{G1}) of 12 μm, a polysilicon window (W_{PW}) of 50 μm, a series MOSFET gate width (W_{G2}) of 4 μm, and typical P-base sheet resistance of 3,000 Ω/sq and typical merged P-base sheet resistance of 10,000 Ω/sq, the IGBT/EST mode transition current density is found to be 8.3 A/cm^2 if a common base PNP transistor current gain of 0.9 and a built-in potential of 0.8 V are assumed.

When the SC-EST structure is operating in the EST mode, most of the anode current arriving at the upper surface can be assumed to flow via the main thyristor path if the entire floating N$^+$ region is injecting electrons. Only a small portion of the anode current then flows via the parasitic thyristor region as illustrated in Fig. 10.14. The hole current is collected by the P-base/N-drift junction (J_4 in Fig. 10.1) of the parasitic thyristor and removed via the cathode contact on the right-hand side. The hole current (I_p) flows via the resistance of the P-base region ($R_{PB,P}$) under the N$^+$ cathode region of the parasitic thyristor before reaching the cathode contact via the P$^+$ region. This produces a voltage drop that creates a forward bias across the junction J_4. The highest forward bias occurs at point B located furthest away from the cathode contact. In a lumped element analysis, the forward bias voltage at point B is given by:

$$V_{\mathrm{B}} = I_{\mathrm{p}}R_{\mathrm{PB,P}} \quad (10.19)$$

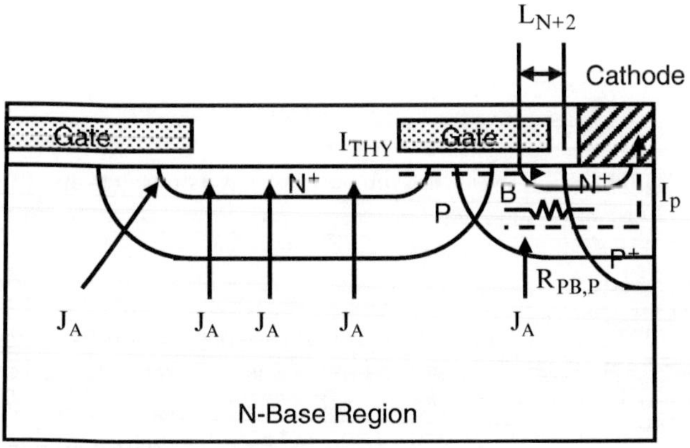

Fig. 10.14 Current flow in the SC-EST structure during the EST mode

The lumped hole current flowing at the parasitic thyristor can be computed by using:

$$I_P = J_A W_{G2} Z \quad (10.20)$$

assuming that the rest of the anode current is flowing via the main thyristor. The lumped resistance of the P-base region at the parasitic thyristor is given by:

$$R_{PB} = \rho_{S,PB} \frac{L_{N+2}}{Z} \quad (10.21)$$

where $\rho_{S,PB}$ is the sheet resistance of the P-base region. The parasitic thyristor in the SC-EST structure latches up when the voltage at point B becomes larger than the junction built-in potential (V_{bi}) because the injection of electrons begins to occur from the N^+ region at point B. Using the above equations with this criterion, the parasitic thyristor latch-up current density is obtained:

$$J_{A,PARA} = \frac{V_{bi}}{L_{N+2} \rho_{S,PB} W_{G2}} \quad (10.22)$$

In the case of an SC-EST structure with a N^+ width (L_{N+2}) of 2 μm, a series MOSFET gate width (W_{G2}) of 4 μm, and typical P-base sheet resistance of 3,000 Ω/sq, the parasitic thyristor latch-up current density is found to be over 3,000 A/cm^2 if a built-in potential of 0.8 V is assumed. This value is much larger than observed in the actual device structure because the entire width of the floating N^+ region does not inject electrons, i.e., the main thyristor only extends over a portion of the floating N^+ region remote from the cathode electrode. In this case, hole current collected across the rest of the region under the floating N^+ region flows via the parasitic thyristor. If only half of the floating N^+ region is assumed to be a part of the main thyristor, the parasitic thyristor latch-up current density is given by:

$$J_{A,PARA} = \frac{2V_{bi}}{L_{N+2} \rho_{S,PB} W_{Cell}} \quad (10.23)$$

For a cell width (W_{Cell}) of 65 μm, the parasitic thyristor latch-up current density is then found to be only 400 A/cm^2.

Simulation Results

The results of two-dimensional numerical simulations for the 5-kV asymmetrical silicon SC-EST structure are described here. The total width (W_{Cell}) of the structure, as shown by the cross section in Fig. 10.1, was 65 μm (area = 6.5 × 10^{-7} cm^{-2}) with a polysilicon window (W_{PW}) of 50 μm in size. The doping profiles for the baseline device structure were already shown in Figs. 10.7–10.9.

In order to understand how closely the SC-EST structure resembles the thyristor structure in the on-state, the on-state characteristics of the 5-kV silicon asymmetric thyristor structure were obtained for the case of various values for the lifetime in the drift region. This thyristor structure was discussed in the previous chapter and its on-state characteristics were shown in Fig. 8.15. The on-state voltage drop at a hole lifetime (τ_{p0}) value of 10 μs is found to be 1.202 V at an on-state current density of 50 A/cm^2 and increases to 4.286 V at a reduced hole lifetime (τ_{p0}) value of 1 μs. The hole distribution in the 5-kV asymmetric thyristor structure was provided in Fig. 8.16. It is worth pointing out that the carrier distribution is symmetric for the thyristor structure.

The on-state characteristics of the 5-kV silicon asymmetric SC-EST structure were obtained by using a positive gate bias voltage of 10 V for the case of various values for the lifetime in the drift region. This device structure has a peak buffer layer doping concentration of 1×10^{17} cm^{-3}. The characteristics obtained from the numerical simulations are shown in Fig. 10.15. The current initially increases exponentially with increasing anode bias when the device is operating in the IGBT mode. The forward drop increases quite rapidly in the IGBT mode due to the low channel density. The main thyristor latches up at a current density ranging from 5 to 10 A/cm^2 as shown by the lower dashed line in agreement with the analytical model. In the EST mode, the on-state voltage drop increases as expected with reduction of the lifetime (τ_{p0}, τ_{n0}) indicated in the figure. The on-state voltage drop at a hole lifetime (τ_{p0}) value of 10 μs is found to be 1.809 V at an on-state current density of 50 A/cm^2 and increases to 6.920 V at a reduced hole lifetime (τ_{p0}) value of 1 μs. The upper dashed line delineates the latch-up of the parasitic thyristor at a current density of 615 A/cm^2.

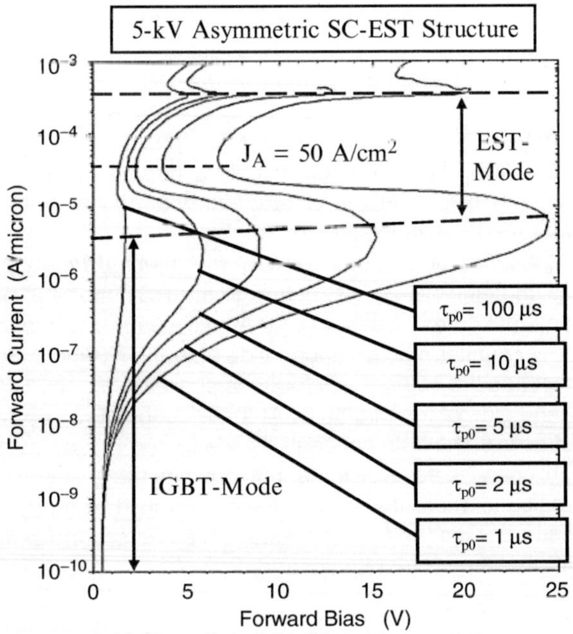

Fig. 10.15 On-state characteristics of the 5-kV asymmetric SC-EST structure: lifetime dependence

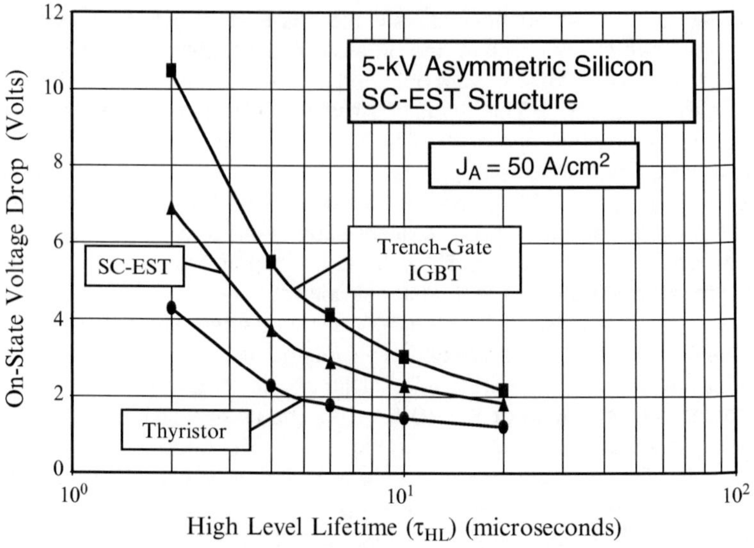

Fig. 10.16 On-state voltage drop for the 5-kV asymmetric SC-EST structure: N-base lifetime dependence

The variation of the on-state voltage drop obtained from the results of the numerical simulation for the 5-kV asymmetric SC-EST structure, as a function of the lifetime in the N-base region, is shown in Fig. 10.16 for the case of an anode on-state current density of 50 A/cm^2. For comparison purposes, the on-state voltage drops for the case of the 5-kV asymmetric trench-gate IGBT structure and for the 5-kV asymmetric thyristor structure are also provided in this figure. It can be observed that the SC-EST structure has a significantly lower on-state voltage drop than the IGBT structure for each lifetime value. This is due to improved carrier distribution in the SC-EST structure with a high free carrier density near the cathode side of the drift region. However, the on-state voltage drop for the SC-EST structure is significantly larger than that for the thyristor structure. This is related to only a portion of the floating N$^+$ region becoming forward biased in the SC-EST structure. The voltage drop across the series MOSFET is small (see Fig. 10.12).

A three-dimensional view of the injected hole concentration within the 5-kV asymmetric SC-EST structure in the EST mode is provided in Fig. 10.17 at an on-state current density of 50 A/cm^2. It can be observed that the hole distribution is very uniform at the anode side of the drift region. However, the hole concentration has the catenary distribution similar to the thyristor structure only on the left-hand side of the structure. At the P$^+$ contact region on the right-hand side of the structure, the hole concentration is forced to zero by the reverse-biased junction. This has the adverse impact of reducing the hole concentration under the cathode region for more than half of the cell width. The reduced hole concentration on the cathode side (and not the voltage drop across the series MOSFET) is responsible for the larger on-state voltage drop observed for the SC-EST structure.

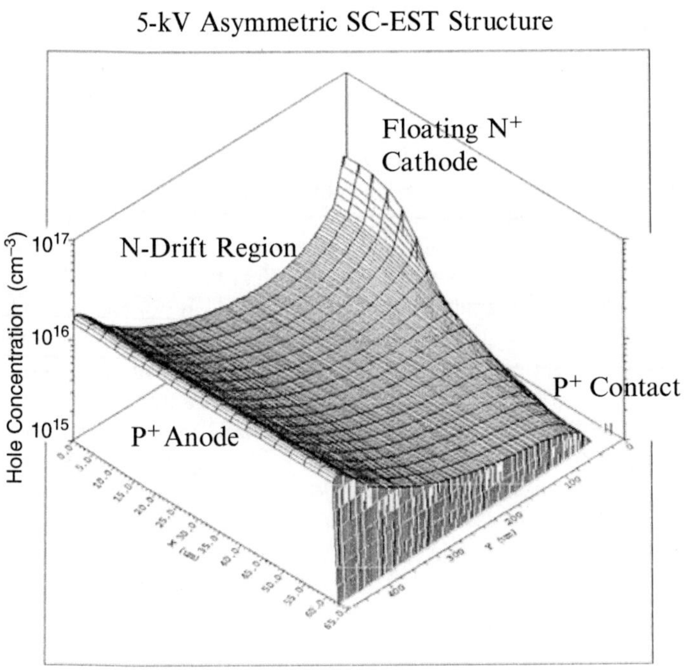

Fig. 10.17 On-state carrier distribution in the 5-kV asymmetric SC-EST structure

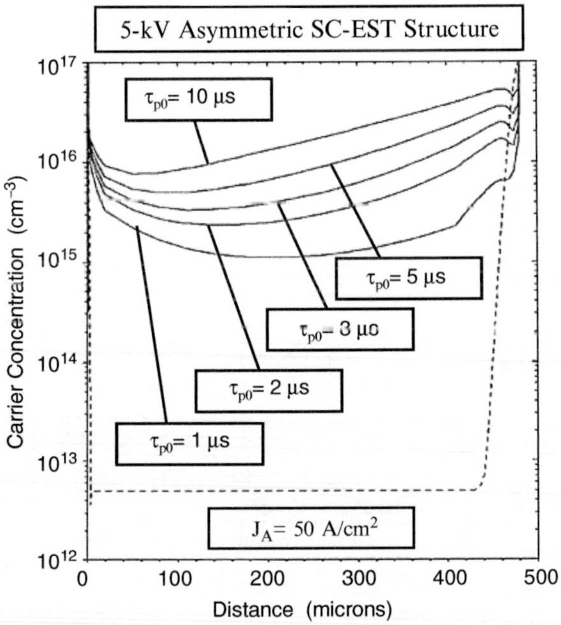

Fig. 10.18 On-state carrier distribution in the 5-kV asymmetric SC-EST structure: lifetime dependence

The on-state voltage drop for the 5-kV asymmetric SC-EST structure is determined by the distribution of carriers injected into the N-base region producing the desired reduction of its resistance. The hole distribution in the 5-kV asymmetric SC-EST structure is provided in Fig. 10.18 for five cases of the lifetime (τ_{p0}, τ_{n0}) in the drift region at $x = 10$ μm. It can be observed that the injected carrier density is four orders of magnitude larger than the doping concentration on the anode side, which is similar to that observed for the IGBT structure (see Fig. 5.15). However, the free carrier concentration on the cathode side of the drift region is much larger in the SC-EST structure. In comparison with the 5-kV thyristor structure (see Fig. 8.16), it can be observed that the hole carrier concentration is similar in magnitude in the drift region on the cathode side for the SC-EST structure at $x = 10$ μm. However, it is much smaller on the right-hand side of the cell as shown in Fig. 10.17. This produces a larger on-state voltage drop for the SC-EST structure when compared with the thyristor structure.

Further insight into the operation of the SC-EST structure can be obtained by examination of the current flow pattern in the various modes. The current flow lines within the 5-kV asymmetric SC-EST structure are shown in Figs. 10.19–10.21 in the IGBT mode, the EST mode, and after latch-up of the parasitic thyristor.

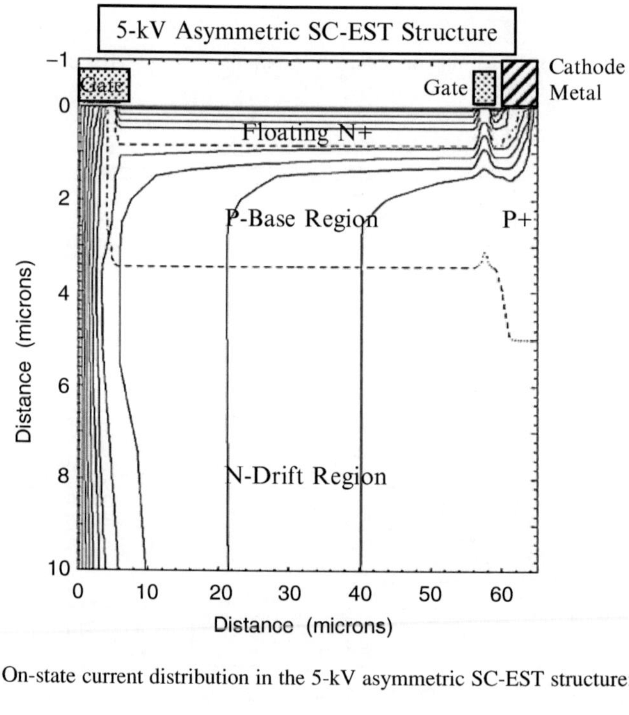

Fig. 10.19 On-state current distribution in the 5-kV asymmetric SC-EST structure: IGBT mode

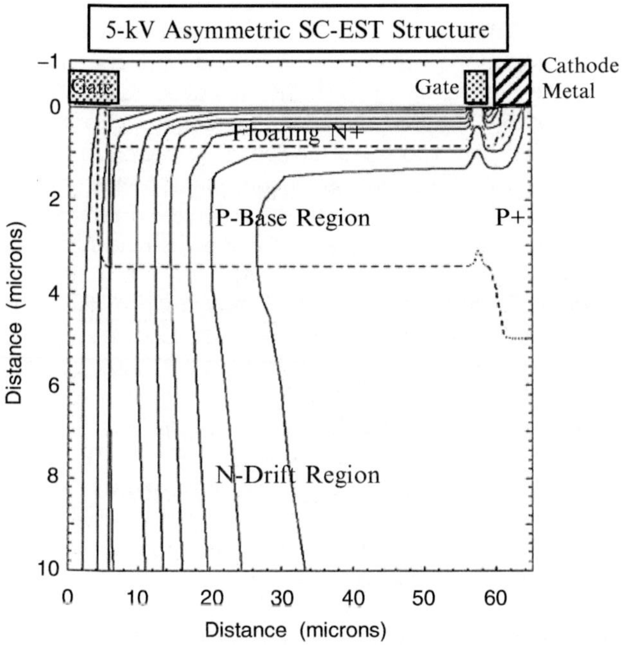

Fig. 10.20 On-state current distribution in the 5-kV asymmetric SC-EST structure: EST mode

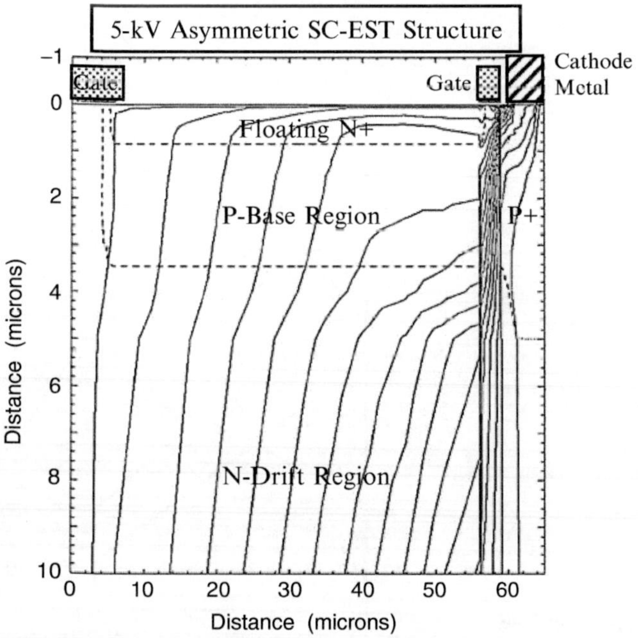

Fig. 10.21 On-state current distribution in the 5-kV asymmetric SC-EST structure: after parasitic thyristor latch-up

In the IGBT mode, a significant amount of current flow occurs via the turn-on MOSFET on the left-hand side (see Fig. 10.19) and no current flow occurs across the junction (J_3) between the floating N^+ region and the P-base region. The hole current flows into the P-base region and is then removed from the cathode contact on the right-hand side. In the EST mode, the current flows via the main thyristor region but is confined to only a portion of the floating N^+ region on the left-hand side. There is insufficient forward bias across the junction J_3 on the right-hand side for injection of electrons. This has an adverse impact on the on-state voltage drop as discussed previously. In addition, a larger amount of the hole current flows via the P^+ contact on the right-hand side because the main thyristor is not turned on over half the length of the floating N^+ region. This greatly reduces the latch-up current density of the parasitic thyristor as discussed for the analytical model. The latch-up current density for the parasitic thyristor in the SC-EST structure is found to be 615 A/cm^2 from the numerical simulations (see Fig. 10.15) in agreement with the second model (see Eq. 10.24). Once the parasitic thyristor latches up, the current concentrates at the parasitic thyristor as observed in Fig. 10.21.

The on-state characteristics of the 5-kV silicon asymmetric SC-EST structure were also obtained as a function of temperature by using a positive gate bias voltage of 10 V for the case of a lifetime of 2 μs in the drift region. The characteristics obtained from the numerical simulations are shown in Fig. 10.22. At low anode current densities (below 0.05 A/cm^2), the on-state voltage drop decreases with increasing temperature while at on-state current densities above 20 A/cm^2, it begins to increase with increasing temperature. The on-state voltage drop increases with increasing temperature at an on-state current density of 50 A/cm^2, which is desirable for allowing paralleling of devices and the prevention of hot spots within the device structure.

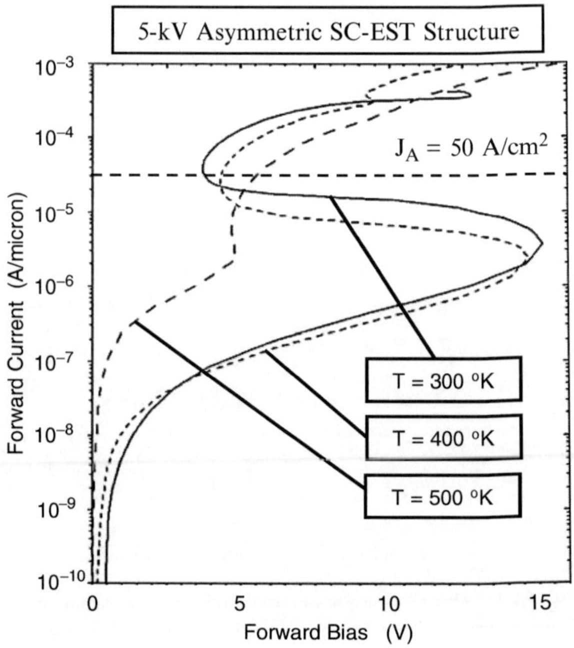

Fig. 10.22 On-state characteristics of the 5-kV asymmetric SC-EST structure: temperature dependence

10.2.3 Turn-Off Characteristics

One of the important advantages of the EST structure, when compared with the GTO structure, is the simplicity of the gate control circuit due to its MOS gate structure. In order to turn off the device, the gate voltage must simply be ramped from the on-state value (nominally positive 10 V) to the off-state value (nominally zero volts) as illustrated in Fig. 5.16 for the IGBT structure. Once the gate voltage falls below the threshold voltage, the conduction path through the channel of the series MOSFET ceases in the EST structure. Consequently, the current flow via the main thyristor path is shut off. In the case of an inductive load, the anode current for the EST structure is then sustained by the hole current flow due to the presence of stored charge in the N-base region. Unlike the GTO structure, there is no prolonged storage time interval for the EST structure during its turn-off. The anode voltage begins to increase in the EST structure almost immediately after the gate voltage reaches zero.

The anode current decreases once the anode voltage reaches the anode supply voltage as shown in Fig. 5.16. For the asymmetric EST structure, the current tail usually occurs in two parts if the anode voltage is insufficient for the space-charge region to extend completely through the N-base region. In this case, there is still some stored charge left in the N-base region near the N-buffer layer after the voltage transient is completed and the anode voltage is equal to the anode supply voltage. During the first part of the anode current decay, the stored charge in the N-base region is removed by recombination, as well the anode current flow. This is a relatively slow decay due to the large high-level lifetime in the N-base region. As the anode current decreases, the hole concentration in the space-charge region decreases allowing the space-charge region to expand even though the anode voltage is constant. Eventually, the space-charge region extends through the entire N-base region when the anode current density becomes equal to the punch-through current density ($J_{A,PT}$). At this point in time, stored charge is present only in the N-buffer layer. The stored charge in the N-buffer layer decreases by recombination at a faster pace due to the smaller lifetime in the N-buffer layer associated with its greater doping concentration than the N-base region. This produces a faster decay of the anode current during the second phase of the current tail as illustrated in the figure.

Based on the above description of the turn-off process in the EST structure, it can be concluded that the turn-off waveforms for the EST structure are similar to those already discussed in Chap. 8 for the MCT structure. The analytical model developed in Chap. 8 for the MCT structure can therefore be applied to the EST structure.

Simulation Example

In order to gain insight into the operation of the asymmetric 5-kV SC-EST structure during its turn-off, the results of two-dimensional numerical simulations for a typical structure are discussed here. The device structure used has the cross section shown in Fig. 10.1 with a cell half-width of 65 μm. The doping profiles for the SC-EST structure used in the numerical simulations were previously provided.

For the typical case discussed here, a high-level lifetime of 4 μs was used in the N-base region. The numerical simulations were performed with an abrupt reduction of the gate voltage from positive 10 V to zero in 10 ns starting from an on-state current density of 50 A/cm^2. The resulting waveforms obtained from the numerical simulations for the anode voltage and current are shown in Fig. 10.23 for the case of an anode supply voltage of 3,000 V. Unlike the GTO structure, there is no storage time associated with the turn-off of the SC-EST structure. The anode voltage increases immediately at the end of the gate voltage transient. The anode voltage increases as the square of the time as predicted by the analytical model until it reaches about 2,000 V. It then increases at a slower rate. This is associated with the onset of avalanche multiplication at high anode bias voltages – an effect not included in the analytical model. The waveforms for the SC-EST structure are very similar to those shown in Chap. 8 for the MCT structure. The anode voltage rise time for the SC-EST structure obtained in the numerical simulations for the case of supply voltage of 3,000 V (0.57 μs) is smaller than that observed for the MCT structure due to the reduced injected hole concentration on the cathode side.

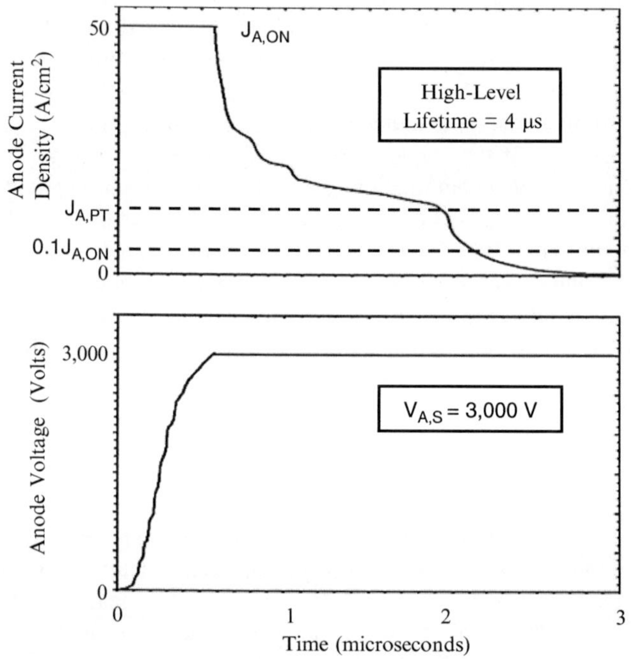

Fig. 10.23 Typical turn-off waveforms for the asymmetric 5-kV SC-EST structure

After the completion of the anode voltage transient, the anode current waveform decays from the initial on-state current density at a rate that decreases with time. The current decays to the punch-through current density (indicated in the figure) in 1.37 μs. A punch-through current density of about 15 A/cm^2 is observed in the numerical simulations. After reaching the punch-through current density, the anode current is observed to decay at a faster rate as described by the analytical model.

A one-dimensional view of the minority carrier distribution in the 5 kV asymmetric SC-EST structure is shown in Fig. 10.24 from the initial steady-state operating point ($t = 0$ μs) to the end of the voltage rise time ($t = 0.57$ μs). These carrier profiles were taken at $x = 10$ μm through the P-base and floating N^+ cathode regions. The initial carrier distribution has the distribution like a P-i-N rectifier. It can be observed from Fig. 10.24 that the carrier distribution in the N-base region near the anode does not change during the anode voltage rise phase. The carrier concentration at the P-base/N-drift junction rapidly reduces to zero within the first 30 ns allowing the junction to support an increase in the anode voltage. This time is shorter than for the MCT structure because the hole concentration on the cathode side is smaller for the SC-EST structure. A significant space-charge region begins to form immediately during the turn-off and expands towards the right-hand side demonstrating that there is no storage phase for the SC-EST structure. At larger anode voltages, the hole concentration in the space-charge region is about 3×10^{13} cm^{-3}, which is consistent with the value for p_{SC} obtained using the analytical model with the carriers moving at the saturated drift velocity and an on-state current density of 50 A/cm^2. The width of the space-charge region can be observed to be about 330 μm when the collector voltage reaches 3,000 V, which is close to that predicted by the analytical model. The electric field profiles in the 5-kV asymmetric SC-EST structure obtained during the voltage rise time from the numerical simulations are similar to those previous shown for the BRT structure. They are omitted from this section in the interest of conserving space.

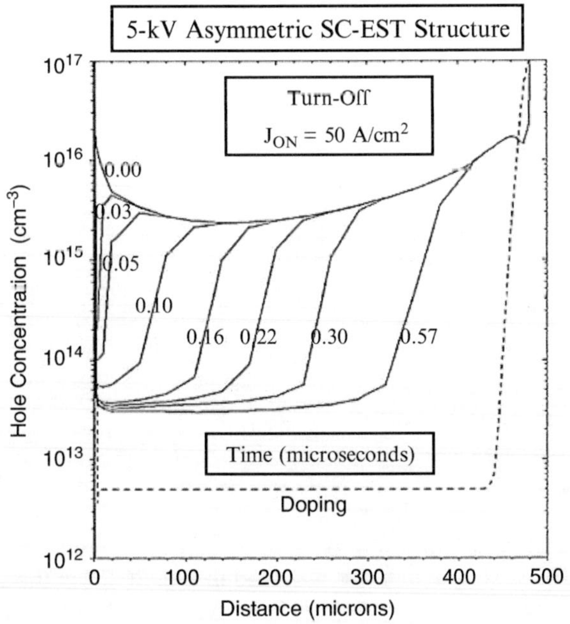

Fig. 10.24 Hole carrier distribution for the 5-kV SC-EST turn-off transient during the voltage rise time

A one-dimensional view of the hole carrier distribution in the 5-kV asymmetric SC-EST structure is shown in Fig. 10.25 during the current tail time. The anode voltage was held constant at the anode supply voltage of 3,000 V during this transient. The hole concentration in the stored charge region begins to decrease immediately after the end of the voltage transient due to the recombination process and the removal of holes and electrons by the anode current flow. At the same time, the space-charge region expands in spite of a constant anode voltage because the hole concentration in the space-charge region reduces. From Fig. 10.25, it can be observed that all the holes in the N-base region have been removed at time $t = 1.99$ μs corresponding to the end of the first phase of the collector current transient (see Fig. 10.23]. Subsequently, the holes remaining in the N-buffer layer are at concentrations well below its doping concentration. Consequently, the recombination of holes in the N-buffer layer during the second part of the anode current transient occurs under low-level injection conditions as assumed in the analytical model. The electric field profiles in the 5-kV asymmetric SC-EST structure obtained during the current fall time from the numerical simulations are similar to those previous shown for the BRT structure. They are omitted from this section in the interest of conserving space.

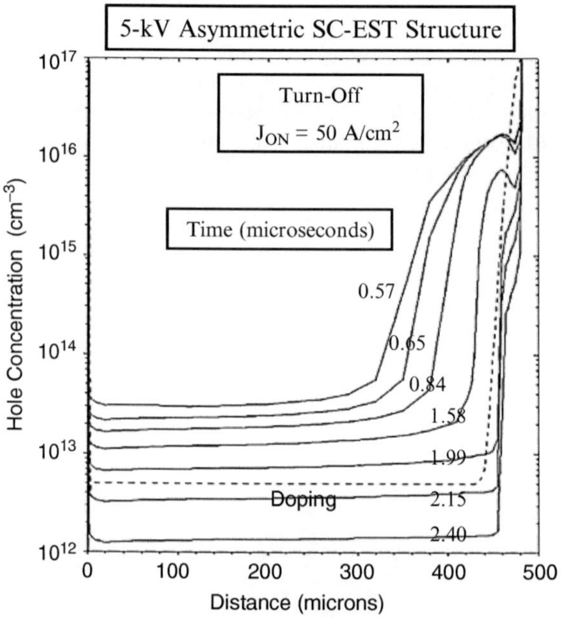

Fig. 10.25 Hole carrier distribution for the 5-kV asymmetric SC-EST turn-off transient during the current tail time

It is instructive to examine the current flow pattern within the EST structure during the turn-off process. The current flow lines within the 5-kV asymmetric SC-EST structure are provided in Fig. 10.26 at the end of the voltage rise time when the anode voltage is 3,000 V while the anode current density is at 50 A/cm^2.

No current flows via the floating N^+ region demonstrating that the main thyristor is not active during the turn-off process. It can be observed that all of the hole current collected by the P-base region flows to the P^+ contact region on the right-hand side. Since the device is supporting a high voltage across the P-base/N-drift junction, the gain of the P-N-P transistor is large under turn-off conditions. Consequently, the resistance of the hole current path through the P-base region below the floating N^+ region and resistance of the merged P-base region must be sufficiently small to keep the voltage across junction J_3 well below the junction built-in potential in order to achieve turn-off in the SC-EST structure. It is worth pointing out that the current density is very uniform in the SC-EST structure during turn-off. This results in an excellent RBSOA for the structure which is close to that derived using one-dimensional analysis. In contrast, the RBSOA for the GTO is degraded by current crowding during the turn-off process.

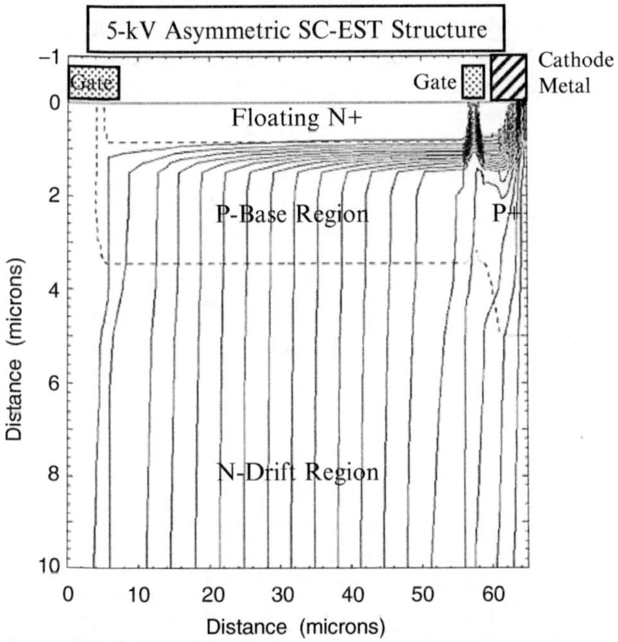

Fig. 10.26 Current distribution for the 5-kV asymmetric SC-EST during turn-off

10.2.4 Lifetime Dependence

The optimization of the power losses for the EST structure requires performing a trade-off between the on-state voltage drop and the switching losses. One approach to achieve this is by adjusting the lifetime in the drift (N-base) region. A reduction of the lifetime in the drift region also alters the lifetime in the N-buffer layer. The impact of reducing the lifetime on the on-state voltage drop was previously shown in Sect. 10.2.2. The on-state voltage drop increases when the lifetime is reduced. The analytical model developed for turn-off of the asymmetric MCT structure presented in Chap. 8 can be used to analyze the impact of changes to the lifetime in the drift region.

Simulation Example

In order to gain insight into the impact of the lifetime in the N-base region on the operation of the 5-kV asymmetric SC-EST structure, the results of two-dimensional numerical simulations for a typical structure are discussed here. The device structure used has the cross section shown in Fig. 10.1 with a half-cell width of 65 μm. The widths of the N-base and N-buffer layer regions are 440 and 30 μm, respectively. The high-level lifetime in the N-base region was varied between 2 and 20 μs. For turning-off the SC-EST structures, the numerical simulations were performed with gate voltage rapidly ramped down from positive 10 V to zero in 10 ns starting from an on-state current density of 50 A/cm^2. The resulting waveforms obtained from the numerical simulations for the anode voltage and current are shown in Fig. 10.27 for the case of an anode supply voltage of 3,000 V.

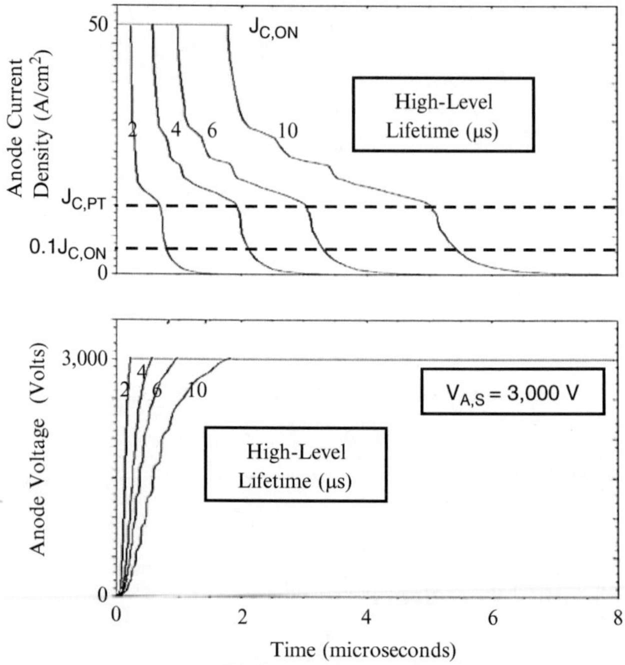

Fig. 10.27 Impact of lifetime on the 5-kV asymmetric SC-EST turn-off waveforms

The numerical simulations show a decrease in the voltage rise time with reduction of the lifetime in the N-base region. This is related to the reduced stored charge in the drift region on the cathode side for smaller lifetime values. The numerical simulations also show a reduction of the anode current fall time during the first part of the decay to the punch-through anode current density ($J_{C,PT}$). The punch-through current density is independent of the lifetime in the N-base region as predicted by the analytical model. A substantial decrease in the anode current fall time is observed when the lifetime is reduced.

10.2.5 Switching Energy Loss

The power loss incurred during the switching transients limit the maximum operating frequency for the EST structure. Power losses during the turn-on of the EST structure are significant but strongly dependent on the reverse recovery behavior of the fly-back rectifiers in circuits. Consequently, it is common practice to use only the turn-off energy loss per cycle during characterization of EST devices. The turn-off losses are associated with the voltage rise-time interval and the current fall-time interval. The energy loss for each event can be computed by integration of the power loss, as given by the product of the instantaneous current and voltage. Since the turn-off switching waveforms for the SC-EST structure have the same shape as those for the MCT structure, the energy loss during the voltage rise-time interval and the current fall-time interval can be computed using the equations previously provided in Chap. 8. For the typical switching waveforms for the 5-kV asymmetric SC-EST structure shown in Fig. 10.23 with an anode supply voltage of 3,000 V, the energy loss per unit area during the anode voltage rise time is found to be 0.028 J/cm^2; and the energy loss per unit area during the collector current fall time is found to be 0.116 J/cm^2. The total energy loss per unit area ($E_{OFF,V}$ + $E_{OFF,I}$) during the turn-off process for the 5-kV asymmetric SC-EST structure is found to be 0.144 J/cm^2.

Using the results obtained from the numerical simulations, the on-state voltage drop and the total energy loss per cycle can be computed. These values are plotted in Fig. 10.28 to create a trade-off curve to optimize the performance of the silicon 5-kV asymmetric SC-EST structure by varying the lifetime in the N-base region. Devices used in lower frequency circuits would be chosen from the left-hand side of the trade-off curve while devices used in higher frequency circuits would be chosen from the right-hand side of the trade-off curve. For comparison purposes, the trade-off curve for the 5-kV trench-gate IGBT structure is also included in the figure. It can be observed from this figure that substantial improvement in the power loss trade-off curve can be obtained by replacing the IGBT with the SC-EST structure. The SC-EST has a sufficiently large FBSOA as shown later to allow replacement of the IGBT with the SC-EST in motor control applications.

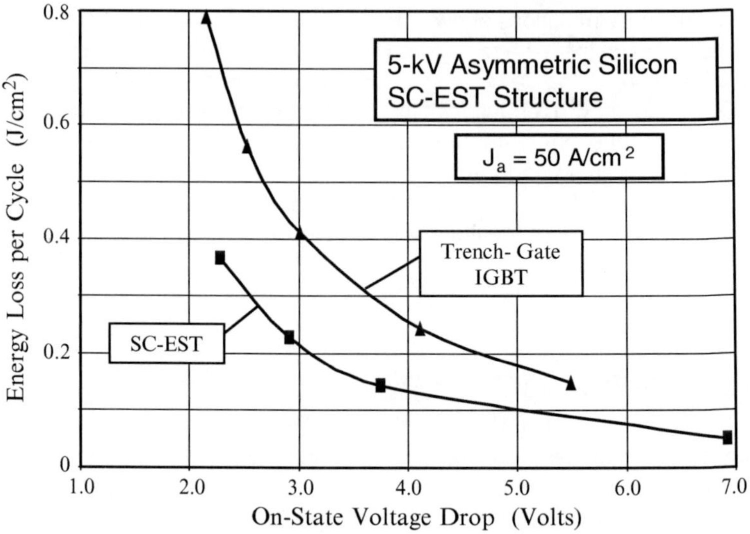

Fig. 10.28 Trade-off curve for the silicon 5-kV asymmetric SC-EST structure: lifetime in the N-base region

10.2.6 Maximum Operating Frequency

High-Level Lifetime (μs)	On-State Voltage Drop (Volts)	On-State Power Dissipation (W/cm²)	Energy Loss per Cycle (J/cm²)	Maximum Operating Frequency (Hz)
10	2.283	57.08	0.365	392
6	2.918	72.95	0.228	558
4	3.746	93.65	0.144	739
2	6.920	173	0.050	540

Fig. 10.29 Power loss analysis for the 5-kV asymmetric SC-EST structure

The maximum operating frequency for operation of the 5-kV asymmetric SC-EST structure can be obtained by combining the on-state and switching power losses:

$$P_{\mathrm{D,TOTAL}} = \delta P_{\mathrm{D,ON}} + E_{\mathrm{OFF}} f \tag{10.24}$$

where δ is the duty cycle and f is the switching frequency. In the case of the baseline asymmetric SC-EST device structure with a high-level lifetime of 4 μs in the N-base region, the on-state voltage drop is 3.75 V at an on-state current density of 50 A/cm^2. For the case of a 50% duty cycle, the on-state power dissipation contributes 94 W/cm^2 to the total power loss. Using a total turn-off energy loss per cycle of 0.144 J/cm^2 in Eq. 10.24 with a total power dissipation of 200 W/cm^2 yields a maximum operating frequency of 739 Hz.

Using the results obtained from the numerical simulations, the on-state voltage drop and the energy loss per cycle can be computed. These values are provided in Fig. 10.29 together with the maximum operating frequency as a function of the high-level lifetime in the N-base region under the assumption of a 50% duty cycle and a total power dissipation limit of 200 W/cm^2. The maximum operating frequency is plotted in Fig. 10.30 as a function of the high-level lifetime in the N-base region for the case of a duty cycle of 50%. It can be observed that the maximum operating frequency can be increased up to 740 Hz by reducing the high-level lifetime to 2 μs. This is much superior to the maximum operating frequency of 150 Hz for the 5-kV GTO structure and 400 Hz for the 5-kV trench-gate IGBT structure.

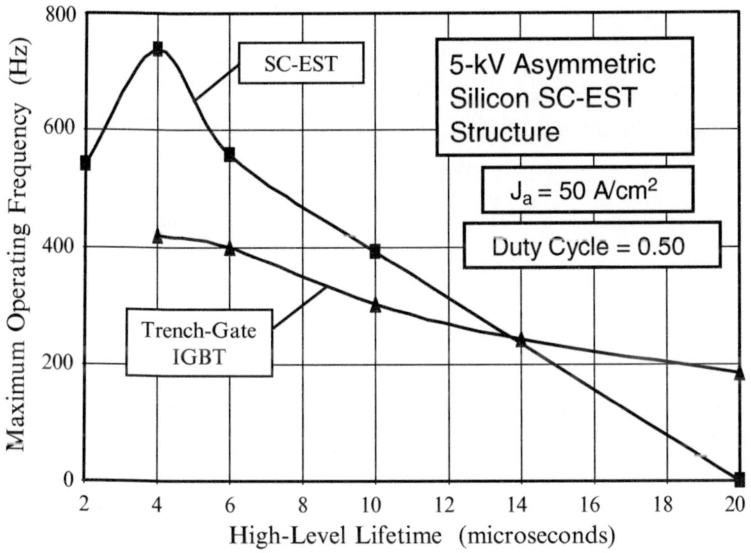

Fig. 10.30 Maximum operating frequency for the 5-kV asymmetric SC-EST structure

10.2.7 Forward-Biased Safe Operating Area

Since the main thyristor current is constrained to flow through a series MOSFET in the SC-EST structure, it exhibits a substantial region of operation where the anode current can be saturated under gate control. This is demonstrated in this section by using the results of numerical simulations for the 5-kV asymmetric SC-EST structure.

Simulation Results

Numerical simulations of the 5-kV silicon asymmetric SC-EST structure were performed for the case of a high-level lifetime of 4 μs with various values for the gate bias voltage while sweeping the anode voltage. The resulting output characteristics are shown in Figs. 10.31 and 10.32. The device was able to

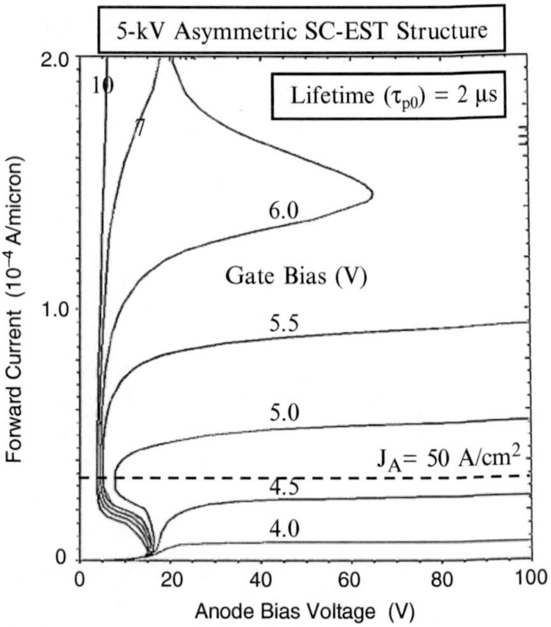

Fig. 10.31 Output characteristics of the 5-kV SC-EST structure

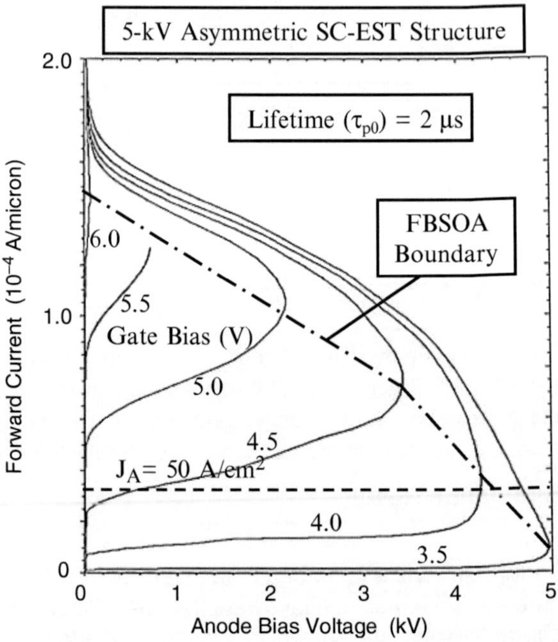

Fig. 10.32 Forward-biased safe operating area of the 5-kV SC-EST structure

saturate the anode current at lower gate bias voltages as expected. At gate bias voltages above 5 V, the SC-EST operates with latch-up of the main thyristor. At gate bias voltages below 5 V, the SC-EST operates in the IGBT mode. Consequently, at these gate bias voltages, it exhibits an excellent FBSOA with current saturation up to nearly 5,000 V. The SC-EST structure has a good FBSOA boundary (although inferior to that of the IGBT – see Figs. 5.73 and 5.74) as indicated by the dot-dashed line in Fig. 10.32.

10.3 5,000-V Silicon DC-EST

The dual-channel emitter switched thyristor (DC-EST) structure was proposed in order to improve the FBSOA of the EST [9]. The DC-EST structure was previously shown in Fig. 10.2. The design and characteristics for the 5,000-V asymmetric DC-EST structure are discussed in this section. The design parameters for the N-base (drift) region required to achieve this blocking voltage are the same as those for the SC-EST. Using the optimum N-base width, the blocking characteristics for the device are obtained. The on-state characteristics for the device are obtained for various lifetime values as well. The gate-controlled turn-off behavior of the silicon DC-EST structure is analyzed including the effect of the lifetime in the drift region.

10.3.1 Blocking Characteristics

The physics for blocking voltages in the first and third quadrants by the asymmetric DC-EST structure is the same as that previously discussed for the SC-EST structure. When a positive bias is applied to the anode terminal of the asymmetric DC-EST structure with zero bias applied to the gate, the P-base/N-base junction (J_2) becomes reverse biased while the junction (J_1) between the P^+ anode region and the N-base region becomes forward biased. The forward blocking voltage is supported across the P-base/N-base junction (J_2) with a depletion layer extending mostly within the N-base region. The electric field distribution within the asymmetric DC-EST structure is essentially the same as that illustrated in Fig. 4.3 for the asymmetric GTO structure. Consequently, the design procedure described in Chap. 4 can be applied to the asymmetric DC-EST structure. From Fig. 4.4, the N-base region width required to obtain a forward blocking voltage of 5,500 V is 470 μm. This width can be slightly reduced when taking into account the voltage supported within the P-base region due to its graded doping profile.

The leakage current in forward blocking mode is produced by space-charge generation within the depletion region. In the case of the asymmetric DC-EST structure in the forward blocking mode, the space-charge generation current at the reverse-biased P-base/N-base junction J_2 is amplified by the gain of the internal P-N-P transistor. Initially, the space-generation current increases with increasing anode bias due to expansion of the depletion region. Concurrently, the current gain (α_{PNP}) of the P-N-P transistor is also a function of the anode bias voltage because the base transport factor increases when the anode bias increases. Prior to the complete depletion of the lightly doped portion of the N-base region, the multiplication factor remains close to unity. It is therefore sufficient to account for the increase in the base transport factor with anode bias as given by Eqs. 4.8 and 4.9.

For the case of the silicon asymmetric DC-EST structure with a width of 450 μm for the lightly doped portion of the N-base region with a doping concentration of 5×10^{12} cm^{-3}, the entire lightly doped portion of the N-base region is completely depleted at a reach-through voltage of 780 V. Once the lightly doped portion of the N-base region becomes completely depleted, the electric field becomes truncated at the interface between the lightly doped portion of the N-base region and the N-buffer layer as illustrated at the bottom of Fig. 4.3. The space-charge generation width then becomes independent of the anode bias because the depletion width in the N-buffer layer is small. Under these bias conditions, the base transport factor also becomes independent of the collector bias as given by Eq. 4.10. Consequently, the leakage current becomes independent of the collector bias until the onset of avalanche multiplication. The leakage currents for the silicon asymmetric DC-EST structure are identical to those provided for the silicon asymmetric IGBT structure in Chap. 5.

Simulation Example

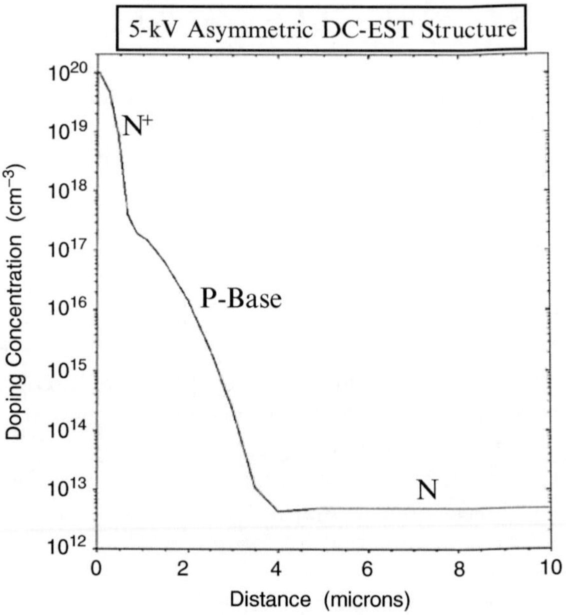

Fig. 10.33 Doping profile for the simulated asymmetric 5-kV DC-EST structure

The results of two-dimensional numerical simulations are described here in order to gain insight into the physics of operation for the 5-kV asymmetric DC-EST structure under voltage blocking conditions. The simulations were performed using a cell with the structure shown in Fig. 10.2 with a width (W_{Cell}) of 30 μm (area = 3.0×10^{-7} cm^{-2}). A floating N$^+$ cathode region width ($W_{PW}/2$) of 20 μm was chosen with a P-base shorting resistance of 1×10^{10} Ω-μm to create a transition from the IGBT mode to the EST mode at a sufficiently low on-state current density as discussed in the next section. The asymmetric DC-EST structure used for the simulations was formed by diffusions performed into a uniformly doped N-type drift region with a doping concentration of 5×10^{12} cm^{-3}. The N-buffer layer was formed by diffusion from the anode side with a depth of 55 μm. The doping profile in the vertical direction through the N$^+$ cathode region is the same as that shown in Fig. 10.7 for the SC-EST structure indicating the net width of the lightly doped portion of the N-base region is 440 μm after accounting for the diffusions. The peak doping concentration of the N-buffer layer is 1.0×10^{17} cm^{-3} and its thickness is 40 μm.

The P-base region for the asymmetric DC-EST structure was formed with a Gaussian doping profile with a surface concentration of 5×10^{17} cm^{-3} and a vertical depth of 3.8 μm as can be seen in Fig. 10.33 where the vertical doping profile in the upper 10 μm of the structure is provided. The N$^+$ cathode region was formed with a Gaussian doping profile with a surface concentration of 1×10^{20} cm^{-3} and a depth of 0.7 μm. The P$^+$ region on the right-hand side for suppressing latch-up of the parasitic thyristor was formed with a Gaussian doping profile with a surface concentration of 5×10^{19} cm^{-3} and a depth of 5 μm. The fabrication process for the EST is similar to that used to manufacture IGBT structures and less complex than that used for the MCT structure.

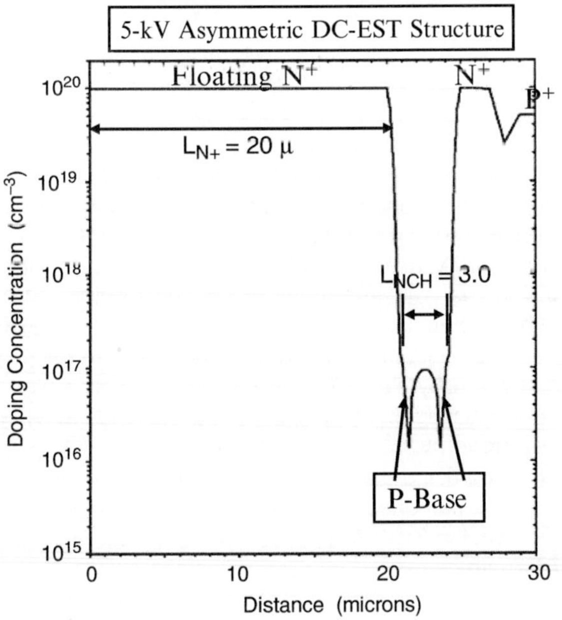

Fig. 10.34 Doping profile across the surface for the simulated 5-kV DC-EST structure

The doping profile across the surface for the 5-kV asymmetric DC-EST structure used for the numerical simulations is provided in Fig. 10.34. This profile was obtained along the horizontal line at $y = 0$ μm. It can be seen that the peak doping concentration of the P-body region of the n-channel MOSFET is 1.0×10^{17} cm^{-3} and its total channel length is 3.0 μm. These values control the on-state voltage drop and current saturation behavior of the DC-EST structure.

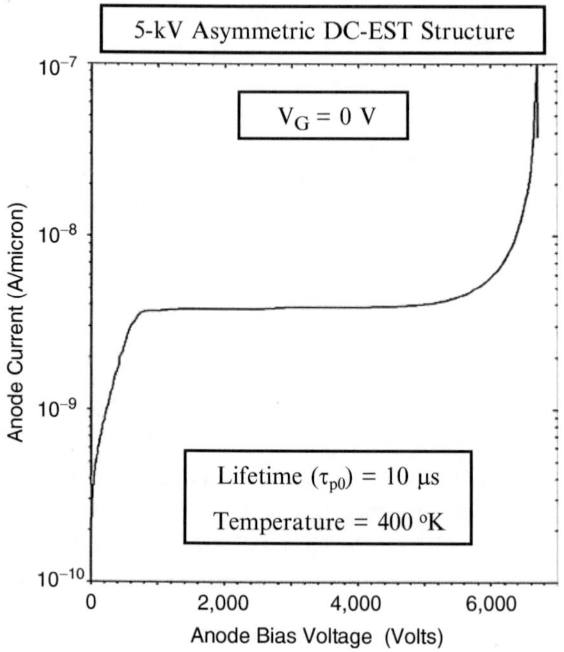

Fig. 10.35 Forward blocking characteristics for the asymmetric DC-EST structure

The forward blocking capability of the silicon asymmetric DC-EST structure was obtained using numerical simulations by increasing the anode bias while maintaining the gate electrode at zero volts. It was found that the device could support more than 6,000 V as shown in Fig. 10.35 at a temperature of 400°K. The leakage current increases rapidly with increasing anode bias voltage until about 780 V as predicted by the analytical model (see Fig. 5.2). This occurs due to the increase in the space-charge generation volume and the increase in the current gain (α_{PNP}) of the open base P-N-P transistor until the anode bias becomes equal to the reach-through voltage obtained using the analytical solution given by Eq. 4.2. The leakage current then becomes independent of the anode voltage until close to the breakdown voltage. This behavior is well described by the analytical model (see Fig. 5.2). The leakage current density obtained using the analytical model is within a factor of 2 of the values derived from the numerical simulations for all cases. The blocking characteristics for the asymmetric DC-EST structure are therefore similar to those for the asymmetric IGBT structure.

The current flow lines within the asymmetric DC-EST structure in the blocking mode are provided in Fig. 10.36 for the case of a gate bias of zero volts. From Fig. 10.36, it can be observed that a significant amount of the anode current flows to the cathode contact via the P^{+} region on the right-hand side. About half of the current collected at the P-base/N-drift region junction J_2 flows via the shorting resistance to the P-base region on the left-hand side and the other half of the current collected at the P-base/N-drift region junction J_2 flows via the MOSFET structure despite the zero gate bias due to the positive bias on the N-drift region (which is the body of a p-channel lateral MOSFET in the DC-EST structure).

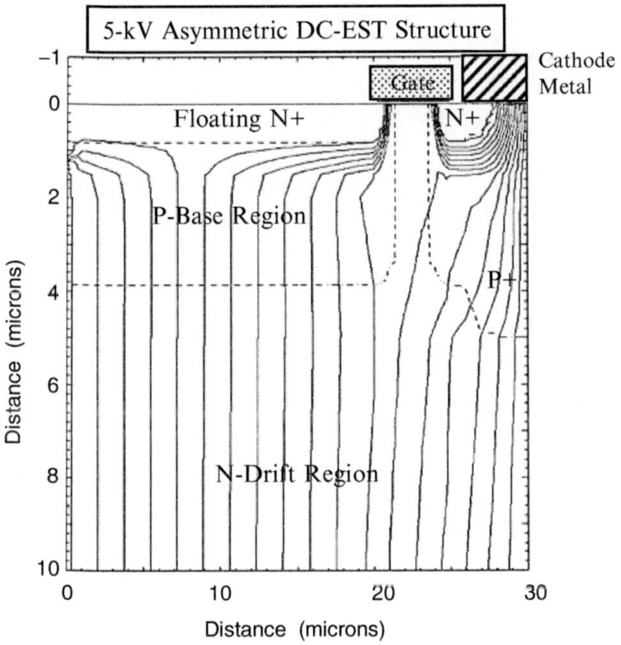

Fig. 10.36 Current flow lines during the blocking mode for the 5-kV asymmetric DC-EST structure: $V_G = 0$ V

As in the case of other asymmetric structures, the anode voltage is primarily supported within the lightly doped portion of N-drift region in the asymmetric SC-EST structure during operation in the forward blocking mode. The electric field profile within the asymmetric DC-EST structure is very similar to that shown previously for the asymmetric IGBT structure and is not included here in the interest of conserving space.

10.3.2 On-State Voltage Drop

The DC-EST structure operates with latch-up of the thyristor structure within the device. Consequently, the free carrier distribution within the N-drift region can be expected to be similar to those for the thyristor structure (see Chap. 2).

However, the incorporation of the n-channel MOSFET in series with the thyristor structure increases the on-state voltage drop. The on-state voltage drop for the DC-EST structure can be analytically calculated using:

$$V_{ON} = V_{THY} + V_{MOSFET} \tag{10.25}$$

The voltage drop across the thyristor (V_{THY}) can be computed using:

$$V_{THY} = \frac{2kT}{q} \ln\left[\frac{J_{A,ON}d}{2qD_a n_i F(d/L_a)}\right] \tag{10.26}$$

where the "d" is half the thickness of the i-region, D_a is the ambipolar diffusion coefficient, n_i is the intrinsic concentration, and L_a is the ambipolar diffusion length in the drift region. The function $F(d/L_a)$ varies with the lifetime in the drift region [20].

In the DC-EST structure, the series MOSFET contains two inversion-mode regions and one accumulation-mode region. The voltage drop across the MOSFET can be computed using:

$$V_{MOSFET} = I_{CH}R_{CH} = I_{CH}\left\{\frac{1}{ZC_{OX}(V_G - V_{TH})}\left[\frac{2L_{iCH}}{\mu_{ni}} + \frac{L_{aCH}}{\mu_{na}}\right]\right\} \tag{10.27}$$

where I_{CH} is the channel current and R_{CH} is the channel resistance. The channel resistance is determined by L_{iCH} the inversion channel length, μ_{ni} the electron mobility in the inversion layer, L_{aCH} the accumulation channel length, μ_{na} the electron mobility in the accumulation layer, C_{OX} the gate oxide capacitance, V_G the gate bias voltage, V_{TH} the threshold voltage [20], and Z the width of the device orthogonal to the cross section. The channel current consists of the current flowing through the main thyristor via the floating N^+ cathode region:

$$I_{CH} = \left(\frac{W_{PW}Z}{2}\right)J_{A,ON} \tag{10.28}$$

Substituting into Eq. 10.27:

$$V_{MOSFET} = J_{A,ON}\left\{\frac{W_{PW}}{2C_{OX}(V_G - V_{TH})}\left[\frac{2L_{iCH}}{\mu_{ni}} + \frac{L_{aCH}}{\mu_{na}}\right]\right\} \tag{10.29}$$

The on-state voltage drop for the 5,000-V asymmetric DC-EST computed by using the above equations is provided in Fig. 10.37 as a function of the high-level lifetime in the drift region. The drift region of the structure consists of a lightly doped portion with a thickness of 440 μm and a buffer layer with thickness of 30 μm. This device structure had a polysilicon window (W_{PW}) of 40 μm, a channel length of 1 μm for each of the inversion and accumulation regions, and a gate oxide thickness of 500 Å.

As in the case of previous structures, the inversion and accumulation layer mobilities were assumed to be 450 and 1,000 cm^2/V-s, respectively. At an on-state current density of 50 A/cm^2, the MOSFET contributes only 0.10 V to the total on-state voltage drop independent of the high-level lifetime. This value is smaller than that for the SC-EST structure in spite of the larger total channel length of the series MOSFET because of the reduced width of the floating N$^+$ region in the DC-EST structure. When the high-level lifetime is reduced below 10 μs, the voltage drop across the thyristor portion begins to increase rapidly resulting in an increase in the on-state voltage drop for the DC-EST structure. This analytical model assumes that the entire length of the floating N$^+$ region operates like a one-dimensional thyristor.

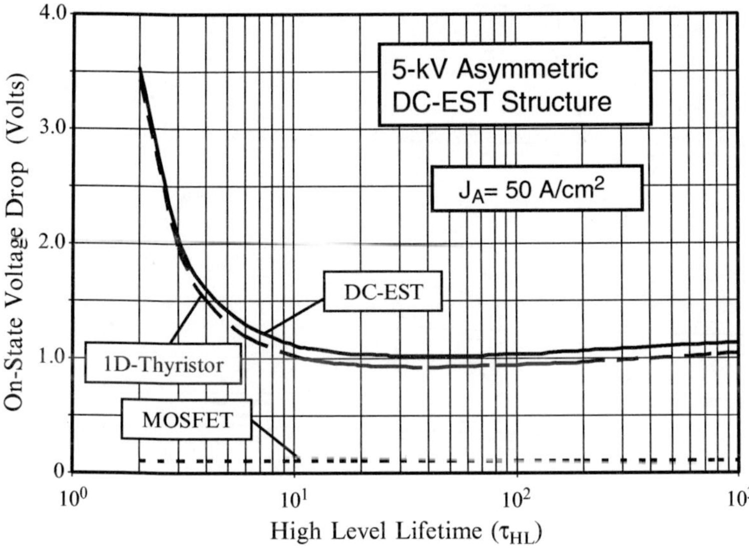

Fig. 10.37 On-state voltage drop for the DC-EST structure

As in the case of the SC-EST structure, the DC-EST has several operating modes in the on-state. At low on-state current densities, the device operates like an IGBT prior to the latch-up of the main thyristor. Once the main thyristor latches up, the device operates in the EST mode. At large on-state current densities, the parasitic thyristor latches up resulting in loss of gate control. The transition point between the IGBT and EST mode in the DC-EST structure cannot be analytically modeled by two-dimensional analysis because the P-base region is shorted to the floating N$^+$ region orthogonal to the cross section. However, the parasitic thyristor latch-up current density for the DC-EST can be analytical modeled by using the same current flow pattern as in the SC-EST structure (see Fig. 10.14).

The parasitic thyristor latch-up current density for the DC-EST structure is given by:

$$J_{A,PARA} = \frac{V_{bi}}{L_{N+2}\rho_{S,PB}W_G} \quad (10.30)$$

In the case of an DC-EST structure with a N$^+$ width (L_{N+2}) of 2 μm, a series MOSFET gate width (W_G) of 5 μm, and typical P-base sheet resistance of 3,000 Ω/sq, the parasitic thyristor latch-up current density is found to be over 2,400 A/cm^2 if a built-in potential of 0.8 V is assumed.

Simulation Results

The results of two-dimensional numerical simulations for the 5-kV asymmetrical silicon DC-EST structure are described here. The total width (W_{Cell}) of the structure, as shown by the cross section in Fig. 10.2, was 30 μm (area = 3.0 × 10^{-7} cm^{-2}) with a polysilicon window (W_{PW}) of 40 μm in size. The doping profiles for the baseline device structure were already shown in Figs. 10.33 and 10.34.

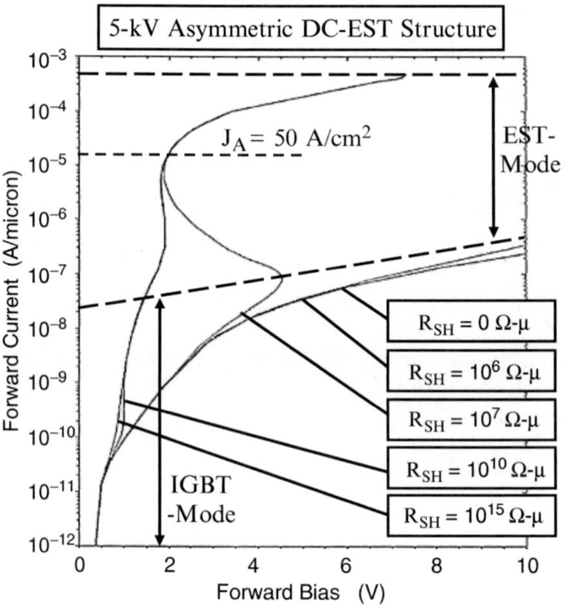

Fig. 10.38 On-state characteristics of the 5-kV asymmetric DC-EST structure: shunting resistance dependence

In order to understand how closely the DC-EST structure resembles the thyristor structure in the on-state, the on-state characteristics of the 5-kV silicon asymmetric thyristor structure were obtained for the case of various values for the lifetime in the drift region. This thyristor structure was discussed in the previous chapter and its on-state characteristics were shown in Fig. 8.15. The on-state voltage drop at a hole lifetime (τ_{p0}) value of 10 μs is found to be 1.202 V at an on-state current density of 50 A/cm^2 and increases to 4.286 V at a reduced hole lifetime (τ_{p0}) value of 1 μs. The hole distribution in the 5-kV asymmetric thyristor structure was provided in Fig. 8.16. It is worth pointing out that the carrier distribution is symmetric for the thyristor structure.

The on-state characteristics of the 5-kV silicon asymmetric DC-EST structure were obtained by using a positive gate bias voltage of 10 V. This device structure has a peak buffer layer doping concentration of 1 × 10^{17} cm^{-3}. A resistance was attached to an electrode connected to the P-base region on the right-hand side of the structure to emulate the shunting resistance for the P-base region. The resistance was varied from zero to 1×10^{20} Ω-μm. The characteristics obtained from the numerical simulations are shown in Fig. 10.38. It can be observed that the main thyristor within the DC-EST structure cannot latch up without the shunting resistance ($R_{SH} = 0$ case). The shunting resistance must be at least 1×10^{7} Ω-μm for obtaining latch-up of the main thyristor. Based on the results shown in Fig. 10.38, a shunting resistance of 1×10^{10} Ω-μm was chosen as appropriate for the DC-EST structure.

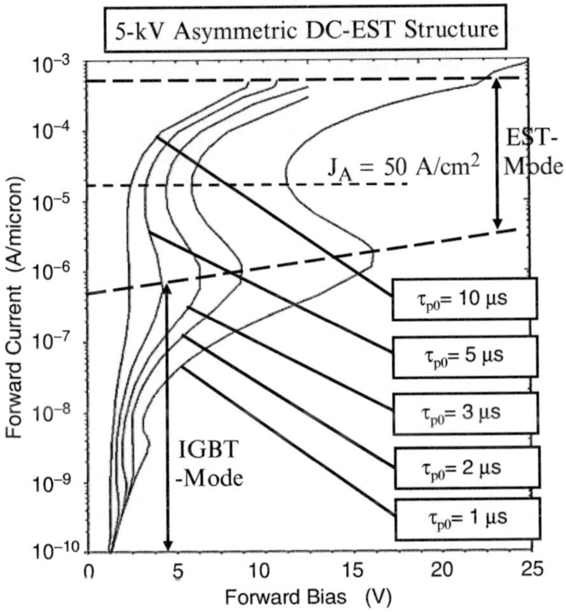

Fig. 10.39 On-state characteristics of the 5-kV asymmetric DC-EST structure: lifetime dependence

The impact of changes in the lifetime in the drift region of the 5-kV asymmetric DC-EST structure was obtained using numerical simulations with a shunting resistance of 1×10^{10} Ω-μm. The resulting on-state characteristics are shown in Fig. 10.39. The current initially increases exponentially with increasing anode bias when the device is operating in the IGBT mode. The forward drop increases quite rapidly in the IGBT mode due to the low channel density. The main thyristor latches up at a current density ranging from 5 to 10 A/cm^2 as shown by the lower dashed line. In the EST mode, the on-state voltage drop increases as expected with reduction of the lifetime (τ_{p0}, τ_{n0}) indicated in the figure. The on-state voltage drop at a hole lifetime (τ_{p0}) value of 10 μs is found to be 1.971 V at an on-state current density of 50 A/cm^2 and increases to 9.146 V at a reduced hole lifetime (τ_{p0}) value of 1 μs. These values are significantly larger than those obtained for the SC-EST structure.

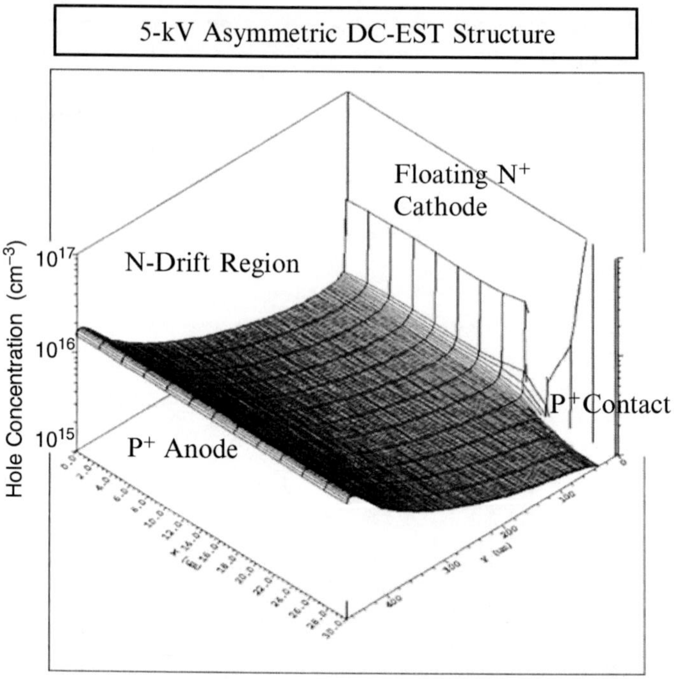

Fig. 10.40 On-state carrier distribution in the 5-kV asymmetric DC-EST structure

The reason for the higher on-state voltage drop observed in the DC-EST structure can be understood by examination of the carrier distribution in the structure. A three-dimensional view of the injected hole concentration within the 5-kV asymmetric DC-EST structure in the EST mode is provided in Fig. 10.40 at an on-state current density of 50 A/cm^2. It can be observed that the hole distribution is very uniform at the anode side of the drift region. However, the hole concentration has the catenary distribution similar to the thyristor structure only on the left-hand side of the structure. At the P$^+$ contact region on the right-hand side of the structure, the hole concentration is forced to zero by the reverse-biased junction. This has the adverse impact of reducing the hole concentration under the cathode region for about half of the cell width. The reduced hole concentration on the cathode side is responsible for the larger on-state voltage drop observed for the DC-EST structure. The current distribution within the 5-kV asymmetric DC-EST structure is provided in Fig. 10.41 during the on-state. It can be observed that almost the entire floating N$^+$ cathode region is active in this structure in contrast to the SC-EST structure. The simple analytical model for the DC-EST is therefore more applicable in this case.

The on-state voltage drop for the DC-EST structure can be reduced by increasing the length of the floating N$^+$ cathode region. This is demonstrated here with numerical simulations performed on a 5-kV asymmetric DC-EST structure B with a total width (W_{Cell}) of the structure of 60 μm (area $= 6.0 \times 10^{-7}$ cm^{-2}) and a

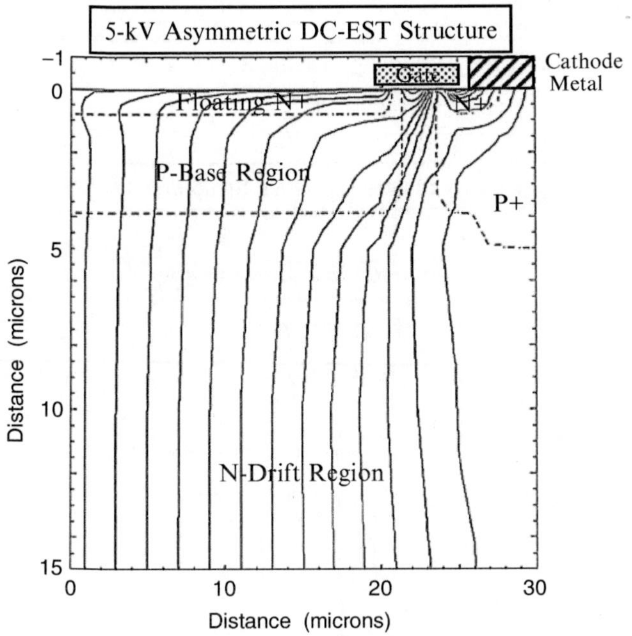

Fig. 10.41 On-state current flow-lines for the 5-kV asymmetric DC-EST structure: $V_G = 10$ V, $J_A = 50$ A/cm^2

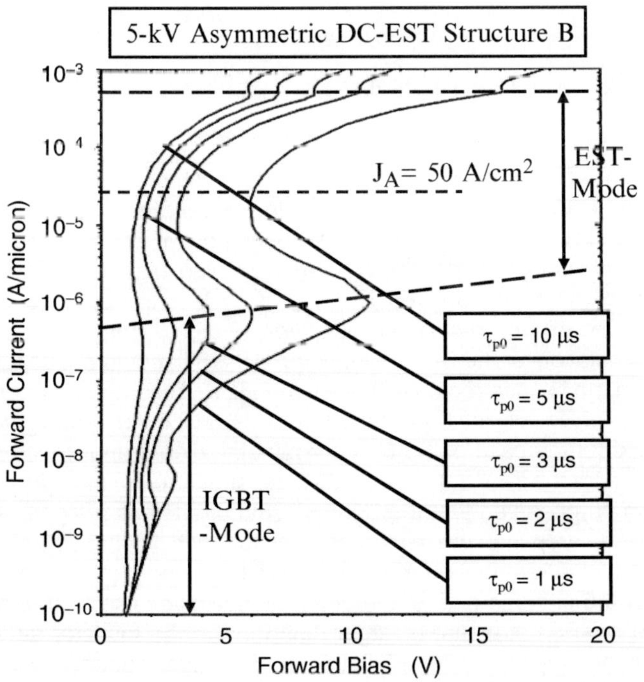

Fig. 10.42 On-state characteristics of the 5-kV asymmetric DC-EST structure B: lifetime dependence

polysilicon window (W_{PW}) of 100 μm in size. The impact of changes in the lifetime in the drift region of this 5-kV asymmetric DC-EST structure was obtained using numerical simulations with a shunting resistance of 1×10^{10} Ω-μm. The resulting on-state characteristics are shown in Fig. 10.42.

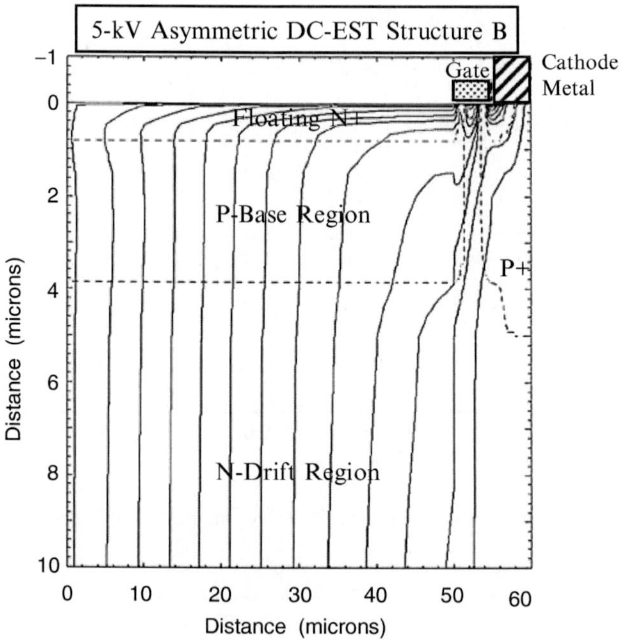

Fig. 10.43 On-state current flow-lines for the 5-kV asymmetric DC-EST structure B: $V_G = 10$ V, $J_A = 50$ A/cm^2

The current distribution within the 5-kV asymmetric DC-EST structure B is provided in Fig. 10.43 during the on-state. In contrast to the previous DC-EST structure EST-A (see Fig. 10.41), it can be observed that an even greater amount of current flows through the floating N$^+$ cathode region in structure B. This results in a reduced on-state voltage drop.

The variation of the on-state voltage drop obtained from the results of the numerical simulation for the 5-kV asymmetric DC-EST structures, as a function of the lifetime in the N-base region, is shown in Fig. 10.44 for the case of an anode on-state current density of 50 A/cm^2. For comparison purposes, the on-state voltage drops for the case of the 5-kV asymmetric planar and trench-gate IGBT structures and for the 5-kV asymmetric thyristor structure are also provided in this figure. It can be observed that the DC-EST structure B has a significantly lower on-state voltage drop than the IGBT structures for each lifetime value. This is due to improved carrier distribution in the DC-EST structure with a high free carrier density near the cathode side of the drift region. However, the on-state voltage drop for the DC-EST structure is larger than that for the thyristor structure. The on-state voltage drop for the DC-EST structure B is much superior to the structure with the shorter floating N$^+$ cathode region (structure EST-A).

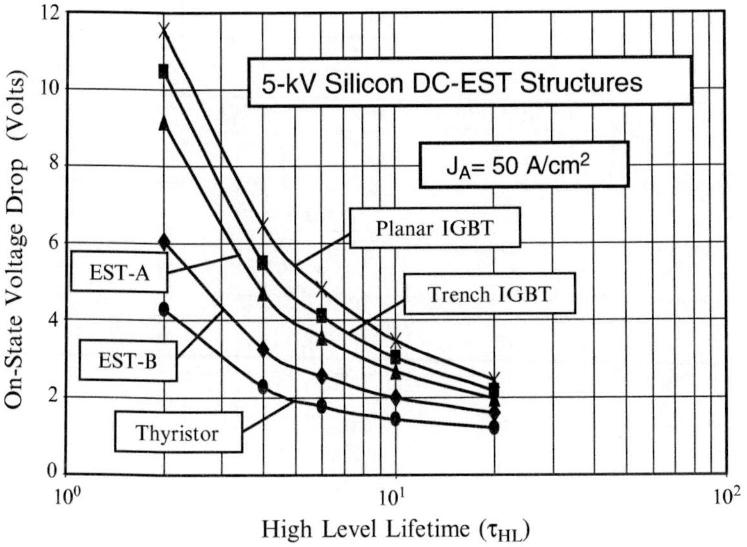

Fig. 10.44 On-state voltage drop for the 5-kV asymmetric DC-EST structures: N-base lifetime dependence

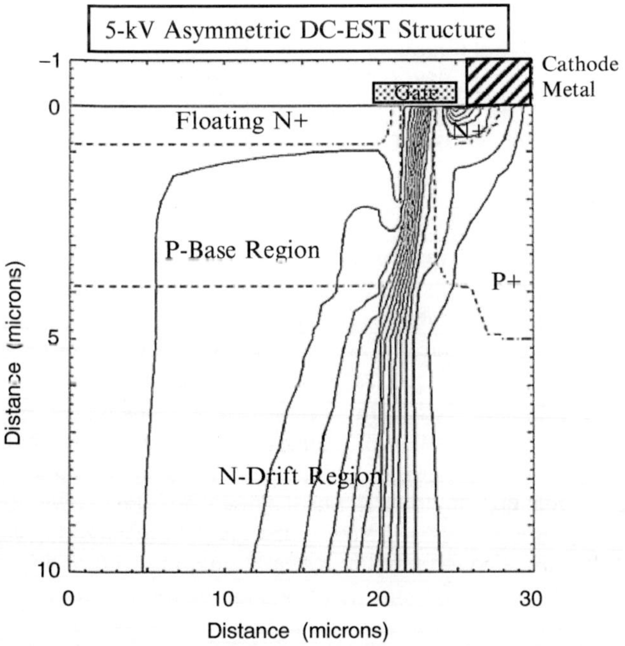

Fig. 10.45 On-state current distribution in the 5-kV asymmetric DC-EST structure A: IGBT mode

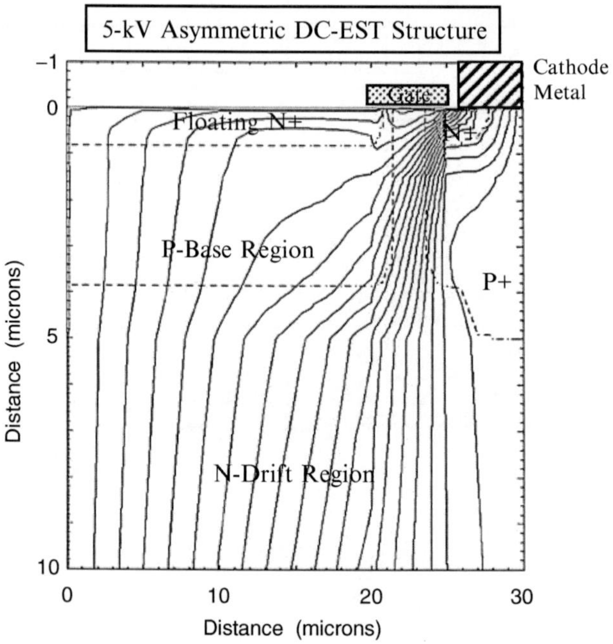

Fig. 10.46 On-state current distribution in the 5-kV asymmetric DC-EST structure A: after parasitic thyristor latch-up

Further insight into the operation of the DC-EST structure can be obtained by examination of the current flow pattern in the various modes. The current flow lines within the 5-kV asymmetric DC-EST structure are shown in Figs. 10.45 and 10.46 in the IGBT mode and after latch-up of the parasitic thyristor for the case of a high-level lifetime of 4 μs. In the IGBT mode at a current density of 0.3 A/cm^2, most of the current flows via the MOSFET on the right-hand side (see Fig. 10.45). Some hole current flows into the P-base region and is then removed from the cathode contact on the right-hand side. Once the parasitic thyristor latches up at a current density of about 1,500 A/cm^2, the current concentrates at the edge of the N$^+$ region on the right-hand side as observed in Fig. 10.46 at a current density of 3,000 A/cm^2. Some of the current can be observed to flow via the main thyristor path as well.

The on-state characteristics of the 5-kV silicon asymmetric DC-EST structure were also obtained as a function of temperature by using a positive gate bias voltage of 10 V for the case of a lifetime of 2 μs in the drift region. The characteristics obtained from the numerical simulations are shown in Fig. 10.47. At low anode current densities (below 0.05 A/cm^2), the on-state voltage drop decreases with increasing temperature while at on-state current densities above 20 A/cm^2, it begins to increase with increasing temperature. The on-state voltage drop increases with increasing temperature at an on-state current density of 50 A/cm^2, which is desirable for allowing paralleling of devices and the prevention of hot spots within the device structure. It can also be observed from Fig. 10.47 that the current density for latch-up of the parasitic thyristor (indicated by the dashed line) decreases as expected with increasing temperature.

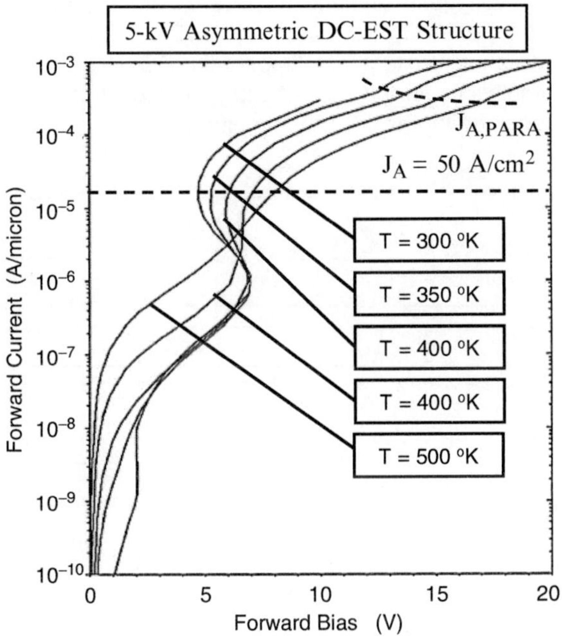

Fig. 10.47 On-state characteristics of the 5-kV asymmetric DC-EST structure A: temperature dependence

10.3.3 Turn-Off Characteristics

The turn-off waveforms for the DC-EST structure are similar to those already for the SC-EST structure in the previous section. The results of numerical simulations are provided here for comparison of the two device structures.

Simulation Example

In order to gain insight into the operation of the asymmetric 5-kV DC-EST structure during its turn-off, the results of two-dimensional numerical simulations for a typical structure are discussed here. The device structure used has the cross section shown in Fig. 10.2 with a cell half-width of 30 μm. The doping profiles for the DC-EST structure used in the numerical simulations were previously provided. For the typical case discussed here, a high-level lifetime of 4 μs was used in the N-base region. The numerical simulations were performed with an abrupt reduction of the gate voltage from positive 10 V to zero in 10 ns starting from an on-state current density of 50 A/cm^2. The resulting waveforms obtained from the numerical simulations for the anode voltage and current are shown in Fig. 10.48 for the case of an anode supply voltage of 3,000 V. Unlike the GTO structure, there is no storage time associated with the turn-off of the DC-EST structure. The anode voltage increases immediately at the end of the gate voltage transient. The anode voltage increases as the square of the time as predicted by the analytical model until it reaches about 2,000 V. It then increases at a slower rate. This is associated with the onset of avalanche multiplication at high anode bias voltages – an effect not included in the analytical model. The waveforms for the DC-EST structure are very similar to those shown for the SC-EST structure. The anode voltage rise time for the DC-EST structure obtained in the numerical simulations for the case of supply voltage of 3,000 V (0.70 μs) is larger than that observed for the SC-EST structure.

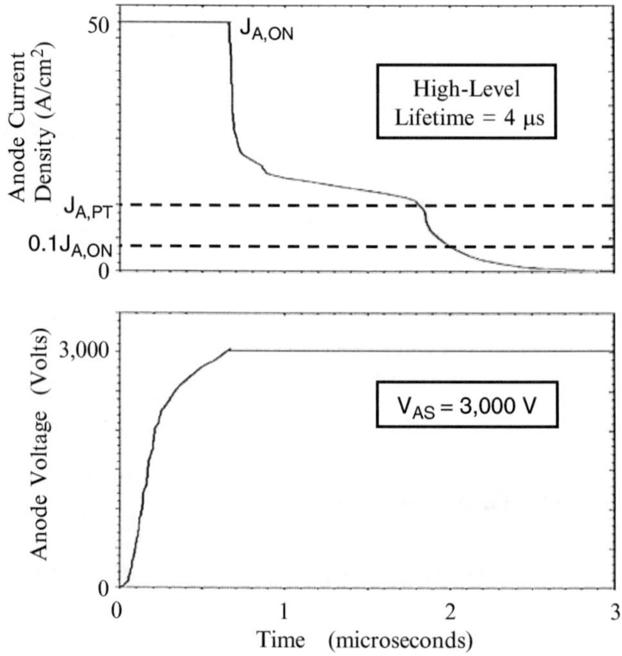

Fig. 10.48 Typical turn-off waveforms for the asymmetric 5-kV DC-EST structure A

After the completion of the anode voltage transient, the anode current waveform for the DC-EST structure decays from the initial on-state current density at a rate that decreases with time. The current decays to the punch-through current density (indicated in the figure) in 1.81 μs. The punch-through current density of about 15 A/cm^2 observed in the numerical simulations. After reaching the punch-through current density, the collector current is observed to decay at a faster rate as described by the analytical model. The current fall time for the DC-EST structure (1.30 μs) is smaller than that observed for the SC-EST structure.

It is instructive to examine the current flow pattern within the DC-EST structure during the turn-off process. The current flow lines within the 5-kV asymmetric DC-EST structure are provided in Fig. 10.49 at the end of the voltage rise time when the anode voltage is 3,000 V while the anode current density is at 50 A/cm^2. No current flows via the floating N$^+$ region demonstrating that the main thyristor is not active during the turn-off process. It can be observed that all of the hole current collected by the P-base region flows to the P$^+$ contact region on the right-hand side via the p-channel MOSFET under the gate electrode. It is worth pointing out that the current density is very uniform in the DC-EST structure during turn-off. This results in an excellent RBSOA for the structure which is close to that derived using one-dimensional analysis. In contrast, the RBSOA for the GTO is degraded by current crowding during the turn-off process.

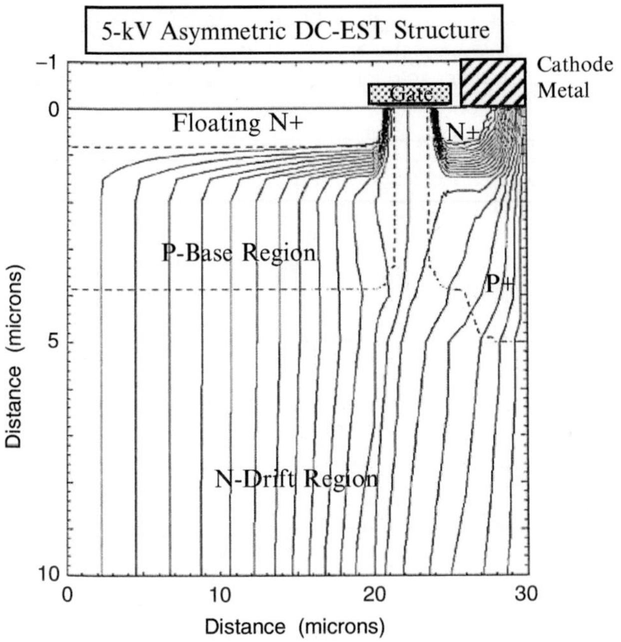

Fig. 10.49 Current distribution for the 5-kV asymmetric DC-EST during turn-off

10.3.4 Lifetime Dependence

The optimization of the power losses for the EST structure requires performing a trade-off between the on-state voltage drop and the switching losses. One approach to achieve this is by adjusting the lifetime in the drift (N-base) region. A reduction of the lifetime in the drift region also alters the lifetime in the N-buffer layer. The impact of reducing the lifetime on the on-state voltage drop was previously shown in Sect. 10.3.2. The on-state voltage drop increases when the lifetime is reduced. The analytical model developed for turn-off of the asymmetric MCT structure presented in Chap. 8 can be used to analyze the impact of changes to the lifetime in the drift region.

Simulation Example

In order to gain insight into the impact of the lifetime in the N-base region on the operation of the 5-kV asymmetric DC-EST structure, the results of two-dimensional numerical simulations for a typical structure are discussed here. The device structure used has the cross section shown in Fig. 10.2 with a half-cell width of 30 μm. The widths of the N-base and N-buffer layer regions are 440 and 30 μm, respectively. The high-level lifetime in the N-base region was varied between 2 and 20 μs. For turning-off the DC-EST structures, the numerical simulations were performed with gate voltage rapidly ramped down from positive 10 V to zero in 10 ns starting from an on-state current density of 50 A/cm^2. The resulting waveforms obtained from the numerical simulations for the anode voltage and current are shown in Fig. 10.50 for the case of an anode supply voltage of 3,000 V.

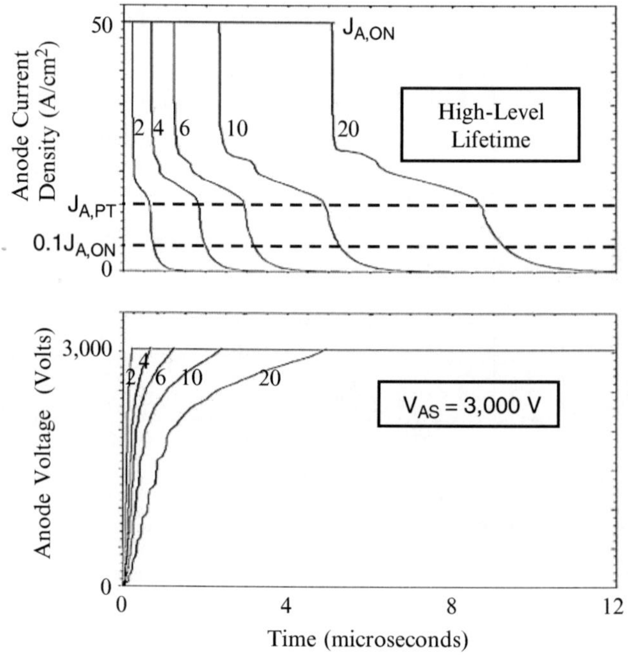

Fig. 10.50 Impact of lifetime on the 5-kV asymmetric DC-EST structure A turn-off waveforms

The numerical simulations show a decrease in the voltage rise time with reduction of the lifetime in the N-base region. The numerical simulations of the 5-kV asymmetrical DC-EST structure also show a substantial decrease in the anode current fall time when the lifetime is reduced. The numerical simulations show a reduction of the anode current during the first part of the decay to the punch-through anode current ($J_{C,PT}$), which is independent of the lifetime in the N-base region as predicted by the analytical model. The turn-off waveforms obtained for the 5-kV asymmetrical DC-EST structure B are not shown here in the interest of space because they are very similar to those shown in Fig. 10.50.

10.3.5 Switching Energy Loss

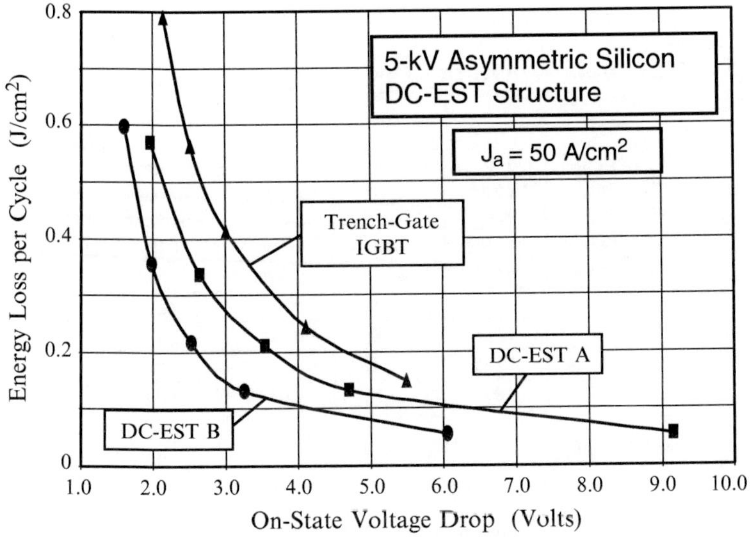

Fig. 10.51 Trade-off curve for the silicon 5-kV asymmetric DC-EST structures: lifetime in N-base region

For the typical switching waveforms for the 5-kV asymmetric DC-EST structure shown in Fig. 10.48 with an anode supply voltage of 3,000 V, the energy loss per unit area during the anode voltage rise time is found to be 0.035 J/cm^2 if the on-state current density is 50 A/cm^2; and the energy loss per unit area during the collector current fall time is found to be 0.098 J/cm^2 if the on-state current density is 50 A/cm^2. The total energy loss per unit area ($E_{OFF,V}$ + $E_{OFF,I}$) during the turn-off process for the 5-kV asymmetric DC-EST structure is found to be 0.133 J/cm^2. The energy loss per cycle for the DC-EST structure B was found to be close to that obtained for structure A.

Using the results obtained from the numerical simulations, the on-state voltage drop and the total energy loss per cycle can be computed. These values are plotted in Fig. 10.51 to create a trade-off curve to optimize the performance of the silicon 5-kV asymmetric DC-EST structure A and B by varying the lifetime in the N-base region. Devices used in lower frequency circuits would be chosen from the left-hand side of the trade-off curve while devices used in higher frequency circuits would be chosen from the right-hand side of the trade-off curve. For comparison purposes, the trade-off curve for the 5-kV IGBT structure is also included in the figure. It can be observed from this figure that substantial improvement in the power loss trade-off curve can be obtained by replacing the IGBT with the DC-EST structures. The DC-EST has a sufficiently large FBSOA as shown later to allow replacement of the IGBT with the SC-EST in motor control applications.

From Fig. 10.51, it can be clearly seen that the DC-EST structure B has a superior trade-off curve because of its lower on-state voltage drop.

10.3.6 Maximum Operating Frequency

High-Level Lifetime (μs)	On-State Voltage Drop (Volts)	On-State Power Dissipation (W/cm²)	Energy Loss per Cycle (J/cm²)	Maximum Operating Frequency (Hz)
20	1.604	40.1	0.598	268
10	1.995	50.1	0.354	424
6	2.536	63.4	0.216	632
4	3.255	81.4	0.131	904
2	6.053	151	0.054	906

Fig. 10.52 Power loss analysis for the 5-kV asymmetric DC-EST structure B

In the case of the baseline asymmetric DC-EST device structure A with a high-level lifetime of 4 μs in the N-base region, the on-state voltage drop is 4.705 V at an on-state current density of 50 A/cm^2. For the case of a 50% duty cycle, the on-state power dissipation contributes 118 W/cm^2 to the total power loss. Using a total turn-off energy loss per cycle of 0.133 J/cm^2 in Eq. 10.24 yields a maximum operating frequency of about 622 Hz. In the case of the asymmetric DC-EST device structure B with a high-level lifetime of 4 μs in the N-base region, the on-state voltage drop is 3.255 V at an on-state current density of 50 A/cm^2. For the case of a 50% duty cycle, the on-state power dissipation contributes 81 W/cm^2 to the total power loss. Using a total turn-off energy loss per cycle of 0.131 J/cm^2 in Eq. 10.24 yields a maximum operating frequency of about 904 Hz. The DC-EST structure B therefore has superior performance.

Using the results obtained from the numerical simulations, the on-state voltage drop and the energy loss per cycle can be computed. These values are provided for the asymmetric DC-EST device structure B in Fig. 10.52 together with the maximum operating frequency as a function of the high-level lifetime in the N-base region under the assumption of a 50% duty cycle and a total power dissipation limit of 200 W/cm^2. The maximum operating frequency is plotted in Fig. 10.53 as a function of the high-level lifetime in the N-base region for the case of a duty cycle of 50% for both structure A and B. It can be observed that the maximum operating frequency for the asymmetric DC-EST device structure B can be increased up to 900 Hz by reducing the high-level lifetime to 2 μs. This is much superior to the maximum operating frequency of 150 Hz for the 5-kV GTO structure and 400-Hz for the 5-kV trench-gate IGBT structure.

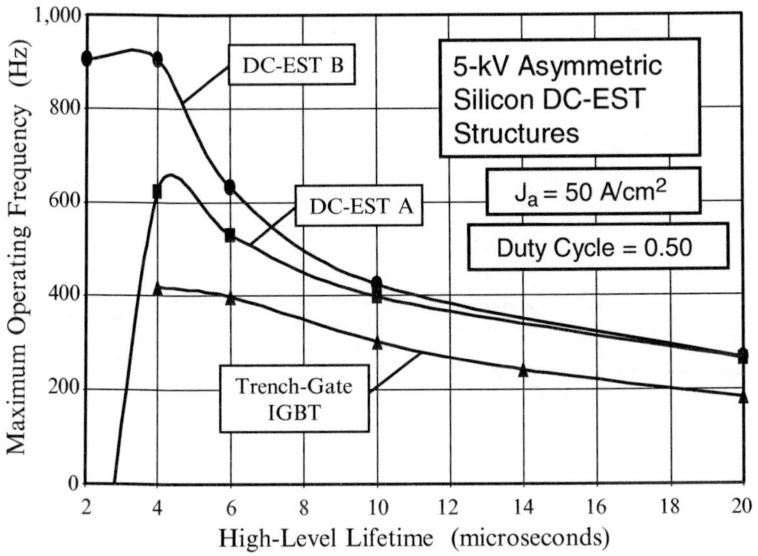

Fig. 10.53 Maximum operating frequency for the 5-kV asymmetric DC-EST structures

10.3.7 Forward-Biased Safe Operating Area

Since the main thyristor current is constrained to flow through a series MOSFET in the DC-EST structure, it exhibits a substantial region of operation where the anode current can be saturated under gate control. This is demonstrated in this section by using the results of numerical simulations for the 5-kV asymmetric DC-EST structures.

Simulation Results

Numerical simulations of the 5-kV silicon asymmetric DC-EST structure A were performed for the case of a high-level lifetime of 4 μs with various values for the gate bias voltage while sweeping the anode voltage. The resulting output characteristics are shown in Fig. 10.54. The device was able to saturate the anode current at lower gate bias voltages as expected. A snapback in the *i–v* characteristics is observed when the device transitions from the IGBT mode to the EST mode with latch-up of the main thyristor. The snapback is reduced when the lifetime is made larger because of the smaller IGBT mode voltage drop.

The output characteristics of the 5-kV silicon asymmetric DC-EST structure B obtained by using the numerical simulations are provided in Fig. 10.55. It was found to saturate the anode current to high voltages as expected and exhibits an excellent FBSOA with current saturation up to 3,000 V. The wide FBSOA boundary obtained by using the numerical simulations for the DC-EST structure is indicated by the dot-dashed lines in Fig. 10.55. Similar wide FBSOA boundaries have been reported for devices with other voltage ratings in the literature [9].

Fig. 10.54 Output characteristics of the 5-kV DC-EST structure A

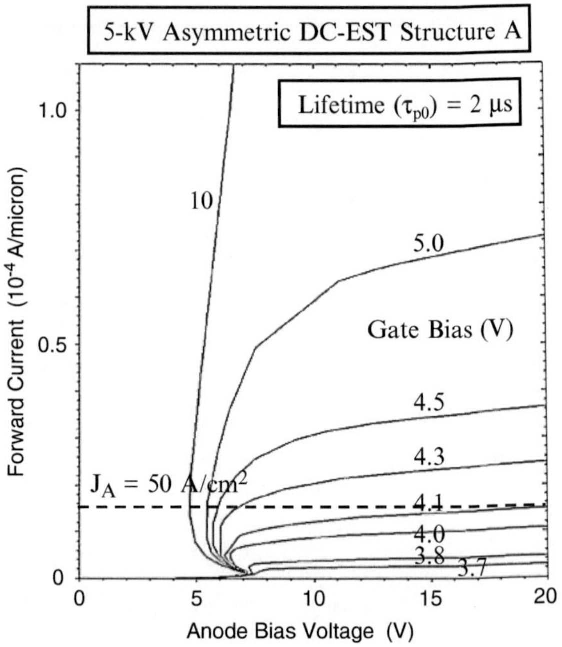

Fig. 10.55 Forward-biased safe operating area of the 5-kV DC-EST structure B

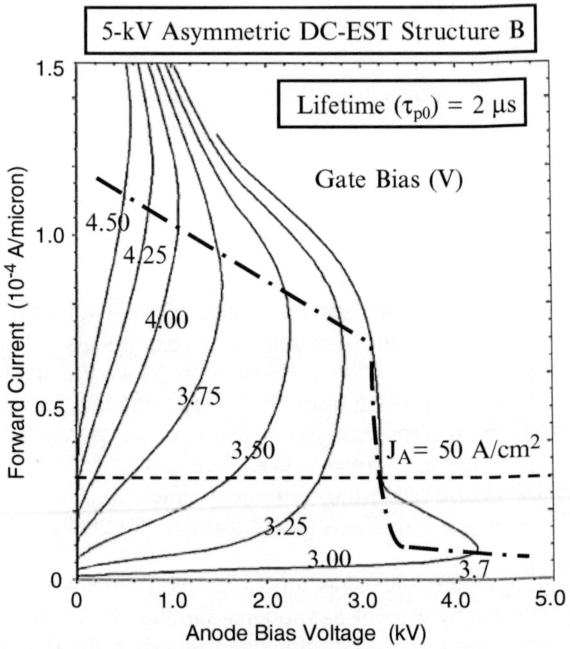

10.4 10,000-V Silicon EST

The 10-kV silicon asymmetric EST structure can be expected to function just like the 5-kV device. However, its design and operation is constrained by the larger blocking voltage capability. The lifetime in the N-base region for the 10-kV device must be larger to maintain a reasonable on-state voltage drop. The larger N-base width results in more stored charge within the structure, which limits the switching frequency.

In Chap. 4, it was demonstrated that the GTO structure has a limited reverse-biased safe operating area (RBSOA) due to influence of the holes in the space-charge region due to current flow. The analysis of the RBSOA for the EST structure is identical to that provided in Sect. 4.4. Using the results shown in Fig. 4.57, it can be concluded that in order to turn off the 10-kV asymmetric EST structure with a collector supply voltage of 6 kV, it is necessary to reduce the collector current density to only 20 A/cm^2. However, one of the merits of the EST structure is the low on-state voltage drop which allows its operation at an on-state current density of 50 A/cm^2. This value will therefore be utilized when determining the on-state voltage drop and switching transients for the 10-kV asymmetric EST structures. Due to RBSOA limitations, the anode supply voltage for the switching transient must be reduced to 5,000 V.

10.4.1 Blocking Characteristics

The electric field distribution within the asymmetric EST structure is essentially the same as that illustrated in Fig. 4.3 for the asymmetric GTO structure. Consequently, the design procedure described in Chap. 4 can be applied to the asymmetric EST structure. From Fig. 4.50, the N-base region width required to obtain a forward blocking voltage of 11,000 V is 1,100 μm. However, the results of the numerical simulation shown in Chap. 4 for the 10-kV GTO structure demonstrate that an N-base width of 800 μm is sufficient.

Simulation Example

In order to gain insight into the physics of operation for the 10-kV asymmetric DC-EST structure under voltage blocking conditions, the results of two-dimensional numerical simulations are described here for a device with N-base width of 850 μm. The simulations were performed using a cell with the structure shown in Fig. 10.2. This half-cell has a width of 60 μm (area = 6.0×10^{-7} cm^{-2}). The asymmetric DC-EST structure used for the simulations was formed by diffusions performed into a uniformly doped N-type drift region with a doping concentration of 2×10^{12} cm^{-3}. All the diffusions in the 10-kV structure had the same parameters as the 5-kV device described in the previous section. The doping profile in the vertical direction through the N$^+$ cathode region is similar to that for the 10-kV BRT structure in Fig. 9.34 indicating that the net width of the lightly doped portion of the N-base region is 825 μm after accounting for the diffusions. The doping profiles for the P-base and N^+ cathode regions are the same as those for the 5-kV asymmetric DC-EST structure.

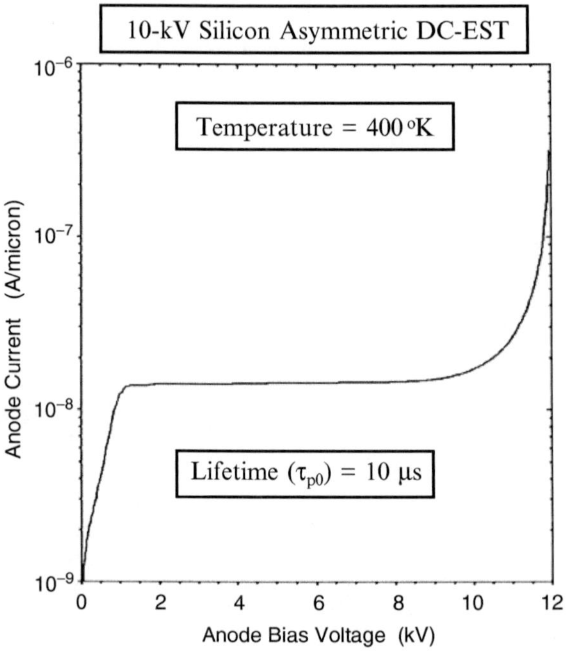

Fig. 10.56 Forward blocking characteristics of the 10-kV asymmetric DC-EST structure

The forward blocking capability of the 10-kV silicon asymmetric DC-EST structure was obtained at 400°K by increasing the anode bias while maintaining the gate electrode at 0 V. The characteristic obtained for a lifetime (τ_{p0}) of 10 μs is shown in Fig. 10.56. The leakage current increases rapidly with increasing anode bias voltage until about 1,000 V. This occurs due to the increase in the space-charge generation volume and the increase in the current gain (α_{PNP}) of the open base P-N-P transistor until the anode bias becomes equal to the reach-through voltage of 1,115 V obtained using the analytical solution given by Eq. 4.2. The leakage current then becomes independent of the anode voltage until close to the breakdown voltage. This behavior is well described by the analytical model. The numerical simulations indicate that a breakdown voltage of 10,500 V is possible with an N-base width of only 825 μm.

The voltage is primarily supported within the lightly doped portion of N-base region in the 10-kV asymmetric DC-EST structure during operation in the forward blocking mode. The electric field profile at various anode voltages are very similar to those previous shown for the 10-kV asymmetric BRT structure in Fig. 9.36.

10.4.2 On-State Voltage Drop

The on-state *i–v* characteristics and on-state voltage drop can be computed using the analytical model discussed in Sect. 10.2.2. In general, a larger lifetime is required in the N-base region for the 10-kV device when compared with the 5-kV device due to the larger width for the N-base region.

Simulation Results

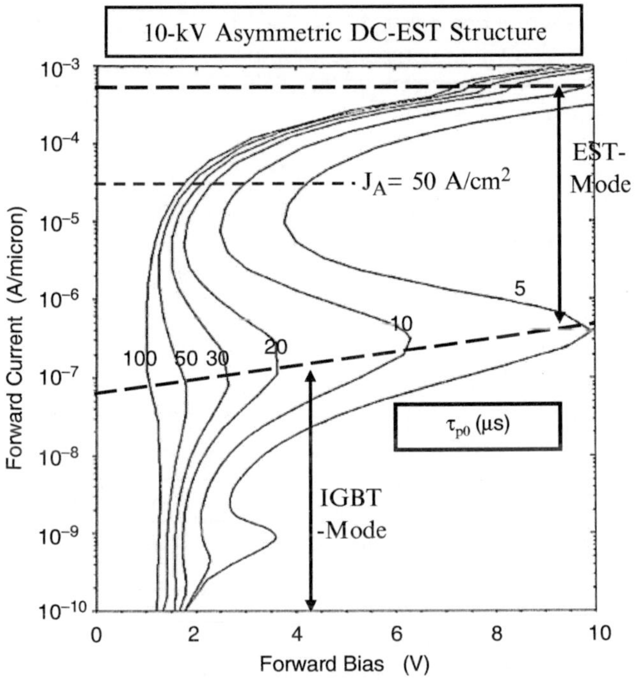

Fig. 10.57 On-state characteristics of the 10-kV asymmetric DC-EST structure

The results of two-dimensional numerical simulations for the 10-kV asymmetrical silicon DC-EST structure are described here. Tho total half-cell width of the structure, as shown by the cross section in Fig. 10.2, was 60 μm (area $= 6.0 \times 10^{-7}$ cm^{-2}). The on-state characteristics of the 10-kV silicon asymmetric DC-EST structure were obtained by using a gate bias voltage of 10 V using various values for the lifetime in the N-base region. The characteristics obtained from the numerical simulations are shown in Fig. 10.57. It can be observed that the on-state voltage drop increases as expected with reduction of the lifetime (τ_{p0}, τ_{n0}) indicated in the figure. The on-state voltage drop for the 10-kV asymmetric DC-EST structure is substantially smaller than that for the 10-kV asymmetric IGBT structure. For this reason, it is possible to operate the 10-kV asymmetric DC-EST structure at a larger on-state current density of 50 A/cm^2 from a power loss stand point. However, due to the limitations of RBSOA, the maximum supply voltage must be reduced to 5,000 V.

The good on-state voltage drop for the 10-kV asymmetric DC-EST structure for larger values of the lifetime in the N-base region is due to the large number of carriers injected into the drift region producing a drastic reduction of its resistance. This is illustrated in Fig. 10.58 where the injected carrier density is shown for five cases of the lifetime (τ_{p0}, τ_{n0}) in the N-base region of the DC-EST structure. It can be observed that the injected carrier density in the drift region on the anode side is more than four orders of magnitude larger than the doping concentration for the case of a lifetime of 100 μs. The injected carrier density is reduced by a factor of 5 times near the anode junction when the lifetime is reduced to 3 μs. There is a significant reduction in the injected carrier density in the middle of the drift region when the lifetime is reduced below 10 μs. This is due to the relatively large width for the N-base region when compared with the 5-kV silicon DC-EST structure. The reduced hole concentration in the drift region produces the observed increase in on-state voltage drop.

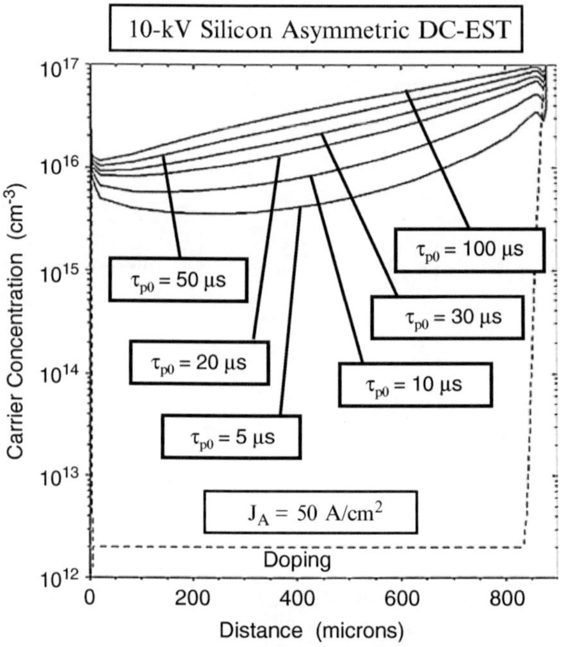

Fig. 10.58 On-state carrier distribution in the 10-kV asymmetric DC-EST structure

The variation of the on-state voltage drop for the 10-kV asymmetric DC-EST structure obtained from the results of the numerical simulations, as a function of the lifetime in the N-base region, is shown in Fig. 10.59 for the case of an anode on-state current density of 50 A/cm^2. For comparison purposes, the on-state voltage drop for the case of the 10-kV asymmetric trench-gate IGBT structure is also provided in this figure. It can be observed that the DC-EST structure has a

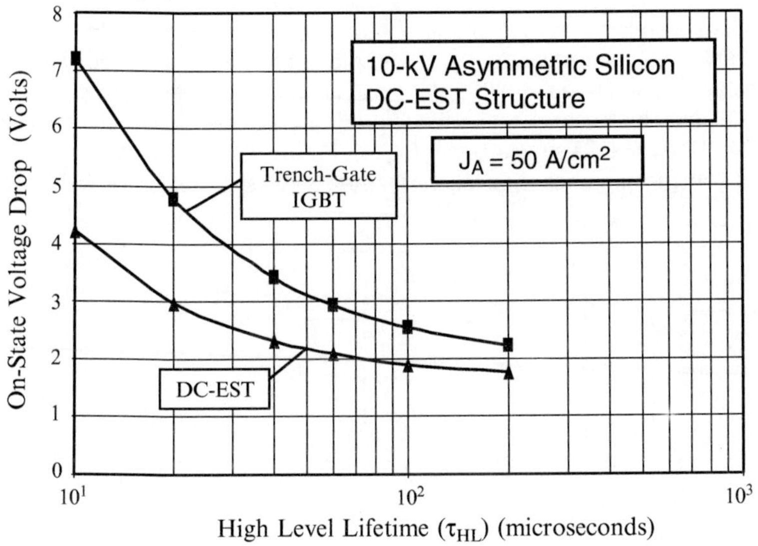

Fig. 10.59 On-state voltage drop for the 10-kV asymmetric DC-EST structure: N-base lifetime dependence

significantly lower on-state voltage drop than the IGBT structure for each lifetime value in spite of the use of the trench-gate structure with high channel density for the IGBT structure. This is due to improved carrier distribution in the DC-EST structure with a high free carrier density near the cathode side of the drift region.

10.4.3 Turn-Off Characteristics

The physics for turn-off of the 10-kV silicon asymmetric DC-EST structure can be expected to be the same as that for the 5-kV device structure. Due to limitations with the RBSOA (as discussed in Chap. 4 for the silicon GTO structure), the 10-kV asymmetric DC-EST structure can be operated at an on-state current density of 50 A/cm^2 only if the anode supply voltage is reduced to 5,000 V. The results of numerical simulations of the 10-kV asymmetric DC-EST structure under these turn-off conditions are discussed here.

Simulation Results

Numerical simulations of the turn-off for the 10-kV asymmetric DC-EST structure with a high-level lifetime of 20 μs were performed by stepping the gate voltage down from positive 10 to 0 V in 10 ns using an on-state current density of 50 A/cm^2. The resulting waveforms obtained from the numerical simulations for the anode voltage and current are shown in Fig. 10.60 for the case of an anode supply voltage of 5,000 V. It can be observed that there is no storage time for the 10-kV asymmetric DC-EST structure. The anode voltage initially increases non-linearly as described by the analytical model during the early stages of the voltage rise but the rate of rise becomes severely reduced at anode voltages beyond 4,000 V due to the onset of significant impact ionization as the RBSOA boundary is approached. The anode voltage almost saturates at 5,000 V indicating operation of the device close to its RBSOA limit. This is consistent with the predictions of the analytical model for the RBSOA of the DC-EST structure (see Fig. 4.57). The voltage rise time is found to be 11.8 μs from the numerical simulations for the 10-kV asymmetric DC-EST structure.

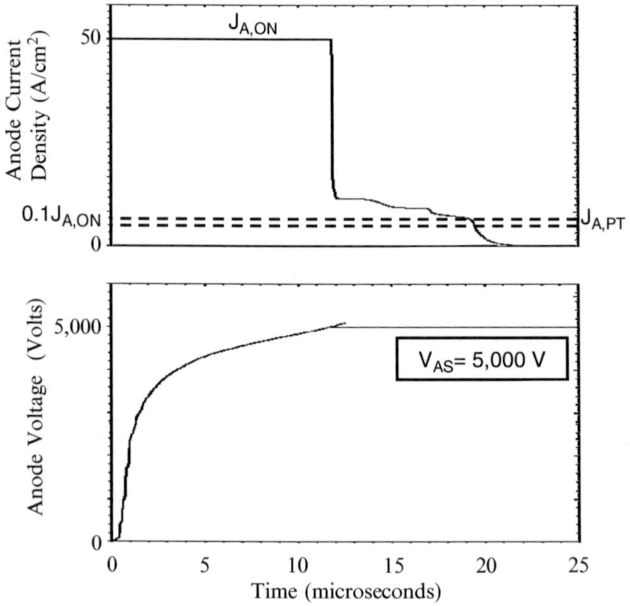

Fig. 10.60 Turn-off waveforms for the 10-kV asymmetric DC-EST

The anode current turn-off occurs with a rapid initial decrease in current to about 10 A/cm^2 followed by a gradual change as expected from the analytical model. This rapid decrease is due to a reduction of impact ionization generated carriers as the anode current decreases (due to recombination of the stored charge) because of the reduction in the electric field in the space-charge region. The punch-through anode current density ($J_{A,PT}$) obtained using the analytical model (Eq. 8.24) is 9 A/cm^2 in agreement with the simulation results. The current fall time is found to be 7.8 μs from the numerical simulations for the 10-kV asymmetric DC-EST structure.

10.4.4 Switching Energy Loss

As discussed previously, the maximum operating frequency for the DC-EST structure is limited by the turn-off losses. The turn-off losses are associated with the voltage rise-time interval and the current fall-time interval. The energy loss for each event can be computed by using the equations previously provided in Sect. 9.2.5. The turn-off energy loss per cycle obtained from the numerical simulations of the silicon 10-kV asymmetric DC-EST structure can be derived from the waveforms in Fig. 10.60. For the case of a high-level lifetime of 20 μs in the N-base region, the energy loss per cycle during the voltage rise time is 1.47 J/cm^2 while the energy loss per cycle during the current fall time is 0.975 J/cm^2 in the case of an on-state current density of 50 A/cm^2 and an anode supply voltage of 5,000 V. The total energy loss per cycle is 2.445 J/cm^2 for the 10-kV silicon asymmetric DC-EST structure.

10.4.5 Maximum Operating Frequency

The maximum operating frequency for the 10-kV asymmetric DC-EST structure is limited by the turn-off losses. The turn-off losses are associated with the voltage rise-time interval and the current fall-time interval. The energy loss for each event can be computed by using the equation previously provided in Sect. 9.2.6. Using this information, the maximum operating frequency for the DC-EST structure can be derived using Eq. 10.24. The data acquired from the numerical simulations of the 10-kV asymmetric DC-EST and IGBT structures are provided in Fig. 10.61 for the case of an on-state operating current density of 50 A/cm^2.

	High-Level Lifetime (μs)	On-State Voltage Drop (Volts)	On-State Power Dissipation (W/cm^2)	Energy Loss per Cycle (J/cm^2)	Maximum Operating Frequency (Hz)
10-kV Asymmetric IGBT Structure	20	4.766	119	2.50	32
10-kV Asymmetric DC-EST Structure	20	2.967	74.2	2.445	51

Fig. 10.61 Power loss analysis for the 10-kV asymmetric DC-EST and IGBT structures with on-state current density of 50 A/cm^2

The maximum operating frequency obtained under the assumption of a 50% duty cycle and a total power dissipation limit of 200 W/cm^2 for the 10-kV asymmetric IGBT and DC-EST structures are found to be 32 and 51 Hz, respectively. The maximum operating frequency for the silicon 10-kV asymmetric DC-EST structure is superior to that for the IGBT structure due to its lower on-state voltage drop.

10.5 Reverse-Biased Safe Operation Area

The analytical solution for the RBSOA for the EST structures can be obtained by using Eq. 4.97 provided for the GTO structure because the physics of operation is similar. However, the GTO structure suffers from current crowding during the turn-off process. This problem does not occur in the EST structures. The RBSOA boundaries for the 5-kV asymmetric EST structures obtained by using numerical simulations are provided in this section.

Simulation Results

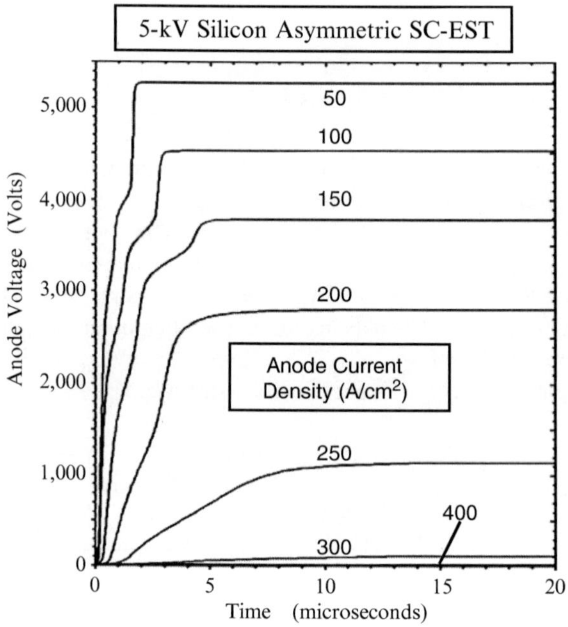

Fig. 10.62 5-kV asymmetric SC-EST RBSOA turn-off waveforms

The RBSOA boundary for the EST structure can be obtained by turning off the structure starting with various on-state current densities. The presence of holes in the space-charge region enhances the electric field at the junction between the P-base region and the drift region. The electric field becomes larger for larger initial on-state current densities. Consequently, the collector voltage at which the on-state current density can be sustained by the impact ionization process becomes smaller. During turn-off, the collector voltage becomes limited as a function of time providing the RBSOA limit for each corresponding on-state current density.

Numerical simulations of the 5-kV silicon asymmetric SC-EST structure were performed for the case of a high-level lifetime of 2 μs with various values for the initial on-state current density. The resulting collector voltage waveforms are provided in Fig. 10.62. At anode current densities below 200 A/cm^2, the collector voltage increases and becomes limited by the on-site of avalanche breakdown. At larger collector current densities, the maximum sustainable current density for the SC-EST structure is limited by turn-on of the parasitic thyristor. Using the anode turn-off waveforms, the RBSOA boundary can be determined as shown in Fig. 10.63. The SC-EST exhibits a RBSOA boundary that is significantly inferior to that observed for the IGBT structure (see Fig. 5.76). However, the RBSOA of the SC-EST structure is sufficiently large for typical motor-control applications.

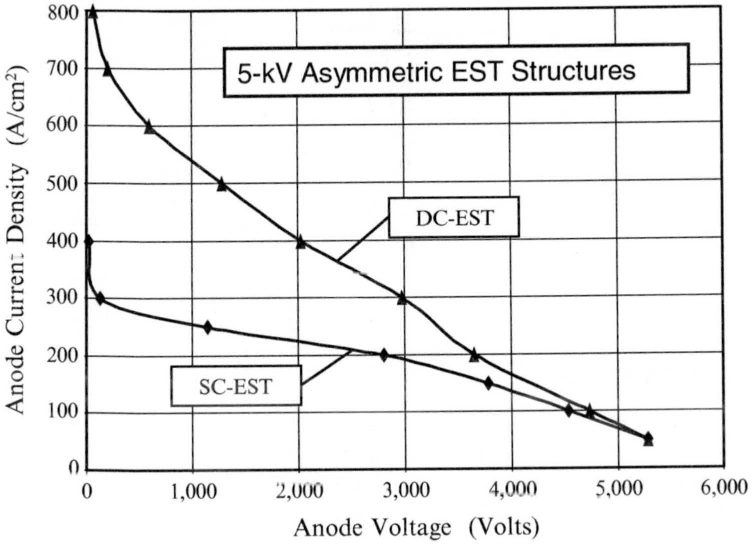

Fig. 10.63 RBSOA boundary for the 5-kV asymmetric SC-EST structure

Numerical simulations of the 5-kV silicon asymmetric DC-EST structure were performed for the case of a high-level lifetime of 2 μs with various values for the initial on-state current density. The resulting collector voltage waveforms are provided in Fig. 10.64. At anode current densities below 400 A/cm^2, the collector voltage increases and becomes limited by the on-site of avalanche breakdown. At larger collector current densities, the maximum sustainable current density for the DC-EST structure is limited by turn-on of the parasitic thyristor. Using the anode turn-off waveforms, the RBSOA boundary can be determined as shown in Fig. 10.63. The DC-EST exhibits a RBSOA boundary that is significantly superior to that of the SC-EST structure but it is much inferior to that observed for the IGBT structure (see Fig. 5.76). The RBSOA of the DC-EST structure is sufficiently large for typical motor-control applications.

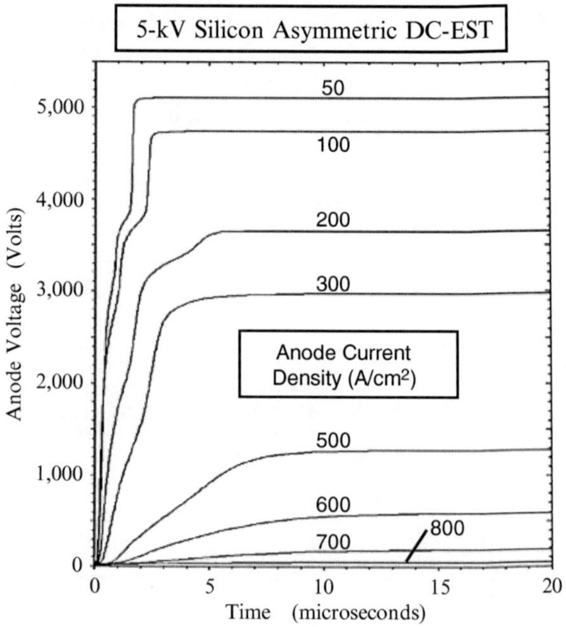

Fig. 10.64 5-kV asymmetric DC-EST RBSOA turn-off waveforms

10.6 Conclusions

The physics of operation and design principles for the silicon EST structures have been described in this chapter. This structure was proposed as an alternative to the MCT structure because of its wide forward-biased safe operating characteristics, which make it compatible with IGBT applications. In addition, the EST structure can be fabricated using a simpler fabrication process that is close to the manufacturing process for IGBT devices.

References

1. B.J. Baliga, "Gated Base Controlled Thyristor", U.S. Patent Number 5,099,300, Filed June 14, 1990, Issued March 24, 1992.
2. B.J. Baliga, "Base Resistance Controlled Thyristor with Single-Polarity Turn-on and Turn-off Control", U.S. Patent Number 5,198,687, Filed July 23, 1992, Issued March 30, 1993.
3. D.N. Pattanayak and B.J. Baliga, "Monolithically Integrated Insulated Gate Semiconductor Device", U.S. Patent Number 5,198,687, Filed May 19, 1987, Issued July 11, 1989.
4. B.J. Baliga, "The MOS-Gated Emitter Switched Thyristor", IEEE Electron Device Letters, Vol. 11, pp. 75–77, 1990.
5. B.J. Baliga, "The MOS-Gated Emitter Switched Thyristor", IEEE International Symposium on Power Semiconductor Devices and ICs, Abstract 4.1.1, pp. 117–121, 1990.

6. M.S. Shekar, et al, "Experimental Demonstration of the Emitter Switched Thyristor", IEEE International Symposium on Power Semiconductor Devices and ICs, Abstract 6.1, pp. 128–131, 1991.
7. N. Iwamuro, M.S. Shekar, and B.J. Baliga, "A Study of EST's Short-Circuit SOA", IEEE International Symposium on Power Semiconductor Devices and ICs, Abstract 3.4, pp. 71–76, 1993.
8. N. Iwamuro, et al, "Comparison of RBSOA of ESTs with IGBTs and MCTs", IEEE International Symposium on Power Semiconductor Devices and ICs, Abstract 5.2, pp. 195–200, 1994.
9. N. Iwamuro, M.S. Shekar, and B.J. Baliga, "Forward Biased Safe Operating Area of Emitter Switched Thyristors", IEEE Transactions on Electron Devices, Vol. 42, pp. 334–339, 1995.
10. S. Sridhar and B.J. Baliga, "Dual-channel EST/BRT: A New High-Voltage MOS-Gated Thyristor Structure", Electronics Letters, Vol. 31, pp. 494–496, 1995.
11. S. Sridhar and B.J. Baliga, "Comparison of Linear and Circular Cell Dual Channel Emitter Switched Thyristors", IEEE International Symposium on Power Semiconductor Devices and ICs, Abstract 8.6, pp. 170–174, 1995.
12. N. Iwamuro and B.J. Baliga, "Reverse Biased Safe Operating Area of Emitter Switched Thyristors", IEEE Transactions on Electron Devices, Vol. 43, pp. 352–357, 1996.
13. S. Sridhar and B.J. Baliga, "Temperature Dependence of the Emitter Switched Thyristor Characteristics", Solid State Electronics, Vol. 39, pp. 769–776, 1996.
14. N. Unten, S. Sawant, and B.J. Baliga, "High Voltage 4kV Emitter Switched Thyristors", IEEE International Symposium on Power Semiconductor Devices and ICs, Abstract 6.3, pp. 221–224, 1997.
15. S. Sridhar and B.J. Baliga, "Output Characteristics of the Dual Channel EST", Solid State Electronics, Vol. 41, pp. 1133–1138, 1997.
16. S. Sawant and B.J. Baliga, "Impact of VLSI Design Rules on High Voltage (2000V) IGBTs/ESTs", IEEE International Symposium on Power Semiconductor Devices and ICs, Abstract 11.13, pp. 253–256, 1998.
17. S. Sawant and B.J. Baliga, "Improved DC-EST Structure with Diode Diverter", Electronics Letters, Vol. 34, pp. 1358–1360, 1998.
18. S. Sawant and B.J. Baliga, "A New Wide SOA DC-EST Structure with Diode Diverter", IEEE International Symposium on Power Semiconductor Devices and ICs, Abstract 7.13, pp. 173–176, 1999.
19. S. Sawant and B.J. Baliga, "Current Saturation Control in Silicon Emitter Switched Thyristors", Solid State Electronics, Vol. 44, pp. 133–142, 2000.
20. B.J. Baliga, "Fundamentals of Semiconductor Power Devices", Springer-Science, New York, 2008.

Chapter 11
Synopsis

The previous chapters have discussed a variety of high-voltage power device structures, based upon both silicon and silicon carbide, for use in high power applications such as mass transportation and power distribution. In this concluding chapter, the performance of these devices is compared and contrasted to provide an overall perspective of the available technologies. The comparison is performed using two categories of voltage ratings, namely 5- and 10-kV blocking voltage capability.

11.1 5-kV Devices

In this section, the performance of power devices with 5-kV asymmetric blocking voltage capability will be compared. The basis for the comparison will be the outcome of the two-dimensional numerical simulations that have been provided for each of the devices in the preceding chapters. The performance attributes that are of prime importance are the on-state voltage drop and the turn-off energy loss per cycle due to their impact on power losses in the applications. In addition, the forward-biased and reverse-biased safe operating area for the various devices will be compared because of their importance in circuit design. Devices with poor safe operating area have required the incorporation of expensive and lossy snubber circuits. With the availability of the IGBTs with excellent safe operating area, power system design has migrated away from the use of snubber circuits allowing major advances in the size and weight of systems. Any modern power device technology must provide similar benefits with significant reduction of power losses in order to have an impact on these applications in the future.

B.J. Baliga, *Advanced High Voltage Power Device Concepts*,
DOI 10.1007/978-1-4614-0269-5_11,

11.1.1 On-State Voltage Drop

All of the high-voltage silicon power devices must rely upon bipolar current flow during the on-state in order to provide sufficiently low on-state voltage drop to be of interest in high power applications. These devices have a nonlinear on-state characteristic with significant current flow commencing only when the voltage exceeds a knee-voltage. It is not appropriate to use an on-resistance to characterize such nonlinear on-state behavior. In contrast, it is usual to describe the on-state operation of the silicon carbide power MOSFETs by using their specific on-resistance. In order to create a unified treatment of the silicon and silicon carbide devices, it is convenient to use the same on-state current density of 50 A/cm^2 for all the structures for obtaining the on-state voltage drop.

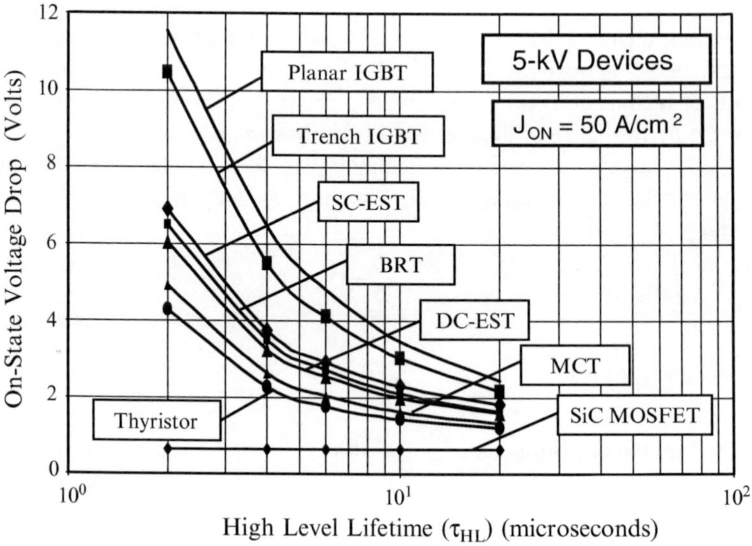

Fig. 11.1 Comparison of the on-state voltage drop of 5-kV devices

The on-state voltage drop for the 5-kV MOS-gated devices discussed in previous chapters can be compared with the aid of Fig. 11.1. In the case of the IGBT structures, the on-state voltage drop increases most rapidly with reduced lifetime in the drift region. The behavior of the planar-gate and trench-gate structures is similar with the on-state voltage drop of the trench-gate devices about 1 V smaller than that for the planar-gate structures for the same lifetime in the drift region. It can be observed from the figure that the on-state voltage drop of even the thyristor structure increases considerably when the high-level lifetime is reduced to 2 μs.

The MCT structure has an on-state voltage drop close to that of the thyristor structure. The BRT, SC-EST, and DC-EST structures have comparable on-state voltage drops which are significantly smaller than that for the IGBT structures but larger than that of the MCT structure. The on-state voltage drop for the silicon carbide inversion-mode power MOSFET can be observed to be independent of the lifetime in the drift region because it is a unipolar device, and much lower than that of all the silicon devices. At the 5-kV blocking voltage rating, the silicon carbide power MOSFET structure is clearly a much superior alternative to the bipolar silicon devices.

11.1.2 Power-Loss Trade-off Curves

The trade-off curve between the on-state voltage drop and the energy loss per cycle is commonly used during the optimization of power devices for applications. This trade-off curve can also be useful for the selection of the best device structure from the power loss point of view. In performing the comparison of the 5-kV devices discussed in this book, a current density of 50 A/cm^2 will be assumed in the on-state and a DC supply voltage of 3,000 V will be assumed in the off-state. The energy loss per cycle corresponds to switching under inductive load conditions.

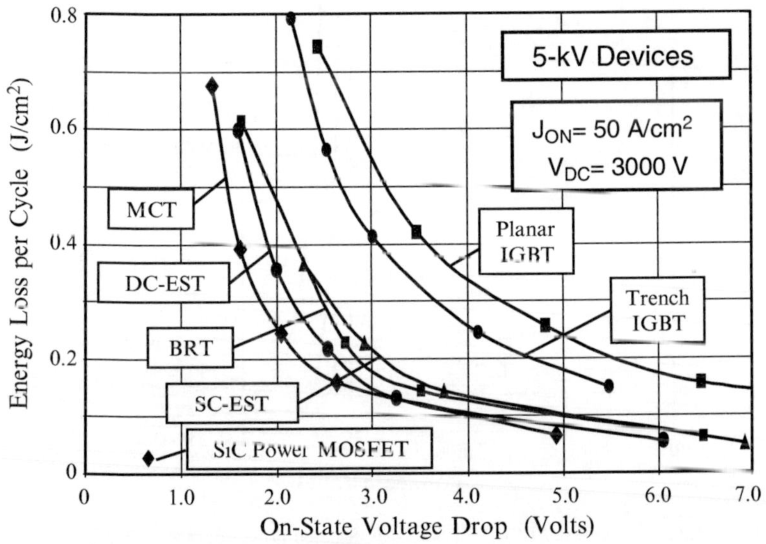

Fig. 11.2 Comparison of the power-loss trade-off curves of 5-kV MOS-gated power devices

The power loss trade-off curve for the 5-kV devices discussed in previous chapters can be compared with the aid of Fig. 11.2. The IGBT structures exhibit the worst trade-off curves with the performance of the trench-gate structure slightly superior to that of the planar-gate structure. The MCT structure has the best trade-off curve among the silicon MOS-gated devices. The DC-EST structure exhibits the next best trade-off curve for power losses. The SC-EST and BRT structure have comparable trade-off curves that are better than those for the IGBT structures but worse than that of the DC-EST and MCT structures. The silicon carbide power MOSFET structure is clearly far superior to the silicon bipolar devices with a low on-state voltage drop and small energy loss per cycle.

11.1.3 Forward-Biased Safe Operating Area

A wide forward-biased safe operating area (FBSOA) is essential for any new MOS-gated power devices that are developed as replacements for the well established IGBT structure. A wide FBSOA allows the controlled turn-on of the device to regulate the rate of current flow in H-bridge circuits used for motor control. If the rate of change of current is uncontrolled or too large, the reverse recovery current of the P-i-N rectifiers used in these applications becomes extremely large, resulting in failure of the diode and the power switch [1].

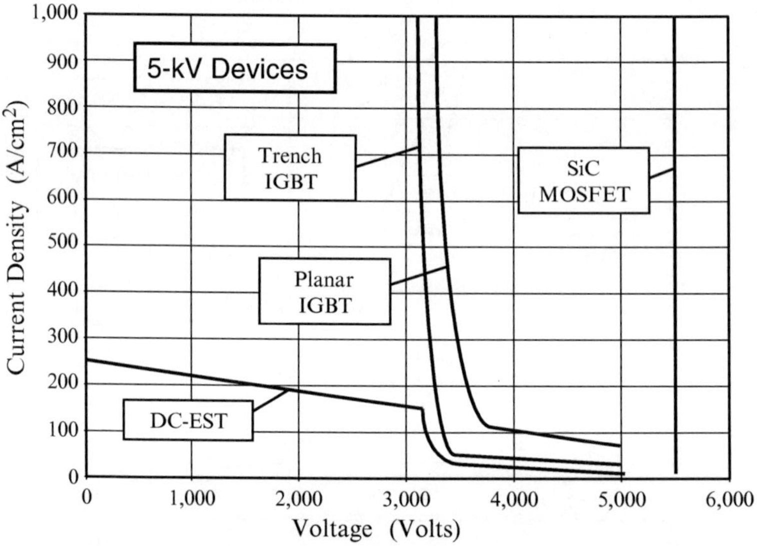

Fig. 11.3 Comparison of the FBSOA of 5-kV MOS-gated power devices

The FBSOA boundary for various devices was obtained by using two-dimensional numerical simulations as described in the previous chapters. These boundaries are shown in Fig. 11.3 for purposes of comparison. The MCT and BRT structures have practically negligible FBSOA, making them unacceptable candidates for hard switching applications commonly used for motor control with IGBTs. The FBSOA of the SC-EST structure is not shown in the figure because it is inferior to that of the DC-EST structure. The FBSOA of the silicon carbide MOSFET structure is a vertical line due to unipolar operation. The trench-gate and planar-gate IGBT structures can be observed to exhibit excellent FBSOA boundaries. The FBSOA boundary for the trench-gate structure is slightly inferior to that of the planar-gate structure. This is associated with the high electric field generated at the corner of the trenches. The FBSOA of the silicon DC-EST structure is much smaller than that of the silicon IGBT structures but may be adequate for applications.

11.1.4 Reverse-Biased Safe Operating Area

One of the merits of the IGBT devices has been their excellent reverse-biased safe operating area (RBSOA). This allows using them for motor control applications without the need for expensive and lossy snubber circuits. A wide RBSOA is essential for any new MOS-gated power devices that are developed as replacements for the well established IGBT structure. A wide RBSOA allows the controlled turn-off of the device in an inductive load circuit. During the turn-off event, the voltage and current at the output terminals of the power switch are simultaneously large resulting in enhanced impact ionization as discussed in Chap. 4.

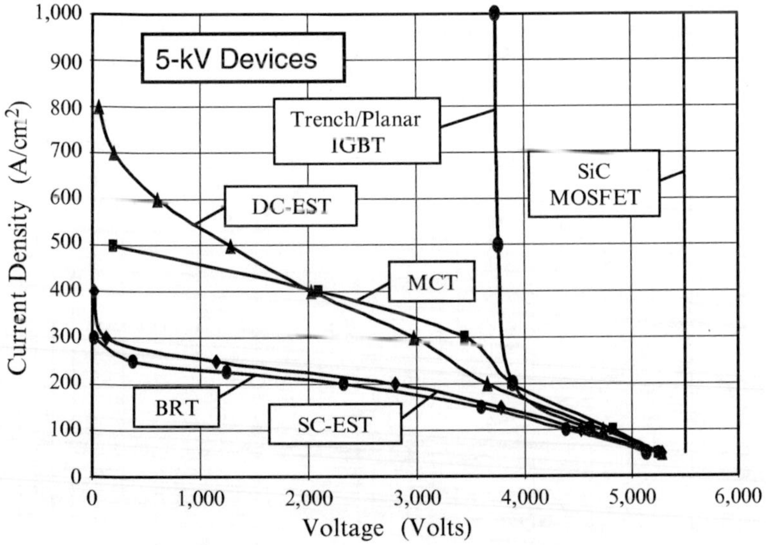

Fig. 11.4 Comparison of the RBSOA of 5-kV MOS-gated power devices

The RBSOA boundary for various devices was obtained by using two-dimensional numerical simulations as described in the previous chapters. These boundaries are shown in Fig. 11.4 for purposes of comparison. The RBSOA of the silicon IGBT is excellent, making it a popular device in motor control applications. The MCT and DC-EST structures have the next best RBSOA among the silicon devices. Their RBSOA, although significantly worse than that of the IGBT structure, is adequate for motor control applications. The silicon BRT and SC-EST structure have an RBSOA that may be marginal from an applications standpoint due to overshoots in current flow as a result of the reverse recovery current from the fly-back diodes. The silicon carbide MOSFET structure has a superior RBSOA when compared with the silicon IGBT due to its unipolar current flow.

11.2 10-kV Devices

In this section, the performance of power devices with 10-kV asymmetric blocking voltage capability is compared. The basis for the comparison is the outcome of the two-dimensional numerical simulations that have been provided for each of the devices in the preceding chapters. The performance attributes that are of prime importance are the on-state voltage drop and the turn-off energy loss per cycle due to their impact on power losses in the applications.

11.2.1 On-State Voltage Drop

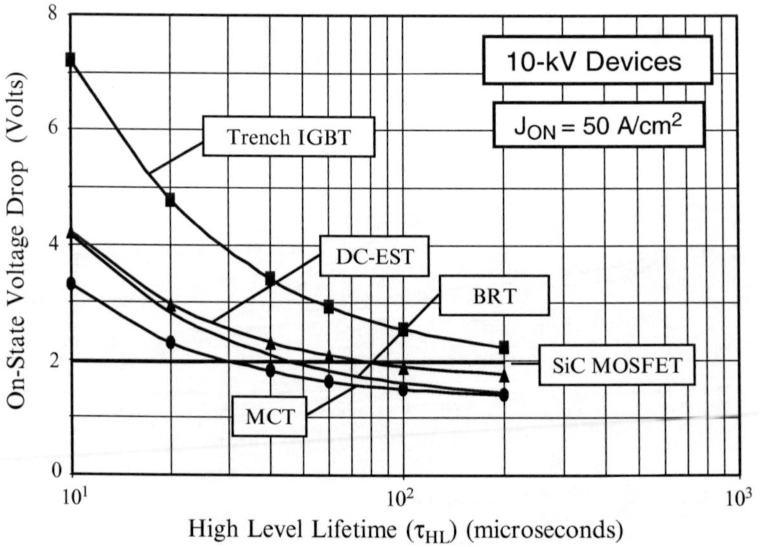

Fig. 11.5 Comparison of the on-state voltage drop of 10-kV MOS-gated power devices

As in the case of 5-kV device structures, in order to create a unified treatment of the silicon and silicon carbide devices, it is convenient to use the same on-state current density of 50 A/cm^2 for all the structures for obtaining the on-state voltage drop. The on-state voltage drop for the 10-kV devices discussed in previous chapters can be compared with the aid of Fig. 11.5. In the case of the IGBT structure, the on-state voltage drop increases most rapidly with reduced lifetime in the drift region.

The silicon MCT structure has the lowest on-state voltage drop among the MOS-gated power device structures. The BRT and DC-EST structures have comparable on-state voltage drops which are significantly smaller than that for the silicon IGBT structures but larger than that of the silicon MCT structure. The on-state voltage drop for the silicon carbide inversion-mode power MOSFET can be observed to be independent of the lifetime in the drift region because it is a unipolar device. In contrast to the case of 5-kV devices, at the 10-kV blocking voltage rating, the silicon carbide power MOSFET structure is not a clearly superior alternative to the bipolar silicon devices from an on-state voltage drop point of view. At larger lifetime values, all the MOS-gated, thyristor-conduction-based, silicon devices outperform the silicon carbide power MOSFET structure in terms of the on-state voltage drop.

11.2.2 Turn-Off Losses

Device Structure	On-State Voltage Drop (Volts)	On-State Power Dissipation (W/cm^2)	Energy Loss per Cycle (J/cm^2)	Maximum Operating Frequency (Hz)
IGBT	4.766	119	2.50	32
MCT	2.304	57.6	2.75	52
BRT	2.819	70.5	2.25	54
DC-EST	2.967	74.2	2.45	51

Fig. 11.6 Power loss analysis for the 10-kV MOS-gated power devices

The optimization of power devices for applications requires a trade-off between on-state power loss and turn-off power loss. In performing the comparison of the silicon MOS-gated 10-kV devices discussed in this book, a current density of 50 A/cm^2 will be assumed in the on-state and a DC supply voltage of 5,000 V will be assumed in the off-state. The energy loss per cycle corresponds to switching under inductive load conditions. The basis for the comparison will be the outcome of the two-dimensional numerical simulations that have been provided for each of the devices in the preceding chapters. For all of the devices, the same high-level lifetime of 20 μs in the N-drift region will be used. It can be seen from Fig. 11.6 that the energy loss per cycle obtained for all of the structures is quite close. As a consequence, the difference in the on-state voltage drop becomes the most significant factor in choosing between the various 10-kV device structures from a power loss point of view.

11.2.3 Maximum Operating Frequency

The maximum operating frequency obtained for the various 10-kV MOS-gated power device structures is provided in Fig. 11.6 for the case of a duty cycle of 50%. For these conditions, the trench-gate 10-kV asymmetric IGBT is limited to a maximum operating frequency of only 32 Hz. All of the other silicon MOS-gated 10-kV asymmetric device structures have a similar maximum operating frequency of about 50 Hz which is considerably greater than that for the IGBT structure. When forward-biased safe operating area is taken into consideration, the 10-kV asymmetric DC-EST structure offers the best performance as an alternative to the IGBT structure.

11.3 Conclusions

In this monograph, the performances of high-voltage (>5 kV) MOS-gated power device structures have been analyzed and compared with each other. This book provides a unified treatment of all the MOS-gated silicon device structures with the same drift region parameters. Based upon the results of numerical simulations, it can be concluded that in the near term the silicon asymmetric DC-EST structure provides the only alternative to the commonly used IGBT structure because of its smaller on-state voltage drop and reasonable forward-biased and reverse-biased safe operating areas. This device can be manufactured using the existing IGBT process technology. On the longer term, the silicon carbide power MOSFET offers excellent performance for a blocking voltage of 5 kV and is competitive with the silicon devices for even the 10-kV blocking voltage rating. For higher blocking voltages, such as 20 kV needed for utility applications, the silicon carbide IGBT structure is an attractive device technology.

Reference

1. B. J. Baliga, "Fundamentals of Power Semiconductor Devices", Springer-Science, New York, 2008.

Author's Biography

Professor Baliga is internationally recognized for his leadership in the area of power semiconductor devices. In addition to over 500 publications in international journals and conference digests, he has authored and edited 16 books (*Power Transistors*, IEEE Press 1984; *Epitaxial Silicon Technology*, Academic Press 1986; *Modern Power Devices*, John Wiley 1987; *High Voltage Integrated Circuits*, IEEE Press 1988; *Solution Manual: Modern Power Devices*, John Wiley 1988; *Proceedings of the 3rd Int. Symposium on Power Devices and ICs*, IEEE Press 1991; *Modern Power Devices*, Krieger Publishing Co. 1992; *Proceedings of the 5th Int. Symposium on Power Devices and ICs*, IEEE Press 1993;

B.J. Baliga, *Advanced High Voltage Power Device Concepts*,
DOI 10.1007/978-1-4614-0269-5, © Springer Science+Business Media, LLC 2011

Power Semiconductor Devices; PWS Publishing Company 1995; *Solution Manual: Power Semiconductor Devices*; PWS Publishing Company 1996; *Cryogenic Operation of Power Devices*, Kluwer Press 1998; *Silicon RF Power MOSFETs*, World Scientific Publishing Company 2005; *Silicon Carbide Power Devices*, World Scientific Publishing Company 2006; *Fundamentals of Power Semiconductor Devices*, Springer Science, 2008; *Solution Manual: Fundamentals of Power Semiconductor Devices*, Springer Science, 2008; *Advanced Power Rectifier Concepts*, Springer Science, 2009. In addition, he has contributed chapters to another 20 books. He holds 120 U.S. Patents in the solid-state area. In 1995, one of his inventions was selected for the *B.F. Goodrich Collegiate Inventors Award* presented at the *Inventors Hall of Fame*.

Professor Baliga obtained his Bachelor of Technology degree in 1969 from the Indian Institute of Technology (I.I.T), Madras, India. He was the recipient of the *Philips India Medal* and the *Special Merit Medal* (*as Valedictorian*) at I.I.T, Madras. He obtained his Masters and Ph.D. degrees from Rensselaer Polytechnic Institute (R.P.I), Troy NY, in 1971 and 1974, respectively. His thesis work involved gallium arsenide diffusion mechanisms and pioneering work on the growth of InAs and GaInAs layers using organometallic CVD techniques. At R.P.I., he was the recipient of the *IBM Fellowship* in 1972 and the *Allen B. Dumont Prize* in 1974.

From 1974 to 1988, Dr. Baliga performed research and directed a group of 40 scientists at the General Electric Research and Development Center in Schenectady, NY, in the area of power semiconductor devices and high voltage integrated circuits. During this time, he pioneered the concept of MOS-Bipolar functional integration to create a new family of discrete devices. He is the *inventor of the insulated gate bipolar transistors* (*IGBT*) which is now in production by many international semiconductor companies. This invention is widely used around the globe for air-conditioning, home appliance (washing machines, refrigerators, mixers, etc) control, factory automation (robotics), medical systems (CAT scanners, uninterruptible power supplies), and electric street-cars/bullet-trains, as well as for the drive-train in electric and hybrid-electric cars under development for reducing urban pollution. The U.S. Department of Energy has released a report that the variable speed motor drives enabled by IGBTs produce an energy savings of 2 quadrillion btus per year (equivalent to 70 GW of power). The widespread adoption of compact fluorescent lamps (CFLs) in place of incandescent lamps is producing an additional power savings of 30 GW. The cumulative impact of these energy savings on the environment is a reduction in carbon dioxide emissions from coal-fired power plants by over one trillion pounds per year. Most recently, the IGBT has enabled fabrication of very compact, light-weight, and inexpensive defibrillators used to resuscitate cardiac arrest victims. When installed in fire-trucks, paramedic vans, and on-board airlines, it is projected by the American Medical Association (AMA) to save 100,000 lives per year in the US. For this work, *Scientific American* magazine named him one of the eight heroes of the semiconductor revolution in their 1997 special issue commemorating the Solid-State Century.

Dr. Baliga is also the originator of the concept of merging Schottky and p-n junction physics to create a new family of power rectifiers that are commercially available from various companies. In 1979, he theoretically demonstrated that the performance of power MOSFETs could be enhanced by several orders of magnitude by replacing silicon with other materials such as gallium arsenide and silicon carbide. This is forming the basis of a new generation of power devices in the twenty-first Century.

In August 1988, Dr. Baliga joined the faculty of the Department of Electrical and Computer Engineering at North Carolina State University, Raleigh, North Carolina, as a Full Professor. At NCSU, in 1991 he established an international center called the Power Semiconductor Research Center (PSRC) for research in the area of power semiconductor devices and high voltage integrated circuits, and has served as its Founding Director. His research interests include the modeling of novel device concepts, device fabrication technology, and the investigation of the impact of new materials, such as GaAs and silicon carbide, on power devices. In 1997, in recognition of his contributions to NCSU, he was given the highest university faculty rank of Distinguished University Professor of Electrical Engineering.

In 2008, Professor Baliga was a key member of an NCSU team – partnered with four other universities – that was successful in being granted an engineering research center from the National Science Foundation for the development of micro-grids that allow integration of renewable energy sources. Within this program, he is responsible for the fundamental sciences platform and the development of power devices from wide-band-gap semiconductors for utility applications.

Professor Baliga has received numerous awards in recognition for his contributions to semiconductor devices. These include two *IR 100 awards* (1983, 1984), the *Dushman* and *Coolidge Awards* at GE (1983), and being selected among the *100 Brightest Young Scientists in America* by *Science Digest* magazine (1984). He was elected *Fellow of the IEEE* in 1983 at the age of 35 for his contributions to power semiconductor devices. In 1984, he was given the *Applied Sciences Award* by the world famous sitar maestro Ravi Shankar at the Third Convention of Asians in North America. He received the 1991 *IEEE William E. Newell Award*, the highest honor given by the Power Electronics Society, followed by the 1993 *IEEE Morris E. Liebman Award* for his contributions to the emerging smart power technology. In 1992, he was the first recipient of the BSS Society's *Pride of India Award*. At the age of 45, he was elected as Foreign Affiliate to the prestigious *National Academy of Engineering*, and was one of only four citizens of India to have the honor at that time (converted to regular member in 2000 after taking U.S. Citizenship). In 1998, the University of North Carolina system selected him for the *O. Max Gardner Award*, which recognizes the faculty member among the 16 constituent universities who has made the greatest contribution to the welfare of the human race. In December 1998, he received the *J.J. Ebers Award*, the highest recognition given by the IEEE Electron Devices Society for his technical contributions to the solid-state area. In June 1999, he was honored at the Whitehall Palace in London with the *IEEE Lamme Medal*, one of the highest forms of recognition given by the IEEE Board of Governors, for his contributions to the development of an apparatus/technology of benefit to society. In April 2000, he was honored by his alma mater as a *Distinguished Alumnus*. In November 2000, he received the *R.J. Reynolds Tobacco Company Award for Excellence in Teaching, Research, and Extension* for his contributions to the College of Engineering at North Carolina State University.

In 1999, Prof. Baliga founded a company, Giant Semiconductor Corporation, with seed investment from Centennial Venture Partners, to acquire an exclusive license for his patented technology from North Carolina State University with the goal of bringing his NCSU inventions to the marketplace. A company, *Micro-Ohm Corporation*, subsequently formed by him in 1999, has been successful in licensing the GD-TMBS power rectifier technology to several major semiconductor companies for worldwide distribution. These devices have application in power supplies, battery chargers, and automotive electronics. In June 2000, Prof. Baliga founded another company, *Silicon Wireless Corporation*, to commercialize a novel super-linear silicon RF transistor that he invented for application in cellular base-stations and grew it to 41 employees. This company (renamed *Silicon Semiconductor Corporation*) is located at Research Triangle Park, N.C. It received an investment of $10 million from Fairchild Semiconductor Corporation in December 2000 to co-develop and market this technology. Based upon his additional inventions, this company has also produced a new generation of power MOSFETs for delivering power to microprocessors in notebooks and servers. This technology was licensed by his company to Linear Technologies Corporation with transfer of the know-how and manufacturing process. Voltage regulator modules (VRMs) using his transistors are currently available in the market for powering microprocessor and graphics chips in laptops and servers.

Index

J

L

M

O

P